DMV Seminar
Band 30

Hermann Weyl's *Raum – Zeit – Materie* and a General Introduction to His Scientific Work

Erhard Scholz
Editor

with contributions by
Robert Coleman and Herbert Korté,
Hubert Goenner, Erhard Scholz,
Skúli Sigurdsson, Norbert Straumann

Springer Basel AG

Editor:

Erhard Scholz
Fachbereich 7
Universität-GH Wuppertal
Gaussstr. 20
42097 Wuppertal
Germany

2000 Mathematical Subject Classification 01A70, 01A60

A CIP catalogue record for this book is available from the
Library of Congress, Washington D.C., USA

Deutsche Bibliothek Cataloging-in-Publication Data
Hermann Weyl's Raum - Zeit - Materie and a general introduction to his
scientific work / Erhard Scholz, ed. With contributions by Robert Coleman ....
          Springer Basel AG
(DMV-Seminar ; Bd. 30)
ISBN 978-3-7643-6476-2   ISBN 978-3-0348-8278-1 (eBook)
DOI 10.1007/978-3-0348-8278-1

© 2001
Originally published by Birkhäuser Verlag, Basel - Boston - Berlin in 2001
Birkhäuser is a member of the BertelsmannSpringer Publishing Group
Printed on acid-free paper produced from chlorine-free pulp. TCF ∞
Cover design: Heinz Hiltbrunner, Basel

ISBN 978-3-7643-6476-2

Preface

This volume arose from a DMV (*Deutsche Mathematiker-Vereinigung*) seminar in the history of mathematics, which took place May 24 to 31, 1992 in the charming environment of Schloss Reisensburg near Günzburg, Southern Germany. It was prepared and run by Jürgen Ehlers, Hubert Goenner, Skúli Sigurdsson, Norbert Straumann and myself. Moreover, we had the joy of an active participation of Matthias Kreck and Peter Slodowy, , indexSlodowy who contributed with their mathematical knowledge and active interest in the history of mathematics. This meeting was an interesting experiment among the DMV seminars which usually treat research topics in mathematics proper. Our workshop aimed at giving a sufficiently concentrated introduction to selected topics in the history of modern mathematical sciences for advanced students and young researchers, working in or coming from nearby fields, as do the seminars of the DMV series in general. In our case that resulted in an unusually diverse background and stimulating exchanges. Participants came from mathematics or physics, having developed interest in the history of mathematics, others came from the history of science with specialization in the history of recent mathematics. We chose to concentrate the seminar on Hermann Weyl's contributions to the rise of general relativity and unified field theories. This topic allows, and in fact even demands, close cooperation and exchange between mathematics, philosophy, physics and their respective histories. Thus, in addition to its main goal, the seminar served to promote interchange across the boundaries of the disciplines involved, their respective histories, the general history of science and intellectual history.

In the early 1990s research on the history of early 20th-century mathematics and physics had gained considerable breadth. Special interest in Weyl's contributions to the interplay between 20th-century mathematics, physics and philosophy had been triggered by different activities at the occasion of the centenary of his birth in 1985, in particular conferences in Zürich and Kiel, although not all of the contributions to these conferences can be considered as contributions to the history of science and mathematics. For our DMV seminar we had to concentrate on a specific topic to make possible productive and focussed presentations and exchanges centered on the history of mathematics. This explains the choice of Weyl's *Raum - Zeit - Materie* (Space - Time - Matter). We looked in particular at the changes in content and shifts in presentation between the different editions of the book from 1918 to 1923, as an indicator for probing the development of Weyl's thought on mathematics, relativity, unified field theory, conception of matter, early turn to probabilism in quantum physics, philosophy and cosmology. The background, context and extensions of the developments formed the pivot of the seminar.

The first part of the proceedings contains papers stemming directly from the seminar. The second part presents an insightful, historically oriented introduction to Weyl's work in mathematics and mathematical physics, written by Robert Coleman and Herbert Korté, which circulated in a preliminary form as preprint already in the late 1980s. This preprint proved useful for the organizers of the seminar and served as background reading material for interested participants. The version published in these proceedings has been revised, updated, and considerably extended.

As seminar organizer responsible for the history of mathematics and as editor of these proceedings I wish to express my thanks to all contributors to the volume and to the coorganizers of the seminar for their work and, not least, their patience. Special thanks, in this respect go to the authors of the second part who, although not having taken part in the seminar, thoroughly revised, updated and extended their 1985 preprint, responding to Skúli Sigurdsson's and the editor's proposals for more attention to historical detail and interconnections. My gratitude also goes to the participants of the seminar, the DMV office, in particular Martin Barner, who besides organizing the seminar so effectively, also shared personal interest in the experiment, and to the staff of Schloss Reisensburg, that provided such pleasant conditions for our work. Last, but not least, I wish to thank the friends from mathematics proper, mentioned above, who encouraged me to undertake the seminar experiment and supported the project with their knowledge and sympathy, thus making it possible to enter this difficult Weylian terrain hoping for a modicum of success. As far as these proceedings are concerned the judgement now rests with the readers.

Wuppertal, August 2000,   *Erhard Scholz*

Contents

General introduction

Erhard Scholz

The investigations of specific issues arising from Hermann Weyl's work, as presented in this book, contribute to an already existing field of rich scholarship in the history of 19th- and early 20th-century mathematics and physics. In particular, the historical investigations of the first part of the book, which are devoted primarily to the historical aspects of Weyl's *Raum - Zeit - Materie*, connect with and contribute to the history of relativity, a field which has seen an increase in number, breadth, and specialization of recent studies. This field has its own regular international meetings, the results of which have been published in the series *Einstein Studies*. In addition, in the recent book *Symbolical Universe: Geometry and Physics 1900 – 1930* edited by Jeremy Gray some related questions have been explored from the point of view of history of mathematics (Gray 1999). The question of how higher dimensional differential geometry and manifolds developed through the second half of the 19th century and became part of the conceptual framework of relativity has been dealt with from various sides: (Reich 1994) for n-dimensional tensor calculus, (Lützen 1995, 1999) for the introduction of higher dimensional geometric thought to rational mechanics during the 19th century, (Scholz 1999b) for an outline of the history of the manifold concept and (Norton 1989) for a discussion of the difficulties which arose trying to apply the manifold concept in the early development of general relativity.

To comprehend how Weyl entered the stage of general relativity one must take the Göttingen milieu into account and in particular its approach to mathematical physics shaped by Klein, Minkowski and Hilbert. Readers interested in these questions will find recent information on these topics in (Rowe 1989, 1997, 1999) and (Sigurdsson 1991, 1994) on Göttingen mathematics and physics in general, (Walter 1999) and (Corry 1997a) on Minkowski, (Corry 1996, 1997b, 1999a), (Sauer 1999) and (Renn/Stachel 1999) on Hilbert's understanding of the role of mathematization in physics and on the interaction of Einstein and Hilbert in 1915. On the development of field theories at the time one can consult (Vizgin 1994) and (Cao 1997). A broader view on Weyl as mathematician and contributor to mathematical physics will be presented in the second part of this volume and can be complemented by diverse contributions to the well-known publications on the occasion of Weyl's centenary birthday (Chandrasekharan 1986) and (Deppert 1988).

The first part of the book begins with an essay by Skúli Sigurdsson addressing difficulties for a proper understanding of Weyl's *Raum - Zeit - Materie* which result from the blending of philosophical, mathematical and physical aspects in Weyl's

presentation. Few of today's readers will have an educational background rich
enough to makes all these features easily accessible. In addition to this our author
insists on the necessity to consciously take into account the deep cultural divide
which separates our time from the 1920s, that is marked by the second world
war and the shift of the center of the scientific enterprise from Europe to North
America.

The chapter discusses Weyl's deep involvement in relativity and matter theories between the classical unified field theories and quantum physics from the
perspective of a creative mind struggling against theories of nature restricted by
the view of classical determinism well suited to an "age of the machine". It makes
us aware of how Weyl's highly complex understanding of matter went far beyond
the borders of mainstream natural sciences and shows how Weyl dealt with the
opposition of freedom (or contingency) and being bound (or lawful determination)
in a dialectical mode which Skúli Sigurdsson characterizes, apparently with some
background smile, as an "idealistic materialist" approach. In particular he warns of
a too narrow reconstruction of Weyl's matter concept from a purely internal point
of view of the physical sciences which might lead to what he calls a "streamlined
lineage" from Weyl to late 20th-century gauge theories. Such "streamlined" lineage
would, in fact, put Weyl's soft and no longer reductive unification approach from
the late 1920s on a par with late 20th-century strive for "grand" or even grandiose
unification programs. From the point of view of underlying natural philosophy,
these recent attempts appear, however, to be much closer in spirit to Weyl's early
unification attempt of 1918 that was given up by Weyl already in the early 1920s.
We should refrain from the temptation to level out such shifts in Weyl's thoughts
which reflect his rich maturing process of the 1920s; otherwise we would, know-
ingly or unknowingly, amputate the historical perception of past and present, and
lose the awareness of their respective potentialities in time.

In another thread of his argument Skúli Sigurdsson follows the problemat-
ics of scientific production as the outcome of individual creativity versus that of
a broader social enterprise, science as a *Kulturgebilde* in Weyl's own words. He
compares the different mixtures of individual invention versus collective weaving
of patterns of knowledge and empirical practices, that were characteristic in the
rise of relativity theory and quantum physics, between which Weyl migrated in
the 1920s (in addition to his mathematical research centering on the reprentation
theory of Lie groups). The chapter thus finishes with an illuminating discussion of
Weyl's self-positioning in the broader enterprise of scientific research, emphasizing the cultural web which gives broader meaning to the creative attempts of the
individual contributors.

The next contribution by Erhard Scholz on Weyl's "purely infinitesimal geometry" studies the links between Weyl's work on differential geometry and his
broader thoughts of that time. The article starts with a short presentation of Weyl's
attempts to supplant the mainstream understanding of the real number continuum and Weyl's shift from (semi)-constructivism to his own version of Brouwer's
intuitionism.These critical analyses of the real continuum were underlying Weyl's

rejection of the usual concept of manifold which is based on classical or — from another point of view — modern set theory.

Weyl's attempts to comprehend in his own way how a (manifold-like) continuum can be constituted and described mathematically led him to his "purely infinitesimal geometry" that is the first geometric structure explicitly conceived as a geometrical gauge field theory. The importance of dialectical philosophy (in Weyl's case of a Fichtean kind) for Weyl's process of clarification is briefly discussed. A more detailed account of this *geistesgeschichtliche* complexity has been presented in (Scholz 1995). The direct link between Weyl's first gauge geometry and Mie type theory of matter is outlined. Scholz argues however, following the hint in (Sigurdson 1991, 204), that Weyl rather early (in late 1920) gave up the ontological claim inherent in a Mie-type approach to matter and thus relativized the importance of his early unified field theory. The background for this shift was, besides philosophical considerations and in a motivational agreement with them, Weyl's early recognition that quantum physics was working towards a new level of materiality that could not be reduced to classical field theory.

In spite of the loss of confidence in his early unified field theory, Weyl continued to be convinced of a deeper lying importance of his gauge geometric ideas. He started a search for some conceptual justification of his gauge geometry, independent from the ontological claim of Mie's theory of matter. This resulted in his famous *Analyse des Raumproblems*, i.e. an analysis of the question which basic assumptions on infinitesimal congruences might be necessary for any reasonable metrical concept of space. In this approach Weyl characterized infinitesimal congruence structures in terms which are notoriously difficult to understand. In his contribution Scholz follows an interpretation of Weyl's characterization of the space problem in terms of fibre bundles, that goes back to proposals for a modern understanding of Weyl presented by Jürgen Ehlers at the DMV-Seminar, supported by discussion by Peter Slodowy during and after the seminar. Scholz considers that interpretation as a reasonable way to present Weyl's *Raumproblem* without essentially departing from the content of Weyl's original formulations and intentions, and presents it as a way to make the latter's considerations more explicit. Readers not acquainted with German language may look for a short résumé of this presentation in (Scholz 1999a). The reader is, moreover, invited to compare this passage with Robert Coleman's and Herbert Korté's presentation of the *Raumproblem*, in which the problem is discussed in much more detail, but from the differing point of view of Finsler metrics, and to compare these two interpretations with Weyl's original texts.

Scholz's article closes with some general observations on shifts in Weyl's methodological conceptions. When Weyl tried to prove the main theorem of the space problem in 1921, he realized that he could only be successful if he admitted the logical principle of the excluded middle for the underlying continuum. Although he considered this proof as an unfinished first attempt, he never came back to it. His analysis of the problem of space thus gave him, at the end of the period here discussed, reasons to reconsider the foundations of mathematics again and to start relativizing his close alliance with Brouwer's program, which he had proclaimed publicly in a famous article published earlier the same year (although written another year earlier).

In the next chapter Hubert Goenner investigates Weyl's role in the early debate on relativistic cosmology. He demystifies, in certain respects, the image of the great mathematician who brings light to the new field by his superior insight of the mathematical structures involved. He places Weyl's early contributions in the first edition of *Raum - Zeit - Materie* and other publications through 1919 in the context of the Einstein – DeSitter debate and shows how Weyl tried to find some sort of compromise between Einstein's static cosmos and a strong version of Mach's principle, which demanded that the Lorentz metric of the cosmos be mass generated, even for a locally flat cosmos, in which case asymptotic boundary conditions were imposed that should represent masses in the distant "horizon" of the cosmos. Weyl proposed, in this respect, to patch a local coordinate region of the DeSitter cosmos to a centrally symmetric fluid solution of the Einstein equations. This led, among others, to the paradoxical effect, that fictitious (coordinate) singularities were cut off by Weyl's procedure, although new real (Lorentz manifold) singularites were introduced and patched into the model.

For some years Weyl participated in the debate about cosmological interpretations of the DeSitter manifold. Hubert Goenner shows how difficulties arose at the time due to the possibility of forming different cosmological interpretations (models) from the DeSitter manifold. He follows Weyl's path through the debate culminating in the proposal that the visible part of the cosmos can be described by the future set of *one* timelike geodesic (Weyl's cosmological hypothesis or principle). Moreover, Weyl argued for a geometrically intriguing selection of spacelike sections and a corresponding orthogonal flow of timelike geodesics, which allowed him to model an early version of an "expanding" universe (inside the DeSitter manifold). In that context he developed a (generally applicable) proposal how to calculate the ensuing redshift. In his final section Hubert Goenner discusses how Weyl's cosmological principle relates to the cosmological debate in the late 1920s and 1930s in which Weyl no longer intervened actively.

In the late 1920s Weyl was confronted with the challenge of relocating his gauge principle in the context of quantum mechanics with its global underdetermination of the complex phase of the wave function. This possibility was noticed by Fritz London and Erwin Schrödinger, but it took Weyl some time to react.[1] In the last chapter of the first part Norbert Straumann gives an introduction to Weyl's "mature" views on the relationship between space-time and matter, taking into account the outcome of the quantum physical "revolution" of the 1920s. Weyl consciously took up the ideas of his earlier gauge-geometric approach and revised his proposal of how to represent symbolically the link between geometry (space-time), electromagnetism and matter in the late 1920s. He considered this new approach explicitly as a counterprogram to Einstein's latest theory of *Fernparallelismus*

---

[1] I thank Skúli Sigurdsson for his insistence on the importance of these proposals.

which Weyl considered as geometrically unnatural and physically unpromising. Therefore this new move has to be taken seriously as a shift — moreover one of utmost importance — in Weyl's thoughts about *Raum - Zeit - Materie*, notwithstanding the fact that it evolved six years after he revised his book with this title for the last time in 1923.

Norbert Straumann argued along these lines at the DMV-seminar and outlines his view in his contribution to this book.[2] He explains how Weyl finally took up the challenge in the context of Dirac's relativistic theory of the electron spin in 1929, for which the reader might like to consult (Kragh 1990). Straumann rereads Weyl's proposal of how to combine general relativity with the Dirac field of the electron in modernized language of gauge field theory. He shows that Weyl introduced a theory of fields and connections in a $SL_2\mathbb{C} \times U(1)$ bundle over spacetime, where Weyl in fact proposed to use the natural representation of $SL_2\mathbb{C}$ and its dual (both of course 2-dimensional) rather than Dirac's 4-dimensional representation. The argument follows Weyl's text closely and may serve readers who are acquainted with basic Yang-Mills theory as a first guide into Weyl's original text which, without such support, makes tremendously difficult reading by similar reasons as in the case of the space problem.[3] Moreover the article shows how Pauli reacted ambivalently (first complete rejection, then moderate acceptance) to Weyl's new theory, and how Weyl himself related the theory to his early gauge theoretic ideas.

The second half of the article gives a brief outline of how Weyl's 2-dimensional spinor representation and additional symmetries of order 2 came to play a major part in elementary particle physics in the 1950s. The additional symmetries of order 2 may be interpreted as representations of parity change (space reflection). These were at first regarded as controversial but were eventually adopted for the representation of neutrino fields. This part of Straumann's paper illustrates the long-range influence of Weyl's ideas by showing the later re-appearance of his work on 2-component spinor fields, albeit in modified form, in elementary particle physics and field theory. More generally, this aspect of Straumann's work illustrates the interesting interaction between mathematics and physics in the 20th century, an important but as yet relatively unexplored domain for future historical research.

In the second part of this volume, Robert Coleman and Herbert Korté give a broad and detailed introduction to Weyl's work in the mathematical sciences and philosophy. Their contribution covers or touches on the whole range of Weyl's mathematical and physical interests: real analysis, complex function theory and Riemann surfaces, elementary ergodic theory, foundations of mathematics, differential geometry, general relativity, Lie groups, quantum mechanics, and number theory. In face of the wide range of topics covered it would not make much sense to give a general overview here, the reader will find a detailed table of contents of this

---

[2] Compare for a broader picture also his recent common article with L. O'Raifeartaigh (O'Raifeartaigh/Straumann 2000).

[3] Weyl explored here and there a symbolical terrain which later would be expressed in terms of fibre bundels, but without a comparably well elaborated set of symbolical tools at his disposal.

contribution at the beginning of part II (p. 159f.). However, there are three topics which merit some comment because of their novel and/or controversial character. These are: Weyl's *Raumproblem* (problem of space), his early analysis of the discrete symmetries, namely, charge conjugation (C), parity (P) and time reversal (T), and his contributions to the foundations of mathematics

In the current mathematical and philosophical literature prevails a standard frame of interpretation of Weyl's *Raumproblem*, which is strongly influenced by the work of E. Cartan, E. Scheibe and D. Laugwitz. On this view the problem is to provide a set of philosophically motivated postulates, compatible with the physics of the time (i.e. general relativity), on the basis of which the class of gauged Riemannian (or semi-Riemannian) metrics, regarded as generalized orthogonal group structures, can be singled out from the class of all "infinitesimal congruence structures" in the sense of Weyl. Scholz's discussion in part I of this book keeps to the general outlook of this perspective and deals with the question of how such infinitesimal congruences can be understood in the light of later refinements of terminology. Coleman and Korté do not share this interpretation of Weyl's *Raumproblem*. Rather, they view the problem as that of providing a set of physically motivated postulates on the basis of which the class of gauged Riemannian metrics can be singled out from the class of all gauged Finsler metrics. Here they present their Finsler-metric approach for the first time in some detail. In this way they achieve a clear picture of the geometric part of the subject. Moreover they critically discuss the other interpretations.

Coleman and Korté give a detailed account of Weyl's early discussion of the discrete symmetries C, P, T and CPT that appeared in the second edition of Weyl's book *Gruppentheorie und Quantenmechanik*. Giving detailed references, they show how Weyl's expressions for these transformations correspond to those used today by comparing Weyl's formulae with those that appear in the first volume of Steven Weinberg's book *The Quantum Theory of Fields* (1995). Their comparison reveals that the correspondence is extremely close. Weyl's analysis differs from the currently accepted one only in that Weyl treats time reversal T as a linear, unitary operator rather than as an antilinear, anti-unitary operator. They point out further that Weyl's analysis was without immediate influence except for the fact that Dirac was motivated by Weyl's analysis of the discrete symmetry C in the context of the second quantized Maxwell-Dirac theory (quantum electrodynamics) to revise his theory of the electron and thereby to predict the existence of the antielectron (positron) and the antiproton.[4]

In their discussion of Weyl's work on the foundations of mathematics, Coleman and Korté present a detailed analysis of Weyl's two major contributions to the subject *Das Kontinuum* (1918a) and "Über die neue Grundlagenkrise der Mathematik" (1921d). In *Das Kontinuum*, Weyl addressed for the first time the question of how much of classical analysis could be constructed on a strictly predicative basis that involves quantification only over the natural numbers. On this basis

---

[4]Compare (Kragh 1990, 90, 102f.).

Weyl was able to construct a countable, "intrinsically complete", ordered field $\mathcal{W}$ and a significant portion of classical analysis, which he initially (and late in his life) hoped to be sufficient for the purposes of mathematical physics. Coleman and Korté note that the structure of Weyl's analysis, which they describe in detail, was strongly influenced by his desire to avoid Russell's theory of ramified types. For this reason, the authors present in some detail an account of the theories of simple and ramified types and show how the theory of simple types is applied in the usual non-predicative construction of the real numbers. Their presentation serves to clarify the difficulties with impredicative definitions as Weyl saw them and to highlight the severe problems that arise if the theory of ramified types is used to avoid them. After briefly pointing out the circumstances and implications of Weyl's conversion by Brouwer to the intuitionist cause, which has recently been dealt with from the Brouwerian perspective by van Dalen (1999), the authors go on to discuss Weyl's major contribution to intuitionism in his essay "Über die neue Grundlagenkrise der Mathematik".

Mathematicians may deplore that they do not find more in this volume about Weyl's work on Lie group theory, in particular his mature work on the classical groups. As additional historical literature on steps into that general field they may consult the recent publications (Hawkins 1999, 2000) and (Slodowy 1999).[5] For our volume they should take into account the professional background of the authors (compare p. 400) and the central goal of this book.

If readers like to consult recent mathematical or physical literature as background and support for the more technical parts of the historical presentations they may consult, e.g., (O'Neill 1983) for semi-Riemannian geometry, (Bleecker 1983), (Marathe/Martucci 1992) or (Schottenloher 1995) for fibre bundles and the mathematical aspects of gauge theories. Our colleagues from physics recommend, in their respective contributions, (Weinberg 1972) for relativity theory and cosmology and (Weinberg 1995, 1996) for quantum field theory. This is, of course, only a small and subjective selection from a plethora of good books which may serve as an introduction to the actual state of knowledge in the respective fields. With respect to notation it should, perhaps, be noted that the Einstein sum convention is widely used in this book, thus e.g. Weyl's "length connection" $\varphi = \varphi_i dx^i$ should be read as $\varphi = \sum_i \varphi_i dx^i$ etc.

Bibliographical references to publications of H. Weyl refer, throughout the whole book, to the *common Weyl bibliography* at the end of this volume. Each item is listed according to the year of first publication and —where it exists— republication in the *Gesammelte Abhandlungen* (GA) of 1968, including the ordinal number of the publication in GA in square brackets. Different editions of books are referred to by their code for the first edition added by year and number (in superscript) of the later edition, e.g. (Weyl 1918c/^{4}1921) for the the fourth edition of *Raum - Zeit - Materie*; in the chapters of the first part Weyl's book

---

[5]T. Hawkins' new book (Hawkins 2000) was not yet available at the time of the final preparation of this book, but will surely be of invaluable help for future investigations relating to this field.

appears also as RZM, thus the fourth edition as (RZM ^{4}1921) etc. References to Weyl's *Selecta* from 1955 are given only, where specific use of this edition is made. Archival material of Weyl and references to all other authors are listed in separate bibliographies at the end of the individual contributions.

The language of direct quotes has been differently treated by the authors of this volume. So one may find them in the original language, mostly German, or in English translation, sometimes both. The choice is not necessarily identical to the choice of language of the contribution. The reader will find a, hopefully, fair combination of German and English. Skúli Sigurdsson gives an explanation for his choice of combination of both languages at the end of his contribution (p. 39), which may be taken, by analogy, for the book as a whole.

References

Bleecker, David 1981. *Gauge Theory and Variational Principles*. London etc.: Addison-Wesley.

Borel, Armand 1986. Hermann Weyl and Lie groups. In (Chandrasekharan 1986, 53–82).

Cao, Tian Yu 1997. *Conceptual Developments of 20th Century Field Theories*. Cambridge: University Press.

Chandrasekharan, Komaravolu (ed.) 1986. *Hermann Weyl 1885–1985: Centenary Lectures Delivered by C. N. Yang, R. Penrose, A. Borel at the ETH Zürich*. New York etc.: Springer.

Corry, Leo 1996. David Hilbert and the axiomatization of physics (1894–1905). *Archive for History of Exact Sciences*, **51**, 83–198.

Corry, Leo 1997a. Hermann Minkowski and the postulate of relativity. *Archive for History of Exact Sciences*, **51**, 273–314.

Corry, Leo 1997b. Hilbert's way to general relativity. In: J. Renn e.a. (eds.) *Alternatives to Einstein's General Relativity*. Basel: Birkhäuser.

Corry, Leo 1999a. Hilbert and physics (1900–1915). In (Gray 1999, 145–188).

Corry, Leo 1999b. David Hilbert between mechanical and electromagnetic reductionism (1910–1915). *Archive for History of Exact Sciences* **53**, 489–527.

Corry, Leo; Renn, Jürgen; Stachel, John 1997. Belated decision in the Hilbert-Einstein priority dispute. *Science* **278**, 1270–1273.

van Dalen, Dirk 1999. *Mystic, Geometer, and Intutionist. The Life of L. E. J. Brouwer.* Oxford: University Press.

Deppert, Wolfgang (ed.) 1988. *Exact Sciences and Their Philosophical Foundations: Exakte Wissenschaften und ihre philosophische Grundlegung.* Vorträge des internationalen Hermann-Weyl-Kongresses, Kiel 1985. Frankfurt/M - Bern etc.: Peter Lang Verlag.

Gray, Jeremy J. (ed.) 1999. *The Symbolical Universe. Geometry and Physics 1890–1930.* Oxford: University Press.

Hawkins, Thomas 1999. Weyl and the topology of continuous groups. In (James 1999, 169–198).

Hawkins, Thomas 2000. *Emergence of the Theory of Lie Groups.* An Essay in the History of Mathematics. Berlin etc.: Springer

James, Ioan (ed.) 1999. *Handbook of the History of Algebraic and Geometric Topology in the 20th Century.* Dordrecht: Kluwer 1999.

Kragh, Helge 1990. *Dirac. A Scientific Biography.* Cambridge: University Press.

Lützen, Jesper 1995. Interactions between mechanics and differential geometry in the 19th century. *Archive for History of Exact Sciences* **49**, 1–72.

Lützen, Jesper 1999. Geometrizing configurations. Heinrich Hertz and his mathematical precursors. In (Gray 1999, 25–46)

Marathe, Kishore B.; Martucci, G. 1992. *The Mathematical Foundations of Gauge Theories.* Amsterdam: North Holland.

Norton, John 1989. Coordinates and covariants: Einstein's view of spacetime and the modern view. *Foundations of physics* **19**, 1215–1263.

O'Neill, Barry 1983. *Semi-Riemannian Geometry.* New York: Academic Press.

O'Raifeartaigh, Lochlainn; Straumann, Norbert 2000. Gauge theory: Historical origins and some modern developments. *Reviews of Modern Physics* **72**, 1–23.

Reich, Karin 1992. Levi-Civitasche Parallelverschiebung, affiner Zusammenhang, Übertragungsprinzip: 1916/17-1922/23. *Archive for History of Exact Sciences* **44**, 77-105.

Reich, Karin 1994. *Die Entwicklung des Tensorkalküls. Vom absoluten Differentialkalkül zur Relativitätstheorie.* Basel etc.: Birkhäuser.

Renn, Jürgen; Stachel, John 1999. Hilbert's foundations of physics: From a theory of everything to a constituent of general relativity. *Preprint* **118**, MPI Wissenschaftsgeschichte Berlin.

Rowe, David E. 1989. Klein, Hilbert, and the Göttingen mathematical tradition. *Osiris* (2) **5**, 186–213.

Rowe, David E. 1997. Perspective on Hilbert. *Perspectives on Science* **5**, 533–570.

Rowe, David E. 1999. The Göttingen response to general relativity and Emmy Noether's theorems. In (J.J. Gray 1999, 189–234).

Sauer, Tilmann 1999. The relativity of discovery: Hilbert's first note on the foundations of physics. *Archive for History of Exact Sciences* **53**, 529–575.

Scholz, Erhard 1995. Hermann Weyl's purely "infinitesimal geometry". *Proceedings International Congress of Mathematicians, Zürich 1994.* Basel etc.: Birkhäuser, 1592–1603.

Scholz, Erhard 1999a. Weyl and the theory of connections. In (Gray 1999, 260–284).

Scholz, Erhard 1999b. The concept of manifold, 1850 – 1940. In (James 1999, 24–64).

Schottenloher, Martin 1995. *Geometrie und Symmetrie in der Physik.* Braunschweig: Vieweg.

Sigurdsson, Skúli 1991. Hermann Weyl, Mathematics and Physics, 1900–1927. *Ph.D. Dissertation* Harvard University. Cambridge, Mass.

Sigurdsson, Skúli 1994. Unification, geometry and ambivalence: Hilbert, Weyl and the Göttingen community. In: Gavroglu, Kostas e.a. (eds.) *Trends in the Historiography of Science.* Dordrecht: Kluwer, 355–367.

Slodowy, Peter 1999. The early development of the representation theory of semisimple Lie groups: A. Hurwitz, I. Schur, H. Weyl. *Jahresbericht DMV* **101**, 97–115.

Vizgin, Vladimir P. 1994. *Unified Field Theories in the First Third of the 20th Century.* Translated from the Russian by J. B. Barbour. Basel etc.: Birkhäuser.

Weinberg, Steven 1972. *Gravitation and Cosmology.* New York: Wiley.

Weinberg, Steven 1995, 1996. *The Quantum Theory of Fields.* 2 vols.: **1** (1995) Foundations; **2** (1996) Modern Applications. Cambridge: University Press.

Walter, Scott 1999. The non-Euclidean style of Minkowskian geometry. In (Gray 1999, 91–127).

Part I

Historical Aspects of Weyl's
Raum - Zeit - Materie

Contents

1 Journeys in spacetime

Skúli Sigurdsson

1.1 Introduction

The happiest thought of Einstein's life was when he imagined falling freely in space, thus fathoming the equivalence principle (Pais 1982, 178). Albert Einstein (1879–1955) was a master at performing thought experiments and performed experiments in the material sense as well, as illustrated by the Einstein/de Haas experiments (Galison 1987, 34–52; also Hughes 1993). Moreover, he calculated and speculated as a theoretical physicist. Thus he satisfies the criteria which the philosopher of science Ian Hacking has proposed to characterize theoretical practice. An ideal theorizer calculates, experiments and speculates. The examples Hacking adduces in *Representing and Intervening* are Newton and Laplace, which nearly fits Hacking's bill (Hacking 1983, 212–215). The mathematican Hermann Weyl (1885–1955) also fits these criteria generously interpreted, for he speculated, calculated and performed thought experiments. In addition, what characterizes many theorizers are their expository gifts which all these scientists have in common, not the least Weyl.

Hacking's book harbored seeds for the awakening of studies of experiments in the natural sciences thus correcting the imbalance in favor of theory in twentieth-century history and philosophy of science, strikingly illustrated by Karl Popper's falsificationist criteria, which bears the imprint of Einstein's relativity theory (Popper [1974] 1992, 37–38). This imbalance was entrenched in modern philosophy of science by the adulation of Einstein and relativity theory by the *Berliner* and *Wiener Kreis* whereby insufficient attention was paid to the "strangeness" of relativity theory as a scientific theory with its dearth of experimental data and the close linkages of the general theory of relativity to cosmology (Hentschel 1990; also Sigurdsson 1992).

Perhaps it is better to describe the kind of data on which relativity theory fed as idealized. Furthermore, it is akin to the striking or golden events that have played an important role in modern high-energy physics (Galison 1987, 19; Galison 1997). Besides the equivalence of inertial and gravitational mass demonstrated in the famous Eötvös experiment, other such effects are the explanation of the perihelion of Mercury, the bending of the path of a light beam by a massive object, and the gravitational red shift of spectral lines. These three additional effects constituted the empirical test cases of the general theory of relativity in the years following its formulation in 1915–1916 by Einstein. It was catapulted into public awareness in 1919 at a dramatic meeting of the Royal Society in London. Confirmation of the tantalizing red shift was sought in vain in the aftermath of the Great War in the visionary Einstein Tower in Potsdam designed by the architect Erich Mendelsohn (Hentschel [1992] 1997). It symbolizes aptly the utopian aspirations and astounding creative fervor after World War I. Regarding such striking physical effects Weyl observed a quarter century later:

> Whereas electric charge is not universally proportional to the inertial mass of bodies their gravity is. This fundamental fact, supported both by daily experience and the most refined experiments, led Einstein to the conception that inertia and gravitation are one (principle of equivalence) and thus to his theory of general relativity. The main reason for my and many others' belief in that theory is the radical explanation it affords for the fact just mentioned. (Weyl 1944c, 213)

The idealized nature and crystalline clarity of such fundamental facts makes them highly amenable to thought experiments while simultaneously highlighting the limitations of a philosophy of science such as that of logical positivism, which focuses on striking events, axiomatic clarity and mathematical rigor. This philosophical movement embraced the quest for unification, the dominant master narrative in twentieth-century physics. This makes even more glaring its inadequacy as a philosophy of science, yet suggests that this master narrative may be a rich resource for the historian of science studying the co-evolution of mathematics, physics and philosophy of science in the early twentieth century (e.g. Vizgin [1985] 1994; Cat, Cartwright, and Chang 1996).

Unity and unification were ideals widely shared by the avant-garde in mathematics, philosophy of science and theoretical physics at that time. They had an aesthetic appeal and hinted at the possibility of having reached a deep understanding of the laws of nature outside the realm of human agency. Thus Weyl wrote in 1918 in the preface to the first edition of *Raum - Zeit - Materie* (hereafter RZM):

> Die Einsteinsche Theorie in ihrem gegenwärtigen Zustande endet mit einem Dualismus von Elektrizität und Gravitation, "Feld" und "Äther"; diese bleiben völlig isoliert nebeneinander stehen. Gerade jetzt eröffnet sich dem Verfasser ein verheißungsvoller Weg, durch eine Erweiterung der geometrischen Grundlage beide Erscheinungsgebiete aus einer gemeinsamen Quelle herzuleiten. So ist die Entwicklung der allgemeinen Relativitätstheorie offenbar noch nicht zum Abschluß gekommen. Es lag aber auch durchaus nicht in der Absicht dieses Buches, das auf dem Feld der physikalischen Erkenntnis heute so besonders kräftig sich rührende Leben an dem Punkt, den es im Augenblick erreicht hat, mit axiomatischer Gründlichkeit in eine tote Mumie zu verwandeln. (Weyl 1918c, vi)

Weyl's words indicate that overcoming the dualism of electricity and gravitation — the two known physical fields at that time — was desirable, that the development of relativity theory was not complete, and that active research in mathematics and physics was associated with freedom and creativity.

The rise of an alternative master narrative in the physical sciences in the late twentieth century suggests that it may be helpful to describe the development of twentieth-century physics as oscillating between the quest for essence, unity and transcendence, and that of novelty, disunity and history. This other narrative is that of anti-reductionism and complexity where solid state physics, not elementary particle physics, is taken to be exemplary (Schweber 1993). The philosophy of science predicated on the theory of relativity occupies a conceptual terrain marked by the poles of essence, unity and transcendence. The story told on the following pages starts in that triangular region in the early years of the twentieth century, yet soon journeys to a region marked by novelty, disunity, interconnectedness and history (Hacking 1996).

These two polygonal regions are embedded in a vast landscape which has become inaccessible to late twentieth-century observers. The principal aim of this chapter is to chart some of the contours of this land, and explore the kinds of creativity and individuality it fostered. As I have noted on an earlier occasion, the historian must regard critically narratives aimed at constructing a streamlined lineage from Weyl's early work on unification to gauge theories in modern physics (Sigurdsson 1994, 364). Erhard Scholz has argued convincingly with reference to Weyl's work at the end of the 1920s that Weyl's espousal of unification should, after the initial enthusiasm in the wake of World War I, be christened moderate or soft. I concur and would hasten to add that Weyl's "softening" is not only characteristic of his ability to free himself of the hold of fixed ideas, but highlights the elusiveness and plasticity of the idea of unity.[1] In an essay on the disunified sciences Hacking lists eleven meanings to the idea of unity of science. The list is not meant to be exhaustive and one might supplement it with the cluster of ideas bearing on metrology and unification, in addition to how the concept unity disaggregates as it circulates between cultural contexts (Hacking 1992, 39–41; also Galison 1998).

Thus in contrast to some contributions to this volume on Hermann Weyl's RZM I am not particularly interested in delineating how this text foreshadowed developments after World War II in elementary particle physics and grand unification, where Weyl's idea of gauge invariance has indeed played a crucial role.[2] That interest is dimmed still further when such attempts are coupled with scanty reference to the already extensive work of professional historians of science on the topic (O'Raifeartaigh and Straumann 2000). The task of unravelling the course of scientific events in the 1910s and 1920s is much too complex to sanction such a cavalier, if not dismissive, attitude to the wealth of pertinent historical and philosophical studies.[3]

My aim in this chapter is to travel back in spacetime and try, like an imaginary geographer or ethnographer, to begin to comprehend the culture of Einstein, Weyl and other grand theorizers. Despite individual differences, they joined hands in securing the general theory of relativity a central place in the history of Western

---

[1] See also E. Scholz' contribution to this volume and (Scholz 2000).

[2] For such a view compare the contributions of N. Straumann and R. Coleman/H. Korté in this volume.

[3] See also (Forman 1991).

thought. This was an interlocked elite cognizant of its own exceptional powers, responsibilities, and place in history. The defining moment for them and their age cohort was the Great War, which marked a fundamental rupture in European and world history, as observed, e.g., by J.L. Garvin editor of *Encyclopedia Britannica.* In a preface to supplementary volumes which appeared in 1926 surveying events in the years 1910–1926 Garvin wrote:

> There have been no more momentous and transforming years in the experience of mankind. Formerly that space of a decade and half would have been called at best a period. We may justly term it an epoch. It crowds into itself more historic drama and social significance, more economic energy and moral ferment, more destructive force, yet more constructive effort and idealism in every sphere, than have been known in most centuries. In wide regions the former political structure and lines of the map have been altered in a manner that would have surpassed all powers of belief if prophesied beforehand. Old Empires and dynasties have vanished; new nations and systems have appeared.

Garvin immediately added regarding science, technology and medicine:

> With this, science and invention have gone forward with accelerating speed to wonderful results. All industrial life is searched by questioning and full of new developments. In this short epoch the former fundamental conceptions of time, space, matter and energy have been dissolved or modified. Speculation on the possibilities of further scientific discovery and of its practical addition to human power never was more daring. Medicine and surgery have made at least an equal advance in their resources for the defence, repair and prolongation of human life. (Garvin 1926, vii)

The remarkable cascade of events in the sciences in the epoch 1910–1926 was discussed, e.g., by Senior Wrangler and Smith's Prizeman Arthur Stanley Eddington (1882–1944) in an entry on the "Universe: Electromagnetic Gravitational Schemes", by Nobel Prizeman and Copley Medallist Einstein on "Space-Time", and by Weyl on the "Modern Conception of the Universe" (Weyl 1926e).

What heightens the challenge of traveling back in time is the fact that the Central European culture which provided such a fertile ground for the rise of modern physics and mathematics, and to which Einstein and Weyl belonged, was destroyed during World War II. Moreover, the historian must not only traverse a terrain marked by two world wars and the Holocaust. A linguistic divide must be crossed which separates the current Anglo-Saxon culture from an earlier era when German still played an important role as the language of modernity while English was in rapid ascendancy (Hobsbawm 1993, 21; Garvin 1926, x). This is especially important in a study of RZM and Weyl's other writings from the period. An analysis of the book and Weyl's work must pay close attention to the references to history, metaphorical possibilities and cultural specificity inherent in his language. Allusions to contemporary events abound and the effect can be uncanny. In the 1919 preface to the third edition of RZM he speaks about the relief looking up from the ruins of the oppressive present towards the stars, harboring an indestructible world of laws (also Schaffer 1989, 153). The 1921 fourth edition having discussed the limits to field physics posed by quantum theory ends on a jubilant note: "Ein paar Grundakkorde jener Harmonie der Sphären sind in unser Ohr gefallen, von der Pythagoras und Kepler träumten" (Weyl [1918c] 1921, 284). Regarding an article on the new foundation crisis in mathematics Weyl subsequently noted: "Nur mit einigem Zögern bekenne ich mich zu diesem Vorträgen, deren stellenweise recht bombastischer Stil die Stimmung einer aufgeregten Zeit wiederspiegelt — der Zeit unmittelbar nach dem ersten Weltkrieg" (Weyl 1921c, 179).

What makes the travels of the imaginary ethnographer even more arduous is the fact that philosophical reflection played a crucial role in scientific practice for Weyl and many of his contemporaries. That is currently less the case. Einstein observed in the late 1940s: "Epistemology without contact with science becomes an empty scheme. Science without epistemology is — in so far as it is thinkable at all — primitive and muddled" (Einstein in Holton [1985] 1986, 164). This sentiment was widely shared, though not unquestioned, by many of Einstein's contemporaries. Thus the theoretical physicist Max Born (1882–1970) idealizing the work of Arnold Sommerfeld (1868–1951) noted that many aspiring theorists learned to calculate under Sommerfeld's guidance; if some of them (e.g. Heisenberg and Pauli) later philosophized that might be ascribed to their contact with Niels Bohr (1885–1962) if external reason had to be sought (Born [1928] 1963, 606).

As seminal teachers of theoretical physics Born and Sommerfeld stressed mathematical prowess and skills, yet did not disparage philosophy. Born, in fact, turned increasingly to philosophy after the late 1920s. Born noted that theoretical physics for Bohr and Einstein, on the other hand, was an area for the application of philosophical principles (ibid.). In the 1920s there existed an implicit division of labor in the blossoming discipline of theoretical physics between calculation and speculation. It was in Copenhagen under Bohr's tutelage that the young masters speculated, philosophized and grappled with problems of foundations (Sigurdsson 1991, 210–219).

Bohr and Einstein are towering figures in twentieth-century physics and despite their marked differences — community and language versus solitude — their conceptual *Lebenswelt* was deeply embedded in the world of neo-Kantian philosophy (Chevalley 1994; Schweber 1995). The stark contrast to the conceptual universe inhabited by theorizers at the end of twentieth century is well illustrated by the Nobel Prize winning physicists Richard P. Feynman (1918–1988) and Steven Weinberg (born 1933) who are steeped in the pragmatic tradition of American science and high energy physics (Schweber 1986; Pestre 1992). Thus Feynman in *The Character of Physical Law* observed that "we have learned from much experience that all philosophical intuitions about what nature is going to do fail." Moreover, Feynman likened philosophy to a disease which had to be beaten by training (Feynman 1965, 53). Likewise, a chapter in Weinberg's recent *Dreams of a Final Theory* is simply entitled "Against Philosophy" (Weinberg 1993; also Holton [1985] 1986).

The work of the historian of science at the dawn of the twenty-first century is insofar challenging as the culture in which he/she has been socialized is purportedly against philosophy or aphilosophical when it comes to scientific practice. It may therefore be important to keep this change in attitude to philosophy in mind when studying RZM. In the preface to the 1918 first edition Weyl noted:

> Zugleich wollte ich an diesem großen Thema ein Beispiel geben für die gegenseitige Durchdringung philosophischen, mathematischen und physikalischen Denkens, die mir sehr am Herzen liegt; dies konnte nur durch einen völlig in sich geschlossenen Aufbau von Grund auf gelingen, der sich auf das Prinzipielle beschränkt. Aber ich habe meinen eigenen Forderungen in dieser Hinsicht nicht voll Genüge tun können: der Mathematiker behielt auf Kosten des Philosophen das Übergewicht. (Weyl 1918c, v)

The three-pronged approach — philosophical, mathematical, and physical — characterizes RZM and although the mathematician may have gained the upper hand it is the partial melting of these prongs which make its study both difficult and so rewarding. Weyl had come of age in Göttingen in the early twentieth century, where philosophy was an integral part of the conceptual universe (Peckhaus 1990) and a belief in a pre-established harmony between mathematics and physics was widespread (Pyenson 1982). Any analysis of RZM must somehow take proper cognizance of that fact. Weyl's tastes in philosophy differed much from that of many contemporary natural scientists, e.g., Einstein's, yet that should not obscure the fact that philosophy was a common medium and source of creative insights for them and their generation. Likewise, though much has been made of the dispute between Einstein and Weyl concerning the latter's idea of gauge invariance and unified field theory, that should not make one overlook the many areas of agreement and mutual respect. In a review of the first edition of RZM Einstein wrote:

> Immer wieder drängt es mich dazu, die einzelnen Teile dieses Buches von neuem durchzulesen; denn jede Seite zeigt die unerhört sichere Hand des Meisters, der den Gegenstand von den verschiedensten Seiten durchdrungen hat. Ich betrachte es als einen glücklichen Umstand, dass ein so ausgezeichneter Mathematiker sich des neuen Gebiets angenommen hat. Er hat es verstanden, mathematische Strenge mit Anschaulichkeit zu verbinden. Der Physiker kann aus dem Buche die Grundlagen der Geometrie und Invariantentheorie, der Mathematiker diejenigen der Elektrizitätslehre und Gravitationstheorie lernen. (Einstein 1918; also Ryckman 1994, 842)

Einstein here pointed to the seminal importance of facilitating a dialog between different scientific disciplines, a task aided by Weyl's considerable expository gifts, and recognized later by Hans Freudenthal, who called Weyl an interpreter between these fields (Freudenthal 1955).

In what follows I discuss Weyl's synthesizing powers and creativity, and the tension between the individual and the collective in Weyl's life.

1.2   Creativity in the age of the machine

The changes in the four successive versions of RZM testify to the rapid advances made on the frontiers of knowledge in the aftermath of World War I. Thus while the fourth edition ended on a utopian note, seeking salvation in the stars, the 1918 first edition ended quite differently. In it Weyl discussed how complicated was the relation between the *Wirklichkeit* each person experienced in daily life and the *objektiven Wesenheiten* with which physics dealt in mathematical symbols and he spoke about the proximity of physics and geometry as exemplified by the close similarity between Maxwell's theory and analytic geometry. The book ended on the following note:

> Ich meine, daß die Physik es nur mit dem zu tun hat, was in einem genau analogen Sinne als formale Verfassung der Wirklichkeit zu bezeichnen wäre. Ihre Gesetze werden ebensowenig in der Wirklichkeit jemals verletzt, wie es Wahrheiten gibt, die mit der Logik nicht im Einklang sind; aber über das inhaltlich-Wesenhafte dieser Wirklichkeit machen sie nichts aus, der Grund der Wirklichkeit wird in ihnen nicht erfaßt. Wenn es der Wahn der scholastischen Methode ist, aus bloß Formalem Wesenhaftes zu deduzieren zu wollen, so ist die Weltanschauung, welche man als Materialismus bezeichnet, nur eine Spielart der Scholastik. (Weyl 1918c, 227)

It would be beyond the scope of this chapter to analyze Weyl's attitude to materialism and its relationship to freedom in depth. Nevertheless, the antagonism to mechanical materialism hints at affinities between Weyl and Victorian physicists which surpass the admiration for the field theory of Faraday and Maxwell which he shared with his Göttingen teachers David Hilbert (1862–1943) and Felix Klein (1849–1925). Victorians such as Maxwell who opposed scientific naturalism and shunned materialism elevated metrology into a theological enterprise (Schaffer 1991, 172), and turned the country house laboratory into a production utopia for sustaining "a symbolic universe in which material technology and spiritual value could, temporarily and delusively, be reconciled" (Schaffer 1998, 177).

In the concluding section of the 1918 first edition of RZM Weyl had also written: "Die Physik, das stellt sich damit heraus, handelt gar nicht von dem Materiellen, Inhaltlichen der Wirklichkeit, sondern was sie erkennt, ist lediglich deren *formale Verfassung*. Sie hat für die Wirklichkeit die gleiche Bedeutung wie die formale Logik für das Reich der Wahrheit." (Weyl 1918c, 227, emph. in orig.)

Thus materiality for Weyl constituted the content of reality, if reality is equated with *Wirklichkeit*. If correct, this interpretation of Weyl's writing may shed light on his engagement with probability as a basic feature of quantum theory and be helpful for untangling his epistemological views. Matter resided outside the realm of field physics although it generated the field effects (also Ryckman 1994, 835).

In 1922 Einstein and Weyl responded in the journal *Wissen und Leben* to a series of questions as to whether the public sensation caused by the theory of relativity was justified. Einstein answered in the negative whereas Weyl answered by repeating his view about the meaning of field physics for reality (Sigurdsson 1991, 180–186; also Einstein 1922). What to Weyl justified some of the sensationalism was not relativity theory, but recent findings in atomic physics. Field physics was strictly lawlike, but statistics reigned in the atomic realm. He wrote:

> Die Materie erscheint nach der hier geschilderten Auffassung als ein *Agens, das seinem Wesen nach jenseits von Raum und Zeit liegt*; dieses aus unzählbaren an sich verbindungslosen Individuen bestehende Agens nennen wir "Materie", sofern wir es als Ursache der im Felde sich ausbreitenden, die Individuen zu einer Welt zusammenknüpfenden Wirkungen betrachten. Seiner inneren Beschaffenheit nach mag es ebensowohl schöpferisches Leben und Wille wie Materie sein. (Weyl 1922d, 904; emph. in orig.)

Weyl had already advocated some of the views adumbrated in this passage, which might be called dialectical or idealistic materialism, three years earlier in a lecture on the relationship between the causal and statistical point of view in physics (Weyl 1920a). Then it had been a minority opinion but in 1922 Weyl claimed, with some exaggeration, that most physicists no longer believed in the deterministic Laplace metaphor (Forman 1971; Sigurdsson 1991, 180–206).

The public sensation caused by relativity theory was noted by the editor of *Encyclopedia Britannica* and enshrined in twentieth-century culture in Einstein idolatry (Barthes [1957] 1972). Yet it was research on the constitution of the atom along the lines of the old quantum theory which captured the attention of the German-speaking physics community. In the demoralized Central European context it was science and especially atomic physics which served as *Machtersatz* for a large segment of the academic community (Forman 1973, 161–165; Heilbron 1986, 86, 92–93). This was the case although atomic physics had received impetus somewhat earlier from attempts at constructing a unified field theory explaining the structure of matter, namely from the work of Gustav Mie (1868–1957), Hilbert and Weyl (Hendry 1984, 6–23).

Weyl was not a member of the physics community strictly speaking, as he acknowledged to Sommerfeld in 1922 (Sigurdsson 1991, 250). Yet, his research on unified field theory and epistemological grappling with causality and quantum theory contributed significantly to the construction of quantum mechanics in the years 1925–1927. A less direct but no less significant contribution of Weyl's stemmed from the value of RZM for a young generation of theoreticians (e.g. Schweber

1990, 987). Weyl's realization of the probabilistic character of quantum theory highlights his ability to weave together strands from different realms of inquiry and make the resulting fabric reveal novel insights. He wove together reflections on the constitution of the mathematical continuum, the relation of infinitesimal geometry to physics and on celestial and statistical mechanics stemming from work done on the eve of the Great War (Sigurdsson 1991, 56–60; von Plato 1994, 61–70, 106–108).

Another instance of such weaving was Weyl's research on group theory, which stemmed partly from the problems raised by RZM (Scholz 1994, Hawkins 1998). This ability to synthesize was celebrated in the mid-1920s by Hilbert in a report on Weyl's geometrical work for the Lobachevsky Prize. Hilbert stressed how Weyl had linked electromagnetism and gravity, i.e. "[sie] zu einer organischen Einheit zu verschmelzen" (Hilbert 1927, 66). Whether the hypothesis that electromagnetism and gravitation followed from Weyl's geometry was borne out by reality could not be decided by current knowledge of physics. Weyl's geometry was the most general infinitesimal geometry, and he had freed Riemannian geometry from its Euclidean past. Hilbert observed furthermore: "In diesen gruppentheoretischen Arbeiten entfaltet sich aufs glänzendste die erstaunliche Fähigkeit Weyl's zur einheitliche [sic] Gestaltung eines weitverzweigten und verschieden zusammengesetzten Stoffes" (ibid., 70).

This ability to synthesize and give a coherent overview of a complex field characterizes Weyl's writings. His book on symmetry contains, in aesthetic garb, numerous deep insights about group theory and the essential dialectic of law and contingency in nature. Weyl illustrated the latter by referring to an intuitive fact: "If nature were all lawfulness then every phenomenon would share the full symmetry of the universal laws of nature as formulated by the theory of relativity. The mere fact that this is not so proves that *contingency* is an essential feature of the world." (Weyl 1952, 26; emph. in orig.)

The contrast between contingency and law, freedom and being bound, spontaneity and mummification is a recurrent theme in Weyl's writings. In an article on Riemann's geometrical ideas and their relation to and impact on group theory, written in 1925 for Lobachevsky's Collected Works, Weyl noted: "Der neue Ansatz des Raumproblems beruht, wie man sieht, auf einer Scheidung zwischen dem, was einfürallemal fest und gegeben, und dem, was an sich zufällig und beliebiger Veränderungen fähig in der Wirklichkeit kausal an die Materie gebunden ist" (Weyl 1925/1988, 38). The weaving together of numerous strands, which Weyl accomplished, created a tension in the fabric. The metamorphosis of RZM over its multiple editions with the gradual abandonment of the belief in a unified field theory and the recognition of the irreducible role of chance in quantum physics, demonstrates his ability to maintain for a long time such a creative tension.

It was important for Weyl that the larger picture would remain in sight; and synthesis might facilitate that, compared to proceeding mechanically and in a piecemeal fashion. In the early 1930s Weyl stressed this point when discussing abstract algebra and topology:

> Wir geben uns nicht gerne damit zufrieden, einer mathematischen
> Wahrheit überführt zu werden durch eine komplizierte Verkettung for-
> meller Schlüsse und Rechnungen, an der wir uns sozusagen blind von
> Glied zu Glied entlang tasten müssen. Wir möchten vorher Ziel und
> Weg überblicken können, wir möchten den inneren Grund der Gedan-
> kenführung, die Idee des Beweises, den tieferen Zusammenhang verste-
> hen. (Weyl 1932, 348)

This emphasis on the broader picture, expressed in visual terms, is reminiscent of
the words of William Whewell who in the 1840s preferred geometry to calculation;
geometrical reasoning resembled directing one's course as one tread the ground,
whereas analytical calculation that of being hurled from one end of railroad tracks
to the other (Whewell in Schaffer 1991, 171; also Shapin 1997, 343–344).

The stance taken by this eminent Victorian is a recurrent motif in the history
of mathematics — the yearning for freedom from the yoke of algorithms and an
increasing industrialization of the pursuit of knowledge, which might enslave the
mathematician (also Sigurdsson 1992, 105–106). The advantage of this approach
was a hallmark of Hilbert's style as a mathematician. "A characteristic feature
of Hilbert's method is a peculiarly *direct attack* on problems, unfettered by algorithms; he always goes back to the questions in their original simplicity" (Weyl
1944a, 135; emph. in orig.). Beyond yielding novel insights for the mathematician,
such freedom bespoke of an essential aspect of human existence. Thus in 1925
Weyl wrote:

> Die Mathematik ist nicht das starre und Erstarrung bringende Schema,
> als das der Laie sie so gerne ansieht; sondern wir stehen mit ihr genau
> in jenem Schnittpunkt von Gebundenheit und Freiheit, welcher das
> Wesen des Menschen selbst ist. (Weyl 1925a, 533)[4]

As Paul Forman has observed, the association of German science and technology with German military might, which Felix Klein characterized as late as
June 1918 as a pre-established harmony, led to a backlash and anti-scientific attitude in the Weimar Republic. Mathematics and natural science were linked to
technological rigidity and strict causality (Forman 1971; Sigurdsson 1991, 266–267;
also Pyenson 1982). In the famous snow scene in *Der Zauberberg* (1924) Thomas
Mann spoke of the icy regularity of the snow crystals which, in their perfect symmetry, epitomized death. They were anti-organic and denied freedom and life. In
Symmetry Weyl quoted from this section of Mann's novel (Weyl 1952, 64) which
illustrates how his reflections on law, contingency and freedom were nourished and
sustained by references to Weimar culture.

In his aforementioned discussion of abstract algebra and topology, Weyl had
continued: "Es ist ja mit einem modernen mathematischen Beweis kaum anders als
mit einer modernen Maschine oder einer modernen physikalischen Versuchsanord-

---

[4]I thank Moritz Epple for having drawn my attention to this passage.

nung: die einfachen Grundprinzipien sind eingebettet und dem Blicke fast entzogen durch eine Fülle technischer Details." Furthermore, Weyl referred approvingly to Klein's assessment of the contributions of Bernhard Riemann (1826–1866) to nineteenth-century mathematics, and how mathematics was in a certain sense furthered most by those whose procedures were intuitive rather than utilizing rigorous methods of proof (Weyl 1932, 348). Weyl did not want to be tied to a mechanical and industrialized plan of action, he wanted to roam freely and intuit new ideas, and though he avoided withering in solitude his was an individual approach to intellectual creation (also Shapin 1991).

It might be intriguing to explore further Weyl's self-conception as a mathematician in light of the studies of Herbert Mehrtens on the history of modern mathematics where Hilbert and Klein represent the opposite poles modern and counter-modern (Mehrtens 1990). On that ideal plane Weyl would be located near the counter-modern pole, yet Weyl's beliefs suggest how Mehrtens' scheme needs to be supplemented. Weyl wanted to understand and not merely to produce mechanically like a factory worker. In his 1932 article on algebra and topology he spoke of the difference between explaining and understanding with reference to hermeneutics. Once in exile he returned to this need for understanding, relating his feelings to his close friend and fellow Hilbert student Erich Hecke (1887–1947). He wrote in the spring of 1936 from Princeton:

> I notice around here, that the Americans have completely learned to master the techniques of science; enough and more than enough mathematics is produced in my vicinity. What they are lacking, and where I can perhaps give useful guidance, is with respect to "reflection". (Weyl in Sigurdsson 1991, 232–233; my transl.)

The latter part of the conceptual pair explaining-understanding is expressed here slightly differently, i.e. in terms of reflection (*Besinnung*). He had done so previously in a lecture published as *Die Stufen des Unendlichen* (Weyl 1931a) and would return to it later. In 1931 he wrote: "Die Gefahr des schöpferischen Tuns, wenn es nicht durch Besinnung überwacht wird, ist, daß es dem Sinne entläuft, abwegig wird, in Routine erstarrt; — die Gefahr der Besinnung, daß sie zu einem die schöpferische Kraft des Menschen lähmenden unverbindlichen 'Reden darüber' wird" (Weyl quoted in Weyl 1954, 631–632).

The checks and balances in Weyl's system of self-governance were meant to guard against the twin evils of ossification and vacuous torrents of words. They were meant to secure serious productivity, but by themselves they were unable to prevent individual exhaustion. Looking back on his life in Europe Weyl told Hecke in 1936:

> I was always in danger of losing myself, because of what was expected of me. I believe, that this critical period is behind me, and now I find, that I do almost as much as is in my powers. Princeton is in that respect so good for me, because the external excitation, which "it is

impossible to tolerate", practically vanishes. (Weyl in Sigurdsson 1991, 232; my transl.)

The ceaseless weaving and synthesizing of different strands whereby Weyl tested the limits of received systems of knowledge had taken its toll. Weyl's wife Helene spoke in the early 1940s of Weyl having experienced an almost god-like pleasure of creativity. This creativity was not based on deep inner security and calm, but rather the opposite, for Weyl was in many ways an inwardly insecure person who partly governed himself by testing his own limits (Sigurdsson 1996, 53 and 66; also Gruber 1985). The testing of limits and acts of creativity entailed commitment to a cluster of notions some of which have previously noted such as law, freedom, being bound, beauty, quest for meaning, and explaining-understanding. Speaking of creativity and the process of bonding with a number of basic notions, the cognitive psychologist Howard E. Gruber has noted:

> Much of the time goes into forming a deep enough cathexis with some particular set of natural objects or ideas to permit steady engagement of the person's whole effort. Such love is not formed in an instant. In matters of work the scientist may be polygamous but not promiscuous. Creativity demands commitment. (Gruber [1978] 1981, 138)

Weyl did not easily surrender what he believed were seminal insights, yet he avoided being imprisoned by his own ideas. This is illustrated by the well-known encounter between Weyl and Einstein in 1918 concerning the feasibility of Weyl's bold interpretation of his infinitesimal geometry, namely that it made possible the unification of gravitation and electromagnetism (Ryckman 1996, 168-179).[5] Not only did some of Weyl's contemporaries view the situation as not decisively settled by Einstein's counterargument (Lorentz [1923] 1937, 373-376). Moreover, it was the grappling with Einstein's objection to Weyl's exciting insight of 1918, namely that neither direction nor length should be integrable in a truly infinitesimal geometry, which inspired Weyl's work on the *Raumproblem* and group theory (Scholz 1994; Hawkins 1998). The advent of the new quantum mechanics in the later 1920s and the work of younger theorists such as Fritz London (1900-1954) enabled Weyl at the end of the 1920s to jettison his earlier understanding as speculative and embrace the quantum-mechanical meaning of his gauge principle (Stachel 1999, 454).

Another instance of Weyl's ability to free himself from earlier ideas or intellectual alliances was his gradual abandoning of a belief in Mie's unified field theory without being unappreciative of what progress it had rendered possible. He wrote at the end of the 1921 fourth edition of RZM about the underlying vision of Mie's theory, quantum theory and the still unresolved problems of dialectical or idealistic materialism:

---

[5]See also Straumann (1987) and the contributions of R. Coleman/H. Korté, section II 4.4, and E. Scholz, section I 2.7, in this volume.

So kühner Hoffnungen müssen wir uns freilich jetzt fürs erste entschla-
gen. Die Gesetze des metrischen Feldes handeln weniger von der Wirklichkeit selber als von dem schattenhaften extensiven Medium, das
die Verbindung zwischen Wirklichem vermittelt, und von der formalen
Verfassung dieses Mediums, die es zur Wirkungsübertragung befähigt.
Bereits hat die *statistische Physik*, die Quantentheorie eine tiefere Sicht
der Wirklichkeit angeschnitten, als sie der Feldphysik zugänglich ist;
das Problem der Materie aber liegt noch ganz und gar im Dunkeln.
(Weyl [1918c] 1921, 283-284; emph. in orig.)

Weyl was in regular contact with Klein at this time as illustrated by a paper
on how to fit projective and conformal geometry in his infinitesimal geometry which
arose out of their correspondence (Weyl 1921f). At the end of 1920 Weyl told Klein
how he had freed himself thoroughly from Mie's theory. Field physics was no longer
the key to *Wirklichkeit*. The field or ether was a powerless transmitter of actions
caused by *Realität* or matter which was beyond the field (Sigurdsson 1991, 204).
However, the ability to free himself should not obscure the presence of a number of
permanent notions guiding Weyl's intellectual endeavors. One such was the quest
for freedom. Having discussed the limitations of Mie's theory at the end of the
1921 fourth edition of RZM, Weyl concluded that the reader:

    ... muß von dem Gefühl errungener Freiheit überwältigt werden
    – ein festgefügter Käfig, in den das Denken bisher gebannt war, ist
    gesprengt–; er muß durchdrungen werden von der Gewißheit, daß un-
    sere Vernunft nicht bloß ein menschlicher, allzumenschlicher Notbehelf
    im Kampf des Daseins, sondern ungeachtet aller Trübungen und alles
    Irrtums doch der Weltvernunft gewachsen ist und das Bewußtsein eines
    jeden von uns der Ort, wo das Eine Licht und Leben der Wahrheit sich
    selbst in der Erscheinung ergreift. Ein paar Grundakkorde jener Har-
    monie der Sphären sind in unser Ohr gefallen, von der Pythagoras und
    Kepler träumten. (Weyl [1918c] 1921, 284)

This language with the irrational mayhem of World War I in the background,
perhaps not unusual for educated citizens at that time, is highly unusual for a scientific text. It is also illustrative of a shift in Weyl's thinking via the medieval
mystic Meister Eckehart from a Fichtean approach to theorizing (Brouwer's intuitionism, radical "Cartesian" geometrization of physics) to a more open-ended
approach wherein matter and quantum theory has an irreducible character (Weyl
1954, 646-647; also Scholz 1995).[6]

The desire to escape from the cage *(Käfig)* of existence and be able to roam
freely had been expressed in the preface to the 1918 first edition of RZM, where
Weyl wrote of how Einstein's relativity theory had enabled human thought to
ascend higher and, as if a wall had collapsed which had barred access to truth,

---

[6]See also the contribution of E. Scholz in this volume (section I 2.7).

one could thus see further and deeper (Weyl 1918c, v). The language employed by Weyl is redolent of mountain imagery (e.g. the verb *erklimmen* for ascend or climb). Scaling the heights away from the lowlands hints at a combination of athletics, aesthetics and transcendence in reaching a vantage point from which the trick of seeing from afar, a gaze from nowhere, seemed possible (Haraway 1988). The mountainous imagery struck a vital chord in Central European culture. Thus a student in Zürich, hearing in 1930 that Weyl was leaving for Göttingen, told him that his lectures had been particularly helpful for "you were to a certain extent able to convey to us students a view of the divine infinite, that often appeared to me like a wonderful view from a thick fog." After comparing mathematics and nature the student likened the mathematician to a "lonely mountain climber, who knows how to search for the divine experience beyond the elementary sport motive, that cannot be expressed in words" (quoted in Sigurdsson 1991, 276-277; my transl.).

The view from the mountaintop might be described as panoramic. The same desire to comprehend a mathematical proof as a whole, not merely in atomized parts, is a vital ingredient of this outlook. In the late 1930s, once in exile, Weyl reviewed the second volume of Courant-Hilbert's *Methoden der mathematischen Physik*. He wrote:

> Nowadays many mathematical books do not seem to be written by living men who not only know, but doubt and ask and guess, who see details in their true perspective — light surrounded by darkness — who, endowed with a limited memory, in the twilight of questioning, discovery, and resignation, weave a connected pattern, imperfect but growing, and colored by infinite gradations of significance. The books of the type I refer to are rather like slot machines which fire at you for the price you pay a medley of axioms, definitions, lemmas, and theorems, and then remain numb and dead however you shake them. [Richard] Courant imparts insights into a situation which has manifold aspects and develops methods without disintegrating them into a discontinuous string of theorems; and nevertheless, the essential results stand out in clear relief. (Weyl 1938c, 596)

Weyl's tastes in mathematical writing and constructing a rich tapestry corresponded to his own style. This was observed by Norbert Wiener (1894-1964) reviewing four books on the space problem by Rudolf Carnap (1891-1970), two by Eduard Study (1862-1930) and Weyl's *Mathematische Analyse des Raumproblems* (Weyl 1923c). Wiener emphasized that Weyl wanted to analyze certain general propositions at the basis of geometry, the notion of *affiner Zusammenhang* played a fundamental role in Weyl's analysis and what happened *im Kleinen* was pivotal in Weyl's generalized differential geometry. Although postulationist in spirit, Wiener noted, Weyl

> ... is fortunate, however, in not being bound in the pedantic straight-jacket of independence-proofs and postulate-counting which has strangled a most promising young science almost in its cradle. Weyl is not

behind the postulationists in rigor, but he is far ahead of them in imagination, and he relegates his meticulous dissection of logical minutiae to the place where it belongs, — the back of the book. (Wiener [1924] 1985, 983)

Wiener had noted concerning Carnap's doctoral dissertation *Der Raum* (1922) that Carnap hoped to preserve what was of permanent value in Kantian philosophy in an intermediary space between the formal space of the mathematican and the space of the physicist, which was experimental and inductive. The analysis situs or what later would be called topological properties of this space, the space of intuition, were known according to Carnap as Kant had considered all space to be known, i.e. a priori. Wiener doubted this, for just as relativity theory had cast serious doubt on the connectivity of space *im Grossen*, then it was hard to believe that the properties of the universe *im Kleinen* would not need to be revised. Wiener wrote: "There are signs that the time may not be far distant when the atomicity which the quantum theory recognizes as a basal characteristic of the universe shall be referred to a fundamentally atomic conception of its space-time framework" (ibid., 980; also Friedman 1995).

What is striking about Weyl's review of Courant-Hilbert is how he had adopted so well to the new cultural environment that he referred to a slot machine to express his dislike of what was mechanistic and aleatory. As he had confided to Hecke in 1936 too much mathematics was being produced around him in Princeton, and what was needed was reflection in order to appreciate details in their proper perspective, in clear relief. There is something tactile, visual, concrete and individualistic about Weyl's tastes and stance towards creativity. In the review of Courant-Hilbert he noted with approval that "when one has lost himself in the flower gardens of abstract algebra or topology, as so many of us do nowadays, one becomes aware here once more, perhaps with some surprise, of how mighty and fruitbearing an orchard is classical analysis" (Weyl 1938c, 596).

Weyl, who had contributed much to group theory and algebra, kept a certain distance to recent developments, as if not to be swept along with the tide, and thus preserve his individuality. He relished freedom and had been content with living in Switzerland away from the main academic centers in Germany as he told Hecke in 1925 (Sigurdsson 1991, 219-220). The stress on the panoramic and tangible was linked to a yearning for unity (Daston 1999, 75, 80, 84). He wanted to avoid becoming a mere cog in a machine in a chain of proofs, which furthermore bespoke an elitist and conservative attitude to knowledge production which abhorred it becoming increasingly industrialized, depersonalized and atomized.

1.3   The I, the collective and its memories

The difficult balancing act between the I, the manifold of possibilities and the collective runs like a red thread through many of Weyl's writings and, as he had remarked in 1925, the essence of man was at the intersection of freedom and being bound. In a 1954 autobiographical sketch, "Knowledge and Reflection" (*Erkenntnis und Besinnung. Ein Lebensrückblick*), Weyl claimed that he had reached a certain closure in his intellectual development with his investigations on group representations in the mid-1920s followed by a period of maturity and the process of growing old. He added:

> Natürlich bin ich in späteren Jahren weder an der Umwälzung vorübergegangen, welche bezüglich unseres Naturwissens die Quantenphysik herbeiführte, noch an der in der graussigen Zerrissenheit unseres Zeitalters emporgewachsenen Existenzphilosophie. Die erstere warf neues Licht auf das Verhältnis des erkennenden Subjekts zum Objekt, im Mittelpunkt der letzeren steht nicht ein reines Ich noch Gott, sondern der Mensch in geschichtlicher Existenz, der sich aus eigener Existenz entscheidet. (Weyl 1954, 648)

Weyl had emphasized in his review of the second volume of Courant-Hilbert that living men not only knew but doubted, asked, and sought meaning. His probing of the limits of field physics is a striking example of productive doubting. In 1922, utilizing the language of dialectical or idealistic materialism, Weyl referred to creative life and will when speaking of matter. It was an agent beyond space and time and the source of the disturbances in the electromagnetic field. Three years earlier lecturing to the Assembly of Natural Scientists in Lugano on the relationship between the causal and statistical point of view in physics he had discussed how the elementary constituents of matter were the site of autonomous and causally absolutely mutually independent decisions making it possible to escape from the rigid pressure of natural laws. He added:

> Diese "Entscheidungen" sind das *eigentlich Reale* in der Welt. Die moderne reine Gesetzesphysik läßt nämlich mehr und mehr erkennen, daß ihre Aussagen für die Welt lediglich jene Bedeutung haben, welche früher die (durch die Relativitätstheorie von der Physik verschluckte) *Geometrie* besaß: Festlegung des *Schauplatzes* der wirklichen Geschehnisse (und nicht der wirklichen Geschehnisse selbst). (Weyl 1920a, 122; emph. in orig.; also Sigurdsson 1991, 186)

Weyl ended the lecture by distinguishing between the inorganic and the organic realm. In the latter the correlation between the decisions made possible "eine der Kausalität entrückte selbstherrliche organisierende Potenz, das *Leben.*" If this conception was correct then it distinguished in principle between life and death. Furthermore, "[w]o *Bewußtsein* und zielsetzender, tatbegründeter Wille heraufkommt, gerät die Lebenspotenz in steigendem Maße in die *Gewalt eines rein geistigen Seins*" (ibid.; emph. in orig.).

In the celebration of life and the power (*Gewalt*) of pure spiritual being removed (*entrücken*) from the rigid pressure of the laws of physics, Weyl drew upon cultural resources equating causality with inflexibility and death as exemplified by the snow scene in *Der Zauberberg.* As Paul Forman argued thirty years ago, the longing in Central Europe for an escape from technological rigidity in the aftermath of World War I paved the way for the emergence of the new quantum mechanics (Forman 1971). For Weyl this yearning for freedom, although motivated by external events, furthermore had deep personal roots, namely how to steer the I on the vast manifold of possibilities without losing oneself in doubt and anxiety. Having mentioned Husserl, Descartes and Kirkegaard, doubt and radical despair in his 1954 autobiographical sketch, Weyl wrote:

> Durch den Zweifel hindurch stoßen wir vor zu dem Wissen um die dem immanenten Bewusstsein transzendente reale Welt; in umgekehrter Richtung aber, nicht des Erzeugnisses, sondern des Ursprungs, liegt die Transzendenz *Gottes,* aus dem herfließend das Licht des Bewusstseins, dem der Ursprung selber verdeckt ist, in seiner Selbstdurchdringung sich ergreift, gespalten und gespannt zwischen Subjekt und Objekt, zwischen *Sinn* und *Sein.* (Weyl 1954, 645; emph. in orig.)

There is something quivering, tactile, and very human about Weyl's attitude. One reaches understanding about an aspect of the world through doubting as if by groping and stumbling along in the search for meaning. This attitude was signalled by quoting from Hölderlin's poem *Die Muße* as an epigram to the 1923 fifth edition of RZM, namely which speaks of "Geist der Unruh, der in der Brust der Erd und der Menschen — Zürnet und gärt ...". Ambivalence, restlessness, doubting and a whiff of anxiety was a recurrent motif in Weyl's writings. In a popular article on Einstein's theory of relativity a few years earlier Weyl spoke about the scene of reality which was a four-dimensional fusion of space and time, an objective world which had no history, but simply was. The history of the individual was experienced differently.

> Nur vor dem Blick des in den Weltlinien der Leiber emporkriechenden Bewusstseins "lebt" ein Ausschnitt dieser Welt "auf" und zieht an ihm vorüber als räumliches, in zeitlicher Wandlung begriffenes Bild. (Weyl 1920b, 131; also Sigurdsson 1994, 365)

In a lecture at the Sixth International Congress of Philosophy at Harvard University in September 1926 Weyl utilized this vivid imagery and developed this argument further. Earlier in the lecture he had noted that the immediate experience was subjective and absolute, whereas the objective world was necessarily relative to the choice of number or other symbols used for its representation. He noted regarding this dyadic contrast: "Dieses Gegensatzpaar: subjektiv – absolut und objektiv – relativ scheint mir eine der fundamentalsten erkenntnistheoretischen Einsichten zu enthalten, die man aus der Naturforschung ablesen kann" (Weyl 1927b, 56). By referring to numbers or a symbolic construction Weyl signalled a certain reconciliation with Hilbert's viewpoint after their disagreement concerning the foundations of mathematics where Weyl had sided with L.E.J. Brouwer (1881-1966). Although Weyl did not want to discuss the possibility of metaphysics in

the Harvard lecture, the dyad was a warning signal insofar as metaphysics strived
for the objective-absolute *Sein*. Weyl stressed the epistemological importance of
this contrasting dyad in his article on the philosophy of mathematics and natural
science for the *Handbuch der Philosophie* (Weyl 1927/1966). "Wer das Absolute
will, muß die Subjektivität, die Ichbezogenheit, in Kauf nehmen; wen es zum Objektiven drängt, der kommt um das Relativitätsproblem nicht herum." In passing,
Weyl noted that Max Born had developed nicely and in a lively manner this line
of thought in his book on Einstein's theory of relativity (Weyl 1927/1966, 153).

Many aspects of Weyl's philosophical reflections and the style used to express
them, e.g., a consciousness crawling up along a worldline, are individualistic if
not outright idiosyncratic. Weyl had noted in the lecture at Harvard: "Um das
Verhältnis von Aussenwelt und wahrnehmendem Bewusstsein zu verstehen, ver-
einfache ich meinen Sinnesleib zu einem *Punktauge*. Das Punktauge beschreibt eine
Weltlinie" (Weyl 1927b, 55; emph. in orig.). His was not a Victorian demon, having
descended from Milton's *Paradise Lost*, but a disassociated organ or consciousness.
Thinking of the crawling of the consciousness, like an earthworm or an isolated
organ, however unusual may have been expedient for performing mathematical
thought experiments. Yet, the reference to Born's book hints at affinities with his
contemporaries, e.g. for Born the objective-relative part of the contrasting dyad
made it possible that the grief of the loneliness of the soul might be alleviated
and kindred spirits could form a bridge between one another (Born 1920, 4; also
Sigurdsson 1991, 115-117 and Holl 1996, 82-99).

Born's view of the role of the I was fairly conventional; progress in physics,
astronomy and chemistry meant coming closer to the goal of removing the I, not
from the act of knowing, but in terms of the finished product *Bild der Natur* (ibid.,
2). This view, akin to that of Einstein, was not espoused by Weyl who, incidentally,
rarely employed the notion of a world picture or *Weltbild* except for the picture
brought to life by the crawling of the pointlike eye or consciousness. Although
he might employ mountain imagery with its clear separation of the knower and
nature, the subject and object, the relationship between the two was filled for him
with tension. Although focussing on the I, it is doubtful that Weyl would have
subscribed to the views of the mathematician Konrad Knopp (1882-1957) who
unaffected by the new quantum mechanics, observed in 1933:

> Es ist die grundverschiedene Rolle, die dem Ich, dem Menschen, einge-
> räumt wird, was die Naturwissenschaften von den Geisteswissenschaften
> scheidet. *Für die Naturwissenschaften ist der Mensch nur die flüchtige
> Staffage, die die große Landschaft des Kosmos vorübergehend belebt.
> Für die Geisteswissenschaften ist die Welt nur die Bühne, auf der der
> Mensch handelnd auftritt.* (Knopp 1933, 205; emph. in orig.)[7]

The relationship between the I, Hermann Weyl, and the community was
frought with tension. As Weyl had confided to Hecke in 1936, he had been in

---

[7]I found reference to this article and quote in (Forman 1991, 84).

the danger of losing himself because of what was expected of him in Europe. It had, e.g., been expected of him to succeed Hilbert as professor of mathematics in Göttingen. That was a formidable task as the theoretical physicist Wolfgang Pauli (1900-1958) told Heinrich Heesch (1906-1995), one of Weyl's students in Göttingen, in 1930. Pauli had sent Heesch a list of ten behavioral commandments. He urged Heesch to show the letter to Weyl (even to everybody in Göttingen) and added:

> "Wir alle" (d.h. die ganze jüngere Mathematiker- und Physikergenera-
> tion, die noch unverdorben ist) haben Weyls Mut sehr bewundert, als er
> die Professur in Göttingen angenommen hat, denn es ist ein Posten von
> schwerer Verantwortung; wird er von dem jetzt in Göttingen herrschen-
> den Geist des beschränkten und snobistischen wissenschaftlichen Lokalpa-
> triotismus und des Herumwurstelns in (jedenfalls teilweise) leerlaufenden
> Organisation völlig verschlungen werden oder wird *er* diesen Geist be-
> siegen? "Wir alle" hoffen von Herzen das letzere und glauben auch, daß
> er zu den wenigen gehört, bei denen dieser Kampf nicht von vornherein
> aussichtslos erscheint. (Pauli to Heesch, Sept. 30, 1930 in Pauli 1993,
> 743-744; emph. in orig.; also Bigalke 1988, 69-109)[8]

Weyl neither won nor lost his "battle", for the situation in Germany changed so drastically in the next few years that Pauli's criteria were no longer applicable. Weyl left Göttingen in the fall of 1933 with his family psychologically bruised yet managed to regain some of his strength in exile (Sigurdsson 1996; also Sigurdsson 1991, 261-276). One of those he "battled" before, during and after his tenure in Göttingen was Hilbert. He described him, on the one hand, as one of the two persons he had met which struck him as a man of genius (the other was Brouwer). On the other hand, he inspired such an overweening belief in the power of mathematics that in 1944, in an obituary for Hilbert who had died the previous year, Weyl described him as the pied piper who had seduced so many rats to follow him into the deep river of mathematics (Weyl 1944a, 132; also Sigurdsson 1991, 37).

Weyl was one of the rats and he had been sufficiently confident in his second publication announcing his unified field theory in 1918 to claim that "the totality of physical phenomena could be derived from a single universal law of the highest mathematical simplicity" (Weyl 1918d, 2; my transl.). In the early 1940s Weyl had changed his mind and no longer believed in such unity nor did he consider that merely electromagnetism and gravitation should be unified. Quantum mechanics and what now would be called nuclear and elementary particle physics had led him to a moderate position. This softening process had already begun in the 1920s with the advent of quantum mechanics. In the academic year 1928-1929 Weyl was a visiting professor at Princeton University. He was approached by a journalist for an opinion regarding Einstein's recent theory of distant parallelism. Weyl said that Einstein's proposal made use of some speculative mathematical consequences in

---

[8]I thank Karl von Meyenn for having drawn my attention to this letter.

order to unify gravitation and electromagnetism, but it did not possess the same convincing character as his general theory of relativity. He stressed that in order to seek unity quantum theory, wave mechanics and material waves had to be taken into account as well (Weyl to James Stokley February 3, 1929; also Weyl 1929b, 246). Moreover, in his answer Weyl had emphasized that in contrast to the picture of science presented in *Herald Tribune*:

> Wissenschaft ist eine in sich helle, durch und durch verständliche, auf kritischem Vernunftsgebrauch und Erfahrung beruhende Analyse der Natur und nicht eine Beschwörung mittels sinnloser Zauberformeln, wie jene Herren offenbar meinen und das Publikum glauben machen wollen. Sie ist ferner ein organisches und durch die Beiträge Vieler stetig gewachsendes Kulturgebilde und besteht nicht aus einzelnen, mit Donnerschall zu verkündenden Sensationen. (Ibid.)[9]

Weyl's answer crystallizes the changes in the dynamic of science and his own self-conception in the decade following the publication of RZM. Whereas the general theory of relativity was heralded as the product of a single author, culminating in the Einstein craze and the myth of the twelve wise men who only could understand the theory of relativity, the situation in mathematical and theoretical physics was quite different in the late 1920s (Goldstein and Ritter 2000). The new quantum mechanics did not have a single author, a fact stressed by Weyl when he referred to a *Kulturgebilde*. This was also emphasized by Norbert Wiener commenting on the widespread publicity caused by the announcement of Einstein's theory of distant parallelism which he likened to a "momentary bullmarket on Einsteiniana". Wiener ridiculed the frequently repeated assertion of the twelve wise men who were only able to fathom the recondite structure of relativity theory. Wiener then mentioned that Einstein's theory rendered possible a good relativistic treatment of the Dirac equation as had been realized simulataneously by Eugene P. Wigner, and Wiener and M.S. Vallarta. He added:

> Simultaneous discoveries of this type are rather the rule than exception in scientific work. They appear especially in the new quantum theory, since every mathematical physicist is directing his full powers to the solution of the problems which have arisen. A suggestion in the work of one author immediately leads to a further attack on the problem in many quarters. There is scarcely a stage in the new quantum theory, apart from its initiation by Professor [Werner] Heisenberg himself [in 1925], which is not the work of several people simultaneously and often very remote from one another. (Wiener [1929] 1985, 914)

Wiener, who was a professor of mathematics at MIT, was in a good position to pronounce on recent trends in physics. During the academic year 1925-1926, when Born was a visiting professor at MIT, they worked together on the new

---

[9]I thank Catherine Goldstein and Jim Ritter for having made this letter available to me.

quantum theory and for some years Wiener continued to be active in that field. Wiener stressed the collective or multiple character of scientific discovery. It was formalized as a sociological theory in 1922 by William F. Ogburn and Dorothy S. Thomas. In an essay, "Singletons and Multiples in Science", Robert K. Merton entrenched this notion still further in the sociology of science. He presented an impressive list of authorities supporting this view from 1822 to the year 1922. One of the testimonies adduced by Merton was that of Einstein in 1921 (Merton [1961] 1973, 354; also Schaffer 1994). Despite believing in multiples and trying to resist uncritical adulation of himself, Einstein was rapidly ascending like a film star into the stratosphere at the end of the decade. Therefore Weyl's admonition to journalists in the United States attempting to prevent Einstein idolatry was to no avail (also Hentschel 1990, 192-195).

The situation in quantum and nuclear physics was very different than that in relativity. Bohr put his indelible stamp on the developments of the new quantum mechanics, yet he neither was nor has been associated with it as Einstein was and has been closely linked to relativity theory. Numerous factors account for the difference, e.g., scale, relationship between theory and experiment, and that between physics and mathematics. By the end of the 1920s the size of the physics community was much larger than a decade earlier and considerably more internationalized. Quantum physics had to account for a plethora of experimental and chemical facts in contrast to the dearth of facts, except for a few idealized and recalcitrant ones, in relativity theory. Differential geometry, the mathematics of relativity theory, and its extension by Weyl and others partly co-evolved with the theory itself, whereas that was much less the case for quantum theory, where established mathematical methods could be used effectively. Physicsts even viewed novel mathematical techniques skeptically, if not with hostility, as illustrated by their attitude to group theory at the end of the 1920s.

Weyl continued to distance himself from the adulation of the individual scientist, e.g., on the occasion of the seventieth birthday of Hilbert in 1932. Weyl claimed Hilbert had neither been a youth leader, an organizer, nor an altogether harmonious personality; neither had he put forth world views. On the contrary, Weyl emphasized the simple virtues of the scientific genius who with diligence and persistence pursued his goals (Weyl 1932). Nevertheless, Weyl's admiration for Hilbert shone through and indicates a tension in Weyl's self-perception between the I and the collective, those who led and were led, between humility and grandeur, between the world in which he had been socialized and the one into which he had grown and shaped as a mature scientist. In his 1944 obituary Weyl wrote of Hilbert and Hermann Minkowski (1864-1909), of the remarkable period after Minkowski came to Göttingen in 1902 until his death, and about Klein who was "like a distant but benevolent god in the clouds." Moreover, Weyl quoted from Hilbert's moving memorial address for Minkowski in which Hilbert described their science as a flowering garden, a metaphor which Weyl had employed in his review of the second volume of Courant-Hilbert (Weyl 1944a, 131-132).

Although science was a collective enterprise the question still remained: who set the research agenda? Was it a collective decision or was the direction of research set by a single person? Did the nature of the decision process depend on time and circumstances? In the 1890s Klein, a key organizing spirit of the Göttingen scientific community, spoke at the congress of mathematics and astronomy held in conjunction with the World Exhibition in Chicago in 1893. Referring to the Göttingen tradition and invoking the name of Carl Friedrich Gauß (1777-1855), Klein suggested that what had once been achieved by a single mastermind was now done by united efforts and cooperation (Klein [1893] 1922, 615; also Parshall and Rowe 1994, 304-327). Although the days of the Gaussian genius seemed to be gone, Klein's conception of the scientific enterprise actually left ample room both for the system builder and the mastermind. In a lecture in Vienna a year later on Riemann and his importance for the development of modern mathematics Klein emphasized how mathematics was propelled forward like the natural sciences. He observed: "Und auch dieses ist ein allgemeines Gesetz, daß zwar viele zur Entwicklung der Wissenschaft beitragen, daß aber die wirklich neuen Anregungen nur auf wenige hervorragende Forscher zurückgehen." The legacy of the seminal scientists was not limited to the brevity of their lives, for they were gradually understood better as time went by as was the case with Riemann (Klein [1894] 1922, 497).[10]

Although Weyl recognized the importance of the patient and persistent labor of scientists, he had imbibed the belief in the exceptional individual advocated by Klein. In Weyl's obituary for Henri Poincaré (1854-1912) it reads:

> Sein Geist hat viel Licht über unsern Weg geworfen! Aber immer sind wir noch rings vom Dunkel umgeben, dessen Aufhellung den kommenden Generationen nur durch stetige und mühsame Arbeit gelingen kann. Mögen ihnen dabei solche Führer und Fackelträger, wie Poincaré einer war, niemals fehlen! (Weyl 1912, 392)

This romantic and seemingly non-technologized world of leaders, torchbearers and pied pipers did not survive subsequent convulsions of history and had already received a fatal blow at the beginning of the Weimar Republic.

The tension between the I — the exceptional individual, the manifold of possibilities — and the collective stayed with Weyl. Although, he genuinely tried to dampen Einsteinian enthusiasm and downplay the fact that Hilbert had been a youth leader, Weyl recognized clearly his own exceptionalness and used that fact and a series of offers from other universities so skillfully during his tenure at the ETH that at the end of the 1920s he had become its highest paid professor (Frei and Stammbach 1992, 124). By adopting the mantle of Hilbert in Göttingen he fulfilled the expectations of the community as well as those of Hilbert, as Sommerfeld told Weyl in the summer of 1930 (Sigurdsson 1991, 263). But Weyl had not stayed for long in Göttingen when he felt trapped by Göttingen and its tradition. The

---

[10] I thank Umberto Bottazini for having drawn my attention to this article. See also (Ryckman 1994, 840).

rise of National Socialism disturbed him fundamentally and endangered his wife and two sons, scientific activities in Göttingen lacked their former vitality and catastrophic state finances in the wake of the great depression reduced his income and undermined his faith in the German system. Already late in 1932 he was offered for the first time membership at the Institute for Advanced Study in Princeton and a year later he and his family had left Germany for Princeton (Sigurdsson 1996, 53-54).

Among the founding members of the Institute was Einstein whose achievements and fame were vital assets for the fledgling institution. Weyl was another exceptional intellect who added to the lustre of the Institute while it in turn offered him and his family a refuge from Germany. Weyl had tried to resist Einsteiniana, yet it contributed to his own symbolic capital and to how he was perceived by other intellectuals. A striking illustration can be found in José Ortega y Gasset's *The Revolt of the Masses*. There it says in a footnote:

> Hermann Weyl, one of the greatest of present-day physicists, the companion and continuer of the work of Einstein, is in the habit of saying in conversation that if ten or twelve specified individuals were to die suddenly, it is almost certain that the marvels of physics to-day would be lost for ever to humanity. A preparation of many centuries has been needed in order to accommodate the mental organ to the abstract complexity of physical theory. Any event might annihilate such prodigious human possibilities, which in addition are the basis of future technical development. (Ortega y Gasset [1930] 1932, 52-53)[11]

The mathematician Oswald Veblen (1880-1960) was a driving force in the rise of Princeton to excellence in mathematics and the natural sciences in the 1920s (Feffer 1998). Subsequently, he became a founding member of the Institute for Advanced Study. In a lecture on geometry and physics in 1922 he discussed recent developments on the border of these fields. He spoke of the axiomatization of geometry and how it had entailed separation from philosophy and physics. This had resulted in increased clarity without any loss of contact or support. The heart of the lecture was devoted to a discussion of how the many "brilliant discoveries in physics has been making the abstract point of view a vital issue in that science also" (Veblen 1923, 129). Veblen discussed at great length the general theory of relativity and the implications of Weyl's work for the geometry of paths which he and L.P. Eisenhart were developing. Veblen recognized that an important source of novelty in mathematics was located on the border of mathematics and physics, in this case that of relativity theory. This was not an isolated observation and might, e.g., be described as a reigning doctrine in the early twentieth century in Göttingen (Sigurdsson 1991, 1-28).

---

[11] Helene Weyl (1893–1948) was the translator of the works of José Ortega y Gasset (1883-1955) into German. On the encounter of Hermann Weyl, Einstein and other German natural scientists with Spanish culture in the early 1920s, see (Glick 1988).

Since the 1890s Hilbert had hoped to treat mechanics — and mathematical physics more generally — like geometry, i.e. by the axiomatic method. Thus a certain unity of the mathematical sciences might be achieved which he envisaged as an indivisible whole, or as an organism, as he expressed it in his address in Paris in 1900 to the International Congress of Mathematicians. He expressed the same conviction in his first note on the foundation of physics in 1915 (Sauer 1999, 533-535, 544; also Corry 1997). , indexCorry This celebration of the unity highlights the positive value associated with it, as well as its protean character. What is special in the Göttingen context is how the fusion of geometry and physics was linked to a particular reading of the history of science. In his second note on the foundation of physics Hilbert invoked a familiar theme: relativity theory had made it possible to oust Euclidean geometry from modern physics as an alien action-at-a-distance law. It turned out that "Geometrie und Physik gleichartigen Charakters sind und als *eine* Wissenschaft auf gemeinsamer Grundlage ruhen" (Hilbert 1917, 64; emph. in orig.). In his 1894 lecture in Vienna Klein had asserted that one of Riemann's achievements had been to embrace the infinitesimal viewpoint and to banish action at a distance from mathematics. Klein saw a parallel between what Riemann had done in mathematics and Faraday in physics (Klein [1894] 1922, 484; Darrigol 1996, 254-258).

These references to action at a distance and its banishment indicate that in Göttingen physics was partly viewed as a model for geometry and mathematics. In the second publication announcing his unified field theory in 1918 Weyl said, referring to his infinitesimal geometry, that he had freed Riemannian geometry from its Euclidean past; he had managed to expell from it the remaining action-at-a-distance element (Weyl 1918d, 2). A year later this epistemological-historical interpretation found its way into the third edition of RZM (Weyl [1918c] 1919, 91). Moreover, Weyl relied on it when he edited and annotated Riemann's 1854 *Über die Hypothesen, welche der Geometrie zu Grunde liegen* re-issued in 1919 by Springer, the publisher of RZM, who was moving into mathematics and physics at that time (Holl 1996). Although the abandonment of the parallel postulate had played an important role in the development of non-Euclidean geometry, Riemann had accomplished much more in his *Habilitationsvortrag* according to Weyl:

> Für die Geometrie geschah hier der gleiche Schritt, den Faraday und Maxwell innerhalb der Physik, speziell der Elektrizitätslehre, vollzogen durch den Übergang von der Fernwirkungs- zur Nahewirkungstheorie: das Prinzip, die Welt aus ihrem Verhalten im Unendlichkleinen zu verstehen, gelangt zur Durchführung. (Weyl 1919c, iii; also Ryckman 1994, 838-840)

Weyl saw his own contribution to the development of his infinitesimal geometry to be a successful completion of a process begun by Riemann, i.e. to expel action at a distance to arrive at a truly infinitesimal geometry. In Weyl's first publication announcing his unfied field theory in 1918 he discussed how in Riemannian geometry a residue of action at a distance had been preserved, for no other reason than its *zufällige Entstehung* in the theory of surfaces (Weyl 1918b, 30). In an addendum Weyl answered Einstein's objection to his theory and said, referring to the infinitesimal geometry: "Es wäre merkwürdig, wenn in der Natur statt dieser wahren eine halbe und inkonsequente Nahegeometrie mit einem angeklebten elektromagnetischen Felde realisiert wäre" (ibid., 42). At the end of World War I, at a time of a deep crisis, Weyl was attempting to free geometry from the fetters of history in order to bring it to a successful completion, a utopian project pursued by bringing an historical process to an end (also Schaffer 1991, 163). Tradition and history turned out to be a burden for Weyl and he felt that Hilbert had seduced the rats to follow him into the deep river of mathematics, e.g., seeking a unified field theory. At the end of World War I history was a valuable and productive resource which made it seem possible for Weyl to escape history and the ruins of the oppressive present seeking transcendence (Weyl [1918c] 1919, vi; also Jardine 1997).

At the outset of this chapter I highlighted some of the challenges that the historian or imaginary ethnographer faces when trying to come to grips with the texture of Weyl's various *Lebenswelten* and the broader meaning of RZM. Language constitutes one of the hurdles which separates the present from the world of metaphorical possibilities in which Weyl lived. German is no longer the *lingua franca* of science but was still a viable contender in the early decades of the twentieth century. In order to heighten the linguistic distance between Weyl's past and the Anglo-Saxon present I deliberately refrained from translating from the original German. I would argue that by paying close attention to Weyl's language his growing dissatisfaction with field theory and concomitant realization that statistics plays a fundamental role in nature becomes literally palpable. Therefore it comes as no surprise that philosophy did not merely serve epistemological ends for Weyl, rather it was associated with a whole mode of being, doubting and seeking meaning and freedom. Hence it comes as little surprise that Weyl had little sympathy for logical positivism. Historians of science journeying to this past might profitably free themselves from the *Käfig* of current mainstream philosophy of science and recognize that logical positivism was only a relatively minor strand of philosophy early in the twentieth century. Weyl pulled together different kinds of philosophy, metaphors and history practising his science. Although the burden of history oftentimes nearly paralyzed him, drawing upon the legacy of the nineteenth century and viewing physics partially as a model for mathematics enabled him to gain fecund insights regarding the constitution of infinitesimal geometry. History could at times be paralyzing and at others liberating, to which RZM with its transcendent and utopian aspirations set against the wanton destruction of the Great War is a stark and vivid reminder.

Acknowledgments

This chapter has been a very long time in the making. It is with pleasure that I thank the following institutions for generous support and hospitality: The Verbund für Wissenschaftsgeschichte (Berlin), the Institute for the History of Science (Göttingen), the Alexander von Humboldt Foundation (Bonn), the German Mathematicians's Association, the staff at Schloss Reisensburg during a sunlit week in 1992, the Max Planck Institute for the History of Science (Berlin), the Icelandic Science Foundation, and the Science Institute of the University of Iceland (Reykjavík). Invitations to give talks enabled me to maintain my interest in Weyl and RZM for which I thank Ulrich Majer and Heinz-Jürgen Schmidt, Jeremy Gray, Tinne Hoff Kjeldsen, David E. Rowe, Christa Binder, the Icelandic Mathematical Society, the organizers of the Novembertagungen in the history of mathematics and those of the 1998 Oberwolfach meeting on the history of mathematics. For a chapter whose gestation period long ago exceeded the reasonable, it is impossible to present a complete list; such is mnemonic frailty. Yet, it is a tremendous pleasure to thank the following individuals for re-kindling my interest in Weyl and RZM at various times and thus pushing the chapter-project closer to completion: Michael Becker, Francesca Bordogna, Lorraine Daston, Anne Davenport, Moritz Epple, Paul Forman, Michael Fortun, Gideon Freudenthal, Hubert Goenner, Catherine Goldstein, Einar H. Gudmundsson, Michael Hagner, Karl Hall, Thomas Hawkins, Sigurdur Helgason, Thórdur Jónsson, Mechthild Koreuber, Lorenz Krüger, Andrew Mendelsohn, Everett Mendelsohn, Alexandre Métraux, Dorinda Outram, Lewis Pyenson, Jürgen Renn, Robert J. Richards, Jim Ritter, Nils Röller, Simon Schaffer, Arne Schirrmacher, Silvan S. Schweber, Otto Sibum, Reinhard Siegmund-Schultze, André Wakefield, and M. Norton Wise. Moreover, I have thanked a number of individuals in footnotes for special contributions. Finally, I want to thank Elvira Scheich for patiently bearing with me and Erhard Scholz for his extraordinary skills as an editor and scholar, and especially for his friendship.

References

Barthes, Roland [1957] 1972. The Brain of Einstein. *Mythologies.* New York: Hill and Wang, 68-70.

Bigalke, Hans-Günther 1988. *Heinrich Heesch. Kristallogeometrie, Parkettierungen, Vierfarbenforschung.* Basel etc.: Birkhäuser.

Born, Max: 1920. *Die Relativitätstheorie Einsteins und ihre physikalischen Grundlagen gemeinverständlich dargestellt.* Berlin: Springer.

Born, Max [1928] 1963. Sommerfeld als Begründer einer Schule. *Naturwissenschaften* **16**, 1035–1036. *Ausgewählte Abhandlungen* **2**, Göttingen: Vandenhoeck und Ruprecht, 604–606.

Cat, Jordi; Cartwright, Nancy; Chang, Hasok 1996. Otto Neurath: Politics and the unity of science. In: Galison, Peter; Stump, David J. (eds.); *The Disunity of Science: Boundaries, Contexts, and Power*, Stanford: University Press, 347–369.

Chevalley, Catherine 1994. Niels Bohr's words and the atlantis of Kantianism. In: Faye, Jan; Folse, Henry J. (eds.); *Niels Bohr and Contemporary Philosophy*, Dordrecht: Kluwer, 33–55.

Corry, Leo 1997. David Hilbert and the axiomatization of physics (1894-1905). *Archive for History of Exact Sciences* **51**, 83-198.

Darrigol, Olivier 1996. The electrodynamic origins of relativity theory. *Historical Studies in the Physical and Biological Sciences* **26**, 241–312.

Daston, Lorraine 1999. Die Akademie und die Einheit der Wissenschaften. In: Kocka, Jürgen; Hohlfeld, Rainer; Walther, Peter Th. (eds.); *Die Disziplinierung der Disziplinen. Die Königliche Preußische Akademie der Wissenschaften zu Berlin im Kaiserreich*; Berlin: Akademie Verlag, 61-84.

Einstein, Albert 1918. [Review of the first edition of Hermann Weyl's Raum - Zeit - Materie.] *Die Naturwissenschaften* **6**, 373.

Einstein, Albert [1922] 1994. Über die gegenwärtige Krise der theoretischen Physik. *Kaizo* (Tokyo) **4**, 1–8. Reprinted in: von Meyenn, Karl (ed.) 1994. *Quantenmechanik und Weimarer Republik.* Braunschweig: Vieweg, 233-239.

Feffer, Loren Butler 1998. Oswald Veblen and the capitalization of American mathematics: Raising money for research, 1923-1928. *Isis* **89**, 474-497.

Feynman, Richard 1965. *The character of physical law.* London: British Broadcasting Corporation.

Forman, Paul 1971. Weimar culture, causality, and quantum theory, 1918–1927: Adaptation by German physicists and mathematicians to a hostile intellectual environment. *Historical Studies in the Physical Sciences* **3**, 1–115.

Forman, Paul 1973. Scientific internationalism and the Weimar physicists: The ideology and its manipulation in Germany after world war I. *Isis* **64**, 151–180.

Forman, Paul 1991. Independence, not transcendence, for the historian of science. *Isis* **82**, 71–86.

Frei, Günther; Stammbach, Urs 1992. *Hermann Weyl und die Mathematik and der ETH Zürich, 1913-1930*. Basel etc.: Birkhäuser.

Freudenthal, Hans 1955. Hermann Weyl: Der Dolmetscher zwischen Mathematikern und Physikern um die moderne Interpretation von Raum, Zeit und Materie. In: Schwerte, Hans; Spengler, Wilhelm (eds.); *Forscher und Wissenschaftler im heutigen Europa*, Oldenburg: Stalling, 357–366.

Friedman, Michael 1995. Carnap und Weyl on the foundations of geometry and relativity theory. *Erkenntnis* **42**, 247-260.

Galison, Peter 1987. *How Experiments End*. Chicago: University Press.

Galison, Peter 1997. *Image and Logic: A Material Culture of Microphysics*. Chicago: University Press.

Galison, Peter 1998. The Americanization of unity. *Dædalus* **127**, 45–71.

Garvin, J.L. 1926. Prefatory note. *The Encyclopedia Britannica*, London, 13th ed., I, vii–xii.

Glick, Thomas F. 1988. *Einstein in Spain: Relativity and the Recovery of Science*. Princeton: University Press.

Goldstein, Catherine; Ritter, Jim 2000. The varieties of unity: Sounding unified theories 1920–1930. *Preprint* Paris.

Gruber, Howard E. [1978] 1981. Darwin's "tree of nature" and other images of wide scope. In: Wechsler, Judith (ed.); *On Aesthetics in Science*, Cambridge, Mass.: MIT Press, 121–140.

Gruber, Howard E. 1985. Going to the limit: Toward the construction of Darwin's theory (1832–1839). In: Kohn, David (ed.); *The Darwinian Heritage*, Princeton: University Press, 9–34.

Hacking, Ian 1983. *Representing and Intervening: Introductory Topics in the Philosophy of Natural Science*. Cambridge: University Press.

Hacking, Ian 1992. Disunified sciences. In: Elvee, Richard Q. (ed.); *The End of Science? Attack and Defense*, Saint Peter, Minnesota: Gustavus Adolphus College, 33–52.

Hacking, Ian 1996. The disunities of the sciences. In: Galison, Peter; Stump, David J. (eds.); *The Disunity of Science: Boundaries, Contexts, and Power*, Stanford: University Press, 37–74.

Haraway, Donna J. 1988. Situated knowledges: The science question in feminism and the privilege of partial perspective. *Feminist Studies* **14**, 575–600.

Hawkins, Thomas 1998. From general relativity to group representations: The background to Weyl's papers of 1925–26. *Matériaux pur l'histoire des mathématiques au XXe siècle*. Société Mathématique de France, Séminaire et Congrès **3**, 69–100.

Heilbron, J.L. 1986. *The Dilemmas of an Upright Man: Max Planck as Spokesman for German Science*. Berkeley: University of California Press.

Hendry, John 1984. *The Creation of Quantum Mechanics and the Bohr-Pauli Dialogue*. Dordrecht: Reidel.

Hentschel, Klaus 1990. *Interpretationen und Fehlinterpretationen der speziellen und allgemeinen Relativitätstheorie durch Zeitgenossen Albert Einsteins*. Basel: Birkhäuser.

Hentschel, Klaus [1992] 1997. *The Einstein Tower: An Intertexture of Dynamic Construction, Relativity Theory, and Astronomy*. Stanford: University Press.

Hilbert, David 1917. Die Grundlagen der Physik (Zweite Mitteilung). *Nachrichten der Königlichen Gesellschaft der Wissenschaften zu Göttingen* (Math.-phys. Klasse), 53–75.

Hilbert, David 1927. Referat über die geometrischen Schriften und Abhandlungen Hermann Weyls, erstattet der Physiko-Mathematischen Gesellschaft an der Universität Kasan. *Bulletin de la Société Phys.-Math de Kazan* **2**, 66–70.

Hobsbawm, Eric J. 1993. Homesickness. *London Review of Books*, April 8, 20–21.

Holl, Frank 1996. *Produktion und Distribution wissenschaftlicher Literatur. Der Physiker Max Born und sein Verleger Ferdinand Springer 1913-1970*. Frankfurt/Main: Buchhändler-Vereinigung.

Holton, Gerald [1985] 1986. Do scientists need a philosophy? In: *The Advancement of Science, and Its Burdens: The Jefferson Lecture and Other Essays*, Cambridge: University Press, 163–178.

Hughes, Thomas P. 1993. Einstein, inventors, and invention. *Science in Context* **6**, 25–42.

Jardine, Nicholas 1997. The mantle of Müller and the ghost of Goethe: Interaction between the sciences and their histories. In: Kelley, Donald R. (ed.), *History and the Disciplines: The Reclassification of Knowledge in Early Modern Europe*; Rochester: University Press, 297–312.

Klein, Felix: [1893] 1922. The Present state of mathematics. *Mathematical Papers International Mathematical Congress, Chicago 1893*, vol. 1, New York: Macmillan. *Gesammelte mathematische Abhandlungen* **2**, 613–615.

Klein, Felix [1894] 1923. Riemann und seine Bedeutung für die Entwicklung der modernen Mathematik. *Gesammelte mathematische Abhandlungen* **3**, 482–497.

Knopp, Konrad 1933. Der Einfluss der Naturwissenschaft auf das moderne Bildungsideal. In: *Die Universität. Ihre Geschichte, Aufgabe und Bedeutung in der Gegenwart*, +Stuttgart: Kohlhammer, 187–217.

Lorentz, Hendrik Aanton [1923] 1937. The determination of the potentials in the general theory of relativity, with some remarks about the measurement of lengths and intervals of time and about the theories of Weyl and Eddington. *Verhandlingen Akademie Wetenschapen Amsterdam* **29** (1923), 383ff. *Collected Papers* **5** Hague: Martinus Nijhof, 363–382.

Mehrtens, Herbert 1990. *Moderne, Sprache, Mathematik. Eine Geschichte des Streits um die Grundlagen der Disziplin und des Subjekts formaler Systeme.* Frankfurt am Main: Suhrkamp.

Merton, Robert K. [1961] 1973. Singletons and multiples in science. In: Storer, Norman W. (ed.), *The Sociology of Science: Theoretical and Empirical Investigations*, Chicago: University Press, 343–370.

O'Raifeartaigh, Lochlainn; Straumann, Norbert 2000. Gauge theory: Historical origins and some modern developments. *Reviews of Modern Physics* **72**, 1–23.

Ortega y Gasset, José [1930] 1932. *The Revolt of the Masses.* New York: W.W. Norton and Company.

Pais, Abraham 1982. *'Subtle is the Lord ...': The Science and the Life of Albert Einstein.* Oxford: University Press.

Parshall, Karen Hunger; Rowe, David E. 1994. *The Emergence of the American Mathematical Research Community, 1876-1900: J.J. Sylvester, Felix Klein, and E.H. Moore.* Providence/London: American Mathematical Society/London Mathematical Society.

Peckhaus, Volker 1990. *Hilbertprogramm und Kritische Philosophie: Das Göttinger Modell interdisziplinärer Zusammenarbeit zwischen Mathematik und Philosophie.* Göttingen: Vandenhoeck und Ruprecht.

Pauli, Wolfgang 1993. *Wissenschaftlicher Briefwechsel mit Bohr, Einstein, Heisenberg u.a. III: 1940-1949.* Edited by Karl von Meyenn. Berlin etc.: Springer.

Pestre, Dominique 1992. The decision-making processes for the main particle accelerators built throughout the world from the 1930s to the 1970s. *History of Technology* **9**, 163–174.

Popper, Karl [1974] 1992. *Unended Quest: An Intellectual Autobiography.* London: Routledge.

Pyenson, Lewis 1982. Relativity in late Wilhelmian Germany: The appeal to a preestablished harmony between mathematics and physics. *Archive for History of Exact Sciences* **27**, 137–155

Ryckman, Thomas A. 1994. Weyl, Reichenbach and the epistemology of geometry. *Studies in the History and Philosophy of Science* **25**, 831–870.

Ryckman, Thomas A. 1996. Einstein *agonists*: Weyl and Reichenbach on geometry and the general theory of relativity. In: Giere, Ronald N.; Richardson, Alan W. (eds.); *Origins of Logical Empiricism.* Minnesota: University of Minnesota Press, 165–209.

Sauer, Tilman 1999. The relativity of discovery: Hilbert's first note on the foundations of physics. *Archive for History of Exact Sciences* **53**, 529–575.

Schaffer, Simon 1989. The nebular hypothesis and the science of progress. In: Moore, James R. (ed.); *History, Humanity and Evolution: Essays for John C. Greene.* Cambridge: University Press, 131–164.

Schaffer, Simon 1991. Utopia limited: On the end of science. *Strategies* **4/5**, 151–181.

Schaffer, Simon 1994. Making up discovery. In: Boden, Margaret A. (ed.), *Dimensions of Creativity*, Cambridge, Mass.: MIT Press, 13–51.

Schaffer, Simon 1998. Physics laboratories and the Victorian country house. In: Smith, Crosbie; Agar, John (eds.); *Making Space for Science: Territorial Themes in the Shaping of Knowledge*, Houndmills, Basingstoke: Macmillan Press, 149–180.

Scholz, Erhard 1994. Hermann Weyl's contribution to geometry, 1917–1923. In: Dauben, Joseph; Sasaki, Chikara (eds.); *The Intersection of History and Mathematics*, Basel etc.: Birkhäuser, 203–230.

Scholz, Erhard 1995. Hermann Weyl's "purely infinitesimal geometry". *Proceedings of the International Congress of Mathematicians, Zürich, Switzerland 1994*, Basel etc.: Birkhäuser, 1592–1603.

Scholz, Erhard 2000. Hermann Weyl on the concept of continuum. In: Hendricks, Vincent F.; Pedersen, S.A.; Frovin, K. (eds.); *Proof Theory: History and Philosophical Significance*, Dordrecht: Kluwer, 213–237.

Schweber, Silvan S. 1986. The empiricist temper regnant: Theoretical physics in the United States 1920–1950. *Historical Studies in the Physical and Biological Sciences* **17**, 55–98.

Schweber, Silvan S. 1990. Gregor Wentzel 1898-1978. *Dictionary of Scientific Biography* **18**, 986–988.

Schweber, Silvan S. 1993. Physics, community and the crisis in physical theory. *Physics Today*, November, 34–40.

Schweber, Silvan S. 1995. Physics, community and the crisis in physical theory. In: Gavroglu, Kostas; Stachel, John; Wartofsky, Marx W. (eds.); *Physics, Philosophy, and the Scientific Community: Essays in the Philosophy and History of the Natural Sciences and Mathematics. In Honor of Robert S. Cohen*, Dordrecht: Kluwer, 125–152.

Shapin, Steven 1991. "The mind is its own place": Science and solitude in seventeenth-century England. *Science in Context* **4**, 191–218.

Shapin, Steven 1997. Signs of the times. *Social Studies of Science*, **27**, 335–349.

Sigurdsson, Skúli 1991, Hermann Weyl, Mathematics and Physics, 1900–1927, *Ph.D. Dissertation* Harvard University.

Sigurdsson, Skúli 1992. Einsteinian fixations. *Annals of Science* **49**, 577–583.

Sigurdsson, Skúli 1992. Equivalence, pragmatic platonism, and discovery of the calculus. In: Nye, Mary Jo; Richards, Joan L.; Stuewer, Roger H. (eds.); *The Invention of Physical Science: Intersections of Mathematics, Theology and Natural Philosophy Since the Seventeenth Century. Essays in Honor of Erwin N. Hiebert*, Dordrecht: Kluwer, 97–116.

Sigurdsson, Skúli 1994. Unification, geometry and ambivalence: Hilbert, Weyl and the Göttingen community. In: Gavroglu, Kostas; Christianidis, Jean; Nicolaïdis, Efthymios (eds.); *Trends in the Historiography of Science*, Dordrecht: Kluwer, 355–367.

Sigurdsson, Skúli 1996. Physics, life, and contingency: Born, Schrödinger, and Weyl in exile. In: Ash, Mitchell G.; Söllner, Alfons (eds.) *Forced Migration and Scientific Change: Emigré German-Speaking Scientists and Scholars after 1933*, Cambridge: University Press, 48–70.

Stachel, John 1999. The dawning of gauge theory. *Studies in History and Philosophy of Modern Physics* **30**, 453–455.

Straumann, Norbert 1987. Zum Ursprung der Eichtheorien bei Hermann Weyl. *Physikalische Blätter* **43**, 414–421.

Veblen, Oswald 1923. Geometry and physics. *Science* **57**, 129–139.

von Plato, Jan 1994. *Creating Modern Probability: Its Mathematics, Physics and Philosophy in Historical Perspective*. Cambridge: University Press.

Vizgin, Vladimir P. [1985] 1994. *Unified Field Theories in the First Third of the 20th Century*. Translated from the Russian by J. B. Barbour. Basel etc.: Birkhäuser.

Weinberg, Steven 1993. *Dreams of a Final Theory: The Search for the Fundamental Laws of Nature*. London: Vintage.

Wiener, Norbert 1924. Four books on space. *Bulletin American Mathematical Society* **30**, 258–262. *Collected Works with Commentaries* **4**, Cambridge, Mass.: MIT Press, 979-983.

Wiener, Norbert 1929. Einsteiniana: Facts and fancies about Dr. Einstein's famous theory. *Technical Rec.* **32**, 403–404. *Collected Works with Commentaries* **4**, Cambridge, Mass.: MIT Press, 913–914.

Archival Material

Weyl, Hermann, February 3, 1929. Letter to James Stokley. Copy in hands of Catherine Goldstein and Jim Ritter.

2 Weyls Infinitesimalgeometrie, 1917 – 1925

Erhard Scholz

2.1 Kontinuum, Grundlagenfragen und kulturelle Zweifel

Als Weyl sich ab Herbst 1916 der gerade neu entstehenden Relativitätstheorie zuwandte, befand er sich in einem tiefgreifenden und vielschichtigen Klärungsprozeß über die Rolle der Mathematik in der Naturerkenntnis und über mögliche Wege einer zufriedenstellenden Begründung der Mathematik des Unendlichen. Als einen ersten Teilschritt dieser Klärung verfaßte er seine im November 1917 abgeschlossene Studie *Das Kontinuum*. In dieser stellte er die Grundzüge einer *konstruktiven* Begründung der Analysis vor, die das Russell-Poincarésche Verbot "imprädikativer Definitionen" einhielt.[1] In Rücksichtnahme auf diese Kritik ließ Weyl hier lediglich solche Definitionen zu, die Existenz- und Allaussagen nur über den natürlichen Zahlen (oder über n-Tupeln natürlicher Zahlen) verwendeten. Modernisiert gesprochen, beschränkte er sich also auf definitorische Terme, die in Prädikatenlogik erster Stufe über IN formulierbar sind. Damit erschien Weyl insbesondere die Konzeptualisierung der *Gesamtheit der reellen Zahlen*, also der Menge IR, als unzulässig. An die Stelle des mathematischen Begriffs der reellen Zahlen setzte er stattdessen ein philosophisches Konzept des (reellen) *Kontinuums*, das als ein nur potentiell Unendliches den intuitiv-philosophischen Rahmen für die mathematischen Theoriebildungen abgeben sollte. In diesem Kontinuum spielte sich für Weyl der mathematische Prozeß der geregelten symbolischen Erzeugung von Objekten und der Formulierung gesetzmäßiger Zusammenhänge ab. Dabei experimentierte er in verschiedenen Stadien seiner Arbeit mit unterschiedlichen Regelsystemen und skizzierte *verschiedene* mögliche Rahmungen des mathematischen Kontinuumsbegriffs.

Von 1916/17 bis 1919 dachte Weyl an die Verwendung einer eingeschränkten prädikatenlogischen Sprache zweiter Stufe über IN. All- und Existenzaussagen erster Stufe galten ihm bis 1919 als völlig akzeptabel; dagegen akzeptierte er Quantorisierungen zweiter Stufe nur eingeschränkt. Aus Weyls Sicht war der Allgemeinbegriff der Prädikate über IN *nicht umfangsdefinit*, also nicht im Sinne einer "sinnvollen" Mengenbildung extensional abschließbar. Quantorisierende Aussagen erschienen ihm in diesem Bereich damit als "evident sinnlos" (Weyl, 1919a, 45). Erst nach einer konstruktiven Spezifizierung (und damit Einschränkung) erschienen sie ihm als zulässig. Einen möglichen Weg zu einer solchen Spezifizierung beschrieb er in *Das Kontinuum* (Weyl 1918a) durch Einführung einer rekursiv aufgebauten Sprache formalisierbarer arithmetischer Ausdrücke. Mit diesen

---

[1] Siehe dazu den Beitrag von R. Coleman und H. Korté in diesem Band.

Ausdrucksmitteln ließ sich eine Teilklasse der reellen Zahlen charakterisieren, die hier als W bezeichnet werden soll und grob gesprochen die durch arithmetisch-konstruktiv angebbare Dedekindsche Schnitte charakterisierbaren reellen Zahlen umfaßte. Das Weylsche arithmetisch definierbare Kontinuum W von 1917/18 besaß die bemerkenswerte Eigenschaft einer relativen "inneren" Vollständigkeit: jede arithmetisch definierbare Cauchyfolge[2] in W besitzt in W einen Grenzwert. Dies ließ W für eine reduzierte Form der Analysis geeignet erscheinen. Reduziert blieb Weyls Kontinuum insbesondere deswegen, weil nicht jede beschränkte Menge ein arithmetisch-konstruktiv definierbares Supremum besitzt.[3]

Die von Weyl 1918 vorgeschlagenen symbolischen Ausdrucksmittel waren aufgrund ihrer Sparsamkeit und Konstruktivität ein bemerkenswerter und produktiver Beitrag zur Klärung möglicher Begründungsstrukturen der reellen Analysis. Weyl selber war jedoch trotz der erreichten Teilerfolge aus philosophischen Gründen schon zum Zeitpunkt der Niederschrift der Arbeit mit seinem Ansatz nicht wirklich zufrieden.

Unter dem Einfluß seines Philosophen-Kollegen in Zürich, Fritz Medicus, entwickelte er nämlich zu dieser Zeit starkes Interesse an der dialektischen Philosophie Fichtes, das seine ältere Ausrichtung an der Husserlschen Phänomenologie ergänzte und radikalisierte. Schon seine an Husserl angelehnten Überlegungen zur "phänomenalen Zeitanschauung" gaben ihm Anlaß, das Kontinuum als nicht in seine einzelnen Bestimmungsweisen, die Punkte, auflösbar zu denken. Fichte bestand in seiner dialektischen Entfaltung des Konzepts des Raumes darauf, auch diesen aus "unendlich kleinen Teilen" aufzubauen, die selber jeweils wieder ein Kontinuum bilden sollten. Jeder solche "unendlich kleine Teil" sollte nach seiner Ansicht vom "Nicht-Ich" als "Sphäre der Wirksamkeit" durch "Intensitäten" oder "Kräfte" hervorgebracht werden und in dieser Weise begrifflich vom "Ich" bestimmt werden. Der Raum wurde von Fichte damit als der "notwendige" Zusammenhang verschiedener "Sphären der Wirksamkeit" charakterisiert, in dem die "freie Wirksamkeit von Ich und Nicht-Ich" in Wechselwirkung treten (Fichte 1802, 194–207).

Weyl hatte offenbar schon länger Husserls Einwände dagegen ernst genommen, eine auf die transfinite Mengenlehre gestützte Theorie des mathematischen Kontinuums auf das Zeitkontinuum anzuwenden. Die Fichtesche Dialektisierung des Raumbegriffs legte nun ganz ähnliche Zweifel auch für das räumliche Kontinuum nahe.[4] Die daraus resultierende Skepsis und Kritik am vorherrschenden Denken der modernen Mathematik wurde für Weyl ein wichtiges Motiv bei der Ausarbeitung seiner "rein infinitesimalgeometrischen" Auffassung der Differentialgeometrie noch im Laufe des Jahres 1918.

Hinsichtlich der Grundlagenfragen der Analysis äußerte sich Weyl noch ein

---

[2]Die Folgenzuordnung selber, nicht nur die Werte, sind dabei als arithmetisch definierbar vorausgesetzt.

[3]Eine genauere Darstellung findet sich im Beitrag von Coleman/Korté §6.3, in diesem Band. Siehe auch (Feferman 1988) und S. Fefermans Beiträge in (Hendricks 2000).

[4]Erste Hinweise auf Fichte als Orientierungspunkt Weyls finden sich in (Weyl 1918a, 2, 70ff.). Zum Fichteschen Einfluß auf Weyl siehe auch (Scholz 1995).

Jahr später in seinem als (Weyl 1919a) publizierten Brief an Otto Hölder im Sinne seines Ansatzes der Kontinuumsschrift von 1917/18. Brouwer, der 1918 begonnen hatte, sein Programm der intuitionistischen Mengenlehre auszuarbeiten und zu publizieren (Brouwer 1918/19), wurde von Weyl zu diesem Zeitpunkt nicht einmal erwähnt. Das änderte sich allerdings während des Jahres 1919 einschneidend.

Im Sommer 1919 trafen sich Weyl und Brouwer im Engadin, und Weyl entdeckte in Brouwers intuitionistischem Konzept der "freien Wahlfolgen" einen Zugang zu einer mathematischen Theorie des Kontinuums, der seinen philosophischen Überzeugungen entgegenzukommen schien.[5] Diese neugewonnene Überzeugung machte er zum Gegenstand seines bekannten polemischen Artikels über die "neue Grundlagenkrise der Mathematik", dessen Manuskript er schon im März 1920 an Brouwer sandte. Im Spätsommer 1920 kam es auf der Bad Nauheimer Tagung der *Gesellschaft deutscher Naturforscher und Ärzte* vom 19. bis 25. September 1920 zu einer erneuten persönlichen Begegnung zwischen Brouwer und Weyl. Auf dieser Tagung gab es für Weyl zwei einschneidende Ereignisse. Das erste war die von Lenards Attacken auf Einsteins Relativitätstheorie ausgelöste Debatte, in die Weyl unter Verwendung seines Konzeptes des (gravito-inertialen) *Führungsfeldes* auf der Seite Einsteins eingriff, um zur Klärung der Scheinparadoxa beizutragen, die Lenard gegen Einsteins Theorie ins Feld führte.[6] Gleichzeitig propagierte er bei dieser Gelegenheit seine eigene feldtheoretische Verallgemeinerung der Relativitätstheorie.

Das zweite für Weyl einschneidende Ereignis war Brouwers Beitrag, in dem dieser sein Programm einer intuitionistischen Mengenlehre und Analysis einer breiteren Wissenschaftsöffentlichkeit vorstellte. Auch in den darüberhinausgehenden persönlichen Gesprächen versuchte Brouwer auf diesem Treffen zunächst erfolgreich, Weyl in Sachen Grundlagenfragen auf seine Seite zu ziehen. Er überreichte Weyl einen Ruf an die Universität Amsterdam und schlug vor, auf diese Weise eine enge Kooperation für eine intuitionistische Neubegründung der Mathematik aufzunehmen. Weyl war für kurze Zeit von dieser Vorstellung einer gemeinsamen Arbeit mit Brouwer sehr eingenommen, wie seinem Brief an den Präsidenten der ETH zu entnehmen ist (Frei/Stammbach 1992, 48ff.).[7]

Mathematisch war Weyl insbesondere von Brouwers Vorschlag fasziniert, die einzelnen Bestimmungsweisen im Kontinuum durch "freie Wahlfolgen" zu repräsentieren, weil dies eine relativ große Offenheit der zulässigen Symbolbildungen mit sich brachte. Dies kam dem Fichteschem Motiv, das räumliche Kontinuums als etwas zu fassen, in dem das "Ich mit Freiheit die Bestimmung" des Zusammenhangs der infinitesimalen Wirkungssphären herstellt (Fichte 1802, 199), jedenfalls schon ein Stück näher als als sein eigenes arithmetisch-konstruktives Verfahren von

---

[5] Siehe dazu (van Dalen 1995, 147ff.) und (van Dalen 1999, 306ff.).

[6] Siehe (Pais 1986, 32), (Goenner 1993).

[7] Für eine breitere kulturhistorische Einordnung des Wechsels von Weyls arithmetischem Konstruktivismus zur Orientierung an Brouwers intuitionistischer Begründung der Mathematik siehe (Mehrtens 1990); eine ausführliche Untersuchung der Rezeption von Brouwers Intuitionismus findet sich bei (Hesseling 2000).

1917/18. In Weyls Sprache lag Brouwers großes Verdienst darin, das Kontinuum als "Medium freien Werdens" in die mathematische Symbolproduktion einzuführen. Das machte in seiner Sicht den Brouwerschen Ansatz seinem eigenen "radikal atomistischen" von 1917/18 überlegen, bei dem "die Begriffe nur ein starres Sein zu erfassen vermögen" (Weyl 1921c, 159).

Diese "Freiheit" war allerdings verbunden mit einer Abschwächung der klassischen logischen Schlußweisen der Mathematik, in denen ja das *tertium non datur* im Sinne der gerade entstehenden Prädikatenlogik erster Stufe eine unaufhebbare Rolle spielte. Brouwers Forderung einer Bindung der logischen Schlußregeln an eine intuitionistische Verifizierung quantorisierter Aussagen machte eine solche Abschwächung notwendig. Weyl übernahm diese Einschränkung ab 1920 und machte zur Semantik dieser Einschränkung einige sehr aufschlußreiche Anmerkungen mit starken Anlehnungen an das Kreditwesen. Darin ließ Weyl, wie auch an andere Stellen seiner Argumentation, zur Motivierung mathematischer Postulate soziale Metaphorik in die legitimierende Erläuterung symbolischen Vorgehens in der Mathematik einfließen:[8] Eine Allaussage interpretierte er als "Urteilsanweisung" vergleichbar einer stets einlösbaren Schatzanweisung, und eine (nicht konstruktive) Existenzaussage als "Urteilsabstrakt", vergleichbar einem "Papier, welches das Vorhandensein eines Schatzes anzeigt, ohne zu verraten, an welchem Ort" (Weyl 1921c, 156ff.).[9]

Andererseits gibt es keinen Hinweis darauf, daß sich Weyl die spezifischen Züge der Brouwerschen intuitionistischen Analysis mit ihren Konsequenzen etwa für den Funktionsbegriff ernsthaft zu eigen gemacht hätte. Seine Bewertung der Brouwerschen Errungenschaften als Verbindung "höchster intuitiver *Klarheit* mit *Freiheit*" (ebda., 179), dementierte er in ungewollter Schärfe noch in demselben Artikel durch das offenherzige Eingeständnis:

... mit vielen seiner (Brouwers, E.S.) Aussagen gelingt es mir nicht, einen Sinn zu verbinden. (ebda., 170)

Diese eklatant widersprüchlichen Bewertungen in ein und demselben Artikel resultierten nur teilweise aus dem bis dahin noch relativ unentwickelten Stand der Brouwerschen intuitionistischen Analysis; sie deuteten auch mögliche, noch kaum sichtbare Differenzen zwischen Brouwer und Weyl bezüglich Konzept und Methode des Aufbaus einer intuitionistischen Analysis (Brouwer) gegenüber einer Analysis des "reinen Kontinuums" an, wie sie Weyl vorschwebte. Vielleicht lassen sie sich sogar als erste, wenn auch noch ganz und gar unausgereifte Anzeichen einer sich erst später voll entfaltenden, der intuitionistischen Begründungsstrategie inhärenten methodologischen Problematik lesen. Dieser Fragestellung kann ich hier nicht

---

[8]Auch in dieser Hinsicht gibt es eine gewisse Parallele zwischen Weyl und der dialektischen Philosophie des deutschen Idealismus. — Ein anderes Beispiel für den Einsatz sozialer Metaphorik zur Legitimierung mathematischer Postulate findet sich in Weyls *Analyse des Raumproblems* (Weyl 1923c, 46), siehe p. 92.

[9]Vergleiche dazu (Majer 1988).

weiter nachgehen.[10]

Hier müssen wir es bei der bloßen Beobachtung belassen, daß Weyl selbst im Zeitraum stärkster Annäherung an Brouwer (in den Jahren 1919/20 bis 1922) dessen intuitionistische Analysis keineswegs im ganzen übernahm, obwohl er vehement eine intuitionistisch inspirierte symbolische Auffassung innerhalb der Mathematik vertrat. Auf den Sachverhalt, daß er mit seiner Konzentration auf die "freien Wahlfolgen" und die Übernahme der Kritik am *tertium non datur* eine durchaus eigenständige Adaptation lediglich ausgewählter Züge der Brouwerschen Analysis vornahm, verwies er dabei ganz explizit (1921c, 158, 170f.).

Dies war, aus späterer Sicht beurteilt, noch immer nicht mehr als ein Durchgangsstadium in der langsamen und schrittweisen Herausbildung von Weyls "reifer" Auffassung über die Grundlagenfragen der Mathematik.[11] Nach 1923 begann er, Hilberts Weg einer axiomatischen Vorgehensweise formalsymbolischer Begründung der Analysis und der Mathematik eine gewisse Berechtigung einzuräumen. Dabei unterschied er deutlich zwischen der logischen Möglichkeit einer solchen Fundierung der Mathematik, der er ab Mitte der 1920-er Jahre eine gewisse Erfolgschance einräumte, und der Problematik einer Sicherung ihrer kulturellen Bedeutung, die er an eine tiefe Beziehung zwischen Mathematik und Naturwissenschaft geknüpft sah (Weyl 1925a, 540ff.). Seine Bereitschaft, Hilberts Beweistheorie zumindest als eine formale Absicherung auf dem Weg aus der "Grundlagenkrise" zu akzeptieren, wurde allerdings durch Gödels Resultate wieder zutiefst erschüttert. So plädierte er ab den 1930-er Jahren grundsätzlich für eine Pluralität der Begründungsmethoden, in denen er neben der Hilbertschen Beweistheorie und dem Brouwerschen Intuitionismus auch dem von ihm 1918 eingeschlagenen, aber unter Brouwers Einfluss zeitweise ganz aufgegebenen Weg einer konstruktiven Sicherung mathematischer Theorien wieder eine nahezu gleichberechtigte Stellung einräumte (Weyl 1946, 277ff.). Diese späte Wiederaufnahme des konstruktiven Ansatzes von 1918 durch Weyl wird von historisch interessierten Vertretern einer konstruktiven oder prädikativen Begründung der Mathematik aus naheliegenden Gründen gerne zu Weyls "eigentlicher" Position erklärt. Aus historischer Sicht liegt darin allerdings eine Stilisierung. Die Auffassung des reifen Weyl ist treffender durch das Bewußtsein einer anhaltenden Krise einer bloß formalen Absicherung der Mathematik gekennzeichnet, auf die er vorzog, durch Pluralität der formalen Begründungsmethoden bei gleichzeitiger aktiver Sorge um die kulturelle Verankerung der Mathematik zu reagieren (Weyl 1946, 1948). Der Aspekt der methodischen Pluralität als grundlegender Zug der Mathematik ist auch das Hauptthema des späten, aus den 1950-er Jahren stammenden Manuskriptes zu *Axiomatic versus constructive procedures in mathematics*, das posthum als (Weyl 1985) publiziert worden ist.

Die kulturelle Verankerung der Mathematik suchte Weyl bis in die 1940-er

---

[10]Insbesondere erwies sich das Konzept der "intuitionistischen Verifizierung" problematischer, als von Brouwer und Weyl ursprünglich erwartet. Siehe dazu (Epple 1999).

[11]Für eine Rahmenbeschreibung der verschiedenen Phasen in Weyls Auffassung der Grundlagenfragen siehe auch (Beisswanger 1965). Für die ersten beiden Phasen vergleiche auch (Leupold 1961).

Jahre mit großer innerer Überzeugung in einer engen Verbindung mit der Philosophie, der Naturtheorie und insbesondere der Schwesterwissenschaft Physik. Mit letzterer wünschte er sich sogar die Mathematik, so weit sie "ernsthafte Kulturangelegenheit" und nicht nur formales Spiel sein soll, zeitweise sogar bruchlos zu verschmelzen (Weyl 1925a, 540f.). Die Erfahrungen des zweiten Weltkrieges und die Beteiligung der mathematischen Physik an der Konstruktion der Atombombe erschütterte jedoch wiederum schon wenige Jahre später seinen "Glauben" (ebda.) an die Sinnhaftigkeit der kulturellen Verankerung der Mathematik über die Verbindung zur Physik in der Geschichte der Moderne. Nun stellte er nicht nur den Glauben an die Zweckmäßigkeit ungehemmter und unspezifizierter "symbolischer Konstruktion der Wirklichkeit" durch die mathematischen Wissenschaften infrage, sondern sogar deren moralische Zulässigkeit.

Eine der deutlichsten Formulierungen dieses Zweifels des späten Weyl findet sich in einem Nachlaßmanuskript über *Entwicklungslinien der Mathematik seit 1900*. In diesem 1949 (oder später) auf englisch gehaltenen Vortrag zog Weyl eine bemerkenswerte Linie von Hardy zu Aristoteles und wieder zurück zu den gerade erlebten Schrecken seiner Zeit, die ich hier etwas ausfürlicher zitieren möchte.

> ... Here some words of Aristotle come to my mind which, to be true, refer to metaphysics rather than mathematics. Stressing its uselessness as much as Hardy does in his apology of mathematics, but at the same time its divinity, he says (Metaphysics 982b): 'For this reason its acquisition might justly be supposed to be beyond human power; since in many aspects human nature is servile; in which case, as Simonides says, 'God alone can have this privilege', and man should only seek the knowledge which is of concern to him ($\tau\grave{\eta}\nu\ \varkappa\alpha\vartheta$' $\alpha\mathring{\upsilon}\tau\grave{o}\nu\ \mathring{\epsilon}\pi\iota\sigma\tau\acute{\eta}\mu\eta\nu$). Indeed if the prets are right and the Deity is by nature jealous, it is probable that in this case they would be particularly jealous and all those who step beyond ($\pi\acute{\alpha}\nu\tau\alpha\varsigma\ \tau o\grave{\upsilon}\varsigma\ \pi\epsilon\rho\iota\tau\tau o\acute{\upsilon}\varsigma$) are liable to misfortune.'
>
> I am not so sure whether we mathematicians during the last decades have not 'stepped beyond' the human realm by our abstractions. Aristotle, who actually speaks about metaphysics rather than mathematics, comforts us by hinting that the envy of the Gods is but a lie of the prets ('prets tell many a lie' as the proverb says).
>
> For us today the idea that the Gods from which we wrestled the secret of knowledge by symbolic construction will revenge our $\H{\upsilon}\beta\rho\iota\varsigma$ has taken on a quite concrete form. For who can close his eyes against the menace of our self-destruction by science; the alarming fact is that the rapid progress of scientific knowledge is unparalleled by a congruous growth of man's moral strength and responsibility, which has hardly chance in historical time. (Weyl, Hs 91a:72, p. 6f.)

In diesen Zweifeln des späten Weyl am kulturellen Sinn und sogar der Zulässigkeit des eigenen wissenschaftlichen Tuns, reflektierten sich ähnlich schon wie in seinen Beiträgen zu den Grundlagenfragen der Mathematik nach Ende des ersten Weltkrieges die Erfahrungen eines Mathematikers in den "Kataklysmen des kurzen 20. Jahrhunderts" (Hobsbawm). Hier ist jedoch nicht der Ort für eine eingehendere Untersuchung dieser gesellschafts- und kulturhistorischen Einflüsse auf Weyls Auffassung der Mathematik[12] oder der verschiedenen von Weyl jeweils ein Stück weit verfolgten nichtklassischen Begründungsansätze der reellen Analysis.

Wichtig bleibt für unseren Kontext, daß Weyl in seinen Arbeiten zur Differentialgeometrie in den hier interessierenden Jahren (1917 – 1925) auf keine ausgearbeitete mathematische Theorie des reellen Kontinuums zurückgreifen konnte, die seinen philosophischen Intentionen in ausreichendem Maße hätte genügen können. Philosophische Konzepte waren für ihn aus diesem Grunde von besonders grundlegender Bedeutung. Dieser Rahmen prägte auch seinen spezifischen Blick auf Riemann und speziell dessen Konzept der Mannigfaltigkeiten und der Differentialgeometrie.

2.2 Mannigfaltigkeitsbegriff

Es waren wohl gerade diese Offenheiten und Ungeklärtheiten im Konzept des Kontinuums, die Weyl zu tiefen und produktiven Überlegungen über Begriff und Struktur differenzierbarer Mannigfaltigkeiten anregten. Schon 1912, als er als junger Privatdozent seine rasch bekannt gewordene Vorlesung über Riemannsche Funktionentheorie hielt, beteiligte er sich an der Ausarbeitung eines präzisen Begriffs der stetigen und differenzierbaren Mannigfaltigkeit, ohne allerdings den begrifflichen Unterschied dieser beiden Aspekte hervorzuheben. So stellte er in Anlehnung an Hilberts Zusatzbemerkung zur Axiomatik der euklidischen Ebene von 1902 eine axiomatische Charakterisierung zweidimensionaler Mannigfaltigkeiten durch zwei Postulate auf (Weyl 1913a, 17f.):[13]

Eine zweidimensionale Mannigfaltigkeit oder Fläche $\mathcal{F}$ wird durch folgende Bestimmungen gegeben: Angabe einer "Gesamtheit von Dingen, die *Punkte der Mannigfaltigkeit* $\mathcal{F}$ heißen" sowie zu jedem Punkt p Auszeichnung von Mengen als Umgebungen U_p von p. Zu jeder Umgebung U_p sei weiterhin eine Bijektion φ_p von U_p auf eine (offene) Kreisscheibe D in der euklidischen Ebene mit Zentrum O gegeben, wobei $\varphi_p(p) = O$. Für dieses System gelte:

(1) Ist q in U_p und U_q Umgebung von q mit $U_q \subseteq U_p$, so liegt $\varphi_p(q)$ im Inneren von $\varphi_p(U_q)$.

(2) Ist q in U_p, $q' = \varphi_p(q)$ und $D' \subseteq D$ eine weitere Kreisscheibe mit Zentrum q', so gibt es eine Umgebung U_q von q mit $\varphi_p(U_q) \subseteq D'$.

Weyl führte, gestützt auf seine Postulate (1) und (2), topologische Konzepte in $\mathcal{F}$ im Sinne der Punktmengentopologie des $\mathbb{R}^n$ ein: Häufungspunkte, isolierte

---

[12] Siehe dazu etwa (Sigurdsson 1991, chap. II) und seine Ausführungen in Kap. 1 diese Bandes.

[13] Siehe auch den Beitrag von R. Coleman und H. Korté, Abschnitt 3.2.

Punkte, innere Punkte einer Teilmenge, Gebiete etc. Dabei unterstellte er ohne weitere logische Analyse, daß die im Sinne der Postulate (1) und (2) ausgezeichneten Umgebungen und Koordinatenabbildungen φ_p auf $\mathcal{F}$ in kohärenter Weise auf topologische Konzepte führen.[14] Er deutete lediglich durch Bezug auf die von Brouwer für bijektive und stetige Abbildungen des $\mathbb{R}^n$ kurz zuvor bewiesenen Gebietsinvarianz an (Brouwer 1911a), daß er von den in den Postulaten (1), (2) ausgezeichneten Koordinatenabbildungen φ_p erwartete, daß sie "umkehrbar eindeutige und umkehrbar gebietsstetige" Abbildungen (also Homöomorphismen)[15] auf das Innere der Kreisscheibe seien (ebda., 19f.). Weiter verwies er ausdrücklich auf die Möglichkeit, ein- und dieselbe Fläche $\mathcal{F}$ durch verschiedene Umgebungssysteme zu charakterisieren, wenn nur die zugrundeliegende "Gesamtheit der Dinge p" gleich ist und — modern gesprochen — die Umgebungsbasen äquivalent sind im Sinne einer wechselseitigen Einschachtelungsmöglichkeit.

Schon aus Gründen des Entwicklungsstandes der Punktmengentopologie ist es kaum verwunderlich, daß Weyl im Wintersemester 1911/12 keinen Anlaß hatte, den Mannigfaltigkeitsbegriff in einen ausgearbeiteten punktmengentopologischen Rahmen zu stellen. Er hielt seine Vorlesung ja zwei Jahre vor der Veröffentlichung von Hausdorffs Axiomatik — wenn auch in einer überraschenden Koinzidenz fast zeitgleich mit Hausdorffs ersten Schritten zur Ausarbeitung von dessen Axiomen.[16] Aber die Distanz zu Hausdorff war nicht nur zeitlicher Natur. Bei Abfassung von *Das Kontinuum* waren Hausdorffs *Grundzüge* weit genug bekannt, daß Weyl die logische Analyse des Mannigfaltigkeitsbegriffs in diese Richtung hätte weiterführen können. Das paßte aber offensichtlich nicht zu seinem konstruktiven Bild der Mathematik und dem Wunsch, einen genetischen Kontinuumsbegriff auch bei der Charakterisierung von Flächen oder Mannigfaltigkeiten symbolisch möglichst treu nachzubilden.

So finden wir zwar einen expliziten Verweis auf Hausdorffs Umgebungsaxiome (Weyl 1918a, 79), aber keineswegs als positiv akzeptierte Modifikation von Weyls

---

[14]Das wäre auf zwei Weisen zu erreichen gewesen: (i) durch die Stetigkeitsforderung an die sich für $U_p \cap U_q \neq \emptyset$ ergebenden Übergangsabbildungen $\varphi_{p,q}$ aus der euklidischen Kreisscheibe D in diese; oder (ii) die Verschärfung der Forderungen an das Umgebungssystem so, daß dieses als Basis der Topologie eines Hausdorffraumes dienen kann. Die letztgenannte Modifikation nahm Weyl erst für die dritte Auflage des Buches im Jahr 1955 durch entsprechende Ergänzungen vor. (Zu zwei Umgebungen von p im Sinne von Postulat (1), (2) gibt es immer eine, die in beiden enthalten ist; sowie Forderung des Hausdorffschen Trennungsaxioms.) Den Hinweis auf diese (bei Hilbert 1902 nicht auftretende) Lückenhaftigkeit der Weylschen Axiomatik von 1912/13 verdanke ich R. Remmert.

[15]Weyl verwendete die Bezeichnung "gebietsstetig" für offene Abbildungen.

[16]Hausdorff machte im März 1912 (genauer in einem datierten Manuskript vom 2. 3. 1912) beim Studium Riemannscher Flächen erste Schritte einer logischen Analyse von Umgebungsbeziehungen. Dies geschah offenbar unabhängig von Weyl. Seine Axiomatik führte Hausdorff in vorläufiger Form während seiner Vorlesung im Sommersemester 1912 bei einer logischen Analyse von Umgebungssystemen im $\mathbb{R}^n$ ein. Bei der (verallgemeinerten) Publikation der Endfassung seiner Axiomatik des topologischen Raumes in seinen *Grundzügen der Mengenlehre* verwies er in einem Nachtrag auf die Parallele zu Weyls (1913a) (Hausdorff 1914, 457). Es spricht aber nichts für einen gegenseitigen Austausch der beiden Mathematiker oder einen Einflusses von Weyl auf Hausdorff. Siehe dazu auch (Scholz 1996, 120ff.)

eigenen, logisch nicht ganz scharfen Ausführungen von 1912/13, sondern lediglich als ergänzenden Hinweis zum Vergleich mit seinem eigenen Zugang. Weyl sprach den Grund für diese distanzierte Haltung gegenüber dem punktmengentopologischen Ansatz am Beispiel der Einführung einer Fläche als Punktmenge deutlich aus:

> Wie aber den stetigen Zusammenhang zwischen diesen Punkten fassen, der sie zur zweidimensionalen Fläche eint? Nachdem wir das Kontinuum in isolierte Punkte zerrissen haben, fällt es jetzt schwer, den auf der Unselbständigkeit der einzelnen Punkte beruhenden Zusammenhang nachträglich durch ein begriffliches Äquivalent wiederherzustellen. Ich schlage im wesentlichen dasselbe Verfahren ein, das ich in dem Analysis-situs-Teil meines Buches *Die Idee der Riemannschen Fläche* befolgt habe. (Weyl 1918a, 79)

Weyl führte an dieser Stelle die Idee einer konstruktiven Vorgabe von Umgebungssystemen ein Stück weiter aus. Mit jedem Punkt P einer Fläche $\mathcal{F}$ ist eine Relation $U(P,Q;n)$ gegeben, "deren Bestehen für zwei Flächenpunkte P, Q und die natürliche Zahl n durch die Worte ausgedrückte wird: Q liegt in der n-ten Umgebung von P" (ebda, 80). Für die zugehörigen Umgebungssysteme $U_{P,n}$ gelten die Postulate:

(1) P liegt in jeder Umgebung $U_{P,n}$ und $U_{P,n+1} \subseteq U_{P,n}$.

(2) Für jeden Punkt P von $\mathcal{F}$ gibt es eine stetige und umkehrbar stetige Bijektion φ_P von $U_{P,1}$ auf das Innnere des Einheitsquadrats Q der "Zahlenebene".[17] Stetigkeit wird dabei in naheliegender Weise durch das geschachtelte Umgebungssystem $\mathcal{U}_P = \{U_{P,n} | n \in \mathbb{N}\}$ präzisiert.

(3) Es gibt ein abzählbares Netz $X \subseteq \mathcal{F}$, das durch eine konstruktiv gedachte injektive Funktion $\lambda : \mathbb{N} \longrightarrow \mathcal{F}$, mit $Bild(\lambda) = X$ und X dicht in $\mathcal{F}$, gegeben wird ("dicht" präzisiert durch die Umgebungssysteme).

Weyl sah diese Definition weiterhin als provisorisch an. Er kritisierte dabei insbesondere den "Übelstand" der Willkürlichkeit der Auswahl der Umgebungssysteme und der Problematik der Spezifizierung der Abbildung λ und schränkte daher ein:

> Wir nehmen diese Willkür (ebenso wie die Willkür der Gegenstände, die wir als Flächenpunkte figurieren lassen) in Kauf, weil hier offenbar doch noch keine reinliche Lösung der Frage vorliegt, wie das Band zwischen dem Gegebenen und dem Mathematischen in klarer Weise zu knüpfen sei (ebda., 82).

---

[17] "Zahlenebene" darf hier nicht bedenkenlos als $\mathbb{R}^2$ gelesen werden, sondern ist je nach Kontext zu verstehen, hier im Sinne der Weylschen reellen Kontinuums der arithmetisch-konstruktiven Theorie von 1917/18, also in unserer Notation als $\mathcal{W}^2$.

Mitte der 1920-er Jahre, als Weyl sich schon der Darstellungstheorie der Lie-gruppen als seinem neuen mathematischen Hauptarbeitsgebiet zugewandt hatte, wurde er von D. F. Egorov im Namen der Russischen Akademie der Wissenschaften für die geplante Herausgabe der Werke N.I. Lobatschewskis um einen Übersichts-artikel über die Geometrie gebeten. Weyl nahm dies zum Anlaß, im Jahre 1925 einen Bericht über *Riemanns geometrische Ideen, ihre Auswirkungen und ihre Ver-knüpfung mit der Gruppentheorie* zu schreiben (Weyl 1925/1988).[18] In dieser Zeit hatte er seine radikalste Phase in den Grundlagen der Mathematik schon hinter sich und betrachtete die Hilbertsche Auffassung der Mathematik und die Mengen-lehre mit größerer Offenheit; aber noch immer stand er der formal-axiomatischen Methode mit einer gewissen, nun aber eher semantisch und kulturell begründeten Skepsis gegenüber. In diesem Essay formulierte er ein ausgewogenes Panorama des Ausarbeitungsstandes des Begriffs der topologischen und differenzierbaren Man-nigfaltigkeit um die Mitte der 1920-er Jahre, ohne dabei seine grundsätzliche Kritik an einer mengentheoretischen Fundierung des Kontinuumsbegriffs zu verschweigen.

Weyl ließ keinen Zweifel an der zentralen Rolle des Mengenbegriffs für die moderne Mathematik. Nach einer knappen Paraphrase von Riemanns Charak-terisierung einer *Mannigfaltigkeit* als Allgemeinbegriff mit verschiedenen Bestim-mungsweisen, gleich ob mit stetigen oder diskreten Übergängen, kommentierte er ohne Zögern aus der Sicht des 20. Jahrhunderts:

> In moderner Formulierung fällt der Begriff der Mannigfaltigkeit (ab-gekürzt: Mf) geradezu mit dem der Menge zusammen; die Elemente der Menge mögen fortan als Punkte bezeichnet werden. (Weyl 1925/1988, 3)

Die in der Ausarbeitung der Mengensprache bis hin zur zeitgenössischen Cantor-Fraenkel-Zermeloschen Mengenlehre zum Ausdruck kommende Neigung zur Präzisierung mathematischer Begriffe durch extensionalisierende Formulierun-gen, war für Weyl also so weit erst einmal unbestritten, gleich welche Auffassung man in den Grundlagenfragen einnahm.[19] Sogar seine Meinung zur Frage, ob ohne eine konstruktive Spezifizierung der Darstellungsmittel sinnvoll ("umfangsdefinit") von der Gesamtheit der Teilmengen einer Menge, also schon etwa von $\mathcal{P}(\mathbb{N})$, ge-sprochen werden kann und damit zumindest indirekt, welche Mengenbegriffe als Kandidaten für im Weylschen Sinne sinnvolle Mannigfaltigkeiten zur Verfügung stehen, hielt er an dieser Stelle zurück. Auffällig ist dagegen die direkt folgende Erklärung des Stetigkeitskonzepts:

> Die Mf ist stetig, wenn die Punkte so miteinander verwachsen sind, daß es unmöglich ist, einen Punkt für sich herauszuheben, vielmehr immer nur zusammen mit einem vag begrenzten Hof, mit einer "*Umgebung*". Mathematisch ist das so zu fassen ....

---

[18] Dieser Artikel ist nicht Teil der *Gesammelte Abhandlungen* Weyls und wurde erst im Jahre 1988 durch K. Chandasekharan publiziert.

[19] Zur Verbreitung der Mengenkonzepte in der Mathematik und der Rolle Riemanns in diesem Prozess siehe (Ferreiros 1999).

Weyl erläuterte diese intuitive Formulierung, daß ein Punkt nicht "atomistisch" sondern immer mit einem kontinuumskonstituierenden "Hof" zu denken sei, in mehreren Stufen. Er begann mit einer kurzen Diskussion von abzählbaren Umgebungsbasen zu jedem Punkt der Mannigfaltigkeit M in Anspielung auf seine Vorgehensweise in *Das Kontinuum* von 1917/18, einschließlich der dort erwähnten Kritik am "Übelstand", daß der Umgebungsbegriff nicht willkürfrei eingeführt sei. Mittlerweile hatte H. Tietze die axiomatische Charakterisierung topologischer Räum durch Systeme offener Mengen fomuliert (Tietze 1923). Weyl wählte eine entsprechende Vorgehensweise, allerdings ohne Tietze explizit zu erwähnen.

> Man vermeidet diese Willkür, wenn man als Grundbegriff den des *Gebietes* statt der Umgebung wählt. Er muß den folgenden Forderungen genügen: Der Durchschnitt zweier Gebiete ist wieder ein Gebiet; zu zwei verschiedenen Punkten P, Q existieren stets punktfremde Gebiete, deren je eines P, deren anderes Q enthält. (ebda. 4)

Auffälligerweise vewendete Weyl hier weiterhin die von Hausdorff 1914 vorgeschlagene Terminologie von *Gebieten* anstelle der von Tietze vorgezogenen Formulierung der *offenen Mengen*. Das ist ein relativ starkes Indiz dafür, daß Weyl mit seiner hier zitierten Beobachtung direkt an die entsprechende Diskussion der Fundamentaleigenschaften offener Mengen ("Gebiete") bei Hausdorff (Hausdorff 1914, 215ff.) anschloss, ohne Tietzes Ausführungen zu kennen.[20]

Weyl arbeitete die Rolle der Stetigkeitsforderung an die Koordinatenwechsel in einem vorurteilslosen und genau formulierten historischen Verweis auf Hilbert (1902), Hausdorff (1914) und seinen eigenen Beitrag von (1913) — nicht in einer formellen Definition — heraus: Hilbert habe diese schon 1902 (für die Ebene) gefordert, allerdings sei erst durch seinen und Hausdorffs Beitrag möglich geworden, die Stetigkeit auch der Koordinatenabbildung selber zu formulieren.

Diese Beobachtung ergänzte er durch einen hochinteressanten physikalisch-semantischen Kommentar. Der Koordinatenbegriff werde durch diese Verallgemeinerung auf beliebige stetige und stetig invertierbare Koordinatenwechsel von allen speziellen Konstruktionen der klassischen Geometrie gelöst und abstrakter gefaßt. Vom Standpunkt der Relativitätstheorie bedeute das eine Lösung der Koordinaten von physikalischen Meßprozessen, und eine Setzung als Struktur, die "a priori willkürlich in die Welt hineingelegt" werde, um relativ dazu die physikalischen

> ...Zustandsfelder einschließlich der metrischen Struktur zahlenmäßig charakterisieren zu können. ...Deutlicher tritt dadurch der Raum als Form der Erscheinungen seinem realen Inhalt gegenüber: der Inhalt wird gemessen, nachdem die Form willkürlich auf Koordinaten bezogen ist. (ebda. 4f.)

---

[20] Hausdorff hatte deutlich hervorgehoben, daß seine Terminologie von der in der Analysis durch Weierstrass eingeführten Bezeichnung insofern abwich, als er für seine "Gebiete" keinen Zusammenhang forderte. Er motivierte diese Abweichung durch die treffende Bemerkung: "Der Begriff einer Menge ohne Randpunkte ist aber so fundamental, daß er entschieden ein eigenes Substantiv verdient." (Hausdorff 1914, 215, Anm. 1)

Daß der "a priori in die Welt hineingelegte" Raumbegriff als Form der Erscheinungen durch mathematische Begriffsschöpfung hervorgebracht wird, die so allgemein sein muß, daß — wie Riemann es formulierte — die Naturerkenntnis "nicht
durch die Beschränktheit der Begriffe gehindert und der Fortschritt im Erken-
nen des Zusammenhangs der Dinge nicht durch überlieferte Vorurtheile gehemmt
wird" (Riemann 1854, 286), hieß für Weyl jedoch *keineswegs*, daß die größtmögliche logische Verallgemeinerung für den zugrunde gelegten Begriff zu wählen sei.
Weyl räumte ein, daß man bei der Abstraktion von Meßprozessen im Sinne der
zeitgenössischen modernen Mathematik durchaus noch weiter gehen könne als in
seiner Auffassung der Relativitätstheorie. Er machte jedoch solchen weiteren Verallgemeinerungen gegenüber einen grundsätzlichen Zweifel geltend:

> Die Mengenlehre, kann man sagen, geht (...) noch weiter; sie reduziert
> die Mf auf eine Menge schlechthin und betrachtet auch den stetigen Zu-
> sammenhang schon als ein in ihr bestehendes Feld. Es ist aber wohl
> sicher, daß sie dadurch gegen das Wesen des Kontinuums verstößt, als
> welches seiner Natur nach gar nicht in eine Menge einzelner Elemente
> zerschlagen werden kann. Nicht das Verhältnis von Element zur Menge,
> sondern dasjenige des Teils zum Ganzen sollte der Analyse des Konti-
> nuums zugrunde gelegt werden. (ebda., 5)

Dies waren deutliche Worte, die zum Ausdruck brachten, daß Weyl im Unterschied zur Mehrheit der zeitgenössischen Mathematiker weiterhin an der Legiti-
mität metaphysischer Überlegungen ("Wesen des Kontinuums") zur Bestimmung
tiefliegender Strukturen innerhalb der mathematischen Begriffsbildung ("stetige
Mannigfaltigkeit") festhielt. Darüberhinaus griff seine Kritik ja nur bei Annahme
des unausgesprochenen methodologischen Prinzips, daß die jeweils allgemeinste
innermathematische Begriffsebene bei der Begründung von Raumkonzepten als
"'Form der Erscheinung" zu dienen habe und nicht vielleicht schon eine komple-
xere Gestalt der mathematischen Begrifflichkeit (topologischer Raum statt un-
strukturierter Menge im Beispiel). Daß andere zeitgenössische Mathematiker —
etwa Hausdorff, um nur einen zu nennen — mit einem solchen methodologischen
Postulat keineswegs einverstanden gewesen wären, sei hier nur angemerkt.

Wie nun andererseits das Postulat, das Kontinuum nicht "in eine Menge
einzelner Elemente zu zerschlagen", mathematisch auszuformulieren sei, konnte
Weyl in der Mitte der 1920-er Jahre nur provisorisch andeuten. So diskutierte er
die Frage, ob in den Untersuchungen zur Dimensionstheorie bei Poincaré, Brouwer
und Menger Ansätze für eine Theorie des Kontinuums im Sinne des von ihm
aufgestellten Postulats zu sehen seien, kam aber zu einem negativen Ergebnis. Er
schloß die Diskussion mit dem Hinweis ab:

> Zu einem vollständigeren Resultat scheint man nur gelangen zu können,
> wenn man das Kontinuum nicht als Gesamtheit einzelner Punkte faßt,
> zwischen denen die Umgebungsbeziehungen bestehen, sondern als ein
> Gebilde, das fortgesetzter Teilungen fähig ist, ohne daß man je auf
> letzte, nicht mehr teilbare Elemente stößt. (ebda., 7)

Eine Präzisierung dieses Gedankens erhoffte sich Weyl durch Rückgriff auf Methoden der kombinatorischen Topologie. Die Idee dazu hatte er schon in seinem 1921-er Artikel *Über die neue Grundlagenkrise der Mathematik* im Sinne einer Auffassung einer Mannigfaltigkeit als *kombinatorische Mannigfaltigkeit* zusammen mit einer unendlichen Folge baryzentrischer Unterteilungen angedeutet. Dabei hatte er Brouwers kombinatorische Definition von Mannigfaltigkeiten und die simpliziale Approximation stetiger Abbildung als Paradigma vor Augen.

Bei der Überarbeitung seines Kommentars zu Riemanns *Hypothesen* für die zweite Auflage pries er zwei Jahre später in diesem Sinne Brouwers Vorgehensweise in (Brouwer 1911b) als "Modell für eine genetische Konstruktion durch fortgesetzte Teilung, bei welcher das Kontinuum nicht mehr atomistisch, als ein System einzelner diskreter Elemente aufgefaßt wird".[21] Weyl ging also im kombinatorisch-topologischen Sinne vom Schema eines simplizialen Komplexes aus, das die — von ihm in abgeschwächter Form angegebene — Sternbedingung erfüllt.[22]

Dabei verschwieg er natürlich nicht die Schwierigkeit, für die Sternbedingung wiederum ein rein kombinatorisches Kriterium anzugeben. Er schlug einen pragmatischen Weg vor, diesem Problem zunächst durch Angabe notwendiger kombinatorischer Bedingungen für Homöomorphie zur $(n-1)$-Sphäre einerseits, Auflistung der bekannten kombinatorischen Schemata der $(n-1)$-Sphären andererseits aus dem Wege zu gehen. Diese Bedingungen sollten in Zukunft "nach oben" und "nach unten" ergänzt werden, bis dies "hoffentlich eines Tages" zu einer eindeutigen Begriffsbestimmung führe (ebda., 109). Er nannte diese Vorgehensweise einen "axiomatischen Weg". Das ist aufschlußreich für Weyls Methodenverständnis, insbesondere die Rolle, die er Axiomen im Forschungsprozeß beimaß. Für genauere Ausführung dieses Gesichtspunktes verwies er auf seine spanischen Vorlesungen (Weyl 1923a, 1924a) sowie auf (Veblen 1922).[23]

Ein *Punkt* der Mannigfaltigkeit erschien Weyl dann als nichts anderes als eine Idealisierung einer unendlichen Folge ineinander geschachtelter Sterne in einer unendlichen Folge baryzentrischer Unterteilungen.

> Unter einem *Punkt der Mannigfaltigkeit* ist eine unendliche Folge solcher Sterne zu verstehen, in welcher jeder Stern ganz im Innern des nächstvorhergehenden enthalten ist; der Sinn dieser Einschachtelungsbedingung zwischen zwei Sternen ist leicht zu formulieren. (Weyl 1921c, 178)

Erscheint diese Charakterisierung aus Sicht der mengentheoretisch fundierten Topologie als recht — vielleicht sogar unnötig — verwickelt, so hatte sie doch aus

---

[21](Weyl 1919c, [2]1923, 23); in der ersten Auflage fehlte dieser Hinweis noch.

[22]"...es ist zu fordern, daß diejenigen Elemente niederer Stufe, welche ein Element n^{ter} Stufe begrenzen, mögliches Teilungsschema nicht einer beliebigen (n-1)-dimensionalen Mf, sondern insbesondere einer $(n-1)$-dimensionalen *Kugel* im n-dimensionalen euklidischen Raum bilden" (Weyl 1925/1988, 10). Allgemeiner hätte er fordern können, daß dies für jeden *Simplex-Stern* einer Teilung gilt, das heißt der Vereinigung aller derjenigen Simplices, die mit einem gegeben 0-dimensionalen Simplex inzidieren.

[23]Mehr dazu in (Scholz 1999b).

Weyls Perspektive den großen Vorzug, als Nachbildung eines Prozesses gelesen werden zu können, bei dem ein Beobachter Netze sich beliebig verfeinernder Bestimmungsweisen für Raumzeitereignisse in die Welt legt, *ohne* dazu einen *physikalischen Meßprozeß* zu benötigen, sondern lediglich qualitative Lokalisierungsprozeduren. In diesem Sinne wäre der Begriff der Analysis-Situs-Mannigfaltigkeit so gefaßt, daß die Ortsbestimmung innerhalb der Mannigfaltigkeit mit symbolischen Mitteln erfolgt, die zumindest im Prinzip einen metrikfreien semantischen Bezug zur Raumzeit-Mannigfaltigkeit andeuten. Weyl war sich dabei natürlich bewußt, daß die Spezifizierung der "beliebigen Verfeinerung" eine mathematische Idealisierung ist, die eine direkte physikalisch realistische Interpretation immer noch unmöglich macht — siehe unten.

Für *differenzierbare* Mannigfaltigkeiten verwies er mathematisch auf den Aspekt der notwendigen Einschränkung des Koordinatenbegriffs, so daß die Übergangsfunktionen differenzierbar sind (Weyl 1925/1988, 12).[24] Dies ergebe die Möglichkeit, in jedem Punkt P der Mannigfaltigkeit einen Tangentialraum an die Mannigfaltigkeit so anzuheften, daß die *infinitesimalen Umgebungen des Ursprungs* des Tangentialraums mit der *infinitesimalen Umgebung von* P identifiziert werden.

> Zu jedem Punkt P einer Mf gehört der zentrierte affin-lineare Raum der Vektoren im Punkte P; die von P ausstrahlenden Linienelemente sind die unendlichkleinen Vektoren. Hierdurch bringen wir lediglich zum Ausdruck, daß in zwei Koordinatensystemen die Differentiale an der Stelle P durch homogene lineare Transformation miteinander zusammenhängen. Die Vektoren bilden, wie man auch sagen kann, den tangierenden linearen Raum, dessen Zentrum den Punkt P deckt und von welchem die unendlichkleine Umgebung des Zentrums mit der unendlichkleinen Umgebung des Punktes P der gegebenen Mf affin zur Deckung gebracht ist. (Weyl 1925/1988, 24)

Diese nicht wesentlich weiter präzisierte Unterscheidung von endlichem Tangentialraum $T_p M$ an eine Mannigfaltigkeit M im Punkt p und infinitesimaler Umgebung von p in M und 0 in $T_p M$, die wiederum in der Angabe der differenzierbaren Struktur von M miteinander "zur Deckung gebracht" werden, war für Weyls Verständnis der Differentialgeometrie entscheidend und spielte schon 1918 bei der Ausarbeitung seiner "reinen Infinitesimalgeometrie" eine wichtige Rolle. Wir werden darauf zurückkommen.

Aus heutiger Sicht verweisen Weyls Formulierungen auf die komplizierte Beziehung zwischen topologischer, kombinatorischer und differenzierbarer Struktur von Mannigfaltigkeiten, die Mitte der 1920-er Jahre noch weitgehend im Dunkeln lagen. Während diese Beziehung für die Dimensionen $n = 2, 3$ in dem Sinne unproblematisch ist, daß jede topologische Mannigfaltigkeit dieser Dimensionen eine kombinatorische Struktur besitzt — somit auch aus Weyls Sicht eine begriffliche

---

[24] Veblen und Whitehead gaben ihre systematische Begründung des Mannigfaltigkeitsbegriffes durch Auszeichnung von Atlanten und Übergangsfunktionen erst gut fünf Jahre später; vgl. dazu etwa (Scholz 1999b).

Existenzberechtigung hat — und in eindeutiger Weise "glättbar" ist, also mit einer differenzierbaren Struktur versehen werden kann, gilt dies für höhere Dimensionen nicht. Deutlich wurde das zum erstenmal in den späten 1950-er Jahren durch Milnors Entdeckung "exotischer" differenzierbarer Strukturen auf der 7-Sphäre (Milnor 1956). Die Klärung der Existenz kombinatorischer und differenzierbarer Strukturen auf Mannigfaltigkeiten der Dimension $n \geq 4$ machte erst seit dieser Entdeckung rasche Fortschritte und führt auf eine lebendige und unabgeschlossene Fragestellung der Topologie der Mannigfaltigkeiten. Anfang der 1980-er Jahre trat schließlich durch die Arbeiten von Freedman und Donaldson die überraschende Vielfalt des Auseinanderfallens von differenzierbarer und topologischer Struktur in der Dimension 4 hervor.

Es kann hier nur angemerkt werden, daß die großen Fortschritte der 4-dimensionalen Differentialtopologie durch Rückgriff und Ausbau modernisierter Fassungen von Eichfeldstrukturen (Yang-Mills Theorie) möglich wurden, die in ihrem begrifflichen Kern auf Weyls "reine Infinitesimalgeometrie" und E. Cartans Theorie der differentialgeometrischen Zusammenhänge zurückgehen. Die historische Erforschung der dazwischen erfolgten Neu- und "Um-" Entdeckungen steht noch aus.[25]

Ähnlich wie Riemann verwies Weyl darauf, daß im Fall der physikalischen Raumzeit die Differenzierbarkeitsannahme in ein höchst problematisches Feld führt, weil diese im Unendlichkleinen die Gültigkeit der klassischen Elementargeometrie voraussetze. Da die Begriffe in ihrer Anwendung schon im endlichen Bereich einen "gewissen Grad von Vagheit und die Tatsachen nur approximative Geltung besitzen" schien es für Weyl unter Bezug auf die Quantenphysik wiederum nahe zu liegen, "jene Vagheit durch einen Wahrscheinlichkeitsansatz wiederzugeben". Wie allerdings differenzierbare und quantenstochastische Struktur untereinander begrifflich und beide zusammen wiederum mit der physikalischen Empirie in Verbindung zu bringen sind, mußte er natürlich offen lassen. So endeten Weyls Ausführungen zum Mannigfaltigkeitsbegriff in seiner Lobatschewski-Schrift mit einer skeptischen Reflexion über die Beziehung von infinitesimaler Struktur der Mannigfaltigkeiten zur Mikrophysik.

> Es bleibt ein Problem, wie dieser Sachverhalt (die Gültigkeit der "Elementargeometrie im Unendlichkleinen", E.S.) in seiner realen Bedeutung päzis zu formulieren ist. Es muß zugegeben werden, daß für diese Frage nach der Bedeutung der Differentialrechnung in ihrer Anwendung auf die Wirklichkeit noch fast nichts geleistet ist. (ebda., 12)

Weyl zielte mit dieser Bemerkung auf den Übergang von quantenphysikalischen Strukturen zur differenzierbaren Struktur der Raumzeit-Mannigfaltigkeit. Obwohl mittlerweile hier ein vielfältiges Forschungsfeld entstanden ist, erscheint Weyls drastische Bemerkung auch aus heutiger Sicht weiterhin höchst aktuell. Wir kehren

---

[25]Für eine erste historische Information über die topologischen Entwicklungen siehe (James 1999), bezüglich der "Neu"-entdeckung der Eichfeldtheorien in der Physik unter anderen (Hoddeson e.a. 1997), (Cao 1997), (O'Raifearteagh 1995) und (O'Raifearteagh/Straumann 2000).

aber hier nach dieser Exkursion zum Mannigfaltigkeitskonzept erst einmal wieder zu Weyls Überlegungen zur Differentialgeometrie um 1918 zurück.

2.3 "Rein infinitesimalgeometrische" Metrik

Nach Abschluß seiner Kontinuumsschrift von 1917/18 wollte Weyl auch die mathematische Strukturierung der Differentialgeometrie grundsätzlich neu durchdenken. Wie dargelegt, erschien es ihm als sinnvoll, die differentialgeometrischen Strukturen strikt aus den "unendlich kleinen Teilen des Raumes" als den eigentlich "Sphären der Wirksamkeit" (Fichte) aufzubauen. Aus diesem Blickwinkel schien ihm in Riemanns Charakterisierung der differentialgeometrischen Metrik ein Mangel darin zu liegen, daß ein Längenvergleich von Vektoren an verschiedenen Stellen der Mannigfaltigkeit *direkt* möglich ist. Er sah darin eine methodische Inkonsequenz, weil sie ein unzulässiges "ferngeometrisches Element ... ohne jeden sachlichen Grund" in die Basisstruktur einbaute, und erklärte dies aus der "zufälligen Entstehung dieser Geometrie aus der Flächentheorie" (Weyl 1918b, 30). Demgegenüber forderte er:

> *Eine wahrhafte Nahegeometrie darf jedoch nur ein Prinzip der Übertragung einer Länge von einem Punkt zu einem unendlich benachbarten kennen,* und es ist dann von vornherein ebensowenig anzunehmen, dass das Problem der Längenübertragung integrabel ist, wie sich das Problem der Richtungsübertragung als nicht integrabel herausgestellt hat.
> (Weyl 1918b, 30; Hervorhebung im Original)

Mit der letzten Bemerkung spielte Weyl auf die von Levi-Civita 1917 eingeführte Konzeption der Parallelverschiebung in Riemannschen Mannigfaltigkeiten an,[26] die Weyl nach den Nachforschungen von Skúli Sigurdsson schon während seiner Vorlesung über Relativitätstheorie an der ETH im Sommersemester 1917 als grundlegende konzeptionelle Struktur diskutiert hatte, weil sie sich ganz in den Rahmen seines Strebens eines Aufbaus der Differentialgeometrie aus "rein infinitesimalem" Gesichtspunkten einfügte (Sigurdsson 1991, 154). So stellte er schon in der ersten Auflage von *RZM* [27] die Levi-Civitasche Parallelverschiebung in Riemannschen Mannigfaltigkeiten ohne Bezug auf die Einbettung in einen höherdimensionalen euklidischen Raum vor. Er verwendete dabei die kovariante Ableitung eines Vektorfeldes $\xi = (\xi^i)$, unter Verwendung der Einsteinschen Summenkonvention:

$$\lambda_k^i := \frac{\partial \xi^i}{\partial x^k} + \Gamma^i{}_{jk} \xi^j \tag{2.3.1}$$

Er argumentierte, daß ξ längs eines Weges $x(t) = (x^i(t))$ als *parallel verschoben* angesehen werden müsse, wenn die gewöhnliche infinitesimale Veränderung des Vek-

---

[26]Siehe dazu (Reich 1992) und (Bottazzini 1999).

[27]"RZM" steht hier und im folgenden als Abkürzung für das Buch *Raum Zeit Materie* (Weyl 1918c). Die einzelnen Auflagen werden durch Jahresangabe mit vorangestelltem Superskript für die Auflagennummer gekennzeichnet, Beispiel (RZM [4]1921).

torfeldes $\delta\xi = (\delta\xi^i)$ für jede infinitesimale Ortsveränderung δx längs des Weges gerade vom (Kovarianz herstellenden) Korrekturterm aufgehoben wird (RZM [1] 1918, 100f.):

$$\delta\xi^i := \frac{\partial\xi^i}{\partial x^k}\delta x^k = -\Gamma^i{}_{jk}\xi^j\delta x^k \tag{2.3.2}$$

Das stimmte mit Levi-Civitas Interpretation überein und ermöglichte Weyl, den Blick auf die intrinsische infinitesimale Struktur der Mannigfaltigkeit zu richten. Wenig später löste Weyl diese Struktur aus ihrer Bindung an die Riemannsche Metrik, formulierte sie als eigenständigen Begriff des "affinen" Zusammenhangs einer Mannigfaltigkeit, entwickelte seine verallgemeinerte metrische Struktur (Weylsche Metrik) und baute darauf seinen Vorschlag für eine einheitliche geometrische Feldtheorie von Elektromagnetismus und Gravitation auf. Er publizierte diese Ideen zunächst in den beiden Aufsätzen *Reine Infinitesimalgeometrie* und *Gravitation und Elektrizität* (Weyl 1918d, 1918b). und übernahm sie dann bei der Überarbeitung für die dritte Auflage in *Raum Zeit Materie*.

Ein *affiner Zusammenhang* Γ einer differenzierbaren Mannigfaltigkeit M bestand für Weyl aus einer (von p differenzierbar abhängigen) linearen Zuordnung infinitesimaler linearer Transformationen $d\gamma$ im endlichen Vektorraum an jedem Punkt $p \in M$ zu jeder infinitesimalen Translation dx in der infinitesimalen Umgebung von p. In Koordinatendarstellung sei etwa $dx = (dx^i)$, dann liegt $d\gamma$ in der Lie-Algebra der linearen Gruppe $gl(n,\mathbb{R})$, und die lineare Abhängigkeit von dx konnte Weyl notieren als

$$d\gamma^i_j = \Gamma^i{}_{jk}dx^k \tag{2.3.3}$$

mit endlichen Koeffizienten $\Gamma^i{}_{jk}$, wobei die Summationskonvention als gültig unterstellt wird. Die Transformation der $\Gamma^i{}_{jk}$ bei Koordinatenwechsel unterliegt dabei Konsistenzbedingungen, die mit denen im Fall des Levi-Civita-Zusammenhangs formell identisch sind. Weiter forderte Weyl die Existenz *geodätischer Koordinatensysteme* im Sinne der Möglichkeit, bei vorgegebenem Punkt p_0 die Koordinaten so zu wählen, daß *in diesem Punkt* die Koeffizienten des Zusammenhangs verschwinden $\Gamma^i{}_{jk}(p_0) = 0$. Dies führte ihn auf die (dazu äquivalente) Symmetriebedingung für den Zusammenhang in beliebigen Punkten p

$$\Gamma^i_{jk}(p) = \Gamma^i_{kj}(p) \tag{2.3.4}$$

In modernisierter Sprache führte Weyl also mit seinem *affinen Zusammenhang* einen torsionsfreien linearen Zusammenhang im $GL(n,\mathbb{R})$-Hauptfaserbündel der Tangentialstruktur ein.

Damit konnte Weyl analog zum Fall des Levi-Civita-Zusammenhangs die *kovariante Ableitung* von Vektorfeldern bezüglich Γ sowie die Konzepte *Parallelverschiebung, geodätische Line* und *Krümmungstensor* definieren (Weyl 1918d, 6–12, RZM [3] 1919, 117–121). Für Weyl erschien die rein mathematisch erklärte Struktur so tief mit einer physikalischen Semantik vequickt, daß er auch in seiner an eine

mathematische Leserschaft gerichtete Einführung des affinen Zusammenhangs die
Erläuterung anschloß:

> In physikalischer Ausdrucksweise ist ein affin zusammenhängendes Kon-
> tinuum als eine Welt zu bezeichnen, in der ein *Gravitationsfeld* herrscht.
> Die Größen Γ_{rs}^i sind die Komponenten des Gravitationsfeldes. (Weyl
> 1918d, 8)

Bei Überarbeitung von *RZM* zur vierten Auflage im Sommer 1920, in der
Zeit der Bad Nauheimer Naturforscherversammlung, ging Weyl zur Bezeichnung
Führungsfeld über, um die grundlegende Bedeutung der affinen Struktur bei der
Diskussion relativitätstheoretischer Fragestellungen noch deutlicher zu kennzeich-
nen.

Dem Programm eines Aufbaus der Differentialgeometrie aus dem "rein Infinitesimalen" folgend, sah Weyl die konforme Struktur $[g]$ zu einer semi-riemann-
schen Metrik g (gegeben durch eine nichtentartete quadratische Form $g_{ij}dx^idx^j$)
begrifflich als wesentlich grundlegender an als die semiriemannsche Metrik selbst.
Es gab für ihn zwei Gründe, sein verallgemeinertes Konzept einer Metrik darauf
aufzubauen. Zum einen war er ja schon bei seinen Studien der Riemannschen Funktionentheorie auf die Wichtigkeit der konformen Struktur gestoßen und bemerkte
nun, daß sich diese auch sehr gut in sein "rein infinitesimales" Programm einfügte:

> Machen wir keine weitere Voraussetzung (als die einer konformen Struk-
> tur, E.S.), so bleiben die einzelnen Punkte der Mannigfaltigkeit in me-
> trischer Hinsicht vollständig gegeneinander isoliert. Ein metrischer Zu-
> sammenhang von Punkt zu Punkt wird erst dann in sie hineingetragen,
> wenn ein *Prinzip der Übertragung der Längeneinheit von einem Punkte
> P zu seinem unendlich benachbarten* vorliegt. Statt dessen machte RIE-
> MANN die viel weitergehende Annahme, daß sich Linienelemente nicht
> nur an derselben Stelle, sondern auch an irgend zwei endlich entfernten
> Stellen der Länge nach miteinander vergleichen lassen. *Die Möglichkeit
> einer solchen "ferngeometrischen" Vergleichung kann aber in einer rei-
> nen Infinitesimalgeometrie durchaus nicht zugestanden werden.* (Weyl
> 1918d, 14, Hervorhebung im Original)

Zusätzlich zu dieser methodischen Überlegung eines "rein infinitesimalen"
Aufbaus der Differentialgeometrie wird für Weyl die Kausalstruktur der allgemeinen Relativitätstheorie eine vergleichbar wichtige Rolle gespielt haben. Schon in
der ersten Auflage von *RZM* diskutierte er die Kausalstruktur in einer Lorentz-
mannigfaltigkeit im Sinne der 4-Parameterschar von Doppelkegeln "der aktiven
Zukunft und der passiven Vergangenheit", einschließlich der jedenfalls rein theoretisch denkbaren Rekurrenz der "Weltlinie meines Leibes" mit der surrealen Kon-
sequenz eines "radikaleren Doppelgängertums ... (surrealer) als es je ein E.T.A.
Hoffmann ausgedacht hat".[28] Selbst wenn Weyl die Kausalstruktur unter kon-
formem Gesichtspunkt zum erstenmal in seinem Brief an F. Klein vom 28. 12.

---

[28] (RZM ^{1}1918, 220). Siehe auch den Beitrag von H. Goenner in diesem Band.

1920 *explizit* diskutierte, ist anzunehmen, daß der Ansatz zu diesem Gedanken ihm auch 1918 schon bekannt war.[29] Eine darauf aufbauende metrische Eichung der konformen Struktur legte dann auch aus dieser Sicht die Einführung einer "infinitesimalen Längenübertragung" nahe.

Die Wahl eines Repräsentanten der gegebenen konformen Struktur lieferte zusammen mit der Metrik in infinitesimalen Umgebungen $g_{ij}dx^i dx^j$ auch eine — wie Weyl formulierte — "pythagoreische Metrik" $g_{ij}\xi^i\xi^j$ im endlichen Tangentialraum. Dies entsprach einer durch die Punkte der Mannigfaltigkeit parametrisierten Schar von Skalierungsprozessen in geometrischen Größenbereichen, vergleichbar der Einführung einer Skalierung im euklidischen Raum durch Descartes, nun aber punktabhängig in sämtlichen endlichen Tangentialräumen. Für die Übertragung der Skalierung von einem Punkt p mit Koordinaten x^i zu einem infinitesimal benachbarten p' mit Koordinaten $x^{i'} = x^i + dx^i$ führte Weyl nun analog zum affinen Zusammenhang einen *Streckenzusammenhang* (von ihm auch häufig als *metrischer Zusammenhang* bezeichnet) ein.

Ein *Streckenzusammenhang* wird, abhängig von der Wahl des Repräsentaten der konformen Struktur, durch eine lineare Differentialform φ, in Koordinaten also $\varphi_i dx^i$, gegeben. Für Weyl bestand die konzeptionelle Funktion dieser Form darin, die infinitesimale Veränderung δl quadratischer Längen von Vektoren $l(\xi) := g_{ij}\xi^i\xi^j$ für Vektoren $\xi = (\xi^i)$ längs einer Kurve $x^i(t)$ durch $p \in M$ messen. Das erreichte er durch die Definition:

$$\delta l := \frac{\partial l}{\partial x^i}dx^i + l\varphi_i dx^i \qquad (2.3.5)$$

Damit definierte er "kongruenten Längentransport" (vergleichbar zur Parallelverschiebung beim affinen Zusammenhang) durch die Bedingung

$$\delta l = 0 \Longleftrightarrow dl := \frac{\partial l}{\partial x^i}dx^i = -l\varphi_i dx^i \qquad (2.3.6)$$

Bei Wechsel der Repräsentation der konformen Struktur durch $\tilde{g}_{ij} = \lambda g_{ij}$ ergab sich aus Konsistenzgründen eine Transformation der zugehörigen linearen Differentialfom

$$\tilde{\varphi}_i dx^i = \varphi_i dx^i - d\log\lambda, \qquad (2.3.7)$$

die Weyl suggestiv als *Eichtransformation* bezeichnete.
Somit ergibt sich zusammenfassend:

Definition 1 *Eine "rein infinitesimalgeometrische Metrik"* (Weylsche Metrik *oder* Eichmetrik) *auf einer differenzierbaren Mannigfaltigkeit M besteht aus Angabe einer konformen Klasse semiriemannscher Metriken $[g]$, zusammen mit einer Klasse $[\varphi]$ linearer Differentialformen $\varphi = \varphi_i dx^i$, die sich gemäß Gleichung (7) bei Wechsel des Repräsentanten der konformen Struktur umeichen. Diese Klasse heißt* Streckenzusammenhang *der Metrik (oder auch kurz* metrischer Zusammenhang).

---

[29]Der mathematische Teil des Briefes wurde als (Weyl 1921c) publiziert.

Aus Sicht der Faserbündel läßt sich Weyls Vorgehen als Arbeit in einem Hauptfaserbündel $L(M)$ mit Gruppe $L = (\mathbb{R}_+, \cdot)$ beschreiben, das zu jeder Wahl eines Repräsentanten g der konformen Metrik eine Trivialisierung $M \times L$ besitzt. Einem Repräsentantenwechsel $\tilde{g}_{ij} = \lambda g_{ij}$ korrespondiert die Übergangstransformation durch Multiplikation mit λ. Ein Zusammenhang in $L(M)$, repräsentiert durch eine reelle 1-Form $\varphi = \varphi_i dx^i$, erfährt dabei die zugehörige Eichtransformation: [30]

$$\tilde{\varphi} = \lambda \cdot d(\lambda^{-1}) + \lambda \varphi \lambda^{-1} = \varphi - d\log\lambda \qquad (2.3.8)$$

Dies entspricht dem von Weyl ermittelten Verfahren von Gl. (7).

Weyl betrachtete nun naheliegenderweise einen affinen Zusammenhang Γ^i_{jk} als *kompatibel* mit einer Weylschen Metrik $([g], [\varphi])$, falls die zugehörige Parallelverschiebung im Sinne der konformen Metrik Winkel erhält und quadratische Längen von Vektoren überträgt wie der Streckenzusammenhang. Er entdeckte und bewies dann Existenz und Eindeutigkeit eines affinen Zusammenhangs, die ihm als die "Grundtatsache der Infinitesimalgeometrie" erschienen:

Satz 1 *Zu einer Weylschen Metrik $([g], [\varphi])$ auf einer Mannigfaltigkeit M gibt es genau einen kompatiblen affinen Zusammenhang.* [31]

Die Existenz und Eindeutigkeit eines "affinen" Zusammenhangs galt ihm nun und im weiteren, etwa bei der *Analyse des Raumproblems*, ein entscheidendes Kriterium für die Akzeptierbarkeit einer infinitesimalgeometrischen Struktur.

2.4 Krümmungskonzepte, Eichkovarianz, Normaleichung

Schon für den Fall des affinen Zusammenhangs Γ^i_{jk} entwickelte Weyl eine schöne infinitesimalgeometrische Charakterisierung der Krümmung durch Parallelverschiebung eines Vektors ξ^i in der infinitesimalen Umgebung eines Punktes p längs zweier infinitesimaler Vektoren dx^i und δx^i. Die iterierte infinitesimale Parallelverschiebung ist ersichtlich nichtkommutativ in den infinitesimalen Translationen. Die Differenz ist ein infinitesimaler Vektor (infinitesimal von 2. Stufe) $\Delta \xi^i$, der linear von ξ^i und alternierend bilinear von den Translationen abhängt (Weyl 1918d, 11):

$$\Delta \xi^i = R^i_{jkl} dx^k \delta x^l \xi^j \text{ mit } R^i_{jkl} = -R^i_{jlk} \text{ und} \qquad (2.4.1)$$

$$R^i_{jkl} = \frac{\partial \Gamma^i_{jl}}{\partial x^k} - \frac{\partial \Gamma^i_{jk}}{\partial x^l} + \Gamma^i_{k\nu}\Gamma^\nu_{jl} - \Gamma^i_{l\nu}\Gamma^\nu_{jk}$$

Es handelte sich dabei also um eine direkte Verallgemeinerung der Riemannschen Krümmung auf allgemeine affine Zusammenhänge. Speziell im Fall der (eindeutig determinierten) affinen Krümmung zu einer Metrik berechnete Weyl die Krümmung F^i_{jkl} der Weylschen Metrik aus den Werten von $(g_{ij}, \varphi_i dx^i)$.

---

[30]Vergleiche z.B. (Bleecker 1981, 30f.) oder (Schottenloher 1995, 257).
[31](Weyl 1918d, 16f.), (RZM ³1919, 111f.)

In ganz analoger Weise bestimmte Weyl die Krümmung des Streckenzusammenhangs. Dabei ergibt sich als Differenz der iterierten infinitesimalen Verschiebung

$$\Delta l = \left(\frac{\partial \varphi^i}{\partial x^j} - \frac{\partial \varphi^j}{\partial x^i}\right) dx^i \delta x^j$$

und damit die *Streckenkrümmung* f als äußere Ableitung

$$f = d\varphi, \quad f_{ij} = \left(\frac{\partial \varphi^i}{\partial x^j} - \frac{\partial \varphi^j}{\partial x^i}\right) \tag{2.4.2}$$

Weyl identifizierte die Streckenkrümmung als Kriterium des Abweichens von einer Riemannschen Metrik:

Satz 2 *In einer Weylschen Mannigfaltigkeit sind folgende Aussagen äquivalent:*

(i) Der Streckenzusammenhang läßt sich wegunabhängig integrieren.

(ii) Die Weylsche Metrik entsteht (zumindest lokal) durch Umeichen aus einer semiriemannschen Metrik.

(iii) Die Streckenkrümmung verschwindet identisch $f_{ij} \equiv 0$.[32]

Zusätzlich zu Koordinatentransformationen betrachtete Weyl das Transformationsverhalten von Tensoren bei Umeichung durch $\tilde{g}_{ij} = \lambda g_{ij}$. Transformiert sich ein Tensor, etwa $t^i{}_j$ bei dieser Umeichung gemäß $\tilde{t}^i{}_j = \lambda^k t^i{}_j$, so sprach Weyl von einer *eichkovarianten* (Tensor-)Größe des *Eichgewichts* k, und im Fall $k = 0$ natürlich von *Eichinvarianten*. In diesem Sinne ist der Metriktensor g_{ij} einer Weyl-Mannigfaltigkeit per definitionem vom Eichgewicht 1, die Weylkrümmung F^i_{jkl} und die Weyl-Ricci-Krümmung $F_{jk} = F^\nu_{k\nu l}$ eichinvariant, dagegen etwa der dualisierte Metriktensor g^{ij} und damit auch die Weyl-Skalarkrümmung $F = F^\nu_\nu = g^{\nu\mu} F_{\mu\nu}$ vom Eichgewicht -1. Diese Eichkovarianz beziehungsweise -invarianz erschien Weyl bei seiner Suche nach geeigneten Lagrange-Funktionen für eine einheitliche Feldtheorie von Elektromagnetismus und Gravitation als wichtiges — und im Vergleich gegenüber Hilbert — neues Kriterium.[33]

Eine andere naheliegende Beobachtung Weyls bezog sich auf Eichmetriken mit nirgends verschwindender Skalarkrümmung, die für Raum-Zeit-Modelle von besonderer Bedeutung waren. In diesem Fall gibt es eine Eichung der Metrik, für die die Skalarkrümmung konstant wird $\tilde{F} \equiv \pm 1$. Normiert man nämlich ausgehend von irgendeiner Eichung g_{ij} durch

$$\tilde{g}_{ij} = |F| g_{ij} \tag{2.4.3}$$

so ergibt sich die gewünschte Normierung $\tilde{F} = |F|^{-1} F$ aus dem Eichgewicht -1 für F. Speziell für semiriemannsche Manngfaltigkeiten nirgends verschwindender Skalarkrümmung gab es aus Weyls Sicht damit *zwei* naheliegende Eichungen:

---

[32](Weyl 1918d, 16, 20; RZM ^{3}1919, 111)
[33]Vgl. Abschnitt 6.

— die Riemannsche Normaleichung g_{ij}, mit verschwindendem Streckenzusammenhang $\varphi_i \equiv 0$, Krümmungstensor R^i_{jkl} und Skalarkrümmung $R \neq 0$,[34]

— die Weylsche Normaleichung $\tilde{g}_{ij} = Rg_{ij}, \tilde{\varphi}_i = -d\log R$, mit gleichem Weyl-Krümmungstensor $\tilde{R}^i_{jkl} = R^i_{jkl}$, aber konstanter (Weyl-)Skalarkrümmung $\tilde{R} \equiv \pm 1$.

Die eigene Normaleichung spielte für Weyl in der Diskussion um Einsteins Einwände gegen die physikalische Interpretation seiner Geometrie eine wichtige Rolle. Die Wegunabhängigkeit atomarer Uhren versuchte er nämlich mit einer nichtverschwindenden Streckenkrümmung der Raumzeit durch die Hypothese in Übereinstimmung zu bringen, daß die Oszillatoren durch Einstellung auf die Skalarkrümmung reguliert werden und sich auf diese Weise wegunabhängig normieren (RZM [5]1923, 299ff.). Diesen (hypothetischen) Prozess setzte er in Analogie zur Richtungseinstellung eines Kompasses durch Übertragung (etwa beim Kreiselkompass) gegenüber der Einstellung beim Magentkompass oder dem "Sternenkompass" (RZM [5]1923, 270f.), verwies aber auf den ad-hoc Charakter dieser Annahme, solange nichts genaueres über Messprozesse im Mikrobereich bekannt war.

2.5 Bemerkungen zum Wechselspiel von konformer und projektiver Struktur

Weyl interessierte sich in den hier betrachteten Jahren außer für die Kontinuumsstruktur von Mannigfaltigkeiten, affine Zusammenhänge und verallgemeinerte metrische Geometrie auch für konforme und projektive infinitesimalgeometrische Strukturen. Letztere diskutierte er in einem Brief an F. Klein vom 28. 12 1920.[35] Der mathematische Teil des Briefes wurde in den Göttinger Nachrichten publiziert (Weyl 1921e). Ein Anlaß, sich etwas eingehender speziell mit der konformen Struktur zu beschäftigen, hatte sich daraus ergeben, daß Weyl von F. Klein um seine Meinung über einen Konflikt zwischen Brouwer und Schouten gebeten worden war.[36] Das brachte ihn dazu, sich ein Stück weit in die Schoutenschen Arbeiten einzulesen.

Schouten war es gelungen, in seinem Symbolismus nachzuweisen, daß für Mannigfaltigkeiten M der Dimension $n \geq 4$ das Verschwinden der Konformkrümmung hinreichend und notwendig dafür ist, daß M (lokal) konform euklidisch ist.[37] Weyl hatte selber in seiner eigenen Arbeit (1918d, 21) lediglich die Notwendigkeit dieses Kriteriums zeigen können. Nun reproduzierte er das Schoutensche Ergebnis in seiner Symbolik und ergänzte es — nach Einführung der projektiven Krümmung einer affin zusammenhängenden Mannigfaltigkeit — durch ein

---

[34]Siehe z.B. (RZM [5]1923, 124).
[35]UB Göttingen, Codes MS Klein, 297
[36]Klein an Weyl, 6. 10. 1920, UB Göttingen, Codex Ms Klein, 296. Vgl. (van Dalen 1999, 295ff.).
[37]Publiziert in (Schouten 1921). Diese Arbeit stand Weyl im Manuskript zur Verfügung.

analoges Kriterium für die projektive Ebenheit einer affin zusammenhängenden Mannigfaltigkeit der Dimension $n \geq 3$ (Weyl 1921e, 201).

Die projektive Struktur definierte Weyl durch Abstraktion aus der affinen, wenn zwei affine Zusammenhänge Γ^i_{jk} und $\tilde{\Gamma}^i_{jk}$ dieselben Geodätischen (bis auf Umparametrisierung) besitzen.[38]

In unserem Zusammenhang ist jedoch ein anderes Ergebnis Weyls als diese lokalen Flachheitsbedingungen von noch weiter reichender Bedeutung.

Satz 3 *Eine Weylsche Metrik ist durch Vorgabe der projektiven und der konformen Struktur eindeutig charakterisiert.*[39]

Weyl verwies ausdrücklich auf die Bedeutung dieser Ergebnisse für die Physik, da in der Relativitätstheorie "projektive und konforme Beschaffenheit eine unmittelbare Bedeutung" haben. Die projektive Struktur setzte er hier mit der Wirkung des gravito-inertialen Führungsfeld gleich und die konforme mit dem "Wirkungszusammenhang der Welt, durch den bestimmt wird, welche Weltpunkte miteinander in möglicher kausaler Verbindung stehen" (ebda., 196), also der Kausalstruktur. Dies war ein Ansatz für eine mögliche begriffliche Tieferlegung der Begründung der Relativitätstheorie, der fast ein halbes Jahrhundert später von J. Ehlers, F. Pirani und A. Schild (1972) (und daran anschließend von R. Coleman und H. Korté) zu einer Fundierung der Relativitätstheorie im Rahmen der Weylschen geometrischen Begrifflichkeit ausgebaut wurde. Weyl selber verfolgte Anfang der 1920-er Jahre diesen Gedanken allerdings nicht weiter. Zu diesem Zeitpunkt glaubt er noch, einem wesentlich ehrgeizigeren Ziel, einer einheitlichen geometrischen Theorie von Gravitation und Elektromagnetismus nahe zu sein. Mehr zur Weiterführung des Weylschen Ansatzes findet man in Abschnitt 4.6 von Teil II dieses Bandes.

2.6 Einheitliche Feldtheorie und dynamistische Materieauffassung

Weyl blieb auch in Zürich gewissermaßen ein externer Vertreter der Göttinger Mathematik und mathematischen Physik. Derem Arbeitsdiskurs mitsamt mathematischen Vorlieben und Ansprüchen auf Beitrag zur Welterklärung blieb er in 1920-er Jahren eng verbunden. Dabei brachte er ein eigenes philosophisches Profil hervor, geprägt durch die ständige Auseinandersetzung mit Husserls Philosophie, die für ihn auch in Zürich durch Helene Weyl sehr persönlich präsent war. Hinzu trat sein enger Kontakt mit F. Medicus, der an der ETH Philosophie und Pädagogik lehrte und ein hervorragender Kenner und engagierter Vertreter der Philosophie des deutschen Idealismus, speziell der Fichteschen Philosophie, war.

---

[38] Diese Äquivalenzbedingung läßt sich auch formulieren als $\tilde{\Gamma}^i_{jk} - \Gamma^i_{jk} = \delta^i_j \psi_k + \delta^i_k \psi_j$ für ein Kovektorfeld ψ_i. Siehe auch den Beitrag von Coleman/Korté in diesem Band, Abschnitt 4.2.

[39] (Weyl 1921c, 196)

D. Hilbert war in der ersten seiner beiden Mitteilungen zu den *Grundlagen der Physik* (Hilbert 1915, 1917) selbständig und aus einer anderen Perspektive als Einstein zu einer Formulierung kovarianter Gleichungen der Gravitationstheorie vorgedrungen. Dies geschah parallel zu Einsteins letztem und entscheidenden Schritt bei der Aufstellung der Feldgleichungen der allgemeinen Relativitätstheorie.[40] Dabei ging es Hilbert gar nicht darum, eine kovariante Verallgemeinerung der klassischen Gravitationstheorie zu finden. Sein Hauptziel war, Gustav Mies elektromagnetische Materietheorie aus dem Kontext der speziellen Relativitätstheorie herauszulösen und durch ein allgemein kovariant formuliertes Variationsproblem zu charakterisieren. Auf diesem Weg ergab sich auf recht natürliche Weise eine Euler–Lagrange–Differentialgleichung, in der die Raumzeit-Metrik und die Variationsableitung der Lagrangefunktion L des elektromagnetischen Feldes in Beziehung gesetzt wurden (Hilbert 1915, 404):[41]

$$\frac{\partial}{\partial g^{\mu\nu}}(\sqrt{g}R) + \frac{\partial}{\partial g^{\mu\nu}}(\sqrt{g}L) = 0 \qquad (2.6.1)$$

Dabei ist R = Skalarkrümmung, $(g^{\mu\nu}) = (g_{\mu\nu})^{-1}$ und $g = \det(g_{\mu\nu})$ — die letzten beiden im Unterschied zur heute üblichen Notation (Hilbert 1915, 396).

Der erste Term war leicht als spurbereinigte Ricci-Krümmung zu identifizieren,

$$\frac{\partial}{\partial g^{\mu\nu}}(\sqrt{g}R) = \sqrt{g}(R_{\mu\nu} - \frac{1}{2}Rg_{\mu\nu}) \qquad (2.6.2)$$

wie Hilbert allerdings erst im Dezember in die Druckfassung seiner Mitteilung einfügte.[42] Nun ließ sich (2) nach Interpretation von $-\frac{\partial}{\partial g^{\mu\nu}}(\sqrt{g}L)$ als Repräsentant des Energie-Impulstensors des elektromagnetischen Feldes als eine (spezielle) Fassung der Einstein-Gleichungen lesen. Nicht mehr als das behauptete Hilbert tatsächlich an dieser Stelle seines Artikels.[43]

Hilbert verfolgte also — anders als Einstein — ein theoretisches Programm, das formal auf Axiomatisierung der Physik zielte, dabei aber inhaltlich zutiefst der Göttinger Tradition einer naturphilosophisch inspirierten mathematischen Physik

---

[40]Diese Bewertung der Hilbertschen Arbeiten vom Spätsommer/Herbst 1915 bleibt auch nach den Ergebnissen von L. Corry, J. Renn und J. Stachel zur Publikationsgeschichte der Hilbertschen Fassung der Einstein-Gleichungen gültig (Corry/Renn/Stachel 1997), wie (Sauer 1999) ausführlich begruendet.

[41](Sauer 1999, 546, 556) diskutiert ausfürlich die Stellung dieser Gleichung in den ursprünglichen Druckfahnen des Artikels vor der von Corry/Renn/Stachel (1997) monierten Abänderung im Dezember 1915.

[42]Siehe (Corry/Renn/Stachel 1997, 1272). Die dort geäusserte Vermutung, Hilbert habe diese Änderung von Einstein abgekupfert, ist allerdings eine wenig plausible Unterstellung. Der Schritt von Gl. (1) nach Gl. (2) erscheint auch ohne ausgeführte Rechnung durchaus in Hilberts Reichweite.

[43]"Die so zu Stande gekommenen Differentialgleichungen zu Gravitation sind, wie mir scheint, mit der von Einstein in seinen späteren Abhandlungen aufgestellten großzügigen Theorie der allgemeinen Relativität im Einklang." (Hilbert 1915, 405) Vergleiche auch die entsprechende Diskussion bei (Sauer 1999, 564).

verpflichtet war.[44] Natürlich war Hilberts Herleitung dabei gleich in mehrfacher Hinsicht von Einstein abhängig. Zunächst profitierte Hilbert indirekt von Einsteins langwierigen physikalischen Untersuchungen zur Verallgemeinerung der Relativitätstheorie, wobei er durch die Gespräche im Juli 1915 ganz aktuell über Einsteins Stand der Überlegungen informiert wurde, einschließlich dessen damaliger Blockade in Sachen einer allgemeinen kovarianten Formulierung der Theorie. Hinzu kam Anfang Dezember 1915 die Möglichkeit und Herausforderung seine Variationsgleichungen an die Einsteinschen Gleichungen der Gravitation anzuschließen.

Vermutlich hatte Hilbert aus Sicht seines Axiomatisierungsprogramms schon in den Gesprächen mit Einstein im Sommer 1915 die Ansicht vertreten, daß auf der grundlegenden Theorieebene Gravitation und Elektromagnetismus kovariant formuliert werden sollten (Axiom II in (Hilbert 1915)).[45] Möglicherweise lieferten aus Einsteins Sicht die Diskussionen im Juli und die daran anschließende Korrespondenz im November 1915 zwischen den beiden Forschern einen Beitrag zur Auflösung von dessen Blockade.

Hilbert glaubte allerdings in der Mieschen Theorie einen vielversprechenden Ansatz vorzufinden, durch deren Fortentwicklung er zu grundlegenden Einsichten für die Theorie der Konstitution der Materie kommen könnte.[46] Auf Grundlage seiner kovarianten Formulierung der Mieschen Theorie inklusive eines aus Einsteins Überlegungen herausgezogenen Gravitationsanteils glaubte er, die Masse des Elektrons elektromagnetisch erklären und gleich den gesamten Energie-Impus-Tensor der Gravitationsgleichungen als rein elektrodynamisch bedingt ansehen zu können. Und selbst bei dieser Reduktion blieb er nicht stehen. Er kündigte darüberhinaus noch an, er habe nun endlich eine

> ...einfache und sehr überraschende Lösung des Problems von *Riemann*, der als der Erste theoretisch nach dem Zusammenhang zwischen Gravitation und Licht gesucht hat. (Hilbert 1915, 398)

Der Hintergrund dieser spektakulären Ankündigung war, daß Hilbert zu diesem Zeitpunkt über die Annahme eines rein elektromagnetisch konstituierten Energie-Impus-Tensors hinaus sogar davon ausging, daß das elektromagnetische Feld umgekehrt durch das Gravitationsfeld bestimmt werden könnte. L. Corry hat in einer vor kurzem erschienenen Studie gezeigt, wie Hilbert erst wenige Jahre früher, im Laufe des Jahres 1913, einen "Paradigmenwechsel" von einem rein mechanistischen Ansatz der Axiomatisierung der Physik zur Mieschen Theorie im Sinne eines elektrodynamischen Reduktionismus vorgenommen hat (Corry 1999b). Die von Hilbert im Winter 1915 erhoffte wechselseitige Bestimmung von Elektromagnetismus und

---

[44] Siehe (Corry 1999b) und (Vizgin 1994).

[45] Dieses Prinzip blieb von der Einschränkung in den von Hilbert zunächst zugelassenen "physikalischen Koordinaten" der Mitteilung von 1915 unberührt, die von J. Renn und J. Stachel als vermeintlich grundsätzlich hervorgehoben wird (Renn/Stachel 1999, 31ff.). Darauf weist im übrigen auf seine Weise schon Sauer ganz deutlich hin (1999, 546f.).

[46] Siehe dazu (Corry 1999c).

Gravitation läßt sich aus dieser Perspektive als ein Versuch verstehen, die beiden von ihm jeweils mit großem Enthusiasmus verfolgten Reduktionsansätze zu einem zusammenzuziehen.

Mathematisch beruhte Hilberts Ankündigung allerdings auf einer überzogen kühnen Schlußfolgerung aus einem (ohne Beweis) angekündigten Theorem: Von den n Euler-Lagrange Differentialgleichungen eines Variationsproblems $\delta \int J\sqrt{g}dx$ auf einer (semi-) Riemannschen Mannigfaltigkeit M der Dimension 4, mit skalarer Lagrangefunktion J, die außer von $x \in M$ noch von n weiteren (möglicherweise ko- oder kontravarianten) "Größen" abhängt, sind stets 4 von den anderen funktional abhängig.

Diese — später in leicht veränderter Fassung von E. Noether — bewiesene Beziehung (Noether 1918) wendete Hilbert 1915 auf den Fall einer Lagrangefunktion $J(g_{ij}, \varphi_k)$ für die Gravitationspotentiale g_{ij} und elektromagnetische Potentiale φ_k an, die im Sinne eines verallgemeinerten Hamiltonprinzips die gemeinsame Wirkung von Gravitation und Elektromagnetismus codieren sollte. Die Variation nach den g_{ij} lieferte ihm Feldgleichungen der Gravitation, die Variation nach den φ_k bei geeignetem Bau von $J(g_{ij}, \varphi_k)$ die 4 allgemein kovariant formulierten Maxwell-Gleichungen, und aus dem Theorem glaubte er folgern zu können, daß die verallgemeinerten Maxwell-Gleichungen aus den 10 Lagrange-Gleichungen der Gravitationspotentiale strikt ableitbar seien. So kam er zu dem überraschenden Schluß,

> *daß in dem bezeichneten Sinne die elektrodynamischen Erscheinungen Wirkungen der Gravitation sind* (Hilbert 1915, 397, Hervorhebung im Original).

Damit brachte Hilbert den alten Traum der Dynamisten des 19. Jahrhunderts von einer einheitlichen Theorie von Gravitation und Elektromagnetismus und einer daraus erwachsenden fundamentalen Theorie der Materieerscheinungen in eine neue, allerdings drastisch reduktionistische Form. Die von Hilbert behauptete Verbindung von Gravitation und Elektromagnetismus stieß jedoch schon im engsten Göttinger Kreis der Kollegen und der einen Kollegin auf äußerste Skepsis. Sie wurde bald aus mathematischer Sicht von F. Klein, E. Noether und H. Weyl kritisiert und als unhaltbar nachgewiesen.[47]

In der überarbeiteten Fassung von Hilberts *Grundlagen der Physik* (Hilbert 1924) tauchte diese Passage nicht mehr auf. Hilbert blieb aber bei einer Verbindung von Gravitationstheorie und Miescher Theorie mit einer gemeinsamen Hamilton-Lagrange-Funktion $\mathcal{H} = R + L$ für gravitationelle (R) und elektromagnetische (L) Wirkung und einer Darstellung des Energie-Impulstensors, in die allein der Anteil L eingeht und genau "mit dem Mieschen elektromagnetischen Energietensor" übereinstimmt (Hilbert 1924, 266).[48]

---

[47]Siehe dazu (Rowe 1999); vgl. auch (Vizgin 1994, 56ff.).

[48]Diese abgeschwächte Form findet sich auch — ohne Hinweis auf die reduktionistische Fassung von 1915 — in Hilberts *Gesammelten Abhandlungen*.

Weyl blieb zunächst vorsichtiger. In seine Vorlesung vom Sommersemester 1917 und die erste Auflage von *RZM* nahm er zwar eine begeisterte Darstellung der Mieschen Theorie im Rahmen der speziellen Relativitätstheorie auf (RZM [1]1918, §25). Von Hilberts reduktionistischem Coup de force einer allgemein kovarianten Verallgemeinerung, verbunden mit dem Anspruch, *schon dadurch* zu einer universellen Feld- und Materietheorie zu kommen, distanzierte er sich dagegen selbst in der abgeschwächten Form. Er kommentierte:

> Wieder ist die Physik, heute die Feldphysik, auf dem Wege, die Gesamtheit der Naturerscheinungen auf *ein einziges Naturgesetz* zurückzuführen, ein Ziel, dem sie schon einmal, als die durch Newtons Principia begründete mechanische Massenpunkt-Physik ihre Triumphe feierte, nahe zu sein glaubte. Doch ist auch heut dafür gesorgt, daß unsere Bäume nicht in den Himmel wachsen. (RZM [1]1918, 170)

Weyl vermied, bei seiner Diskussion einer möglichen Hamilton-Funktion $\mathcal{H}$ der allgemeinen Relativitätstheorie unter Einschluß der Maxwell-Lorentz-Wirkung, ein "exakt gültiges Weltgesetz" vorzuschlagen und formulierte lediglich ein Prinzip, das "so weit trägt, als unsere Kenntnis der Materie sicher reicht" (ebda, 216). Schon bei der Kommentierung der Mieschen Theorie im speziell relativistischen Sinne hatte er angemerkt:

> Wir wissen ja nicht, ob wir mit denjenigen Zustandsgrößen, welche der Mieschen Theorie zugrunde liegen, zur Beschreibung der Materie ausreichen, ob sie tatsächlich rein 'elektrischer' Natur ist. Vor allem aber hängt die dunkle Wolke aller jener Erscheinungen, mit denen wir uns heute notdürftig vermittels des Wirkungsquantums auseinandersetzen, über dem Land der physikalischen Erkenntnis, wer weiß, welch neuen Umsturz drohend. (RZM [1]1918, 170)

Dementsprechend schlug Weyl an dieser Stelle lediglich eine allgemeine Form der Hamiltonfunktion $\mathcal{H} = H + L + S_m + S_e$ vor. Dabei sollten die ersten beiden Terme die Feldwirkungen der Gravitation und des Elektromagnetismus darstellen, im wesentlichen $H = R$ (Riemannsche Skalarkrümmung), $L = F_{ik}F^{ik}$, mit (allgemein relativistischer) Maxwell-Feldstärke F_{ik}; die beiden letzten Terme S_m, S_e sollten ad-hoc postulierte "Substanzwirkungen" der Gravitation und der Elektrizität repräsentieren.

Seine früh im Jahre 1918 gefundene Lösung des mathematisch-philosophischen Problems einer rein infinitesimalgeometrischen Charakterisierung des Kontinuums führte Weyl jedoch dazu, den von Hilbert eingeschlagenen Weg einer Vereinheitlichung von Gravitation und Elektromagnetismus und einer darauf aufbauenden fundamentalen, rein feldtheoretischen Materietheorie auf seine eigene Weise doch noch ein Stück weit fortzusetzen. Weyls naturphilosophische Motivlage war dabei gar nicht weit entfernt von Hilberts. Beide einte der gemeinsame Bezug auf Riemann und das Anliegen einer "dynamistischen", das heißt einer einheitlichen, feldtheoretischen Materieerklärung. Weyl bereinigte diese allerdings um die — bei Hilbert allzu offensichtliche — reduktionistische Einfärbung und trug sie mathematisch subtiler und sprachlich-literarisch gekonnter vor, als dies seinem ehemaligen Lehrer möglich war.

Weyl stellte zunächst Riemanns programmatische Bemerkung, daß im Fall einer kontinuierlichen Raummannigfaltigkeit der Grund für die Metrik in den innerhalb des Kontinuums wirkenden Kräften gesucht werden müsse (Riemann 1854, 286) in einen neuen Kontext. Einstein hatte aus dieser Sicht schon den ersten Schritt getan und die Bestimmung der semiriemannschen Metrik g mit gravito-inertialen Kräften in einen gemeinsamen Begründungszusammenhang gebracht.[49] Für Weyl eröffnete sich nach seiner Verfeinerung des Konzeptes der infinitesimalgeometrischen Metrik durch Angabe einer konformen Struktur $[g]$ und eines Eichfeldes $[\varphi]$ ein neuer Weg, auch das elektromagnetische Feld in den gemeinsamen Bestimmungszusammenhang der verallgemeinerten Metrik einzubeziehen. Er interpretierte nämlich schlichtweg die Komponenten des Streckenzusammenhangs φ_i als Potentiale des elektromagnetischen Feldes.

Das elektromagnetische Feld $f := (f_{ij})$ ergab sich dann durch äußere Ableitung $f = d\varphi$, in Komponenten $f_{ij} = \frac{\partial \varphi_j}{\partial x_i} - \frac{\partial \varphi_i}{\partial x_j}$, entsprach also in Weyls geometrischer Sprache genau der Streckenkrümmung der Weylschen Metrik. Das erste System der Maxwell-Gleichungen folgte damit aus formalen Gründen unmittelbar

$$df = dd\varphi \;\; = \;\; 0, \tag{2.6.3}$$

bzw. $\quad \dfrac{\partial f_{ij}}{\partial x_k} + \dfrac{\partial f_{jk}}{\partial x_i} + \dfrac{\partial f_{kj}}{\partial x_i} \;\; = \;\; 0 \quad$ in Komponenten.

Weyl sah in dieser schon aus begrifflichen Gründen erfüllten Identität über die bloße Strukturanalogie hinaus ein Indiz dafür, daß seine Theorie eine der Situation angemessene logisch-mathematische Form besaß. Das erste System der Maxwell-Gleichungen nahm nämlich seiner Auffassung nach im Theorieaufbau eine relativ-apriorische Rolle ein; und das spiegelte sich in der aus strukturellen Gründen erfüllten Identität (3) wider.[50] Dagegen konnte das zweite System der Maxwell-Gleichungen

$$\frac{\partial h^{ij}}{\partial x_j} = s^i \tag{2.6.4}$$

(mit einer Tensordichte h^{ij} als elektromagnetische Felddichte und Viererstromdichte s^i) zwar der Form nach in seinen Theorierahmen eingefügt werden; die konkrete Ausfüllung, das heißt die Bestimmung der h^{ij} und s^i aus der Hamiltonfunktion

---

[49] Diese vorsichtige Formulierung vermeidet bewußt eine verengende Leseweise des Riemannschen Satzes im Sinne des späteren Machschen Prinzips, dem zwar Einstein für kurze Zeit anhing, Weyl allerdings nicht, oder höchstens als kurzes Durchgangsstadium; vgl. H. Goenners Beitrag in diesem Band.

[50] "Das 1. System der Maxwellschen Gleichungen ... ist, wenn unsere Auffassung vom Wesen der Elektrizität zutrifft, ein Wesensgesetz, dessen Gültigkeit noch völlig unabhängig davon ist, welche Naturgesetze den Wertverlauf der physikalischen Zustandsgrößen in der Wirklichkeit beherrschen." (RZM [3]1919, 244f.), ungeändert in (RZM [5]1923, 301). In ähnlicher Weise auch in (Weyl 1919b, 66f.)

$\mathcal{H}$ war jedoch davon abhängig, welche Annahmen für die Hamiltonfunktion und damit die konkrete Füllung des symbolischen Rahmens der Weltgesetze gemacht wurden (Weyl 1919b], 75; RZM ^{3}1919, 247–251).[51] Die Suche nach geeigneten Hamiltonfunktionen und damit physikalisch geeigneten symbolischen Repräsentanten für die Felddichte h^{ij} und Viererstromdichte s^i erschien in seiner Theorie als offenes Forschungsproblem, zu dessen Lösung er zunächst nur provisorische Vorschläge machen konnte.

Weyl war allerdings für etwa zwei Jahre davon überzeugt, daß die Struktur seiner Theorie aufgrund ihres konsequenten begrifflich-geometrischen Aufbaus besser als jede vorhergehende geeignet sein müsse, die fundamentalen Naturgesetze zu verstehen. Insbesondere legte er auf die in seinem Kontext naheliegende Forderung der Eichinvarianz für die Hamiltonsche Wirkungsdichte $\mathcal{W} = \mathcal{H}\sqrt{g}$ bei der Formulierung des Variationsproblems

$$\delta \int \mathcal{W}dx = \delta \int \mathcal{H}\sqrt{g}dx_1 \ldots dx_n \qquad (2.6.5)$$

großen Wert.

Da $\sqrt{g}$ vom Eichgewicht 2 ist, waren in Weyls Theorie Lagrangefunktionen vom Eichgewicht -2 von bevorzugtem Interesse; denn nur in diesem Fall ergaben sich eichinvariante Lagrangedichten. Damit erschien aber der von Hilbert gewählte und später auch von Einstein übernommene Ansatz einer Herleitung der Gravitationstheorie mit einem Feldwirkungsterm der Skalarkrümmung R (Eichgewicht -1) für Weyl aus grundsätzlichen Erwägungen bestenfalls als vorläufige Annäherung. Weyl schlug die Erforschung quadratischer Lagrangefunktionen (in R) als aussichtsreicheren Weg vor, obwohl dies

> ... wohl auf die Maxwellschen elektromagnetischen, nicht aber auf die Einsteinschen Gravitationsgleichungen führt; an ihre Stelle treten Differentialgleichungen 4. Ordnung. In der Tat ist es aber auch unwahrscheinlich, dass die Einsteinschen Gravitationsgleichungen streng richtig sind ... (Weyl 1918b, 40)

Etwa im Fall

$$\mathcal{W} = -\frac{1}{4}R^2\sqrt{g} + \beta f_{ij}\mathbf{f}^{ij} \qquad (2.6.6)$$

mit der elektromagnetischen Feld*dichte* $\mathbf{f}^{ij} = \sqrt{g}f^{ij}$, von dem Weyl nicht behaupten wollte, "daß es in der Natur realisiert" sei (RZM ^{3}1919, 253) erhielt er

$$\frac{\partial \mathbf{f}^{ij}}{\partial x_j} = s^i \text{ mit } s^i = c\sqrt{g}\varphi^i, c \text{ normierbare Konstante.}$$

Variation nach den g_{ij} lieferte ihm in diesem Fall Gravitationsgleichungen, die von den Einstein-Gleichungen um Terme abweichen, die im Grenzfall einer Raumzeit mit unendlichem Volumen gegen Null gehen (RZM^{3}1919, 254f.).

---

[51]So nicht mehr in der 5. Auflage von RZM.

Obwohl diese Kompatibilität im Grenzfall nur eine notwendige Bedingung für die Akzeptabilität seiner Theorie war, schloss sich Weyl nun in leichter Abwandlung der von Hilbert ebenso enthusiastisch wie vorschnell vertretenen Hoffnung an, den Schlüssel zu einer endgültigen umfassenden Erklärung der Materie und ihrer fundamentalen Bewegungsgesetze in der Hand (oder besser im Kopf und auf dem Papier) zu haben. Hilbert hatte gegen Ende seiner ersten Note über die *Grundlagen der Physik* proklamiert:

> ...ich bin auch der Überzeugung, daß durch die hier aufgestellten Grundgleichungen die intimsten bisher verborgenen Vorgänge innerhalb des Atoms Aufklärung erhalten werden und insbesondere allgemein eine Zurückführung aller physikalischer Konstanten auf mathematische Konstanten möglich sein muß — wie denn überhaupt damit die Möglichkeit naherückt, daß aus der Physik im Prinzip eine Wissenschaft von der Art der Geometrie werde: gewiß der herrlichste Ruhm der axiomatischen Methode ..." (Hilbert 1915, 407)[52]

Weyl war zwar nicht von Hilberts Weg im konkreten überzeugt, glaubte aber, in seinem Rahmen dem *Problem der Materie* eben doch rein feldtheoretisch auf die Spur kommen zu können.

Er verwies — paradigmatisch — auf die Rolle der Konstanten β in seinem Ansatz für die Hamiltonsche Wirkungsdichte (5) mit dem Kommentar, es habe etwas "Abstruses", wenn "dem Weltbau gewisse reine Zahlen von zufälligem numerischen Wert zugrunde liegen sollen" (Weyl 1919b, 83). Er hoffte darauf, ein "Weltgesetz" von ähnlichem Typ wie (5) zu finden, bei dem β nicht festgelegt ist, aber sinnvoll nur Werte in einer diskreten Menge annehmen kann. Die Restriktion für β erhoffte er sich durch eine Variationsbedingung an $\mathcal{W}$. Für jede (außerhalb eines kompakten Weltgebiets verschwindende) Variation von $([g], [\varphi])$, für die der von der Streckenkrümmung resultierenden Anteil der Variation von $\mathcal{W}$ gleich Null ist (in (5) $f_{ij}\mathbf{f}^{ij}$), verschwindet auch die Variation des von der Weyl-Riemannkrümmung herkommenden Anteils (R^2).

Die so ausgezeichneten Werte von β würden dann Weltzustände parametrisieren, in denen elektromagnetische und graviationelle Wirkungen getrennt werden können und jeweils für sich "stabil" sind.

> ...Dadurch würde das Problem der Materie zu einem "Eigenwert"-Problem: nur zu gewissen diskreten Werten von β gehören reguläre Lösungen. Ihnen entsprechen mögliche Korpuskeln, die aber doch alle neben- oder ineinander, sich gegenseitig feine Modifikationen der inneren Struktur aufzwingend, in derselben Welt existieren. Merkwürdige Konsequenzen für die Organisation des Weltalls scheinen da aufzudämmern und die Möglichkeit einer Erklärung seiner Ruhe im großen, Unruhe im kleinen. (Weyl 1919b, 83)[53]

---

[52] Nicht mehr in (Hilbert 1924).
[53] ähnlich in (RZM ^{3}1919, 260f.), so nicht mehr in (RZM ^{5}1923).

Aus mathematischer Sicht spielte Weyl hier indirekt auf seine erfolgreichen Arbeiten als junger Analytiker zu den Eigenschwingungen einer Membran an.[54] Wer dafür offen war, konnte weit darüber hinausgehend die Hoffnung heraushören, einer naturphilosophisch wie mathematisch-physikalischen gleichermaßen überzegende Metapher für die Ordnung einer auch existentiell als zufriedenstellend erfahrbaren Welt nahe zu sein.[55]

An der mathematisch-physikalischen Durchführung des Programms beteiligte sich zu Anfang der 1920-er Jahre zunächst eine beträchtliche Zahl von Physikern mit großem Engagement, aus der Sommerfeld-Schule vor allem der junge W. Pauli. Weyls ursprüngliche Hoffnung stellte sich allerdings im Laufe weniger Jahre als nicht erfüllbar heraus — jedenfalls nicht auf dem von Weyl ursprünglich ins Auge gefaßten Weg. Zu dieser Einschätzung trug das steigende Wissen im Bereich der Quantenphysik entscheidend bei.[56]

Einstein hatte Weyls Theorie von vornherein zwar zu einer mathematisch genialen, aber physikalisch belanglosen Erfindung erklärt. Seiner Ansicht nach verdarb die Wegabhängigkeit der Streckenübertragung von vornherein jede physikalisch sinnvolle Interpretation der Weylschen Theorie. Die Frequenzstabilität atomarer Oszillatoren sei ein unabweisbarer Beleg für eine globale Eichbarkeit der physikalischen Metrik. Damit sei im Weylschen Kontext formuliert die Streckenkrümmung Null. Außerdem sei physikalisch die Beschränkung auf den (semi-)Riemannschen Fall schon allein deshalb vorzuziehen, weil dort die Metrik in der Erfahrung verankert werde.[57]

Weyl erklärte in einem aufschlußreichen Briefwechsel mit Einstein im Laufe des Jahres 1918, warum ihn bei allem Respekt vor Einsteins physikalischer Intuition dessen Ablehnung seiner neuen Theorie aus methodologischen Gründen nicht überzeugen konnte.[58] Von seiner Theorie erhoffte Weyl sich eine Aufklärung der Tiefenstruktur der Materiekonstitution im Mikro- und im kosmologischen Makrobereich. Das schloß seiner Meinung nach eine Nivellierung der Streckenkrümmung im Mesobereich keineswegs aus. Weyls Überlegung war in diesem Punkt ganz analog der Riemannschen Bemerkung gegen Ende des Habilitationsvortrages, daß die Euklidizität des Raumes im mittleren Bereich keineswegs schwache Krümmung im Großen oder starke, aber sich bei Mittelbildung ausgleichende Krümmungsschwankungen im Kleinen ausschließt. Dies müsse man sogar annehmen, falls sich so "die

---

[54]Vgl. dazu den Beitrag Coleman/Korté in diesem Buch, Abschnitt 2.3.

[55]Siehe dazu wiederum Skúli Sigurdsson in Kapitel 1.

[56]Zur Reaktion auf Weyl siehe (Sigurdsson 1991). Das breite Spektrum der Versuche. einer einheitlichen Feldtheorie zum Durchbruch zu verhelfen, in dem Weyls Ansatz wiederum nur eine Stimme in einem ganzen Orchester darstellte, ist Gegenstand der Untersuchungen von J. Ritter und C.Goldstein (Goldstein/Ritter 2000). Zur Theoriegeschichte der einheitlichen Feldtheorie vgl. (Vizgin 1994). Eine ironische Pointe kann man aus zeithistorischer Sicht in gewissen Parallelen zu den aktuell virulenten, hochgesteckten Hoffnungen auf eine einheitliche Materietheorie durch Stringmodelle sehen, wobei natürlich aus verschiedenen Gründen die Parallele nicht zu eng gezogen werden darf.

[57]Vgl. (Straumann 1987).

[58]Die Korrespondenz ist mittlerweile in (Einstein 1998) voll publiziert.

Erscheinungen auf einfachere Weise erklären liessen" (Riemann 1854, 285).

Tatsächlich schlug Weyl sogar eine erste ad-hoc Hypothese vor, die zeigen sollte, daß es im Rahmen seiner Theorie für atomare Oszillatoren sehr wohl möglich
ist, sich ortsabhängig aber trajektorienunabhängig zu stabilisieren und sich damit
strukturkonsistent zu eichen. Er nahm dazu eine Einstellung des Oszillationspro-
zesses auf die Weylsche Normaleichung mit Skalarkrümmung $F = c$ konstant, etwa
mathematisch normiert mit $c = 1$, an. Eine Entscheidung darüber, ob seine Theorie fruchtbar sei oder nicht. könne lediglich *nach* Auswickelung der Konsequenzen
seines Ansatzes bis hin zu einer kompletten *Beobachtungstheorie* erfolgen (RZM
^{3}1919, 253f.) Dies war eine methodologisch kohärente Gegenposition zu Einstein,
deren explikatorische Fruchtbarkeit zur Zeit allerdings bloß erhofft war und völlig
von der weiteren Theorieentwicklung abhing.

Interessanterweise begann Weyl jedoch schon zwei Jahre nach seiner Debatte
mit Einstein, ein Jahr nach der ersten Einarbeitung seiner Feldtheorie in RZM,
massiv an der Durchführbarkeit seines ursprünglichen Programm zu zweifeln. Diese Wendung zeichnete sich ab, bevor alle theorieimmanenten Möglichkeiten zur
Durchführung seines Ansatzes ausgeschöpft waren. Sie war Bestandteil einer breiteren intellektuellen Reorientierung.

2.7 Beginn einer intellektuellen Reorientierung ab Ende 1920

Die Jahre 1918 bis 1920 waren für Weyl die Phase der stärksten Öffnung der eigenen mathematischen und physikalisch-theoretischen Arbeit für philosophische
Motive. In dieser Zeit versuchte er, Fichtes "radikalen Konstruktivismus" hinsichtlich Raum-, Zeit- und Materie-Begriff so weit zu folgen, wie es ihm sinnvoll
und mathematisch-physikalisch umsetzbar erschien, und konvertierte zeitweilig zu
Brouwers Intuitionismus bezüglich der Grundlagenfragen der Analysis. Der Höhe-
punkt der Konversion fand seinen Niederschlag in Weyls vielzitiertem Beitrag *Über
die neue Grundlagenkrise in der Mathematik*, in dem er seinen eigenen konstruktiven Ansatz einer arithmetischen Begründung der Analysis von *Das Kontinuum* (1918a) aufgab und Brouwers intuitionistisches Programm zur "Revolution"
für das "Staatswesens der Analysis" erklärte (Weyl 1921c).[59] Ähnlich dramatisch
äußerte er sich in einem Brief vom 27. 9. 1920 an den Präsidenten der ETH Zürich,
in dem er behauptete, gemeinsam mit Brouwer "an einer Revolutionierung und
Neubegründung der Analysis" zu arbeiten (Frei/Stammbach 1992, 49). Aber so
heftig Weyl in der zweiten Jahreshälfte 1920 nach außen hin höchst subjektive
Proklamationen über den Zustand der Mathematik abgab, so eindeutig sind auch
die Hinweise darauf, daß sich schon im Laufe des Jahres 1920 seine Sicht der
Beziehung zwischen "freier" symbolproduzierender Aktivität und vorgeordneter
Wirklichkeit innerlich zu transformieren begann, insbesondere bezüglich der be-
grifflichen Reduziblität der äußeren materiellen Welt, die in Weyls Sprache als ein

---

[59]Siehe dazu (Hesseling 2000).

Teil der "transzendenten Realität" angerufen wurde.[60]

Schon diese Art der Sprache verweist auf die große Nähe dieser Seite der der Symbolkonstruktion vorgeordneten Wirklichkeit zu metaphysischen und religiösen Spekulationen. Noch Jahrzehnte später berichtete Weyl von der engen Beziehung, die in diesen Jahren für ihn zwischen philosophisch-wissenschaftlicher Grundlagenklärung und religiöser Selbstfindung bestand. Letztere gipfelte für ihn nach einer ausführlichen Lektüre der Schriften des Meister Eckehart in einer langen meditativen religiösen Erfahrung in einem "herrlichen Engadiner Winter" im Jahre 1922, durch die er zur definitiven Annahme einer der symbolischen Konstruktion vorgeordneten "transzendenten Realität" geführt worden sei. Das meinte Weyl hier im Sinne einer "unmittelbaren" mystischen Gotteserfahrung (Weyl 1954, 646f.). Es kann aber keinen Zweifel darüber geben, daß sich Weyl schon *vor* der Selbstvergewisserung dieser etwas schwierigen und nie intersubjektivierbaren Seite der "transzendenten Realität" dazu durchgerungen hatte, die Nichtlösbarkeit des "Problems der Materie" durch freie symbolische Produktion anzuerkennen.

Bei der Einarbeitung seiner Eichfeldtheorie und die darauf aufbauende Materieerklärung in die 3. Auflage von RZM hatte Weyl ähnlich wie Hilbert, wenn auch philosophischer formuliert, in Aussicht gestellt, daß eine Assimilation der Physik an die Geometrie nun fast erreicht sei:

> Wir hatten erkannt, daß Physik und Geometrie schließlich zusammenfallen, daß die Weltmetrik eine, ja vielmehr die physikalische Realität ist. Aber letzten Endes erscheint so diese ganze physikalische Realität doch als eine bloße Form; nicht die Geometrie ist zur Physik, sondern die Physik ist zur Geometrie geworden. (RZM ³1919, 263)

Wenig später spitzte er die Aussage dahingehend zu, die Physik habe "für die Wirklichkeit keine weitergehende Bedeutung wie die formale Logik für das Reich der Wahrheit" (ebda.). Das schien von der Figur her zu einigen Äußerungen Hilberts, der in seinen programmatischen Vorstellungen über die Beziehung von Mathematik und Physik in ganz ähnlicher Rhetorik davon gesprochen hatte, daß mit *seiner* Theorie "... die Möglichkeit näherrückt, daß aus der Physik im Prinzip eine Wissenschaft von der Art der Geometrie werde: gewiß der herrlichste Ruhm der axiomatischen Methode ..." (Hilbert 1915, 407).

Weyl und Hilbert standen sich also in den Jahren um 1919 und 1920 in ihrem Streben nach einer rationalistischen Materieerklärung und damit gleichzeitig der Propagierung einer reduktionistischen Beziehung zwischen Mathematik und Physik in nichts nach, sondern wetteiferten miteinander auf diesem Wege. Ironischwerweise hielten sie sich dabei wechselseitig den Reduktionismus vor, der sie am jeweils anderen, nicht aber bei sich selbst störend auffiel. Hilberts Kritik an Weyl findet sich am schärfsten in seinen von D.Rowe herausgegebenen Vorlesungen

---

[60]Siehe zum Beispiel Weyls Sprachverwendung bei der Dikussion der Beziehung zwischen mathematischer Symbolwelt und der äußeren materiellen Welt in *Die Erkenntnislage in der heutigen Mathematik* (Weyl 1925a), 540).

von 1919 ausgesprochen, in denen er seinem ehemaligen Schüler vorwarf "Hegel-sche Physik" zu betreiben (Hilbert 1992, 99f.) Weyls Kritik an Hilbert findet sich in seinem Briefwechsel mit Klein und wissenschaftsöffentlich in seinem sonst sehr respektvollen Nachruf auf seinen Lehrer (Weyl 1944a 171).

Schon während des Jahres 1920 entdeckte Weyl jedoch philosophische, mathe-matische und physikalische Gründe, die gegen diese Art einer immer noch äußerst reduktionistischen Auffassung des physikalischen Naturbegriffs sprachen. In einem kleinen an ein breiteres Publikum gerichteten Aufsatz *Das Verhältnis der kausa-len zu statistischen Betrachtungsweise der Physik* (Weyl 1920a) stellte er die Halt-barkeit eines durchgängig klassisch deterministischen Naturbildes in Frage und äußerte sich dabei, wie er meinte, bewußt im "Gegensatz zu der heute herrschen-den Ansicht" (ebda., 121). Dabei ging es ihm darum, die klassische Kausalität als eine Idealisierung zu verstehen, die theorieimmanent und bereichsbezogen ihre Bedeutung hat, der jedoch eine fundamentale Bedeutung für die ontische Konstitu-tierung der Naturprozesse nicht (mehr) zugesprochen werden könne. Ein Teilschritt seines Arguments für diesen Blickwechsel war die Anerkennung einer ontologisch und mathematisch "selbständigen Rolle" der Statistik. Hinzu trat eine phanta-siereiche Spekulation über eine mögliche Verbindung einer quasi-relativistischen Kausalitätsstruktur der Raumzeit mit einer lediglich statistischen Determinierung der Prozesse im Kleinen, symbolisch repräsentiert durch die in der zeitlichen Ord-nung offenen Bestimmungen eines ganz Weylisch verstandenen intutionistischen Kontinuums. Er setzte darauf, daß zeitgenössische Mathematik und Physik den Rahmen eines mathematischen Naturbildes abstecken könnten, in dem Determi-nierung und Zukunftsoffenheit zusammengehen und ineinandergreifen. So hoffte Weyl, daß sich schließlich ein naturtheoretischer Rahmen abzeichnet, in dem exi-stentiell erfahrene und gewollte Freiheit *innerhalb* der Natur kohärent gedacht und ohne kulturelle Spaltungen möglich gemacht wird.[61]

Jede ernsthafte Kritik an dieser Spekulation sollte in Rechnung stellen, daß hier ein philosophisch produktiver junger Mathematiker versuchte anzudenken, wie eine tief in die europäische Moderne eingeschriebene Spaltung zwischen natur-wissenschaftlich-mathematischem Denken und breiterer, unter anderem geistes-wissenschaftlich geprägter, Kultur *ohne Regression* gegebenenfalls auflösbar sein könnte. Dabei war sich Weyl des vorläufigen und spekulativen Charakters seiner Überlegungen völlig bewußt. Sein Tonfall hielt sich weit entfernt vom Stil, in dem die "endlich gefundene Lösung" eines alten Problems vorgestellt wird. Ihm ging es eher um Ahnung und Anreiz, überkommene Denkblockaden aufzulösen.

Unabhängig von der Frage ihrer weiteren wissenschaftlichen Verwirklichung trugen solche metaphysischen Spekulationen offenbar dazu bei, daß Weyl die Kri-tik seitens der Physiker an seiner "rein feldtheoretischen" Materieerklärung bereits

---

[61]Dieser Aufsatz wird insofern mit gutem Grund, wenn auch in unglücklicher, durch den "Opportunismus"-Vorwurf verzerrender Interpretation, von P. Forman in seiner Darstellung des Ausbruchs aus dem klassischen Determinismus des physikalischen Naturbildes in der "Weimarer Kultur" angeführt (Forman 1971, 1980). Trotz dieser (und anderer Verzerrungen) bleibt diese Arbeit durch ihre radikale kulturelle Orientierung verdienstvoll und von bleibender Wirkung.

zu einem Zeitpunkt anzunehmen bereit war, als es für ihn aus innertheoretischen Gründen durchaus möglich gewesen wäre, seine Theorie als ein in seinen Potentialen noch unausgeschöpftes Forschungsprogramm zu verteidigen.[62] Das von ihm gegen Einstein angeführte methodologisches Argument, daß eine physikalische Bewertung seiner Theorie erst *nach* voller Ausarbeitung einer Beobachtungstheorie in dessen Rahmen möglich sei, war ja 1920 nicht weniger gültig als zwei Jahre vorher.

Tatsächlich brachte Weyl jedoch bei seiner Überarbeitung von RZM für die 4. Auflage eine merkliche — und für ihn völlig neue — Distanz bei der Diskussion des Realitätsgehaltes seiner Feld- und Materietheorie zum Ausdruck. Von der Vorstellung einer rein feldtheoretischen Erklärung der Materiekonstitution nahm er nun sichtlich Abschied. Die Aussagekraft seiner Theorie wollte er nur noch auf den reinen Feldverlauf beschränken. Er ging nun davon aus, daß die inneren Konstituierungsprinzipien der Materie gar nicht mehr direkt durch Feldgesetze repräsentierbar seien, sondern die Materie als Resultat vorauszusetzen sei und in der Theorie nur noch gewissermaßen als singuläre Randbedingung auftrete. Die Materie sei in den Feldsingularitäten abgebildet und werde nicht durch die Feldgesetze determiniert, sondern trage eher umgekehrt zur Determinierung des Feldverlaufes bei.

Weyl sah die in dieser Theoriestruktur enthaltene (Determinierungs-) Lücke, die bisher lediglich provisorisch von der (alten) Quantentheorie ausgefüllt wurde, aus naturphilosophischen Gründen als durchaus wünschenswert an (RZM [4]1921, 276). Von der zukünftigen wissenschaftlichen Entwicklung erhoffte er sich eine besser fundierte Ausfüllung der identifizierten Lücke, jedoch in einer noch nicht gefundenen Theorieform jenseits des klassischen Determinismus. Ähnlich wie in seinem Aufsatz zur *kausalen und statistischen Betrachtungsweise in der Physik* wies er auf eine logisch und ontologisch irreduzible Rolle statistischer Überlegungen hin, die geeignet sein könnten, mathematisch den Bruch mit dem klassisch-deterministischen Naturbild auszudrücken.

> Es muß einmal klipp und klar gesagt werden, daß die Physik bei ihrem heutigen Stande den Glauben an eine auf streng exakten Gesetzen beruhende geschlossene Kausalität der Natur gar nicht mehr zu stützen vermag. (RZM [4]1921, 283)

Weyl unterschied damit deutlicher als mancher Naturwissenschaftler seiner Zeit (und vielleicht sogar deutlicher als noch manche wissenschaftshistorische Beschreibung unserer Zeit) die Ebenen einer theorieimmanenten Kausalstruktur — wie er sie etwa für die Relativitätstheorie an anderer Stelle des Buches ausführlich und

---

[62]Für Weyl spielte dabei anscheinend die Kritik derjenigen Physiker, die anfangs seiner Feldtheorie gegenüber sympathisierend eingestellt waren eine wichtige Rolle. Speziell zu nennen ist dabei W. Pauli, der Weyls Theorie nicht erst mit seinem Beitrag zur *Enzyklopädie der mathematischen Wissenschaften* (Pauli 1921) — im Manuskript fertiggestellt Dezember 1920 — sondern auch schon in einem Diskussionsbeitrag auf der Bad Nauheimer Naturforscherversammlung grundsätzlich kritisierte (Pauli 1920).

in großer Klarheit ausführte — und die eines metaphysischen "Glaubens an eine ...geschlossene Kausalität der Natur". Letztere, so brachte Weyls Formulierung im übrigen völlig klar zum Ausdruck, hatte durch die Physik noch nie wirklich begründet, sondern bestenfalls (oder genauer schlimmstenfalls) lediglich "gestützt" werden können.[63] Seine eigene Theorie von 1918 war für ihn nun zu eng mit einem solchen Bild einer strukturell deduzierbaren "geschlossenen Kausalität" der Natur verbunden. Die Quantenphänomene schienen ihm eine Öffnung des Blicks notwendig und möglich zu machen.

Gegen Ende des Jahres 1920 sprach Weyl seine Verabschiedung von der Mieschen Theorie seinem (neben Hilbert) zweiten mathematischen Lehrer, Felix Klein, gegenüber knapp und deutlich aus:

> Endlich habe ich mich gründlich von der Mie'schen Theorie losgemacht und bin zu einer andern Stellung zum Problem der Materie gelangt. Die Feldphysik erscheint mir keineswegs mehr als der Schlüssel zu der Wirklichkeit; sondern das Feld, der Äther, ist mir nur noch der in sich selbst völlig kraftlose *Übermittler* der Wirkungen, die Materie aber eine jenseits des Feldes liegende und dessen Zustände verursachende Realität. Mit dem "Weltgesetz" (Hamiltonsches Prinzip), das die Wirkungsübertragung im Äther regelt, wäre noch gar wenig für das Verständnis aller Naturerscheinungen gewonnen. (Weyl an Klein, 28. 12. 1920)

Weyls Phase einer klassisch-feldtheoretischen Materieauffassung war also zeitlich eng auf die Jahre 1918 bis 1920 begrenzt. Nun erst war er so weit, den Blick auf eine, wie er es wenig später nannte, *Agenstheorie der Materie* zu richten, in der er sowohl die beiden denkbaren Varianten des Reduktionismus von stofflichen und wirkfähigen Aspekte in den klassischen Materiebegriffen (Substanz und Dynamismus) und sogar deren Voraussetzung, die dualistische Trennung dieser beiden Aspekte, hoffte vermeiden zu können.

Bei diesem nächsten Entwicklungsschritt zur Agenstheorie der Materie orientierte er sich anfangs vorwiegend an Leibniz (Weyl 1924f), bezog er sich allerdings schon wenig später, bei der Ausarbeitung seiner Erstauflage der *Philosophie und Naturwissenschaften*, ausdrücklich auch auf Schellings Naturphilosophie, in der er Ansätze vorfand, die sehr gut mit seiner neuen Haltung bezüglich Agens und Feld korrespondierten.

> Schelling hat, zum Teil beeinflußt von Leibniz, in seinem Ersten Entwurf der Naturphilosophie (1799, Sämtliche Werke, Cotta,1858, III) die moderne Entwicklung vorahnende Gedanken geäußert. 'Es müßte also', heißt es auf S. 21, 'in der Erfahrung etwas vorkommen, das, obgleich selbst nicht im Raum, doch Prinzip aller Raumerfüllung wäre.'

---

[63] Eine entsprechende Ebenentrennung mit einem höchst engagierten Plädoyer für die Formulierung eines offenen Naturbegriffs, war ja schon eines der bekannteren Fichteschen Themen gewesen (Fichte 1800/1879), mit denen sich Weyl in den Jahren vorher intensiv beschäftigt hatte.

Diese 'Naturmonade' ist nicht selbst Materie, sondern Aktion, 'für die man kein Maß hat als ihr Produkt selbst'. Auf Grund des Satzes, daß das Streben aller ursprünglichen Tendenzen auf Erfüllung des Raumes geht', kommt es dann zur Konstruktion der gestaltlosen Flüssigkeit, an deren Stelle wir heute das Feld setzen müßten. (Weyl 1927b, 136)

Die "Naturmonaden" waren für Weyl nun die Agensstrukturen, die aus der Perspektive der deterministischen Feldstruktur gewissermaßen nur von "außen" in deren "Lücken" eintraten, das heißt die Parameter der Singularitäten bestimmten. Aus einer anderen mathematischen Perspektive erschienen sie als irreduzibel stochastisch, physikalisch als Gegenstand der sich herausbildenden Quantenphysik.

Schellings Naturphilosophie zu begegnen, konnte man im Züricher intellektuellen Freundeskreis von F. Medicus zu dieser Zeit wohl kaum vermeiden. Medicus lehrte schwerpunktmäßig und regelmäßig die Philosophie des deutschen Idealismus und zeigte dabei, obwohl selbst Fichte–Experte, Einfühlungsvermögen und Respekt auch für die "konkurrierenden" Philosophien Hegels und Schellings. So hielt Medicus unter anderem im SS 1919 an der ETH eine Vorlesung über "Die nachkantische Spekulation (Fichte, Schelling, Hegel)". Eine spätere Aufzeichnung eines Hörers (Hansjörg Dändliker) aus dem WS 1943/44 zeigt, wie pointiert Medicus die Differenzen in der nachkantischen dialektischen Philosophie herausarbeitete: Zerstörung des alten Gegenstandsbegriffs, als vom Subjekt getrennten, durch Kant und ansatzweise Überwindung des (alten) Gegenstandsverhältnisses des Subjekts in der *Kritik der Urteilskraft*. Durch Fichte und Schelling sei im Anschluss daran die "Welt des Erlebens" zum Mittelpunkt der Philosophie gemacht worden.[64] Dieser Weitung des philosophischen Blickes korrespondierte bei Weyl eine steigende Anerkennung der Rolle quantenphysikalischer Erklärungsmuster, die auch mehr und mehr in das Zentrum seiner physikalischen Untersuchungen rückte (Sigurdsson 1991, 221ff.).

Die mathematisch-physikalisch und philosophisch produktive Wendung in der nun vorgestellten "Agenstheorie der Materie" war allerdings wieder einmal ganz Weylisch und gut geeignet, die traditionellen Grenzvereinbarungen zwischen den naturwissenschaftlichen und kulturtheoretisch-philosophischen Diskursen aufzukündigen. Und tatsächlich finden sich auch in der Naturphilosophie unseres Jahrhunderts Spuren einer sich fortentwickelnden und weiterhin offenen Rezeption der Weylschen Beiträge. Zum Beispiel knüpfte der alte Ernst Bloch in seinen Tübinger

---

[64]Wie Medicus den Gegensatz Fichte und Schelling charakterisierte, geht aus den Aufzeichnungen stichwortartig und dadurch sehr pointiert hervor:
"Fichte keine Philosophie d. Natur. also keine lebendige Natur gegenständliche Natur. — Schelling davon durchdrungen, dass Natur lebendig, kann desh., nicht mit chem. u. physikalischem Erkenntnissen verst. werden. Wenn die Natur z. Erlebnis wird, fühlen wir uns ihr verwandt. ... — Fichte: Natur ist da als Widerstand, damit es sittliche Aufgaben geben kann. Nat. ist notwendige Vorstellungs Welt. — Fichte hat eine Art Naturhass: Gegen Naturtriebe. — Schelling liebt die Natur. Sieht das Ethische nicht nur als im Gegensatz zur Natur. nähert Natur- u. Ethische- Notwendigkeit einander. — F: Sittliches Bewusstsein will, dass der Mensch nicht seinen Neigungen folge, sondern den Pflichten. — S: Sittliche Freiheit liegt dort, wo er wenn er richtig handelt, zugleich seiner Natur folgt. ..." (ETH Zürich, Codex Ms. Medicus 1117a:7)

Vorlesungen zur Naturphilosophie begeistert an Weyls Agenstheorie der Materie an. Er versuchte, sie mit Zurückhaltung und Vorsicht (die nicht typisch für Blochs Diktion sind) als Anzeichen dafür zu interpretieren, daß nun möglicherweise — anders als zur Zeit der Philosophen des 19. Jahrhunderts — die "Dialektik vor der Tür" eines offenen, die spezialisierten Kultursegmente übergreifenden, Naturbegriffes stünde (Bloch 1978, 230).

Weyls Abrücken von einer rein dynamistischen — und dabei klassisch deterministischen — Materietheorie während des Jahres 1920 bedeutete nicht, daß er zur selben Zeit etwa auch von seinen "rein infinitesimalgeometrischen" Auffassungen innerhalb der Geometrie Abstand genommen hätte. Zunächst einmal löste er sich lediglich von seiner anfangs, also in den Jahren 1918/19, geradezu überbordenden Neigung, den mathematischen Objekten der Eichgeometrie *unmittelbare* physikalische Bedeutung zuzusprechen. Begleitend zu der soeben skizzierten intellektuellen und metaphysischen Umorientierung begann Weyl jedoch, seine eichgeometrische Struktur durch eine neue, nun neokantianisch eingefärbte, philosophisch-mathematische Analyse zu untermauern. Dazu entwarf er apriorisch vertretbare Konzepte, die jeder Kongruenzgeometrie im Infinitesimalen zugrunde liegen sollten.

2.8 Analyse des Raumproblems

Schon 1919 stellte Weyl anläßlich der Neuherausgabe von Riemanns Habilitationsvortrag (Weyl 1919c) die Frage nach einer begrifflichen Analyse derjenigen Bestimmungen geometrischen Denkens, die Riemanns metrische Struktur innerhalb der Klasse der (später so genannten) Finsler-Metriken in einem relativ apriorisch Sinne auszeichnen. Riemann hatte selber schon darauf verwiesen, daß metrische Begriffe in einer Mannigfaltigkeit auch anders als durch eine positiv definite Differentialform $ds^2 = \sum g_{ij} dx^i dx^j$ eingeführt werden könnten; er hatte jedoch die genauere Untersuchung solcher Alternativen aus eher pragmatisch Gründen unterlassen (Riemann 1854, 278).[65] Weyl stellte nun im Jahre 1918 die Frage aus Sicht seiner Vorstellungen zur "reinen Infinitesimalgeometrie" neu.

Dabei brachte er zwei zunächst noch unverbundene Aspekte ins Spiel: die Betrachtung freier infinitesimaler Beweglichkeit in jedem Punkt der Mannigfaltigkeit als einen Teilaspekt der Helmholtz-Lieschen Untersuchungen und die Existenz und Eindeutigkeit eines mit der Metrik kompatiblen affinen Zusammenhangs. Noch konnte er zwischen diesen beiden Aspekten keine ausgearbeitete Beziehung vorschlagen, aber er wagte unter Verweis auf die zentrale Rolle, die der affine Zusammenhang in seiner eigenen Verallgemeinerung der metrischen Differentialgeometrie sowie bei Levi-Civita (1917), Hessenberg (1917) und Schouten (1919) spielte, eine Vermutung.

---

[65] "Die Untersuchung dieser allgemeinern Gattung würde zwar keine wesentlich anderen Principien erfordern, aber ziemlich zeitraubend sein und verhältnismässig auf die Lehre vom Raum wenig neues Licht werfen, zumal da sich die Resultate nicht geometrisch ausdrücken lasse; ..." (Riemann 1854, 278)

Bei der fundamentalen Bedeutung, die nach den neueren Untersuchungen (...) dem affinen Grundbegriff der infinitesimalen Parallelverschiebung eines Vektors für den Aufbau der Geometrie zukommt, erhebt sich insbesondere die Frage, ob die Mannigfaltigkeiten der Pythagoreischen Raumklasse die einzigen sind, welche die Aufstellung dieses Begriffs ermöglichen und welche dementsprechend nicht bloß eine Metrik sondern auch affinen Zusammenhang besitzen. Die Antwort lautet wahrscheinlich bejahend, ein Beweis dafür ist aber bisher nicht erbracht worden. (Weyl 1919c, 27)

Diese Formulierung läßt natürlich verschiedene Präzisierungen zu; dabei hängt viel von der Interpretation des Terms *Mannigfaltigkeiten der Pythagoreischen Raumklasse* ab. Mit "pythagoreisch" bezeichnete Weyl im allgemeinen nichtausgeartete quadratische Formen, hier wäre dann von semiriemannschen Metriken in einer Mannigfaltigkeit die Rede;[66] im Kontext dieser Anmerkung, dem Kommentar zu Riemanns Habilitationsvortrag, verwendete Weyl die Bezeichnung allerdings durchweg auf positiv definite quadratische Formen eingeschränkt. Das zeigen alle von Weyl zur Päzisierung angeführten Formeln in diesen Anmerkungen.

Weiter bezeichnete Weyl zwei allgemeine positive (vom Grad 1) homogene Metriken f_1 und f_2, die jeweils also die Finsler- Bedingung

$$f(\lambda x_1, \ldots, \lambda x_n) = |\lambda| f(x_1, \ldots, x_n) \qquad (2.8.1)$$

erfüllen, als zu derselben *Klasse* gehörig, falls $f_2(x) = f_1(Ax)$ mit $A \in \mathrm{GL}(n, \mathbb{R})$. Er forderte von einer "natürlichen" Finslermannigfaltigkeit, daß die (infinitesimalen) Metriken f_p für alle Punkte $p \in M$ einer Klasse angehören.

Es wird natürlich sein, vorauszusetzen, daß sich die verschiedenen Punkte der Mannigfaltigkeit nicht schon hinsichtlich der in jedem von ihnen herrschenden Maßbestimmung unterscheidet; das formuliert sich analytisch dahin, daß die den verschiedenen Punkten P entsprechenden Funktionen f_P all aus einer, f durch lineare Transformation der Variablen hervorgehen. (Weyl 1919c, 26f.)

Darüberhinaus induziert die angegebene Klasseneinteilung der Finslermetriken eine Einteilung der "natürlichen" Finslermannigfaltigkeiten in infinitesimalgeometrische *metrische Raumklassen*. Zwei Finslermannigfaltigkeiten (M, f) und (M', f') gehören dabei genau dann derselben *metrischen Raumklasse* an, wenn beide im eben genannten Sinne "natürlich" sind und darüberhinaus f_p und f'_q für alle $p \in M$ und alle $q \in M'$ derselben Klasse angehören, also linear ineinander transformiert werden können. Das oben angeführte erste Zitat enthält also die

Vermutung 1 (Weylsches Raumproblem von 1919) *Unter den metrischen Raumklassen der im Weylschen Sinne "natürlichen" Finslermannigfaltigkeiten sind*

---

[66]Diese Sprachverwendung zieht sich durch alle späteren Arbeiten zur Analyse des Raumproblems.

die Riemannschen Mannigfaltigkeiten dadurch ausgezeichnet, daß diese und nur diese jeweils genau einen kompatiblen affinen Zusammenhang besitzen.

Kurz nach Aufstellung dieser Vermutung veränderte Weyl die Formulierung des Raumproblems mit Blick auf seine verallgemeinerte Eichgeometrie und löste es in dieser veränderten Fassung. Die weitere Diskussion um das "Raumproblem" schloß zunächst ausschließlich an die spätere Formulierung an. So dauerte es fast vier Jahrzehnte bis D. Laugwitz im Jahr 1958 darauf hinwies, daß Weyl in seinen Kommentaren zu Riemann eine speziellere Vermutung gewissermaßen als Vorform seiner späteren "Analyse des Raumproblems" aufgestellt hatte. Laugwitz gelang es, Weyls Vermutung durch differentialgeometrische Methoden, weitgehend unabhängig von den gruppentheoretischen Überlegungen zu beweisen, die für die Untersuchung zur späteren Fassung entscheidend wurden (Laugwitz 1958).[67]

Für Weyl spielte die Vermutung zur Begründung der Riemannschen Struktur schon 1919 nur eine untergeordnete Rolle. Für ihn war eine analoge konzeptionelle Begründung für seine eigenen verallgemeinerten (Weylschen) Metriken viel interessanter. Auch hier suchte er nach einer Charakterisierung einer "natürlichen" Klasse allgemeinerer Raumstrukturen, unter denen dann die Weylschen Mannigfaltigkeiten durch die Bedingung der Existenz und Eindeutigkeit eines kompatiblen affinen Zusammenhangs ausgezeichnet sein könnten. In diese Richtung setzte er seine Überlegungen zur "Analyse des Raumproblems" im folgenden Jahr bei der Ausarbeitung der 4. Auflage von RZM fort und nahm einen neuen Abschnitt zu diesem Thema auf. Darin begann er seine geometrische Struktur durch Verbindung gruppentheoretischer Argumente mit der Untersuchung strukturkompatibler affiner Zusammenhänge abzustützen (RZM [4]1921, §18 "Gruppentheoretische Auffassung der Raummetrik").[68]

Weyl nahm dabei Motive aus der Helmholtz-Lieschen Behandlung des klassischen Raumproblems auf, modifizierte sie aber aus "rein infinitesimalgeometrischer" Sicht und dem Gesichtspunkt einer Verträglichkeit mit Prinzipien der allgemeinen Relativitätstheorie. Eine Schlüsselstellung räumte er der Idee ein, daß

---

[67]H. Korté und R. Coleman stellen in ihrem Beitrag in diesem Band eine eigene Interpretation vor, in der der Unterschied zwischen der Weylschen Vermutung von 1919 und der späteren ausgereiften Formulierung des Weylschen Raumproblems bestritten wird. Dadurch kommen sie auch zu einer Einschätzung des Laugwitzschens Beitrages, den ich aus historischer Sicht nicht teile.

[68]Tatsächlich berichtete Weyl nach Fertigstellung des Manuskripts von (RZM [4]1921) selbstbewußt an Klein:
"An Neuerungen wird Sie vielleicht interessieren: ein kurzer Beweis im Anhang dafür, daß der Krümmungsskalar im Riemannschen Raum die einzige von den g_{ik}, deren 1. und 2. Ableitungen abhängende und in den 2. Ableitungen lineare Invariante ist (einfacher scheint mir als der Vermeil'sche Beweis); vor allem aber eine neue gruppentheoretische Formulierung des Raumproblems ('woher kommt die quadratische Differentialform?'), die von meinem Standpunkt aus und dem der allgemeinen Relativitätstheorie (welche die Definition der Form nicht annimmt, aber beliebige virtuelle Veränderungen der metrischen Fundamentalform zulässt) anstelle des Helmholtz-Lie'schen Raumproblems tritt. Bislang habe ich aber nur die Formulierung in Händen und den Beweis für die niedrigsten Dimensionszahlen; ich bemühe mich augenblicklich, das vermutete allgemeine Resultat sicherzustellen." (Weyl an Klein 28. 12. 1920)

eine infinitesimalgeometrische Kongruenzstruktur in jedem Falle einen eindeutig bestimmten kompatiblen affinen Zusammenhang zu besitzen habe.

Eine infinitesimale Kongruenzstruktur sollte nach Weyl auf die Auszeichnung punktgebundener (endlichdimensionaler) "Drehungsgruppen" G_p auf dem "Vektorkörper" (dem endlichen Tangentialraum) in jedem Punkt p in M aufbauen, wobei alle G_p untereinander und zu einer Lie-Untergruppe $G \subset SL(n, \mathbb{R})$ isomorph und zueinander konjugiert sind, $G_p = h_p{}^{-1}Gh$.[69] G sollte nach Weyl die "ein- für allemal feste Natur der Metrik", die Abhängigkeit der G_p von p die "verschiedenen Orientierungen" der Drehungsgruppen in den verschiedenen Punkten zum Ausdruck bringen (Weyl 1923c, 48f.; RZM [5]1923, 139f.).

Mathematisch gesehen trieb Weyl in der vorgenommenen Begriffsanalyse die differentialgeometrischen Konzepte wieder einmal über die Grenzen des Bekannten und zu seiner Zeit formal Präzisierbaren hinaus. Die variierende "Orientierung der Gruppe" (oben als G_p notiert) beschrieb er intuitiv und ohne eindeutige symbolische Fixierung; kein Wunder, daß sie bei seinen mathematischen Lesern zu durchaus verschiedenartigen Interpretationen führte. Weyl tastete sich hier in Richtung einer Struktur vor, die heute in der Terminologie der Faserbündel schärfer zu fassen ist, als es Weyl mit den ihm zur Verfügung stehenden sprachlichen Mitteln möglich war.[70]

Dabei ist — nicht zuletzt mit Blick auf verschiedene spätere vereinfachende Interpretationen der Weylschen Analyse — wichtig, daß Weyl deutlich zwischen einer zugrunde gelegten *Drehungsgruppe* $G \subset SL(n, \mathbb{R})$ und dem zugehörigen Normalisator$(G) =: H$ in $GL(n, \mathbb{R})$, der von ihm so genannten *Ähnlichkeitsgruppe* unterschied.[71] Aus seiner Sicht erschien es Weyl als sinnvoll, bei Koordinatenwechsel und lokalem Trivialisierungswechsel (sobald wir die Weylschen Andeutungen in der Sprache der Hauptfaserbündel zuspitzen) die *Ähnlichkeitsgruppe H*, und nicht die Drehungsgruppe G selber als Transformationsgruppe in Betracht zu ziehen. Bei dieser Unterscheidung hatte er die die konforme Struktur der (Weylschen) Eichmetrik vor Augen.

Bei Ausformulierung dieser Argumentation in Faserbündelsprache verwendete Weyl nur zur Beschreibung der *Ähnlichkeitsstruktur* ein Hauptfaserbündel $H(M)$ über der differenzierbaren Mannigfaltigkeit M mit typischer Faser und Strukturgruppe $H \subset GL(n, \mathbb{R})$ (zusammenhängende Liegruppe), das aus einer Reduktion des tangentialen n-Bein- Bündels auf die Gruppe H entsteht.[72] Zur

---

[69]Die "Drehungsgruppen" sollen Volumen erhalten. Im Laufe der weiteren Analyse stellte Weyl allerdings fest, daß die Determinantenbedingung überflüssig ist.

[70]Die folgende Darstellung stützt sich wesentlich auf die von J. Ehlers während des DMV-Seminars vom 25. bis 29. 5. 1992 vorgetragene Interpretation der Weylschen *Analyse des Raumproblems*. P. Slodowy danke ich für geduldige produktive Kritik an früheren Fassungen dieses Beitrages.

[71]Vgl. Anm. 69.

[72]Da Weyl bei der Analyse des Raumproblems ausschliesslich lokal argumentierte, kann das Tangentialbündel in diesem Kontext als trivialisierbar angenommen werden. Die (implizit angenommene) Reduzibilität war also topologisch unproblematisch und diente lediglich der Analyse der Bedingungen geeigneter infinitesimaler Strukturbildung.

Beschreibung der *infinitesimalen Kongruenzen* betrachtete er für ein $G \subset H$ mit $H = N(G)$ ein Unterbündel des adjungierten Bündels $\mathrm{ad}H(M)$ mit G als typischer Faser, $G(M) \subset \mathrm{ad}H(M)$.[73] Jede lokale Trivialisierung von $H(M)$ über $U \subset M$ führt dann gleichzeitig mit H zu einer "Koordinatendarstellung" der Drehungen $G_p \equiv G$ in jedem Punkt $p \in U$. Trivialisierungswechsel von $G(M)$ wird durch die Konjugation mit punktabhängigen "Ähnlichkeitsabbildungen" $h(p) \in H$ gegeben, $\tilde{G}_p = h^{-1}(p)Gh(p)$. Diese erschienen bei Weyl unter der Bezeichnung "verschiedene Orientierung" der Gruppen G_p bei Wechsel des Punktes p (Weyl 1923c, 47f.). Diese Charakterisierung der "verschiedenen Orientierungen" entsprach der von Weyl intendierten Umparametrisierung im Infinitesimalen (also auf liealgebrawertigen 1-Formen und Zusammenhängen) durch die adjungierte Darstellung von H auf der Liealgebra $\mathbf{g} = \mathrm{Lie}(G)$. In moderner Sprache wird dies durch Charakterisierung der infinitesimalen Eichtransformationen durch Schnitte in $\mathrm{ad}\, H(M)$ zum Ausdruck gebracht.[74] Wenn Weyl seine Verwendung der adjungierten Operation auch nicht ausdrcklich begründete, erschien sie für mit der zeitgenössischen Auffassungen der Lietheorie vertraute Leser wohl durchaus naheliegend.

Eine *infinitesimale Kongruenzstruktur* setzt nun nach Weyl außer der Spezifizierung der "Natur des Raumes" durch Auswahl von $M, G \subset H$, und der "Orientierungen der Drehgruppen" durch $H(M)$ noch die Angabe eines verträglichen linearen (nicht notwendigerweise torsionsfreien) Zusammenhanges Λ als *kongruente Verpflanzung* voraus. Die von Weyl nicht weiter analysierte Verträglichkeit ist dabei aus heutiger Sicht so zu verstehen, daß der lineare Zusammenhang Λ auf das Bündel $H(M)$ reduzierbar ist, also speziell Holonomie in H besitzt (Weyl 1923c, 48).

Zwei kongruente Verpflanzungen Λ und $\tilde{\Lambda}$ charakterisierten dabei für Weyl aus naheliegenden Gründen — nämlich in Analogie zu seinem Streckenzusammenhang von 1918 - ein- und denselben (verallgemeinerten) *metrischen Zusammenhang*, falls ihre Differenz eine Differentialform $A(p)$ mit Werten in $\mathbf{g} = \mathrm{Lie}(G)$ ist, sie also durch eine punktabhängige Schar "infinitesimaler Rotationen" auseinander hervorgehen; in Koordinaten:

$$\tilde{\Lambda}^i_{j,k}(p) = \Lambda^i_{j,k}(p) + A^i_{j,k}(p) \qquad (2.8.2)$$

mit $(A^i_{j,k}(p))_{i,k}$ in $\mathbf{g}$ für $1 \leq j \leq n$. Ein abstrakter *metrischer Zusammenhang* ist also (in einer Trivialisierung betrachtet) nichts anderes als eine Äquivalenzklasse $[\Lambda]$ linearer Zusammenhänge Λ (die auf das Bündel $H(M)$ reduzibel sind), modulo Differentialformen mit Werten in $\mathbf{g} \subset \mathbf{h} = \mathrm{Lie}(H)$ (Weyl 1923c, 49). Bei Berücksichtigung der (von Weyl nicht explizierten) Kohärenzbedingungen bei Trivialisierungswechsel ist dementsprechend der affine Raum aller Zusammenhänge in $H(M)$, der die kovarianten "Basisformen" auf $H(M)$ als Translationsvektorraum besitzt, in die affine Unterräume zu Basisformen mit Werten in $\mathbf{g}$ zu faktorisieren.[75]

---

[73] Wegen Lokalität der Argumentation war die Existenz solcher Unterbündel unproblematisch, vgl. vorige Anmerkung.

[74] Vergl. z.B. (Schottenloher 1995, 272).

[75] Eine liealgebrawertige 1-Form η in $H(M)$ heisst dabei *kovariante Basisform*, wenn sie längs

In leicht interpretierender Zuspitzung können wir damit formulieren:

Definition 2 *Eine* abstrakte infinitesimalgeometrische Kongruenzstruktur, *wie sie von Weyl intendiert wurde, besteht aus*

(i) *einer differenzierbaren Mannigfaltigkeit M der Dimension n, sowie zusammenhängenden Liegruppen $G \subset H = N(G) \subset GL(n, \mathbb{R})$,*

(ii) *einem Hauptfaserbündel $H(M)$ mit Strukturgruppe H, das durch Reduktion aus dem tangentialen $GL(n, \mathbb{R})$-Bündel der n-Beine von M entsteht, sowie assoziiert dazu einem Unterbündel $G(M)$ des adjungierten Bündels $H(M)$ mit typischer Faser G.*

(iii) *und einem linearen, auf das Bündel $H(M)$ reduziblen Zusammenhang Λ, als* kongruente Verpflanzung.

G heißt dabei die *Drehungsgruppe*, H die *Ähnlichkeitsgruppe* der infinitesimalen Kongruenzstruktur, Λ eine *kongruente Verpflanzung* und die zugehörige Äquivalenzklasse $[\Lambda]$ modulo kovarianten Basisformen mit Werten in **g** ein *(abstrakter) metrischer Zusammenhang* der Struktur.

Für Weyl charakterisierten die Daten von (i) die "Natur des Raumes", während durch (ii) die "Orientierungen der Drehungsgruppe" in den verschiedenen Punkten $p \in M$ angegeben wird. In Anlehnung an die spätere (nach-Cartansche) Terminologie beim Cartanschen Raumproblem werde ich im folgenden bei Vorgabe von (i) und (ii) von einer $H - G$-*Struktur auf* M sprechen. Eine Weylsche infinitesimale Kongruenzstruktur bestand in diesem Sinne also aus einer $H - G$-Struktur und einem zugehörigen linearen Zusammenhang. Dabei ist wichtig, daß *im Unterschied zur Theorie der G-Strukturen*, zwar die Werte des Zusammenhanges auf **h** reduziert werden können, die Äquivalenzklassen modulo Differentialformen dann aber mit Werten in **g** gebildet werden.

Zusätzlich zu diesen "rein begriffsexplikatorischen" Bedingungen von Definition 2 ist nach Weyl noch eine "synthetische Forderung" hinzuzufügen, die erst zum Ausdruck bringt, wann eine $H - G$-Struktur dem "Wesen des Raumes" entspricht. Weyl formulierte diese in der Form zweier Postulate.

Postulat 1 (Eindeutigkeit des affinen Zusammenhangs) *Jeder "metrische Zusammenhang", d.h. jede Äquivalenzklasse $[\Lambda]$, der betrachteten $H - G$-Struktur enthält genau einen affinen Zusammenhang Γ. Für jeden auf $H(M)$ reduziblen linearen Zusammenhang Λ gibt es also genau einen affinen Zusammenhang Γ, so daß im Sinne von Gl (2) für alle Punkte $p \in M$ gilt:*

$$\Gamma^i_{j,k}(p) = \Lambda^i_{j,k}(p) + A^i_{j,k}(p)$$

---

des vertikalen Bündels verschwindet (Weyl betrachtete die Formen sowieso nur auf der Basis M) und sich bei Trivialisierungswechsel mit $h(x)$ richtig, d.h. durch die adjungierte Operation transformiert $\bar{\eta} = \mathrm{ad}_{h^{-1}(x)}\eta$, nach Trivialisierung also $\bar{\eta} = h^{-1}\eta h$; vgl. z.B. (Schottenloher 1995, 266).

Postulat 2 (Freiheit) *In einem vorgegebenen Punkt $p_0 \in M$ kann jedes System $L^i_{j,k} \in \mathbb{R}^{3n}$ als Wert einer "kongruenten Verpflanzung" $\Lambda^i_{j,k}(p_0)$ eines "metrischen Zusammenhangs" $[\Lambda]$ auftreten:*

$$L^i_{j,k} = \Lambda^i_{j,k}(p_0)$$

Postulat 1 galt Weyl als *kohärenzsicherndes Prinzip* für die betrachtete infinitesimale Kongruenzgeometrie. Es verallgemeinerte in naheliegender Weise die Rolle des Levi-Civita– (beziehungsweise des Weyl–) Zusammenhanges in einer Riemannschen (Weylschen) Mannigfaltigkeit.

Motivierende Argumente für die Doppelstruktur seines Postulates ("Kohärenz" bzw. Eindeutigkeit und "Freiheit") suchte Weyl auf verschiedenen Ebenen. So verwies er darauf, daß auf hoher Abstraktionsebene eine Analogie zum Postulat der freien Beweglichkeit in der klassischen Analyse des Raumproblems durch Helmholtz und Lie vorliege, wenn auch in der Form den Bedingungen der infinitesimalen Kongruenzgeometrie angepaßt und modifiziert. Der Freiheit der Bewegung (*Transitivität* der Bewegungsgruppe auf den Fahnen in der Sprache der linearen Geometrie) entsprach hier die größtmögliche Variationsmöglichkeit des "metrischen Feldes" (Λ) im Sinne des Freiheitspostulats. An die Stelle der Eindeutigkeit der Bewegung bei Vorgabe von Punkt und Richtungen (*einfache* Fahnentransitivität) setzte Weyl hier die Eindeutigkeit des affinen Zusammenhangs in jeder Äquivalenzklasse des "metrischen Zusammenhangs".

Als Weyl im Frühjahr 1922 in Barcelona und Madrid zum erstenmal ausführlich über seine Formulierung und die Lösung des Raumproblems vortrug, ging er bei der Begründung seiner Postulate über die Analogie zur klassischen Geometrie und Kinematik weit hinaus. Er argumentierte ausführlich mit einer *sozialen Metapher* für die Sinnhaftigkeit seines Raumpostulates. Ähnlich wie die Individuen mit dem sozialen Ganzen im Rahmen einer geeigneten Staatsverfassung miteinander so in Verbindung treten, daß die größtmögliche Freiheit der Individuen mit der Bewahrung des "Wohls des Ganzen" in Hand gehe, so sei in einer vernünftigen synthetischen Bestimmung der Raumstruktur zu gewährleisten, daß größtmöglicher Spielraum in den Punkten für die Verteilung und Kraftwirkung der Materie besteht und sich dennoch bei jeder aposteriorischen Bestimmung des "metrischen Zusammenhangs" eine strukturelle geometrische Kohärenz durch die eindeutige Auszeichnung eines affinen Zusammenhangs einstellt. Der größtmöglichen Freiheit der Individuen entspreche die Ausstattung der metrischen Raumstruktur mit größtmöglichen Freiheitsgraden für die metrikbestimmende Materierfüllung. Der affine Zusammenhang stehe als Kohärenzbedingung an der Stelle der Sicherung des "Wohls des Ganzen" (Weyl 1923c, 46f.).

Selbst wenn die entsprechende Textpassage unmittelbar eher motivierend als im logischen Sinne begründend gemeint war, kann man der Dichte der sozialen Rhetorik entnehmen, daß das zugehörige semantische Feld für Weyl mehr darstellte als lediglich ein nachträglich angehefteter Kontext seiner Begriffsbildung. Sie sollte als Indiz dafür genommen werden, *daß der soziale Kontext von Weyl in die*

Bildung seiner symbolischen Strukturen der Infinitesimalgeometrie als konstituierendes Moment mit einbezogen wurde.[76]

Aus mathematischer Sicht lieferte Postulat 2 in Verbindung mit dem Eindeutigkeitspostulat unmittelbar eine Dimensionsbedingunge für die betrachteten "Drehungsgruppen" G. Da nämlich

$$\Lambda^i_{j,k} = \Gamma^i_{j,k} - A^i_{j,k}$$

folgt wegen $\Gamma^i_{j,k} = \Gamma^i_{k,j}$ für dim $g = N$

$$n^3 = n \cdot \frac{n}{2}(n+1) + nN$$

und damit

$$N = \frac{n}{2}(n-1) \qquad (2.8.3)$$

Allerdings bemerkte E. Scheibe in einer Untersuchung über dreißig Jahre später, daß diese Dimensionsbedingung schon aus dem Postulat des eindeutig determinierten affinen Zusammenhangs folgt (Scheibe 1957, 196, Hilfssatz 1).[77] Er bewies in die hier verwendete Sprache übertragen:

Lemma 1 (Scheibe) *Für eine $H-G-$Struktur gilt:*

$$Postulat\ 1 \implies Postulat\ 2$$

Unter Berücksichtigung des Scheibeschen Lemmas läßt sich demnach der mathematische Kern der Weylschen Postulate für das "Wesen des Raumes" zusammenfassen als die Einschränkung auf eigentliche $H-G-$Strukturen in folgendem Sinne:

Definition 3 *Eine $H-G-$Struktur über M heiße* eigentlich, *falls in ihr Postulat 1 gilt.*

Für Weyl lag der Reiz seiner "Analyse des Raumproblems" allerdings darin, daß er das Problem keineswegs aus rein mathematischen Erwägungen heraus formuliert hatte. Schon die Herausarbeitung der (eigentlichen) infinitesimalen Kongruenzstrukturen verstand er, wie schon angemerkt, unter explizitem Bezug auf Kant als eine mathematisch-philosophische Bestimmung des "Wesens des Raumes" oder auch unter Verweis auf Husserl als "phänomenologische Wesensanalyse" des Raumes (RZM [4]1921, 133; [5]1923, 139ff.; 1923c, 49). Darüberhinaus bezog Weyl bewußt und in außerwöhnlicher Weise den sozialen Horizont von Handlungserfahrungen in die von ihm vorgenommenen "Wesens"- bestimmungen ein. In methodologischer Anspielung auf Kant legte er insbesondere Wert auf den "rein begriffsexplikatorischen" Charakter von Definition 1, während er die Raumpostulate 1 und 2

---

[76]Vergleiche Anmerkung (8).

[77]Scheibe verwendete dabei eine verwickelte Formulierung, die dem noch relativ unentwickelten Stand der Prinzipalbündel geschuldet zu sein scheint. Entsprechend langwierig erscheint die Formulierung seines Beweises. Dessen Gültigkeit soll durch diese Bemerkung nicht infrage gestellt werden.

ausdrücklich einem "*synthetischen Teil* im Kantischen Sinne" seiner Untersuchung zurechnete.

Im weiteren untersuchte Weyl nun notwendige Bedingungen dafür, daß eine Gruppe G überhaupt in einer eigentlichen $H - G$-Struktur einer Mannigfaltigkeit M der Dimension n auftreten kann. Dazu analysierte er insbesondere die Konsequenzen für die Liealgebra **g** und erhielt

Lemma 2 (Weyl) *Die Liealgebra* **g** *einer Gruppe* $G \subset SL(n, \mathbb{R})$, *die als Drehungsgruppe einer eigentlichen infinitesimalen Kongruenzstruktur einer Mannigfaltigkeit* M *der Dimension* n *auftritt, erfüllt folgende (notwendigen) Bedingungen:*

a) *Spur*$(A) = 0$ *für alle* $A \in$ **g**.

b) *dim* **g** $= \frac{1}{2}n(n - 1)$.

c) *Ist für eine symmetrische bilineare Abbildung* $B \in Hom(\mathbb{R}^n \times \mathbb{R}^n, \mathbb{R}^n)$ *für alle* $x \in \mathbb{R}^n$ $B_x := B(\_, x) \in$ **g**, *so gilt* $B_x = 0$.[78]

Die Spurbedingung a) folgt aus der Forderung $G \subset SL(n, \mathbb{R})$, die Dimensionsbedingung b) folgerte Weyl, wie oben angegeben, aus seinem "Freiheitspostulat" und die Unsymmetriebedingung c) aus der Eindeutigkeitsforderung für "affine" Repräsentanten eines metrischen Zusammenhangs.

Weyl vermutete, daß aus Bedingungen a), b), c) seines Lemmas die Isomorphie von G zu einer verallgemeinerten orthogonalen Gruppe folgt, $G \cong SO(p, q)$ mit $p + q = n$, und daß in diesem Sinne der "pythagoreische Raumtyp" eine gut vertretbare apriorische Rechtfertigung besitzt. Dabei war er von der begrifflichen Kohärenz seiner Überlegungen so überzeugt, daß er seine Analyse und die zugehörige Vermutung schon nach der Verifizierung der Vermutung lediglich für die Dimensionen $n = 2, 3$ in die 4. Auflage von RZM aufnahm (RZM ^{4}1921, §18).

Noch vor Ausführung eines kompletten Beweises kündigte er also seinen Hauptsatz an. Im Laufe der Ausarbeitung des Beweises entdeckte er, daß für Dimension $n \geq 3$ die Forderung der Volumentreue, $G \subset SL(n, \mathbb{R})$, überflüssig ist. In der endgültigen Fassunge lautete seine Satz dann:

Satz 4 (Hauptsatz der Analyse des Raumproblems) *Ist* **g** *Liealgebra einer Lieuntergruppe von* $GL(n, \mathbb{R})$ *und erfüllt die Bedingungen* b), c) *von Lemma 2, sowie im Fall* $n = 2$ *auch die Bedingung* a), *so ist* **g** *Liealgebra einer verallgemeinerten orthogonalen Gruppe* $G = SO(p, q)$ *(mit* $p + q = n$).

Noch während des Jahres 1921 gelang ihm ein mühseliger Beweis für die Kernaussage des Hauptsatzes; jedoch machte dieser ursprüngliche Beweis viele Fallunterscheidungen erforderlich.[79] Bei der Ausarbeitung seiner Vorlesungen über das

---

[78] (Weyl 1923c, 50f. etc.). Weyl sprach dabei von B als einer "symmetrischen Doppelmatrix".

[79] Angekündigt in (Weyl 1921f), ausgeführt in (Weyl 1922b). "Kernaussage" bedeutet: Prämisse des Satzes noch mit der Spurbedingung a) in allen Dimensionen.

Raumproblem in Barcelona (publiziert 1923) vereinfachte er den Beweis beträchtlich und gab dem Satz die angegebene allgemeine Form. Der Beweis verwendete allerdings immer noch umfangreiche Rechnungen im Rahmen der Weierstraßschen Elementarteilertheorie.[80] Damit war Weyl der Nachweis gelungen, daß eine infinitesimale Kongruenzstruktur, die "dem Wesen des Raumes" entspricht (das heißt die Weylschen Raumpostulate erfüllt), eine verallgemeinerte orthogonale Gruppe $SO(p,q)$ als Drehungsgruppe besitzt.

Daraus zog er sogar, scheinbar beiläufig, noch das

Korollar 1 *Zu einer vorgegebenen eigentlichen infinitesimalen Kongruenzstruktur (d.h. eigentlichen $H - G-$Struktur auf einer differenzierbaren Mannigfaltigkeit M samt ausgezeichnetem linearen Zusammenhang Λ) gibt es eine Weylsche Metrik ($[g], [\varphi]$), deren infinitesimale Kongruenzstruktur mit der vorgegebenen $H - G-$Struktur zusammenfällt und deren affiner (Weyl-) Levi-Civita Zusammenhang Γ mit Λ im Sinne von Gl (2) äquivalent ist.*

Weyl skizzierte den Beweis dieser Aussage wie folgt: Nach dem Hauptsatz der Analyse des Raumproblems ist $G \cong SO(p,q)$ und $H \cong \mathbb{R}_+ * SO(p,q)$. Die vorgegebene $H - G-$Struktur gestattet es, eine differenzierbar vom Punkt p abhängige Orthogonalbasis des Tangentialraumes $T_p(M)$ und damit eine konforme Struktur $[g]$ der Signatur (p,q) auf M zu bestimmen. Denn lokale Trivialisierung liefert in jedem Punkt $p \in M$ eine lineare Darstellung der Operation von $SO(p,q)$ auf T_pM und durch Integration eine invariante Metrik $g(p)$ in jedem Punkt $p \in M$. Eine Auswahl von $g \in [g]$ zu treffen, "heißt die Mannigfaltigkeit eichen". Der durch Λ bestimmte "metrische Zusammenhang" $[\Lambda]$ enthält genau einen affinen Zusammenhang Γ. Aus Γ und g läßt sich der Streckenzusammenhang $[\varphi]$ in lokalen Koordinaten bestimmen. Weyl gab dafür eine explizite Formel an (Weyl 1923c, 51f.).[81]

Als Weyl im Jahre 1923 die endgültige Fassung seines *Raumproblems* publizierte, hatte E. Cartan die Fragestellung schon aus der französischen Übersetzung der 4. Auflage von RZM aufgenommen und im geometrischen Teil zum Cartanschen Raumproblem umformuliert. Cartan verstand die von Weyl charakterisierte infinitesimale Kongruenzstruktur in der von ihm vorgezogenen Sprache der Zusammenhänge im Sinne begleitender n-Beine. Seine Darstellung trug allerdings der für Weyls geometrische Struktur wichtige Unterscheidung von H und G nicht mehr ausdrücklich Rechnung (Cartan 1922, Cartan 1923, 171–174). Die nächste Generation der Differentialgeometer, Chern, Nomizu, Klingenberg und andere, formten Cartans Beschreibung aus Sicht der Hauptfaserbündel noch einmal um und betrachteten die Vorgabe einer G-Struktur auf M als ausreichend für die geometrische Charakterisierung des Weylschen Raumproblems.[82] Dabei ist eine

---

[80]Beweisskizze in der 8. Vorlesung (Weyl 1923c, 51ff.), Ausführung in Anhang 12 (Weyl 1923c, 88ff.).

[81]Die Angabe von φ anstelle von Γ ist Scheibes Sicht in der 1988 modifizierten Reininterpretation des Weylschen Raumproblems (Scheibe 1988, 69).

[82]Ich halte mich hier an die Darstellung von (Kobayashi-Nomizu 1963, 288).

G-Struktur auf M nichts anderes als eine Bündelreduktion des linearen n-Bein-Bündels in TM auf die Strukturgruppe und typische Faser $G \subset GL(n, \mathbb{R})$.

Aus dieser Sicht bestand dann das *Cartansche Raumproblem* in der Angabe derjenigen Gruppen G, für die jede zugehörige G-Struktur, jeweils genau einen torsionsfreien linearen Zusammenhang besitzt ("eigentliche" G-Struktur). Das Spezifische der Weylschen Eichstruktur im Wechsel zwischen linearem -, Drehungs- und Ähnlichkeitsbündel wurde dabei ausgeblendet.

Es blieb die "gereinigte" Fassung, in der Drehungs- und Ähnlichkeitsstruktur als identisch behandelt wurden. Dadurch ging die für Weyls Infinitesimalgeometrie wichtige Unterscheidung von infinitesimaler Kongruenzgruppe G und einer dazu ergänzenden Eichgruppe H/G und damit *der eigentlich Kern der Weylschen Eichstruktur* zunächst einmal verloren. Auf die im Gefolge Weyls eingetretene Strukturnivellierung wies E. Scheibe in seiner Dissertation wieder hin (Scheibe 1957) und kam später in verschiedenen epistemologischen Beiträgen auf diesen Sachverhalt zurück.[83]

Cartan konnte das so modifizierte Weylsche Raumproblem nach einem geometrischen Zwischenschritt[84] direkt mit seinen Klassifikationsergebnissen der einfachen und halbeinfachen Liegruppen lösen. Diese elegante Methode wurde zu einem der Anlässe für Weyl, sich näher mit der Theorie der Liegruppen und deren Darstellungen zu beschäftigen. Weitere Anlässe und wichtige technische Hilfsmittel für seine Arbeit kamen allerdings darüberhinaus aus der Invariantentheorie.[85]

2.9 Weitere Anlässe zum Umdenken

Noch in anderer Hinsicht erhielt Weyl aus seiner Beschäftigung mit dem Raumproblem Anlaß zum Umdenken. Seine Analyse arbeitete ja mit allgemeinsten begrifflichen Annahmen, weit weg von jedweder konstruktiven Präzisierung. In seiner Kritik an Hilbert sprach er schon aufgrund subtiler methodischer Differenzen gegenüber der klassischen Analysis von der "...inneren Haltlosigkeit der Grundlagen, auf denen der Aufbau des Reiches ruht" (Weyl 1921c, 143). In krassem Gegensatz dazu kamen ihm bei den höchst allgemeinen Begriffs- und Strukturuntersuchungen des Raumproblems nicht die geringsten Anflüge eines Zweifels, obwohl sie dem Anspruch eines konstruktiven oder intuitionistischen Aufbaus der Mathematik keineswegs genügten. Diese überraschende Sicherheit schöpfte Weyl aus einer

---

[83] Im Beitrag von R.Coleman und H. Korté in diesem Band wird E.Scheibes Darstellung des Weylschen Raumproblems ausführlich und (aus meiner Sicht übertrieben) kritisch gewürdigt. Es bleibt zu ergänzen, daß sich in (Scheibe 1988). ein dankenswert deutlicher und klärender Hinweis auf den "historical myth" in der mathematischen Literatur findet, in der bei der Generation nach Cartan üblicherweise die Differenz des Weylschen und des Cartanschen Raumproblems unerwähnt bleibt und stattdessen die Cartansche Formulierung als bruchlose Präzisierung der Weylschen Charakterisierung gehandelt wird. Eine eingehende historische Untersuchung von Cartans Beitrag und möglichen Verschiebungen noch einmal in der Generation nach Cartan steht allerdings aus.

[84] Die Gruppe G einer eigentlichen G-Struktur hat keinen nichttrivialen invarianten Unterraum im $\mathbb{R}^n$.

[85] Siehe (Borel 1986), (Hawkins 1997, 1999) und (Slodowy 1999).

ihm offensichtlich scheinenden und daher gleichsam von vornherein unterstellten "transienten Bedeutung" seiner eigenen symbolischen Produkte. Das ging so weit, daß er nicht einmal zu dem Zeitpunkt Zweifel an der Richtigkeit seines Hauptsatzes $G \cong SO(p,q)$ besaß, als er noch keinen allgemeinen Beweis in Sicht hatte.

> In der 4. Auflage meines Buches 'Raum, Zeit, Materie' (Springer 1921) hatte ich die Stirn, den eben erwähnten übrigens rein algebraischen Satz über infinitesimale Gruppen linearer Transformationen (den Hauptsatz der Analyse des Raumproblems, E.S.) als Vermutung auszusprechen allein auf Grund seiner transienten Bedeutung für das Raumproblem. (Weyl 1921e, 228)

Diese Differenz wurde dadurch weiter pointiert, daß Weyl bei der Ausarbeitung seines Beweises des Hauptsatzes nicht ohne Verwendung des *tertium non datur* im zugrunde gelegten reellen Kontinuum auskam. Am Ende der ersten Publikation seines Beweises gestand er ein;

> ...in der vorliegenden Form ist er (der Beweis, E.S.) nur zwingend, wenn wir uns in einem Zahlbereich bewegen wie z.B. dem der rationalen Zahlen, in welchem eine Zahl entweder $= 0$ oder $\neq 0$ ist. Die Fallunterscheidungen der Elementarteilertheorie sind von diesem Standpunkt aus ein besonders bedenklicher Ausgangspunkt. Darauf wird aber erst zurückzukommen sein, wenn einmal eine neue Analysis in bestimmterer Ausgestaltung vorliegt, als es heute der Fall ist. (Weyl 1922b, 295)

Weyl kam also in diesem "transient" motivierten mathematischen Kontext ohne größere "Gewissensbisse" wieder auf die klassische Logik in den reellen Zahlen zurück. Die Rekonstruktion seiner Ergebnisse im Sinne der Kontinuumsanalysis vertagte er auf unbestimmte Zeit, bis die technischen Voraussetzungen für einen entsprechenden Beweis zur Verfügung stehen würden. Durch die Verschiebung seiner Forschungsinteresse zur Darstellungstheorie der Liegruppen und ihrer Anwendung in der Quantenphysik wurden für ihn jedoch in den nächsten Jahren andere Themen wichtiger als die Ausarbeitung "einer neuen Analysis in bestimmterer Ausgestaltung".

Noch vor Ablauf des Jahres 1920 hatte er die Berufung nach Amsterdam abgelehnt. Die gemeinsame Arbeit mit Brouwer an der "Revolutionierung der Analysis" kam nie zustande. Gegenüber Klein schrieb er im häufig zitierten Brief vom 28. 12. 1920 als Grund, "...das Risiko des holländischen Klimas kann ich nicht übernehmen". Der Hintergrund dieses Satzes liegt in schweren asthmatischen Anfällen, die Weyl schon in der Schweiz immer wieder zur Unterbrechung seiner Lehrtätigkeit und Aufenthalt im Hochgebirge zwangen.[86] So wurde Weyls Beschäftigung mit

---

[86]Ich danke Skúli Sigurdsson für diesen Hinweis. Zusätzlich mag man aber auch die metaphorische Bedeutung dieses Satzes in Rechnung stellen; siehe dazu etwa die Verwickelungen im Streit zwischen Brouwer und Schouten (van Dalen 1999, 295ff.).

dem Raumproblem zu einem Durchgangsstadium für ihn in Richtung auf größere Offenheit sowohl gegenüber der klassischen Analysis als auch einer Anerkennung axiomatischer Begriffsfundierung innerhalb der Mathematik. Letztere sollte aus seiner Sicht allerdings weiterhin eng an über die Mathematik hinausgehende semantische Kontexte gebunden bleiben. Dies war Mitte der 1920-er Jahre sein Angebot an Hilbert, mit dessen "Formelspiel irgendein(en) *Sinn* zu verknüpfen". Zwar blieb er weiterhin parteilich für eine intuitive und kontextuell gebundene Vorgehensweise innerhalb der Mathematik; er war aber bis Mitte der 1920-er Jahre in seiner eigenen Arbeit von der radikalen Haltung der Zeit 1918 bis 1920 hinsichtlich der Notwendigkeit und Möglichkeit einer Neubegründung der Mathematik durch einen Prozeß symbolischer Selbsthervorbringung abgerückt. Nun konnte er versuchen, die *heutige Erkenntnislage in der Mathematik* "sine ira et studio" zu beschreiben (Weyl 1925a, insbesondere 540, 542).

Wir haben gesehen, wie Weyls Infinitesimalgeometrie der Jahre 1917 bis 1925 in vielerlei Hinsicht an Riemanns Motive anknüpfte und dabei insbesondere durchaus bewußt an die in Göttingen gepflegte Riemann-Tradition.[87] Natürlich läßt sich jede Tradition verschieden auslegen; schon Hilbert und Klein beriefen sich beide auf Riemann und hatte doch so verschiedene Mathematikauffassungen und Arbeitsprofile, daß sie von H. Mehrtens als typische entegegengesetzte Vertreter des Spektrums von "Moderne" und "Gegenmoderne" angesehen werden (Mehrtens 1990). Weyl definierte in diesem Spektrum einen eigenen und eigenwilligen Ort. Seine bedeutendste mathematische Leistung, der Beitrag zur Darstellungstheorie der Liegruppen, stand gerade bevor. Weyls auf disziplinäre Querbezüge achtender Arbeitsstil, seine Abneigung gegenüber ihm als bloß formal erscheinenden Theoriebildungen und seine Bemerkungen zu den Grundlagen der Analysis lassen ihn aus Mehrtens' Sicht als einen typischen Vertreter der mathematischen "Gegenmoderne" erscheinen. Weyls Sensibilität und sein aktives Interesse an vielfältigen semantischen Bezügen der von ihm betriebenen Mathematik, also gerade die vermeintlich "gegenmodernen" Züge seiner Arbeitsweise trugen jedoch entscheidend zur Wirksamkeit seiner Ideen innerhalb der "modernen" Mathematik und theoretischen Physik bei.

Schon wenige Jahre nach den in diesem Beitrag vorgestellten Arbeiten eröffneten sich nach dem Übergang zur "neuen" Quantentheorie überraschende neue physikalische Bezüge für Weyls ursprünglich aus philosophischen Fragestellungen heraus entwickelten Konzepte der infinitesimalen Eichgeometrie. Diese waren nun nicht mehr an einen rein feldtheoretischen Ansatz dynamistischer Materieerklärung gebunden wie die Theorie von 1918, sondern griffen mit Diracs Elektronentheorie auf die neu entstehende quantenphysikalische Erklärung der Mikroprozesse zurück.[88] Sie gewannen nach einem komplexen Prozeß der Neu- und Umentdeckung durch die nächste Generation der mathematischen Physiker eine tiefliegende Bedeutung für das Verständnis der offeneren Strukturbildungen des

---

[87]Zur Etablierung dieser Tradition siehe unter anderem (Rowe 1989, 1997a, 1997b).
[88]Siehe dazu den Beitrag von N. Straumann in Kap. I 4 dieses Bandes.

"dynamischen Agens" der Materie. Auch in dieser überraschenden Neuaufnahme kann man gut eine bemerkenswerte Parallele zwischen Weyls Arbeiten zur Infinitesimalgeometrie und der langanhaltenden Wirkung Riemanns durch das gesamte 20-te Jahrhundert sehen.

Literatur

Beisswanger, Peter 1965. Die Phasen in Hermann Weyls Beurteilung der Mathematik. *Mathematisch-Physikalische Semesterberichte* **12**, 132–156.

Bleecker, David 1981. *Gauge Theory and Variational Principles*. London etc.: Addison-Wesley.

Bloch, Ernst 1978. *Die Lehren von der Materie*. Frankfurt/Main: Suhrkamp.

Borel, Armand 1986. Hermann Weyl and Lie groups. In (Chandrasekharan 1986, 53–82).

Bottazzini, Umberto 1999. Ricci and Levi-Civita: from differential geometry to general relativity. In (Gray 1999, 241–259).

Brouwer, Luitzen E. J. 1911a. Beweis der Invarianz der Dimensionszahl. *Mathematische Annalen* **69**, 169–175. CW **2**, 430- -453.

Brouwer, Luitzen E. J. 1911b. Über Abbildungen von Mannigfaltigkeiten. *Mathematische Annalen* **71**, 97–115. CW **2**, 454–476.

Brouwer, Luitzen E. J. 1918/1919. Begründung der Mengenlehre unabhängig vom logischen Satz vom ausgeschlossenen Dritten. *Verhandlingen Kgl. Akademie Wetenschapen Amsterdam*. CW **1**, 150–190.

Brouwer, Luitzen E. J. 1920. Intuitionistische Mengenlehre. *Jahresbericht DMV* **28**, 203–208. CW **1**, 230–235.

Brouwer, Luitzen E. J. 1975/1976 (CW). *Collected Works*. Ed. A. Heyting, 2 vols. Amsterdam: North-Holland.

Cao, Tian Yu 1997. *Conceptual Developments of 20th Century Field Theories*. Cambridge: University Press.

Cartan, Elie 1922. Sur un théorème fondamental de M. H. Weyl dans la théorie de l'espace métrique. *Comptes Rendus Ac. Sci.* **175**, 82–85. *Oeuvres Complètes* **3.1**, Paris 1955 [62], 629–632.

Cartan, Elie 1923. Sur un théorème fondamental de M. H. Weyl. *Journal de Mathématiques* **2**, 167–192. Ouevres Compètes **3.1**, Paris: Gauthier-Villars 1955, 633–658.

Chandrasekharan, Komaravolu (ed.) 1986. *Hermann Weyl 1885–1985: Centenary Lectures Delivered by C. N. Yang, R. Penrose, A. Borel at the ETH Zürich.* New York etc.: Springer.

Corry, Leo 1996. David Hilbert and the axiomatization of physics (1894–1905). *Archive for History of Exact Sciences,* **51**, 83–198.

Corry, Leo 1997. Hilbert's way to general relativity. In J. Renn e.a. (ed.) *Alternatives to Einstein's General Relativity.* Basel: Birkhäuser.

Corry, Leo 1999a. Hilbert and physics (1900–1915). In (Gray 1999, 145–188).

Corry, Leo 1999b. David Hilbert between mechanical and electromagnetic reductionism (1910–1915). *Archive for History of Exact Sciences* 53, 489–527.

Corry, Leo 1999c. From Mie's electromagnetic theory of matter to Hilbert's unified foundation of physics. *Studies in the History and Philosophy of Modern Physics* **30**, 159–183.

Corry, Leo; Renn, Jürgen; Stachel, John 1997. Belated decision in the Hilbert-Einstein priority dispute. *Science* **278**, 1270–1273.

van Dalen, Dirk 1995. Hermann Weyl's intuitionistic mathematics. *Bulletin of Symbolic Logic* **1**, 145–169.

van Dalen, Dirk 1999. *Mystic, Geometer, and Intutionist. The Life of L. E. J. Brouwer.* Oxford: University Press.

Deppert, Wolfgang (ed.) 1988. *Exact Sciences and Their Philosophical Foundations: Exakte Wissenschaften und ihre philosophische Grundlegung.* Vorträge des internationalen Hermann-Weyl-Kongresses, Kiel 1985. Frankfurt/M - Bern etc.: Peter Lang Verlag.

Earman, John; Glymour, Clark 1978. Einstein and Hilbert: Two months in the history of general relativity. *Archive for History of Exact Sciences* **19**, 291–308.

Einstein, Albert 1998. *The Collected Papers of Albert Einstein* Vol **8**: The Berlin Years: Correspondence 1914–1918. Princeton: University Press.

Ehlers, Jürgen; Pirani, Felix A. E.; Schild, Alfred 1972. The geometry of free fall and light propagation. In: O'Raifeartaigh, L. (ed.) *General Relativity, Papers in Honour of J. L. Synge.* Oxford: Clarendon Press, 63–84.

Epple, Moritz 1999. Did Brouwer's intuitionistic analysis satisfy its own epistemological standards? In (Hendricks e.a. 1999).

Feferman, Solomon 1988. Weyl vindicated: "Das Kontinuum" 70 years later. *Atti del Congresso Temi e prospettive della logica e della filosofia della scienza contemporanee.* Cesena 7–10 gennaio 1987. 1. Bologna: CLUEB, 59–93.

Ferreiros, José. *Labyrinth of Thought. A History of Set Theory and Its Role In Modern Mathematics.* Basel: Birkhäuser.

Fichte, Johann Gottlieb 1800/1879. *Die Bestimmung des Menschen.* Unter Berücksichtigung der Ausgaben 1801, 1838 und 1845. Hers. Karl Kehrbach. Leipzig: Reclam 1879.

Fichte, Johann Gottlieb 1802. *Grundriss des Eigenthümlichen der Wissenschaftslehre in Rücksicht auf das theoretische Vermögen.* Jena - Leipzig. SW **1**, 331–411. GA **I.3**, 142–208.

Forman, Paul 1971 Weimar culture, causality, and quantum theory, 1918 – 1927: Adaptation by German physicists and mathematicians to a hostile intellectual environment. *Historical Studies in the Physical Sciences* **3**, 1–116. Deutsche Übersetzung in (von Mayenn 1994, 61–179).

Forman, Paul 1980. Weimar culture, causality, and quantum theory, 1918 – 1927: Adaptation by German physicists and mathematicians to a hostile intellectual environment. In: Colin Chant, John Fauvel (eds.). *Darwin to Einstein: Studies on Science and Belief.* Essex: Longman, 267—302.

Frei, Günther; Stammbach, Urs 1992. *Hermann Weyl und die Mathematik an der ETH Zürich, 1913 - 1930.* Basel usw.: Birkhäuser.

Goenner, Hubert 1993. The reactions to relativity I: The anti-Einstein campaign in Germany in 1920. *Science in Context* **6**, 107–133.

Goldstein, Catherine; Ritter, Jim 2000. The varieties of unity: Sounding unified theories 1920–1930. *Preprint* Paris.

Gray, Jeremy J. (ed.) 1999. *The Symbolical Universe. Geometry and Physics 1890–1930.* Oxford: University Press.

Hausdorff, Felix 1914. *Grundzüge der Mengenlehre.* Leipzig: Veit.

Hawkins, Thomas 1998. From general relativity to group representations: The background to Weyl's paper of 1925-26. Matériaux pur l'histoire des mathématiques au XXe siècle. Société Mathématique de France, *Séminaire et Congrès* **3**, 69–100.

Hawkins, Thomas 1999. Weyl and the topology of continous groups. In (James 1999, 169–198).

Hendricks, Vincent F.; Pedersen, S.A.; Frovin, K. (eds.) 2000. *Proof Theory: History and Philosophical Significance.* Dordrecht: Kluwer.

Hesseling, Dennis E. 2000. *Gnomes in the Fog. The Reception of Brouwer's Intui-tionism in the 1920's*. Basel: Birkhäuser (in Druck).

Hessenberg, Gerhard 1917. Vektorielle Begründung der Differentialgeometrie. *Ma-thematische Annalen* **78**, 187–217.

Hilbert, David 1902. Über die Grundlagen der Geometrie. *Nachrichten Gesellschaft der Wissenschaften Göttingen*, 233-241. Gekürzte Fassung in *Mathematische Annalen* **56**, 381–422. Vollständiger Nachdruck als *Anhang IV* zu *Grundlagen der Geometrie* [2]1903, 121ff.; [7]1930, 178–230.

Hilbert, David 1915. Die Grundlagen der Physik. Erste Mitteilung. *Nachrichten Gesellschaft der Wissenschaften Göttingen*, 395–407 [nicht in GA].

Hilbert, David 1917. Die Grundlagen der Physik. Zweite Mitteilung. *Nachrichten Gesellschaft der Wissenschaften Göttingen*, 395–407 [nicht in GA].

Hilbert, David 1924. Die Grundlagen der Physik. *Mathematische Annalen* **92**, 1–32. GA **3** (1935), 258–289 [*veränderter* Abdruck von (1915, 1917)].

Hilbert, David 1927. Referat über die geometrischen Schriften und Abhandlungen Hermann Weyl's, erstattet der Physiko-Mathematischen Gesellschaft an der Universität Kasan. *Bulletin de la Société Phys. Math. de Kazan* (3) **2**, 66–70 [nicht in GA].

Hilbert, David GA. *Gesammelte mathematische Abhandlungen*. 3 Bände. Berlin: Springer 1932, 1933, 1935. Neudruck New York: Chelsea: 1965.

Hoddeson, Lillian; Brown, Laurie; Riordan, Michael; Dresden, Max (eds.) 1997. *The Rise of the Standard Model*. Particle Physics in the 1960s and 1970s. Cambridge: UP.

James, Ioan (ed.) 1999. *Handbook of the History of Algebraic and Geometric To-pology in the 20th Century*. Dordrecht: Kluwer 1999.

Kobayashi, Shoshichi; Nomizu, Katsumi 1963. *Foundations of Differential Geome-try*. Vol 1. London etc.: John Wiley.

Laugwitz, Detlef 1958. Über eine Vermutung von Hermann Weyl zum Raumpro-blem. *Archiv der Mathematik* **9**, 128–133.

Leupold, Rudolf 1961. Die Grundlagenforschung bei Hermann Weyl. *Dissertati-onsschrift* Universität Mainz.

Levi-Civita, Tullio 1917. Nozione di parallelismo in una varietá qualunque e con-seguente specificacione geometrica della curvatura Riemanniana. *Rendiconto del Circolo Matematico di Palermo* **42**, 173–205.

Majer, Ulrich 1988. Zu einer bemerkenswerten Differenz zwischen Brouwer und Weyl. In: (Deppert e.a. 1988, 543–553).

von Mayen, Karl (Hrsg.)1994. *Quantenmechanik und Weimarer Republik.* Braunschweig-Wiesbaden: Vieweg.

Mehrtens, Herbert 1990. *Mathematik — Moderne — Sprache. Die mathematische Moderne und ihre Gegner.* Frankfurt/Main: Suhrkamp

Milnor, John 1956. On manifolds homeomorphic to the 7-sphere. *Annals of Mathematics* (2) **64**, 399–405.

Noether, Emmy 1918. Invariante Variationsprobleme. *Nachrichten Gesellschaft der Wissenschaften Göttingen*, 235–257.

Pais, Abraham 1986. *"Raffiniert ist der Herrgott ... " Albert Einstein Eine wissenschaftliche Biographie.* Aus dem Englischen von R. U. Sexl, H. Kühnelt, E. Streeruwitz. Braunschweig/Wiesbaden: Vieweg.

Pauli, Wolfgang 1920. Diskussionsbemerkungen zum Vortrag Weyl: Elektrizität und Gravitation. *Physikalische Zeitschrift* **21**, 650 [nicht in Collected Scientific Papers].

Pauli, Wolfgang 1921. Relativitätstheorie. *Encyklopädie der Mathematischen Wissenschaften* **5.2**, Leipzig: Teubner, 539–775.

O'Raifeartaigh, Lochlain 1995. The birth of gauge theory. In B. Gruber (ed) *Symmetries in Science* **7**, New York: Plenum, 433– 443.

O'Raifeartaigh, Lochlainn; Straumann, Norbert 2000. Gauge theory: Historical origins and some modern developments. *Reviews of Modern Physics* **72**, 1–23.

Reich, Karin 1992. Levi-Civitasche Parallelverschiebung, affiner Zusammenhang, Übertragungsprinzip: 1916/17-1922/23. *Archive for History of Exact Sciences* **44**, 77-105.

Reich. Karin 1994. *Die Entwicklung des Tensorkalküls. Vom absoluten Differentialkalkül zur Relativitätstheorie.* Basel etc.: Birkhäuser.

Renn, Jürgen; Stachel, John 1999. Hilbert's foundations of physics: From a theory of everything to a constituent of General Relativity. *Preprint* **118**, MPI Wissenschaftsgeschichte Berlin.

Riemann, Bernhard 1854. Über die Hypothesen, welche der Geometrie zu Grunde liegen. Habilitationsvortrag Göttingen. *Göttinger Abhandlungen* **13** (1867). *Gesammelte Werke*, 272-287.

Rowe, David E. 1989. Klein, Hilbert, and the Göttingen mathematical tradition. *Osiris* (2) **5**, 186–213.

Rowe, David E.1997a. Perspective on Hilbert. *Perspectives on Science* **5**, 533–570.

Rowe, David E. 1997b. Felix Klein and the Göttingen mathematical community. *Preprint* Mainz University.

Rowe, David E. 1999. The Göttingen response to general relativity and Emmy Noether's theorems. In (J.J. Gray 1999, 189–234).

Rowe, David E. 2000. Einstein meets Hilbert: At the crossroads of physics and mathematics. *Preprint* Mainz University.

Sauer, Tilmann 1999. The relativity of discovery: Hilbert's first note on the foundations of physics *Archive for History of Exact Sciences* **53**, 529–575.

Scheibe, Erhard 1957. Über das Weylsche Raumproblem. *Journal für r. u. a. Mathematik* **197**, 162–207.

Scheibe, Erhard 1988. Hermann Weyl and the nature of spacetime. In (Deppert 1988, 61–82).

Schelling, Friedrich Wilhelm J. 1799. *Erster Entwurf eines Systems der Naturphilosophie.* Sämtliche Werke, Hrsg. K. F. A. Schelling. Band **3** Stuttgart: Cotta 1858.

Scholz, Erhard 1995. Hermann Weyl's purely "infinitesimal geometry". *Proceedings International Congress of Mathematicians,* Zürich 1994. Basel etc.: Birkhäuser, 1592–1603.

Scholz, Erhard 1996. Logische Ordnungen im Chaos: Hausdorffs frühe Beiträge zur Mengenlehre. In: E. Brieskorn (Hrsg.). *Felix Hausdorff zum Gedächtnis.* Band **1** *Aspekte seines Werkes.* Wiesbaden: Vieweg, 107–134.

Scholz, Erhard 1999a. Weyl and the theory of connections. In (Gray 1999, 260–284).

Scholz, Erhard 1999b. The concept of manifold, 1850 – 1940. In (James 1999, 24–64).

Scholz, Erhard 2000. Hermann Weyl on the concept of continuum. In: (Hendricks e.a. 2000, 213–237).

Schouten, Jan Arnoldus 1919. Die direkte Analysis zur neueren Relativitätstheorie. *Verhandlingen Kgl. Akademie der Wetenschappen Amsterdam* **12** (6).

Schouten, Jan Arnoldus 1921. Über die konforme Abbildung n-dimensionaler Mannigfaltigkeiten mit quadratischer Massbestimmung auf eine Mannigfaltigkeit mit euklidischer Massbestimmung. *Mathematische Zeitschrift* **11**, 58–88.

Sigurdsson, Skúli 1991. Hermann Weyl, Mathematics and Physics, 1900–1927. *Ph.D. Dissertation* Harvard University, Cambridge, Mass.

Sigurdsson, Skúli 1994. Unification, geometry and ambivalence: Hilbert, Weyl and the Göttingen community. In: Gavroglu, Kostas e.a. (eds.) *Trends in the Historiography of Science.* Kluwer, 355–367.

Sigurdsson, Skúli 1996. Physics, life, and contingency: Born, Schrödinger, and Weyl in exile. In: Ash, Mitchell; Söllner, Alfons (eds.), *Forced Migration and Scientific Change. Emigré German-Speaking Scientists and Scholars after 1933.* Cambridge: Universtiy Press, 48–70.

Slodowy, Peter 1999. The early development of the representation theory of semisimple Lie groups: A. Hurwitz, I. Schur, H. Weyl. *Jahresbericht DMV* **101**, 97–115.

Straumann, Norbert 1987. Zum Ursprung der Eichtheorien bei Hermann Weyl. *Physikalische Blätter* **43** (11), 414–421.

Tietze, Heinrich 1923. Beiträge zur allgemeinen Topologie I. *Mathematische Annalen* **88**, 290–312.

Veblen, Oswald 1922. *Analysis Situs.* New York: American Mathematical Society Publications.

Vermeil, H. 1917. Notiz über das mittlere Krümmungsmass einer n-fach ausgedehnten Riemann'schen Mannigfaltigkeit. *Nachrichten Göttinger Gesellschaft Wissenschaften,* Math.-phys. Kl. 334–344.

Vizgin, Vladimir P. 1994. *Unified Field Theories in the First Third of the 20th Century.* Translated from the Russian by J. B. Barbour. Basel etc.: Birkhäuser.

Archivalia

Hausdorff, Felix, Ms 2. 3. 1912. Analysis Situs. Studien. vom 2. 3. 1912. Universitätsbibliothek Bonn, *Nachlaß Hausdorff,* Kapsel 31, Faszikel 121, Blatt 1.

Medicus, Fritz Hs 1117a:7 Notizen über Schelling. Bibliothek ETH Zürich, *Nachlaß Medicus,* Hs 1117a:7.

Weyl an Klein 28. 12. 1920. Universitätsbibliothek Göttingen *Nachlaß Klein* **12**, 297.

Weyl, Hermann Hs 91a:72. Entwicklungslinien der Mathematik seit 1900. Problems and methods of 20th century mathematics. Gastvorlesung nach 1950, englisch. Bibliothek ETH Zürich, *Nachlaß Weyl,* Hs 91a:72.

3 Weyl's contributions to cosmology

Hubert Goenner

3.0 Introduction

Compared with his outstanding achievements in mathematics and mathematical physics, the notable contributions of Hermann Weyl (1885-1955) to the subfield of physics "cosmology" are less well known. Nevertheless, during the period 1918 to 1930, in five editions of his book *Raum-Zeit-Materie* (four of which differ) and in several articles, he was instrumental for the development of relativistic cosmological modeling by lending creative ideas to it. The most prominent of these ideas is what he himself, in an article in the *Encyclopedia Britannica* (Weyl 1926e), called "Weyl's hypothesis" and which was taken up by other authors as "Weyl's postulate" (Robertson 1933, Narlikar 1979, Raychaudhuri 1979) or as "Weyl's principle" (Weinberg 1972). It is the assumption that there exists a common rest system[1] for the galaxies and that their worldlines originate in a common point in the past (Weyl 1923e). Weyl also found an exact, spherically symmetric static solution of Einstein's field equations with cosmological constant for incompressible matter (RZM, RZM ^{3}1919, 1919d). He also was one of the first scientists calculating an (approximately) *linear* relation between the redshift of galactic spectra and distance (Weyl 1923e).

However, being in good company with Einstein himself and the little band of scientists adressing questions of cosmology at the time, Weyl did not think of "expanding" solutions of the field equations as we are used to call them now. As others, he was not ready to clearly separate coordinate singularities from singularities inherent in the geometry and did not come to a coordinate independent definition of the concepts "static" and "stationary". The task of disentangling questions connected with the freedom in the choice of coordinates from those related to the arbitrariness inherent in a splitting of space- time into space and time was left for and solved by the next generation of cosmologists.

This article is not intended as a historical narrative in its proper sense. There exist interesting books concerning the history of modern cosmology in general which include and comment on Weyl's contributions (Merleau-Ponty 1965, North 1965, Kragh 1996). Also, more specialized monographs (Kerszberg 1989) or articles (Kragh 1987, 1999) are available. My purpose here is to use the hindsight of the present day for describing Weyl's rôle in the development of modern cosmology from the angle of a physicist with an interest in history - somewhat in the spirit

---

[1]An instantaneous common rest system is a reference system in which all galaxies are at rest relative to each other at a certain moment. In the corresponding rest system some time later, again all galaxies are at rest. Their relative distance may have changed, though, in the meantime.

of, but less technical than, for example, Ellis who surveyed the full period of 1917 to 1960 (Ellis 1989).

After a brief exposition of how DeSitter and Einstein developed relativistic cosmology, in sections 3 and 4, I concentrate on Weyl's interpretation of the DeSitter-solution concerning its presumed singularities, material sources and behaviour in time. In Section 5 and 6, I discuss Weyl's principle and his calculation of the cosmological redshift while, in Section 7, some remarks concerning the reception of Weyl's hypothesis by cosmologists are made. A familiarity of the readers with basic concepts of differential geometry and of general relativity including Einstein's field equations is assumed. The appendices are intended to assist those perplexed by the many different coordinate systems employed, and to supply some mathematical concepts necessary to an understanding of current physics literature.

3.1   DeSitter and Einstein introduce relativistic cosmology

The field of relativistic cosmology began with conversations of Einstein (1879-1955) and the astronomer DeSitter (1872-1934) during Einstein's visit in Leiden in 1915 acknowledged by DeSitter in his second paper on Einstein's theory of gravitation and its astronomical consequences (DeSitter 1916b, footnote 155). In this paper, written in September and October 1916 and published in December of 1916, DeSitter considered the gravitational field produced by the fixed stars described as a hollow spherical shell around the Sun, and derived an approximate solution (to first order). From this solution he concluded that light coming from the mass shell should be displaced towards the *violet* as compared to a light source in the hollow space inside the shell. Although this result was in obvious conflict with the fact, also mentioned by DeSitter, that the spectra of the stars showed a small systematic displacement toward the red, he nevertheless used a small violet-shift in order to obtain an estimate for the total mass within a given distance (DeSitter 1916b, 175-177).

After the discussions with DeSitter, in February 1917, Einstein published his famous paper on cosmology (Einstein 1917), in which he enlarged his field equations by the so-called cosmological term containing the cosmological constantΛ:

$$R_{\alpha\beta} - \frac{1}{2}\, Rg_{\alpha\beta} + \Lambda g_{\alpha\beta} = -\kappa\, T_{\alpha\beta}\, , \ \kappa = \frac{8\pi G}{c^4}$$

where $\alpha, \beta = 0, 1, 2, 3$. On the left hand side we find the parts of the curvature tensor algebraically related to the matter tensor $T_{\alpha\beta}$, i.e. the Ricci tensor $R_{\alpha\beta}$ and its trace, the curvature scalar R. Usually, in cosmological modeling, the matter tensor of a perfect fluid with energy density ρ and pressure p is adopted as the source of the cosmological gravitational field.

In his paper, Einstein's point of departure was the instability of a spherically symmetric Newtonian universe of stars — for the case of its gravitational potential tending to a constant finite value at infinity. In general relativity, this led to the question what the proper general boundary condition at (spacelike) infinity for the components of *every* particular metrical field $g_{\alpha\beta}$ should be. By taking up ideas of Mach, Einstein demanded that the inertial mass of a particle infinitely removed from all the other masses vanish. For him, the only possible solution for a stable system of stars led to a gravitational potential growing without limit with increasing distance. However, an infinite potential at infinity would imply that the velocity of stars increase toward infinity, a conclusion not supported by observation. The ensuing importance of the boundary conditions contradicted another Machian idea, for which Einstein coined the name *Mach's Principle*, i.e. that both the gravitational and inertial fields are *uniquely* determined by the matter distribution.[2]

While in this situation, as Einstein noted in his paper, DeSitter resigned to the point of view that no unique general boundary conditions at infinity exist but that they must be prescribed by hand for every particular gravitating system, Einstein cut the Gordian knot by a new hypothesis: boundary conditions at spatial infinity are superfluous anyhow because the universe is spatially closed. He showed that his new field equations admitted a cosmological model [3] with space sections (i.e. hypersurfaces of constant time) of constant positive curvature forming a 3-sphere. The line element for the model, now called *Einstein-cosmos*, is given by

$$ds^2 = c^2 dt^2 - R_0^2 \left[1 + \frac{k}{4}\left(x^2 + y^2 + z^2\right)\right]^{-2} (dx^2 + dy^2 + dz^2), \qquad (3.1.1)$$

where $k > 0$ or, after normalization $k = +1$, describes the constant curvature of the space sections (Riemann 1854). The energy density of gravitating matter ρ generating this solution is constant and connected to the cosmological constant Λ and the "radius" R_0 of the universe by

$$\rho = \frac{c^2 k}{4\pi G R_0^2} = \frac{\Lambda}{4\pi G} \qquad (3.1.2)$$

with the Newtonian gravitational constant G.

The universe pictured by this solution is static, isotropic and homogeneous or, a cosmological model of ultimate regularity. These features corresponded well to the craving for stability and immutability by all those terrorized by the 19th century-idea of an eventual "heat death" of the universe. Although DeSitter had hypothesized about the possibility of a cosmological model whose boundary "would thus be necessarily finite and limited" he clearly thought of an isolated island of matter surrounded by a vacuum solution with the particular boundary conditions Einstein originally had had in mind (DeSitter 1916b, 182).

The one who, in general relativity, had first made the idea of a closed space section work, was Einstein's colleague in the Prussian Academy of Sciences, astronomer Karl Schwarzschild (1873-1916), in his paper "On the gravitational field

---

[2]For a recent exposition of cosmological ideas in Newtonian gravitational theory cf. Norton 1999. For an extended discussion of Mach's ideas cf. (Barbour and Pfister 1995).

[3]The expression "cosmological model" was used only much later, of course.

of a sphere of incompressible fluid according to Einstein's theory" (Schwarzschild 1916). He must have been seriously ill already when writing it, because a note in the Sitzungsberichte of the Preussische Akademie der Wissenschaften of Febr. 24, 1916 (p. 313) says that the paper was presented to the academy by its member Schwarzschild "through the intervention of Mr. Einstein". While the paper was published one month later, Schwarzschild died in May.[4]

The space sections of Schwarzschild's solution exactly form part of a 3-sphere S^3. With the idea of a *spatially* closed universe in the fore, DeSitter, one year after Schwarzschild, worked out an idea which, as he acknowledged, had been suggested originally by Ehrenfest. DeSitter described it as "the idea to make the four-dimensional world spherical in order to avoid the necessity of assigning boundary conditions " (DeSitter 1917a, 1219, footnote 1). It turned out that he could solve Einstein's *vacuum* field equations with cosmological constant, i.e. without assuming any distribution of gravitating matter ($T_{\alpha\beta} = 0$). The line element of the resulting cosmological model is, in the case of a four-dimensional *Euclidean* space

$$ds^2 = dw^2 + \sin^2 w[d\chi^2 + \sin^2 \chi d\Omega^2], \qquad (3.1.3)$$

where $d\Omega^2 = d\theta + \sin^2 \theta d\varphi^2$ is the line element on the 2-sphere. Eq. (1.3) describes the metric of S^4 locally and isometrically embedded into Euclidean five-dimensional space. The line element (1.3) represents a 4-dimensional space with constant Riemannian curvature $K = +1$, and subspaces $w =$ const also with positive constant curvature $k > 0$.[5] The corresponding space-time Lorentz metric describes a 4-dimensional manifold of constant positive or *negative* curvature K. This cosmological model is now called *DeSitter-space* for $K < 0$ and *anti-DeSitter-space* for $K > 0$:

$$ds^2 = \left[1 + \frac{K}{4} \left(c^2 t^2 - x^2 - y^2 - z^2 \right) \right]^{-2} \left(c^2 dt^2 - dx^2 - dy^2 - dz^2 \right) . \qquad (3.1.4)$$

The constant curvature K now relates to the cosmological constant through

$$\Lambda = 3K . \qquad (3.1.5)$$

The Ricci scalar is $R^\sigma{}_\sigma = 4\Lambda$.[6].

Einstein, who at that time still was influenced by ideas of Mach reacted disfavorably to DeSitter's solution: "According to my opinion it would be unsatisfactory, if a thinkable world without matter did exist. The metrical $g_{\alpha\beta}$ -field must be necessitated by matter, must not exist without it" (DeSitter 1917a, postscript, letter from Albert Einstein of March 24, 1917, my translation).[7]

---

[4]Whether Schwarzschild or Einstein took up the idea of a closed space from the discourse carried on by mathematicians since Riemann's time still is to be investigated.

[5]The line element in eq. (1.3) is used by DeSitter for pedagogical reasons. (DeSitter 1917b, p.10)

[6]The sign of K for the two cases depends on sign conventions made in the definition of the Riemann and Ricci tensors

[7]In the following, all translations form German originals are mine.

In a further paper submitted in July and published in November 1917 DeSitter compared his solution to the Einstein-cosmos and worked out a large number of results (DeSitter 1917b). For example, he pointed out that for the space sections besides the topology of S^3 the topology of projective space could be used ("elliptical" space).[8] He also discussed the possibility of ghost images due to the closedness of space. He calculated timelike and spacelike geodesics of his solution given by equation (1.4) and hinted at the existence of what later became known as "particle horizon" (Rindler 1956) (Cf. mathematical appendix I). DeSitter also derived a redshift for signals exchanged between different points and compared it with observational data obtained for three spiral galaxies. For the discussion below, his stress on the observational irrelevance of the coordinates used as, for example, in eq. (1.4) or, in an equivalent form of metric also given by him (DeSitter 1917b, p. 7) for $K < 0$:

$$ds^2 = \cos^2 \chi c^2 d\tau^2 - R_0{}^2(d\chi^2 + \sin^2 \chi d\Omega^2) \qquad (3.1.6)$$

with $(R_0)^2 = -K^{-1}$, is interesting.[9] In particular, for DeSitter *universal* time did not exist in contrast to the case of the Einstein-cosmos. While the Principle of General Covariance had been accepteded together with general relativity, it took a long time until those working in the field ceased to identify coordinates named t and r with measurable time and radius — as had been permissible in pre-relativistic theory.

While in eqs. (1.4) and (1.6) for $K < 0$ the space sections $t = $ const and $\tau = $ const correspond to 3-spaces of positive constant curvature, DeSitter displayed yet another form of his metric

$$ds^2 = (R_0)^2 \left\{ dT^2 - \sinh^2 T(d\chi^2 + \sinh^2 \chi d\Omega^2) \right\} \qquad (3.1.7)$$

in which the hypersurfaces $T = $ const correspond to 3-spaces of negative curvature. From a geometrical point of view, this is not surprising, because the DeSitter-space can be represented as a one-sheeted hyperboloid in a 5-dimensional flat (Minkowski-) space (DeSitter 1917b, 10). Depending on the choice of the timelike direction, the space sections of this (4-dimensional) hyperboloid can have positive, negative, or even zero curvature (Cf. eq. (4.6)). Or, judged from the family of timelike world lines *orthogonal* to these space sections, three *different* such families have been selected in order to obtain the metrical forms of eqs. (1.6), (1.7) and (4.6). Such families of world lines will play a prominent role in the physical interpretation of the De Sitter-space as a cosmological model.

While in eq. (1.6) the metric components do not depend on the time coordinate, they do in (1.7). In the form given by (1.4), the metric depends both on t and the space coordinates. This induced early workers in the field to believe that the corresponding metrics were describing different situations. Another point to

---

[8]In this topology the endpoints of diameters of the sphere are identified.

[9]The corresponding metric for **anti**-DeSitter space is given by $ds^2 = \cosh^2 \chi c^2 d\tau^2 - R_0^2(d\chi^2 + \sinh^2 \chi d\Omega^2)$.

be noted is that the curves $\chi = \theta = \varphi = $ const in (1.7) do form timelike geodesics, while in (1.4) they do *not*. Thus, the world lines ascribed to the cosmological matter ("substrate") are different, something also overlooked in the beginning. In the mathematical appendix II a list of various forms of the DeSitter-cosmos and coordinate transformations relating them is given.

3.2   Under the spell of Einstein's authority: Weyl's attempt to introduce gravitating matter into the DeSitter-cosmos

DeSitter's paper started a lengthy discussion and correspondence with Einstein, Felix Klein, and others, notably Hermann Weyl. The main issues debated were:

- Is the De Sitter-cosmos devoid of matter or do gravitating sources enter via singularities of the metric?

- Is the DeSitter-cosmos a time-independent solution, is it stationary or genuinely dynamic?

- Which topology can be used for the space sections: the spherical or the elliptical one in which antipodes are identified, or both?

The last question, after some differing approaches had been cleared up, was settled first. Weyl, in the first edition of his book *Raum-Zeit-Materie* started his last section "Betrachtungen über die Welt als Ganzes" (Reflections about the world as a whole) with a discussion of topological questions (RZM).[10] In rederiving the Einstein-cosmos he pointed out that both the spherical and the elliptical topologies are acceptable choices. Certainly, Weyl was well aware of the theorem that even-dimensional *projective* spaces are *not* orientable while an odd-dimensional projective space as is the case here, is orientable. Felix Klein (1849-1925), in his paper concerning conservation laws in a spatially closed world (Klein 1918), supported this point of view as had already DeSitter.[11] In talking about the non-orientable elliptic plane, Klein had in mind a four-dimensional projective space-time; otherwise the impossibility of distinguishing past and future in four-dimensional space time would not make sense.

After having derived the Einstein-cosmos Weyl displayed the DeSitter metric in the form of Eq. (B.5) of appendix II, i.e.

$$ ds^2 = \left(1 - \frac{\rho^2}{R^2}\right) c^2 dt^2 - \frac{d\rho^2}{1 - \rho^2/R^2} - \rho^2 d\Omega^2 \, , \qquad (3.2.1) $$

where ρ is a radial coordinate and $d\Omega^2$ the line element on the 2-sphere as introduced after eq. (1.3). Its spatial part $t = $ const leads to a 3-sphere[12]. $\rho = 0$

---

[10]Weyl did not use the modern term topology but spoke of *analysis situs* (the analysis of position).

[11]The postcard referred to by Kerzberg (Kerzberg 1989, 269) which Klein sent to Einstein on April 25, 1918 also is not in conflict with the theorem.

[12]Set $\rho = R \sin \chi$.

corresponds to one of its "poles". The points $\rho = R\left(\chi = \frac{\pi}{2}\right)$ were called "equator" or "spatial horizon" by Weyl. At this horizon one of the components of the metric becomes unbounded. Weyl's conclusion was

> ... one sees, that the possibility of an empty world contradicts the laws of nature which we here adopt as valid. *At least at the horizon there must exist masses.* (RZM, 225)

Weyl's position corresponds exactly to Einstein's when he criticized DeSitter's solution as *unphysical* (Einstein 1918). Einstein based his discussion on the line element of eq. (1.6) and concluded:

> DeSitter's system does solve the [field] equations [...] everywhere except on the surface $\chi = \pi/2$. There, the component g_{44} of the gravitational potential tends to zero as it does in the proximity of a gravitating mass point. It can be assumed that DeSitter's system does in no way correspond to a world without matter, but rather to a world the material content of which is concentrated in the surface $\chi = \pi/2$ [...] (Einstein 1918, 272).

In following this misleading interpretation of the concept of a singularity (apparent versus genuine singularity, cf. below), Weyl went on by cutting out part of the space-time manifold around $\chi = \pi/2$ and replaced it by an exact solution of the Einstein field equations with cosmological constant and generated by a perfect fluid matter distribution with constant energy density. He assumed this solution to model the large scale distribution of the stars and took for it the interior Schwarzschild solution of Einstein's field equations with cosmological constant. The metrical components of both solutions of Einstein's field equations were to be joined smoothly at the boundaries.

The only solution with gravitating incompressible fluid matter known at the time was *Schwarzschild's interior* solution (Schwarzschild 1916);

$$ds^2 = \frac{1}{4}\left[3\sqrt{1 - \left(\frac{R}{\ell}\right)^2} - \sqrt{1 - \left(\frac{r}{\ell}\right)^2}\right]^2 c^2 dt^2 - \frac{dr^2}{1 - \frac{r^2}{\ell^2}} - r^2 d\Omega^2 \quad (3.2.2)$$

with $\frac{1}{\ell^2} = \frac{8\pi G}{3c^4}\mu_0$, where μ_0 is the *constant* energy density of the fluid. The space sections of this solution are also spaces of constant positive curvature. Thus it is easy to join them continuously to those of the DeSitter-cosmos. The metric (2.2) solves Einstein's field equations *with* cosmological constant Λ if μ_0 is replaced by $\mu_0 - \frac{\Lambda c^4}{8\pi G}$ and the (radially varying) pressure $p(r)$ by $p + \frac{\Lambda c^4}{8\pi G}$. Weyl, however, did not avail himself of this simple solution (2.2) for the star fluid filling the zone cut out but derived a more general metric (RZM, 225, 226; Weyl 1919d, 31):

$$ds^2 = \frac{1}{4}\,(1-x_0)^{-2}[3\sqrt{f(R)} - (1+2x_0)\sqrt{f(r)}]^2 c^2 dt^2$$

$$- \left(1 + \frac{2M}{r} - (\frac{r}{\ell})^2\right)^{-1} dr^2 - r^2 d\Omega^2 \qquad (3.2.3)$$

where $f(r) := \left(1 + \frac{2M}{r} - (\frac{r}{\ell})^2\right)$ and $x_0 := \frac{\Lambda c^4}{8\pi G \mu_0}$. The values $x_0 = 0, M = 0$
lead back to (2.2). Eq. (2.3) gives also an exact solution of Einstein's equations
with cosmological constant for perfect fluid matter. (Cf. Appendix III where I
embed Weyl's solution into a larger class.)

Apparently, he came to his solution after having derived the most general
spherically symmetric, static solution of Einstein's vacuum field equations with cosmological constant. It was found also by other people and is called Schwarzschild-
DeSitter, Kottler or Weyl-Trefftz solution:

$$ds^2 = \left(1 + \frac{2M}{\rho} - \frac{\rho^2}{\ell^2}\right) c^2 dt^2 - \frac{d\rho^2}{1 + \frac{2M}{\rho} - \frac{\rho^2}{\ell^2}} - \rho^2 d\Omega^2 . \qquad (3.2.4)$$

In contrast to what is claimed occasionally, Eq. (2.4) does *not* describe a spacetime of constant curvature; its space sections $t = $ const are the same as in (2.3)
and *not* spaces of constant 3-curvature. A physical interpretation of the metric
(2.4) would be that of a gravitating *mass point* within the DeSitter-space. This
solution contains a *genuine* singularity at $\rho = 0$ where the curvature invariant

$$R^{\alpha\beta}{}_{\gamma\delta} R^{\gamma\delta}{}_{\alpha\beta} = 24 \left(\frac{2M^2}{\rho^6} - \frac{2M^2}{\rho^3 \ell^2} + \frac{1}{\ell^4}\right) \qquad (3.2.5)$$

blows up while, for M = 0, it is a constant.[13]

While this causes no harm in the fluid zone replacing the mass horizon, Weyl
went one step further. He also placed a ball of fluid modeling the star matter
around the center $r = 0$ described by the same metric (2.3) and with radius r_0.
This ball was then smoothly joined to (2.4) at $r = r_0$; (2.4) itself was joined to
the fluid zone at $r = r_1 > r_0$.

Thus, instead of removing the *apparent* singularity at $\rho = \ell = R$ ($\chi = \pi/2$)
by joining a fluid zone to the De Sitter-cosmos devoid of matter, Weyl introduced
a genuine singularity at $\rho = 0$. At the time, however, a coordinate-independent
definition of a "singularity" was not known. Nevertheless, it is curious why a
mathematician of Weyl's caliber did not get worried by the fact that a spacetime of constant curvature i.e. a space-time with constant curvature *at each event*,
could not contain any singularity and that he attempted to "cure the imagined

---

[13]In order to distinguish *apparent* from *genuine* singularities we must look at coordinate independent scalars formed from the curvature tensor.

illness (apparent singularity) by real poison (genuine singularity)" .[14] He and other excellent theoreticians at the time seemingly got misled by a special coordinate patch without taking note of the homogeneity and isotropy of a space of constant curvature. Weyl set forth his particular point of view in more detail in a separate paper (Weyl 1919d). He also took over his arguments to the case of his unified field theory combining gravitational and electromagnetic fields. In it, an additional term simulating another cosmological constant enters. In Weyl's words:

> Because a new electromagnetic term accompanies here Einstein's cosmological term, the existence of a material particle becomes possible without need for a mass horizon; the particle is necessarily charged" (RZM [5]1923, 271-272).

However, then "matter therefore is a real singularity of the field" (ibid. 273). Some of the underlying issues were noticed by North who comments:

> The singularity [at the mass horizon] was not removed, but simply exchanged for a point singularity with which, for some reason, Weyl was much happier" (North 1965, 413).

A reason for this happiness might be taken from a remark of Kottler's who interpreted Weyl's view as supporting "the hypothesis of island worlds" still a subject of debate at the time (Kottler 1922, 229, footnote 131).[15] From this angle, the third metric, the ball of perfect fluid representing star matter Weyl had placed into the De Sitter-cosmos at $\rho = 0$, could be seen as modeling not just one but many stars or nebulae scattered over the sky, due to the homogeneity of the model. In fact, Laue followed such a line, but within the *Einstein-cosmos*, (Laue 1923, 248).

Yet another mathematician who was influenced by Einstein's proposal that his field equations with cosmological constant must not allow solutions with vanishing matter tensor, was Felix Klein. Although he showed that the DeSitter-cosmos necessarily leads to a solution with vanishing matter tensor (Klein 1918) and stated that his results "contradict the objection raised against DeSitter by Einstein in his communication of March 1918 and which were then supported by Weyl in his book and, recently, by extended calculations in a special note in the *Physikalische Zeitschrift*" he left the impression that not all was settled:

> It may be noted, however, that the $g_{\mu v}$ [...] become infinite along the fundamental hyperboloid which can be considered as an equivalent to the non-existence of matter at the non-singular points of the world.

On the other hand, he stated clearly that the apparent singularity (at Weyl's mass horizon) was not different from the singularity in the origin of polar coordinates.

---

[14]In the context of a recent paper on Einstein's attitude with regard to singularities (Earman and Eisenstaedt 1999), interesting comments concerning Weyl and the subjects of this and the next section have been made. I read them as supporting my conclusions.

[15]This hypothesis assumed that the Milky Way was not the only galaxy but that all spiral nebulae represented further distant galaxies.

Thus, also Klein could not free himself entirely of Einstein's interpretation of DeSitter's solution. Here, we have a good example of the double-edged influence of an authority in physics even on first-rate mathematicians.

Weyl kept his dubious interpretation also in the third, improved, and the fourth, extended, editions of Raum-Zeit-Materie (RZM [3]1919, [4]1921). He only dropped the special solution for which he had joined three metrics. Pauli, in his famous review (Pauli 1921), reported about the agreement of Einstein and Weyl but added a cautious note "A final answer to the question is not yet available". Also, Laue joined the initiated ones when, in his book on relativity, he claimed that the DeSitter-cosmos possess a singularity and

> ... it probably will have to be interpreted as a mass distribution with finite surface-density such that DeSitter's solution does not hold for a gravitational field without bodies, but for a distribution of all the matter on a surface (Laue 1923, 246).

3.3 The DeSitter cosmos a static or a time-evolving solution? Weyl's gradual move toward cosmological modeling

One slight but not unimportant change in the third and fourth editions of Weyl's book must be noted, though. Now, the conclusion of Weyl was

> ... one sees that the possibility of a *static*, empty world contradicts the laws of nature which we here take for granted ..." (RZM [3]1919, 241; RZM [4]1921, 254, my emphasis).[16]

The insertion of the word "static" shows that Weyl had accepted Klein's arguments. Klein had been quick in noting that the time coordinate t – in contrast to the case of the Einstein-cosmos – was not uniquely determined up to a time translation. He stated that "there are ∞^6 ways to introduce [...] a t " such that the metric remain static[17] (Klein 1918, p. 420). Here, another one of Einstein's credos came into play, as seen by Klein:

> According to Einstein's opinion the starting point must be the remark that the world which we look for should be interpreted as a *statical* system (Klein 1918, 420).

Klein was already aware of the fact that the coordinate systems in which the DeSitter-cosmos looked static were covering only *part* of the whole manifold. Klein used DeSitter's metrical form (1.6) and related it to the 4-dimensional hyperboloid

$$\xi^2 + \eta^2 + \zeta^2 - v^2 + \omega^2 = \frac{R^2}{c^2} \qquad (3.3.1)$$

---

[16]With"laws of nature" Weyl means Einstein's field equations.

[17]The symbol ∞^6 expresses a sixfold infinity, i.e. one generated by six parameters, each of which extends to infinity

embedded into a flat, 5-dimensional space through

$$\left.\begin{array}{l} \xi = \frac{R}{c} \sin\chi\cos\vartheta \ , \ \eta = \frac{R}{c} \sin\chi\sin\vartheta\cos\varphi \\[2mm] \zeta = \frac{R}{c} \sin\chi\sin\vartheta\sin\varphi \ , \ v = \frac{R}{c} \cos\chi\sinh\frac{ct}{R} \\[2mm] \omega = \frac{R}{c} \cos\chi\cosh\frac{ct}{R} \ . \end{array}\right\} \qquad (3.3.2)$$

From (3.2)

$$\tanh\frac{ct}{R} = \frac{v}{\omega} \qquad (3.3.3)$$

and it follows that $-\infty < t < \infty$ is mapped to only the intervall $-1 < \frac{v}{\omega} < +1$, whereas in (3.1) $-\infty < v, \omega < +\infty$ is possible. As eq. (3.3) may be rewritten as

$$t = \frac{1}{2}\frac{R}{c}\,\ell n\,\frac{\omega+v}{\omega-v}, \qquad (3.3.4)$$

to Klein $v - \omega = 0$ and $v + \omega = 0$ corresponded to future and past infinity while $v = \omega = 0$ described Weyl's mass horizon.

Whereas for Klein a metric staying invariant under time translations was considered to be *static*, according to Weyl (RZM, p. 192) the metrical form of a *static* gravitational field must be

$$ds^2 = f^2 dt^2 - \gamma_{ik}dx^i dx^k \ , \qquad (3.3.5)$$

where f and $\gamma_{ik}(i, k = 1, 2, 3)$ are functions of the spatial coordinates x^i only. Thus, Weyl's canonical form (3.5) corresponds, in modern terminology, to the existence of a normal, timelike Killing-vector field $\xi^\alpha = \delta_0^\alpha$.[18] Klein's definition contains also *stationary* metrics for which the timelike Killing vectors are *not* hypersurface-orthogonal. In the fourth edition of his book Weyl included a definition of "stationary" which was, according to him taken from the Italian school and characterized by "all $g_{\alpha\beta}$ being independent of the time coordinate, whereas the 'coefficients on the side' g_{01}, g_{02}, g_{03} need not vanish" (RZM [4]1921, 244).

How difficult the conceptual situation was to the less initiated is exemplified by a remark of the theoretically inclined assistant in experimental physics at the University of Freiburg Cornel Lanczos (1893-1974) (Lanczos 1923, 73). He defined a solution of Einstein's field equations to be *stationary* if all metrical coefficients are independent of time in a coordinate system such that all masses are permanently at rest. For *vacuum* solutions, i.e. those without gravitating matter being present, he erroneously claimed as a necessary and sufficient condition for stationarity that the *time-lines* of the coordinate system be *geodesics*. Thus, by Lanczos' definition the DeSitter-cosmos is not stationary.

Although Weyl had given what later turned out to be the accepted definition of "stationary", he still abstained from distinguishing properties due to the

---

[18]A Killing vector field generates a local isometry. It satisfies $\nabla_{(\alpha} \xi_{\beta)} = 0$, where ∇ denotes the covariant derivative and the brackets symmetrize the expression.

coordinate system from those inherent in the space-time geometry. While in both the third and the fourth editions of his book (RZM [3]1919, 242; [4]1921, 255) he claimed it understandable that "within the actually observed world a static situation obtains as far as the electromagnetic and the gravitational field in the large are concerned", in the fourth edition he remained undecided as to the behaviour of the DeSitter-cosmos. Taking into account Klein's coordinate system and his own he remained puzzled:

> In the first case the world as a whole would not be static, and it would be consistent with the laws of nature that it be empty; this assumption is followed by DeSitter. In the second case we do have a static world, which is impossible without a mass horizon; this assumption, discussed here in more detail, is preferred by Einstein. (RZM [4]1921, 256).

Perhaps it was this apparent discrepancy which lead Weyl to what was called later his "cosmological hypothesis" and which he first hinted at in the fifth edition of his book (RZM [5]1923, 294-295). In a recent paper, the gradual change in the rôle played by the particular congruence of curves selected by Weyl as world lines of the cosmological substrate, from being just a property of the DeSitter cosmos to becoming an additional assumption for cosmological modeling, has been described (Bergia and Mazzoni 1999).

From our present knowledge, we would not characterize the space-time of constant curvature as globally static, despite of the fact that the 10-parameter isometry group of a four-dimensional space-time of constant curvature contains a locally timelike normal Killing vector field. It remains timelike only in an extended coordinate patch, not on the full manifold. The frontier where the space-time character of the vector field changes coincides with the event horizon. This is quite similar to the exterior Schwarzschild solution which is static only *outside* of its event horizon at the Schwarzschild radius. However, modern usage of the word "DeSitter-cosmos" refers only to that part of the full manifold described by the line element

$$ds^2 = c^2 dt^2 - e^{at}(dx^2 + dy^2 + dz^2) \qquad (3.3.6)$$

with a constant a, for which the timelike Killing vector is hypersurface-orthogonal. The corresponding isometry, however, is not a time translation but a time translation followed by a space dilation (Cf. mathematical appendix IV).

That the DeSitter cosmos is not static became very clear to Lemaître: "If coordinates and a corresponding decomposition into space and time respecting the homogeneity of the universe are introduced, one finds that the field is no longer static [...]" (Lemaître 1927, 51).

3.4 Cosmological redshift and Weyl's hypothesis

From hindsight, we have seen that both, in the questions of whether a singularity exists in the DeSitter-cosmos or whether it be a static or a dynamical solution, i.e. one evolving in time, Weyl's contributins did not lead to a breakthrough - up to the fifth edition of his book. However, there he very successfully calculated the cosmological redshift. We recall that DeSitter, in 1917, used the form (1.6) of his metric and determined a redshift for what he considered to be the worldlines of fundamental observers, i.e. the timelike paths $\chi = \theta = \varphi = $ const. They do not describe freely falling observers moving on geodesics, qua definition, but are non-geodesical curves. Observers at $\chi = \chi_0$ keep the *same proper distance*, though. The redshift followed from

$$\frac{\nu_{\text{obs.}}}{\nu_{\text{emit.}}} = \frac{d\tau_{\text{emit.}}}{d\tau_{\text{obs.}}} = \sqrt{\frac{g_{00}(\text{emit.})}{g_{00}(\text{obs.})}} = \left| \frac{\cos \chi_{\text{emit.}}}{\cos \chi_{\text{obs.}}} \right| . \qquad (3.4.1)$$

Placing the observer at $\chi = 0$ we get $\frac{\lambda_{\text{emit.}}}{\lambda_{\text{obs.}}} = |\cos \chi|$ or, with $\Delta \lambda = \lambda_{\text{emit}} - \lambda_{\text{obs}}$ and, to first order,

$$\frac{\Delta \lambda}{\lambda_{\text{obs}}} \approx \chi^2 + 0(\chi^4) . \qquad (3.4.2)$$

Thus, according to DeSitter the redshift grows *quadratically* with increasing (angular) distance $0 \le \chi < \pi/2$.

As remarked before, in the first four editions of his book Weyl did not bother to mention the cosmological redshift, but waited until the fifth. Nevertheless, in a paper on the physical foundation of his generalized theory of relativity (unified field theory) (Weyl 1921d, 478) he commented on the "redshift of spectral lines for stars the distance [...] of which compares to the world radius". According to him an exchange of observer and star position would lead to a violet shift. The increase with growing χ is interpreted by Weyl in the sense that "the redshift must be understood as the result of an approach toward the mass horizon". This means that he considered the redshift to be a *gravitational* one, not a Doppler shift; this is consistent with DeSitter's calculation. As we shall see below in Section 6, with the various possible reasons for a redshift depending on the motion of observers, the radiating stars and the distribution of matter between them, the problem of calculating a *cosmological* redshift was not an easy one at the time.

Now, Lanczos had found an altogether different representation of the DeSitter-cosmos in which the metric components became time dependent (Cf. Appendix II, eq. (B.10):

$$g = c^2 dT^2 - \ell_0^2 \cosh^2 \frac{cT}{\ell_0^2} [d\omega^2 + \sin^2 \omega d\Omega^2] \qquad (3.4.3)$$

(Lanczos 1922b, 539, footnote 3). The redshift may be calculated from (4.3) with the help of the null geodesics (in the subspace $\theta = \varphi = $ const) joining worldlines through $\omega = 0$ and $\omega = \omega_0 \ne 0$. If the signal is emitted from $\omega = 0$ at the time T and arrives at $\omega = \omega_0$ at time T_1, then the redshift will be

$$\frac{\Delta \lambda}{\lambda} = \frac{\cosh cT_1/\ell_0}{\cosh cT/\ell_o} - 1 \simeq 2(T_1^2 - T^2) \frac{c^2}{\ell_0^2} . \qquad (3.4.4)$$

If we introduce a distance between observer and emitting source $D := c(T_1 - T)$ and set $(T_0 + T)/\ell_0 \simeq 1$ in the spirit of Weyl , then

$$\frac{\Delta\lambda}{\lambda} \simeq D/\ell_0 \; ;$$

ℓ_0^{-1} corresponds to what later became known as the Hubble constant. In (4.3) the world lines $\omega = \theta = \varphi =$ const describe *freely falling* test particles the proper distance of which *increases* with time. Lanczos did *not* deal with the redshift in the paper referred to; however, a proper interpretation of (4.4) would have to be in terms of a *Doppler*-effect. Weyl did read Lanczos' paper which criticized his attempt to introduce matter into the DeSitter cosmos. Around that time he did not think it necessary to keep separated the *gravitational* redshift and a redshift caused by relative motion, i.e. the *Doppler* shift of spectral lines:

> According to Einstein's equivalence principle, from the Doppler effect it follows that light oscillates in a slower rythm if it reaches us from a location of lower gravitational potential than that [light] generated by the same atomic phenomenon at the location of the observer; spectral lines appear as shifted towards the red end. Doppler-effect and Einstein-effect are inseparably linked. (Weyl 1923e, 230)

Nevertheless, in comparing both forms (1.6) and (4.3) for the metric of space-time, he might well have come to conjecture that without an a priori selection of a congruence of world lines (describing the observer on a star and the sources of light on all the other stars) a calculation of the redshift remains ambiguous. Of course, this is mere guess-work; we do not know how Weyl arrived at his hypothesis put forward in the *fifth* edition of Raum-Zeit-Materie (RZM [5]1923, 295).

On the other hand, Weyl seemingly moved toward his postulate of 1923 concerning the motion of the cosmological substrate quite independently of Lanczos' paper which had appeared only in December 1922 (Lanczos 1922b). This may be seen from his report on relativity theory as it had been debated during the meeting of German natural scientists and physicians in Bad Nauheim (Weyl 1922c). There, he repeated his definition of a *static* gravitational field as in Section 4 (cf. eq. (3.5)) and stated:

> In a static gravitational field, the frequency of a light wave emitted by a body at rest is the same everywhere in space, measured in the cosmic time t, the time coordinate in the system of the four static coordinates (ibid., 317).

As in the fourth edition of his book, he introduced static coordinates for the DeSitter-cosmos such that in the five-dimensional embedding space its metrical form is

$$g = \frac{z^2}{R^2} \, dt^2 - (dx_1^2 + dx_2^2 + dx_3^2 + dz^2), \qquad (3.4.5)$$

where $z^2 = R^2 - r^2$, $r^2 = x_1^2 + x_2^2 + x_3^2$. He then noted that t in (4.5) is not the only static time but one of many possible ones. Which of them could be essential for the propagation of light? The coordinates used in (4.5) describe only part of the full space-time of constant curvature, a wedge ("Keil", see below).

> If the world consists just of one such wedge, as Einstein assumes, then
> we have to pick the particular t which is uniquely determined up to
> a linear transformation and corresponds to the wedge. If this is in
> harmony with reality, then the propagation of a light wave from the
> moment of its generation on is influenced by the joinder of the world
> as a whole; one should expect, instead, that the light wave can react on
> it only after having run through the whole space of the world. (ibid.,
> 319)

Although this sentence of Weyl is characterized more by elusive poetry than by mathematical clarity, we may take it as a first approach to his concept of the "range of influence" of a star used later in the formulation of the cosmological hypothesis. That Weyl had been thinking intensively about causality at the time may be seen from his essay on the relationship of causal and statistical considerations in physics (Weyl 1920a). Much stronger arguments along this line have been given in the larger setting of a convincing biographical reconstruction of Weyl's life by Skúli Sigurdsson (Sigurdsson 1991, Chapter 5 and 1996, 48-70. Cf. also Forman 1971, p. 74-80).[19]

Klein's geometrical analysis of the space-time of constant curvature of 1918 might also have been instrumental. In contrast to Weyl's, Klein's analysis of the DeSitter-cosmos assumed an *elliptical* topology. For him, the DeSitter-cosmos is represented as a double wedge, the flanks $v + \omega = 0, v - \omega = 0$ (cf. eq. (3.4)) of which correspond to infinite future and infinite past while on the edge $\omega = v = 0$ time remains undefinable. We note that $\omega = v = 0$ corresponds to $\chi = \pi/2$ for the form of the metric given by (1.6); a quick calculation shows that the curves $\chi = \pi/2, \vartheta = $ const, $\varphi = $ const are *null* geodesics. These are boundaries of what later received the name of the DeSitter-cosmos. Klein remarked that "the double wedges which are cut out of the fundamental hyperboloid by different pairs of tangential spaces [different pairs of his "flanks"] share only pieces and reach out from these" (Klein 1918, 422).

In essence, already Klein described the characteristic features of the DeSitter-cosmos as a part of the four-dimensional space of constant curvature. The analysis applied to this cosmological model by Weyl was new in his use of what we now call "conformal structure" i.e. the set of time-oriented null cones in each event. This has been stressed by Ehlers (1988, 95). As Weyl saw it:

> The timelike among these [geodesic] wordlines are branches of hyper-
> bolas reaching from $-\infty$ to $+\infty$. The null cones emanating from such a

---

[19]See also the article of Erhard Scholz in this volume.

geodetic line ℓ and open towards the future are covering only a part of the hyperboloid which is bounded by two straight generating lines parallel to each other. In the downward direction ($-\infty$) the area becomes infinitely narrow in comparison with the full extent of the hyperboloid; in the upward direction it comprises the circumference of the surface almost completely. (RZM [5]1923, 295)

What Weyl describes here is Klein's "wedge" seen as generated by the subsequent null cones from a freely falling star.

This range of influence (Wirkungsgebiet) contains ℓ together with ∞^3 other geodetic lines forming a pencil diverging towards the future (converging towards the past)(ibid. 295).[20]

Klein's comparison of overlapping different double wedges in Weyl's description becomes easy to grasp:

But here, only such matter-elements the world lines of which belong to one and the same pencil are acting upon each other right from the beginning; different such systems start to penetrate each other causally only in the course of their history" (ibid., 295). Thus, Weyl singles out a particular congruence of timelike geodesic curves as the world lines of the cosmological substrate. This choice is the essence of his hypothesis: "It lies near at hand that all known heavenly bodies belong to one single such system. (ibid. 295)

In his subsequent paper Weyl points out

... that this assumption about the 'state of rest' of the stars is missing in DeSitter's papers [...] – the only possible one consistent with the homogeneity of space and time" (Weyl 1923e, footnote, 231).

Thus, he indicates that his hypothesis implies the existence of a common time parameter or, turned around, that the stars have a common instantaneous rest system. This was a necessary prerequisite, if DeSitter's observation referred to above, i.e. that no universal time exists in his solution, was to be circumnavigated. Remember that Einstein had begun the derivation of his cosmological model also with an assumption concerning a common rest system:

Therefore I believe that, for a start, we may use the following approximate assumption in our considerations: a coordinate system exists relative to which matter can be regarded as permanently at rest (Einstein 1917, 148).

---

[20]In my translation of "Wirkungsgebiet" I used Weyl's own formulation in his paper written in English (Weyl 1930, 936, 938). Ehlers (Ehlers 1988) and Bergia (Bergia and Mazzoni 1999) use "domain of action" while Kerszberg (Kerszberg 1989) translates it as "range of action". Weyl is versatile in his use of language; in one paper he uses "Wirkungsbereich" (Weyl 1923e, 230/231); in the fifth edition of his famous book he also speaks of "Wirkungszusammenhang" and "Wirkungswelt" (RZM [5]1923, 295 and 322, respectively)

In DeSitter's solution the *instantaneous* rest system replaces Einstein's *permanent* rest system. Einstein then went on to write down the energy-momentum-tensor of a pressureless fluid adapted to this rest system. Thus, what is really new in Weyls principle, is the *focussing* to the past of the congruence of matter world lines of stars and their causal relationship. In his formulation for *Encyclopedia Britannica*, Weyl's hypothesis reads as:

> On Weyl's supplementary hypothesis the world lines of the stars form a sheaf, which rises in a given direction from the infinitely distant past, and spreads out over the hyperboloid in the direction of the future, getting broader and broader (Weyl 1926a, 910).

Weyl neither in the fifth edition of his book nor in his subsequent paper (Weyl 1923e) did display explicitly a form of the DeSitter metric mirroring his hypothesis, i.e. that his chosen congruence of worldlines also corresponded to the particular choice of a rest-system for the cosmological matter. This was done only in 1925 by Lemaître (Lemaître 1925a,b) who introduced coordinates for the DeSitter-cosmos such that its metric became

$$g = c^2 dt^2 - e^{at}(dx^2 + dy^2 + dz^2) \ . \tag{3.4.6}$$

3.5 Calculation of the cosmological redshift by Weyl, Lanczos, and Silberstein

In 1923, Weyl briefly described his method for the calculation of the redshift in an appendix to his book.[21] If s and σ are the proper times of the light source and the observer, respectively, and we draw the light cone from a point s_0 toward the observer cutting his world line in the point $\sigma(s_0)$, then $d\sigma = \frac{d\sigma}{ds}\big|_{s_0} ds$. According to Weyl, the redshift is determined by

$$\frac{\nu_{\text{observer}}}{\nu_{source}} = \frac{d\sigma(s)}{ds} \ . \tag{3.5.1}$$

In order to calculate $\frac{d\sigma}{ds}$ for the DeSitter-space Weyl worked in the 5-dimensional embedding space of the space-time housing the space of constant curvature, or after a reduction by two coordinates, dealt with a hyperboloid in a 3-dimensional Lorentz space. While not giving the calculation he arrived at the result

$$\frac{\Delta\lambda}{\lambda} = tg \ \frac{r}{R} \ , \tag{3.5.2}$$

where r is the proper distance between light source and observer at the time of the observation.[22] Thus, to lowest order, the redshift is *linear* in distance and Weyl had, in principle, found what was later called Hubble's law. His lengthy and

---

[21] Appendix III of the fifth edition of Raum-Zeit-Materie.

[22] Proper distance is defined by the line element of the subspace at constant time.

somewhat tricky calculation was published shortly after the 5th edition of the book in a separate paper (Weyl 1923e). In the book, Weyl related his formula to the observed redshifts of spiral nebulae:

> It is remarkable that neither the elementary nor the Einstein-cosmos do lead to such a redshift. Naturally, today it is not possible to claim that our explanation hits the mark; in particular because the views concerning the nature and distance of the spiral nebulae will have to be clarified greatly. (RZM [5]1923, 323)

Shortly after Weyl, also Lanczos had submitted a paper on the redshift in the DeSitter-cosmos to Physikalische Zeitschrift. The time span between these two papers was such that Lanczos had seen the new edition of Weyl's book and his subsequent paper before sending back the proof sheets. He acknowledged both publications of Weyl but believed, in a "note added in proof", "that my investigation has not become superfluous even in its special part, because our assumptions on the distribution of the material world lines are not congruent". He then went on to describe the differences between his and Weyl's choice of congruence. In fact, Lanczos used his metric form (4.3) and got a different expression for the cosmological redshift which, although being also linear in the distance, depended on the time τ_0 of light emission

$$\frac{\Delta\lambda}{\lambda} = \cos\frac{a}{R} - \sin\frac{a}{r}\sinh\tau_0 , \qquad (3.5.3)$$

where a is the geodesic distance of light source and observer. This is not surprising because he did not use Weyl's geodesic congruence as world lines for observer and emitter. He interpreted the redshift as fully generated by the Doppler effect of the stars moving away from each other. After a crossing of the "mass-horizon" the spectral shift would however become a shift toward the violet. Thus, it is clear that Lanczos had not yet accepted Weyl's hypothesis.

A theoretical physicist who not only had not accepted Weyl's hypothesis but had misunderstood it as claiming "a necessary feature of DeSitter's world" in disguise, was Ludwik Silberstein (1872-1948) (Silberstein 1924 b, 909), who had already published two books on relativity theory (Siberstein 1922, 1924 a). He criticized Weyl's calculation of the redshift as containing "a number of rather obscure technicalities" and presented a calculation of his own. For it, he again used the form (1.6) of the line element, but took *geodesics* as the fundamental word lines of observer and light source. The result he obtained, if a *special relativistic* Doppler shift is neglected, is

$$\frac{\Delta\lambda}{\lambda} = \sin\frac{r}{R} \qquad (3.5.4)$$

where "r is the observer's distance from the star at the moment receiving its light" (ibid., 912). To lowest order, from (5.4) Weyl's result (5.2) is reproduced

$$\frac{\Delta\lambda}{\lambda} \simeq \pm\frac{r}{R} . \qquad (3.5.5)$$

Silberstein pointed out that the distances to the spiral nebulae were far too uncertain as to use these objects in testing equation (5.5). [23] However, he turned to globular clusters of stars and derived from a table of values for nine such clusters a mean value of the "world radius" of $R \simeq 6.10^{12} a.u. \cong 6.10^{25} cm$. This fitted well to the value derived from the distance measurements to the Magellanic Clouds made by Shapley. Eddington, in the second edition of his book on the mathematical theory of relativity (Eddington 1924, 162) nevertheless used the spiral nebulae to confirm the significance of DeSitter's cosmological model.

Weyl responded to Silberstein's polemics (Weyl 1924d) and defended his hypothesis:

> The hypothesis [...] simply means that the stars of the system are able to act upon one another from eternity. Another hypothesis has been followed by Mr. K. Lanczos, but mine has the great advantage of not introducing a singular initial moment, of conserving the homogeneousness of time. (Moreover, it is the only one which satisfies the requirement.). (Weyl 1924d, 349)

In a different setting he interpreted his hypothesis in the following words:

> According to [the] hypothesis [...] all stars of a system seem to flee away from an arbitrarily chosen central star; their spectral lines are shifted to the red end for an observer on the central star – the more so, the farther they are. Now, the spiral nebulae, which probably are the most distant celestial objects, display strong redshift of their spectral lines, a few exceptions granted. (Weyl 1924c, 483).

North, in his history of cosmology, carefully related the various calculations of authors using different world lines for source and observer, different concepts of distance and different time intervals. He also distinguished possible causes for the redshift and refers to DeSitter's original calculation as the "DeSitter effect" (North 1965, 92-104). A present day summary distinguishes six different causes for redshifts in a cosmological model. In principle, there can be three *gravitational* redshifts due to the inhomogeneity of the matter distribution at the radiating source (z_1), at the observer (z_2) and due to the large-scale inhomogeneities in the universe (z_3). Furthermore, three *kinematical* Doppler-effects due to the motions of the source and observer relative to the rest system of the cosmological substrate (z_4, z_5) plus the effect resulting from the expansion of the universe (z_6). With the DeSitter-cosmos being homogeneous one would expect z_3 to vanish; however, an effect can still occur due to the acceleration of the observer relative to the rest system of matter (Ellis 1989, 374). In DeSitter's *static* frame (1.6), $z_6 = 0$ and $z_3 \neq 0$ with all other contributions equal to zero; in the coordinate frame called (by Ellis) "Robertson-frame", i.e. for the metrical forms (3.6) or (4.3), $z_6 \neq 0$, $z_3 = 0$ with all other terms equal to zero. Thus, for Weyl and Lanczos, the effect arose

---

[23] For the debate about the distances of spiral nebulae at the time cf. (Smith 1982).

from the expansion of the universe, while Silberstein considered also additional motions of the light source.

3.6 The reaction to Weyl's principle

In the same article (Weyl 1924c), Weyl really got excited to an extent that shows clearly how his philosophy of nature did influence the suggestion of his cosmological hypothesis:

> If I think about how the world lines of a star system with a common asymptote climb up on DeSitter's hyperboloid from the infinite past, I'd like to say: the world is born from eternal rest of 'father ether'; but it is stirred up by the 'spirit of unrest' [Geist der Unruhe] (Hölderlin), who is at home in the agens of matter, in 'the breast of earth and man' and 'will never again come to rest'. (Weyl 1924c, 484)[24]

In fact, I find it strange, probably merely due to hindsight, that the discussion of cosmological models at the time centered so much on the DeSitter-cosmos which contained no gravitating matter and was regarded by Eddington only as a limiting case, the circumstances of the actual world being intermediate between Einstein's and DeSitter's cosmological model (Eddington 1924, 160). It would have been more natural to look for generalizations of the Einstein-cosmos, corresponding to pressureless matter. And this indeed was done by Friedman (1888-1925), who's pioneering paper had been published almost a year before Weyl announced his cosmological hypothesis (Friedman 1922). Friedman in St. Petersburg was an outsider who mainly worked in meteorology and fluid dynamics. He had kept the properties of the space sections of the Einstein-cosmos, i.e. of being spaces of positive constant curvature and of being orthogonal to the time lines. However, his results not only were *not* taken into consideration by Einstein, DeSitter, Weyl, Lanczos, Eddington, and Silberstein but first were branded as erroneous by Einstein himself (Einstein 1922).

A little later, Einstein confessed having made a mistake himself and considered Friedman's results as illuminating (Einstein 1923a). However, this did not help Friedman; his second paper in which he extended his results to space-sections with *negative* constant curvature again was met with neglect (Friedman 1924). Friedman, in his first paper had even pointed out that some of his non-stationary solutions could be obtained also with *vanishing* cosmological constant. *Nine* years later this remark finally was taken up by Einstein (Einstein 1931). Note, however that Einstein did not recall Friedman's *second* paper when, in 1932 he discussed the so-called Einstein-DeSitter cosmos (cf. footnote on 78 of Robertson 1933, Einstein-DeSitter 1932).

Weyl, in principle, could have read and absorbed Friedman's first paper which appeared in August 1922; he signed the preface to the fifth edition of his book

---

[24]For the linkage between Weyl's science and philosophy cf. (Sigurdsson 1991, 1996).

in "autumn 1922" (RZM [5]1923, vi). Einstein had reacted to Friedman's paper in mid September. A number of other articles which had appeared in 1922 are quoted by Weyl as, for example, communications of Cartan, Schouten and Struik, Wirtinger, St. John and Sommerfelds book (Sommerfeld 1922). As these names seem to indicate, Weyl perhaps read only papers of well-established colleagues. Weyl also does not mention Friedman in his papers in Physikalische Zeitschrift mailed in January and April 1923 nor in his articles of 1926 and 1930. In the first he refers to Lanczos; in the second to Silberstein, Tolman, and Robertson. True, with the latter paper Weyl wanted to establish priority for his calculation of the cosmological redshift after Hubble had secured it empirically thanks to the newly developed technique of measuring distances with δ-Cepheid stars. Thus, Hubble's law, i.e. equation (5.5) had been put on empirical ground. Friedman never had considered the cosmological redshift. And Weyl did not mention Lemaître as well who, in 1927, had found anew and, seemingly independently of Friedman's papers, an expanding cosmological model as an exact solution of Einstein's equations with gravitating matter (Lemaître 1927). Lemaître, in his paper, referred to "Weyl's hypothesis" in a footnote but did not accept it and followed Lanczos' splitting of space-time into space and time.[25] The first sign that Weyl had noticed Friedman's paper is in 1934, where he adopted Eddington's point of view that

> ...the true solution will lie somehow in between the hyperbolic world and the cylindrical world. [And] such solutions have been given already in 1922 by Friedman, later by Lemaître [...]. (Weyl 1934c, 422)

The paper is the printed version of a lecture Weyl had given in July 1933 for a course during the university-vacations in Göttingen, in front of a general natural-science oriented public.

Whether Weyl did not notice Friedman's paper or whether – with Einstein – he did not judge it as being relevant, in any case, Friedman's solutions did not exist for him until 1933. Einstein-cosmos and DeSitter-cosmos were the only two exact solutions of cosmological significance in his mind. They are the only ones referred to also in his article on the philosophy of mathematics and natural science (Weyl 1926/[3]1966, 77).

However, when Friedman had no readers or followers, Weyl with his hypothesis likewise did not impress many of his few scientific colleagues working in the field of cosmology. We saw already that Lanczos, Silberstein, and Eddington had their own thoughts. Einstein remained inactive on cosmological questions until 1931, more or less.[26] In his appendix "On the cosmological problem" in the third edition of *The meaning of relativity* (Einstein 1946, 104-126) as well as in the later editions of his two German books (Einstein 1960, 1970), he did not mention

---

[25] In his publication of 1927 Lemaître does not mention Friedman. However, in the English translation of his paper which appeared in 1931 in *Monthly Notices* (Friedman 1922) is given as a reference.

[26] He reacted to Friedman and to Selety (Einstein 1923a, b) and discussed the topic in his correspondence, though.

Weyl but just Friedman. Only after Lemaître had found the coordinates displaying that part of the space of constant curvature given in the form (3.6), now named DeSitter-cosmos, and after Robertson had conceptually cleared up what a cosmological model means (Robertson 1933), Weyl's principle was accepted more or less by the mainstream.[27]

Robertson, in a footnote of a paper written in Göttingen, expressed Weyl's hypothesis in this form " ... by assuming that the geodesics concerned form a diverging pencil" (Robertson 1928). Five years later, in his review article, he then stressed Weyl's postulate as guaranteeing the introduction of a cosmic time (Robertson 1933, 65). Richard Tolman (1881-1948) in his influential book discussed Weyl's principle as one hypothesis among others, although a sympathetic one which

> ... has shown more immediate promise of furnishing an account of the observed relation between redshift and distance" (Tolman 1934, 356).

The "more" referred to Tolman's own hypothesis. Again, Otto Heckmann (1901-1983), who published a monograph on cosmology during the Third Reich and referred to Einstein himself more than once, kept silence on Weyl's contributions except for mentioning his book in connection with the Anschaulichkeit to be reached by embedding cosmological models in a five dimensional flat space and, by mentioning in the bibliography that "H. Weyl" had translated Eddington's book on the expanding universe (Heckmann 1942, 48, 105, Eddington 1933).[28] In fact, Weyl's hypothesis had become superfluous and was replaced by the *cosmological principle*, i.e. the hypothesis that, in the space sections, no point and no direction are preferred.

Weyl had contributed himself to this situation, because he had not been able to make clear to cosmologists what the essence of his principle was. Of all things, his distinction at the time between what we call, in present physics language, *test*-matter and *gravitating*-matter had not been convincing. [29] Apparently, for him the gravitating matter was confined to the singularity of the mass horizon while the stars were made of matter of negligible influence to the global properties of the universe. Consequently, his selection of a particular congruence of world lines for test-particles is conceptually different from Friedman's and, later, Robertson's choice of a congruence of freely falling *gravitating* matter-particles. Moreover, he had not explained in any detail why the selected world lines originating in a point in the past and interacting causally with each other must likewise define a common rest system. [30] In fact, as the mathematician Kurt Gödel (1906-1978) later showed,

---

[27]Robertson did not use the term "comological model" but spoke of "types of universes". Tolman employed "Model of Universe" at least since 1931 and "cosmological model" in his book (Tolman 1931, 1934).

[28]Skuli Sigurdsson told me that this translator is not Hermann but Weyl's wife *Helene*

[29]Test-matter is matter *reacting* only to the influence of the gravitational field but not forming its source.

[30]All world lines of the congruence parametrized by one and the same time parameter.

there exist cosmological models with local vorticity for which there is no *cosmic time* defining a common rest-system of matter (Gödel 1949). For some, his principle reflected the *homogeneity* of the DeSitter-cosmos:

> ... that the Weyl hypothesis has the very attractive feature of putting all the particles (nebulae) in the model on the same footing, so that there would be nothing unique about the phenomena observed from any particular nebula" (Tolman 1934, 358).

As mentioned before, for others, the introduction of cosmic time was decisive (Robertson 1933, 65), although this assumption had also been made by Einstein and Friedman without Weyl's hypothesis. Thus it is no wonder, that Weyl's best known contribution to cosmology did not make it into such popular books as Eddington's (Eddington 1933) and Jeans' (1931); there he is noted for his unified field theory, but Friedman, Lemaître, DeSitter, and Einstein are the pioneers in understanding the universe.

3.7 Conclusion

The slow reception of Weyl's hypothesis properly reflects the *relative* importance of Weyl's contributions to cosmology, namely being creative and helpful but not of decisive influence for the eventual evolution of cosmological modeling towards an expanding universe. Weyl's hypothesis ingeniously solved the puzzle of how to calculate the cosmological redshift for the DeSitter universe. Only that DeSitter's was the wrong solution in the sense of not being supported by the empirical data supplied by astronomers using the big telescopes in the United States. Of course, this matter was still open to debate until 1929. But the insight that a realistic cosmological model must account for *gravitating* matter existed already at the time. It is characteristic that Weyl himself seems to have been the foremost promoter of his cosmological hypothesis. What could have become a convincing support of his hypothesis, i.e. his derivation of the red shift, was done in a way too complicated as to be readily absorbed by others. Weyl's stature in mathematics and science may thus explain why the hypothesis still is mentioned in some modern books on gravitation and cosmology, notably by authors not specialized in cosmological research (Cf. Møller 1952, 368; Møller1952/²1972, 516).

In a way, Weyl's and Einstein's positions were similar: both started developments in cosmology but were quickly left behind by the actual course of affairs, i.e. the consequences drawn by others. Both men were guided by philosophical tenets as, for example, a belief in the immutability of the world. From the point of view of a mathematician, the geometry of the DeSitter-cosmos is the most harmonious to be found: a space-time homogeneous both in space and in time. For Weyl, out of this universe, causality, in the form of a swarm of interacting stars emanating from a common origin and diverging toward the future, separated out our part of the world.

Both the gradual accumulation of empirical data and the growth in understanding of the theoretical implications, for cosmology, of general relativity changed this picture to a less perfect world. To some, this may be a desillusioning assessment; however, Weyl's work in other areas of general relativity and gravitation, notably his contributions to the conceptual development of space-time geometry (linear connection, gauge idea)[31] and in the area of (axi-symmetric) exact solutions of Einstein's field equations has retained its preferred place.

Acknowledgements
To the DMV and, in particular, to Erhard Scholz I am grateful for having been invited to this fruitful workshop in such a graceful setting. Both, Erhard Scholz and Skúli Sigurdsson educated me greatly on Weyl through their papers and many delightful discussions. From both Skúli's and Erhard's careful and critical yet encouraging reading of a draft of this article I have profited enormously. I also learned much from the lectures and comments of the other participants of the workshop.

3.8   Mathematical appendices

I. The event horizon of the DeSitter-cosmos

The concept of the event horizon is an important one for discussions of cosmological models and black holes. It seems therefore necessary that it is explained in this cosmological setting. We ask whether there are light signals needing an infinite time to reach the observer. Their geometrical locus is said to define the *event horizon*. (We follow Rindler 1956). In order to calculate the *event* horizon for the standard model of cosmology I restrict the derivation to *radial* null geodesics. Thus, if we use Friedman's form for metrics with flat space sections as the easiest case

$$g = c^2 dt^2 - S^2(t)[dr^2 + r^2 d\Omega^2] , \qquad (A.1)$$

we obtain, for the radial *null* geodesics, the equation

$$0 = c^2 dt^2 - S^2(t) dr^2 . \qquad (A.2)$$

If a signal is emitted at time t_1 at $r_1 > 0$ and travels towards $r < r_1$ to be reached at time t, then after integration:

$$r = r_1 - c \int_{t_1}^{t} \frac{dt'}{S(t)} . \qquad (A.3)$$

The proper distance between initial and final point of the signal in the space section $t = $ const is defined by

$$d(t) = S(t)r = S(t) \left[r_1 - c \int_{t_1}^{t} \frac{dt'}{S(t)} \right] . \qquad (A.4)$$

---

[31] For such contributions of Weyl cf. (Scholz 1994, 1999; Mielke and Hehl 1988).

Let the observer be at $r = 0$ such that $d(t) = 0$. For signals needing an infinite time to reach the observer we must obtain

$$r_1 = c \int_{t_1}^{\infty} \frac{dt'}{S(t)} . \tag{A.5}$$

For the DeSitter-cosmos in the form (3.6)

$$r_1 = \frac{2c}{a} e^{-at_1}$$

follows. The geometrical locus of the event horizon thus is

$$r = \frac{2}{a} ce^{-at} . \tag{A.6}$$

II. The DeSitter-cosmos in various coordinate systems.

In the literature, the DeSitter cosmos is given in many different coordinate systems. As a help for the reader, in this appendix a list of such coordinate patches and the transformations among them is provided.

1. Forms of the DeSitter-metric not depending on the time coordinate.

We start from eq. (1.4) in polar coordinates r, θ, φ, i.e. from

$$g = \left[1 + \frac{K}{4} (c^2 t^2 - r^2) \right]^{-2} (c^2 dt^2 - dr^2 - r^2 d\Omega^2) . \tag{B.1}$$

To simplify notation we assume $K = 1$ and introduce new coordinates t', r' by

$$
\begin{aligned}
2t &= tg \frac{t' + r'}{2} + tg \frac{t' - r'}{2} \\
2r &= tg \frac{t' + r'}{2} - tg \frac{t' - r'}{2}
\end{aligned}
$$

and obtain

$$g = (4 \cos^2 r')^{-1} (dt'^2 - dr'^2 - \sin^2 r' d\Omega^2) . \tag{B.2}$$

A further coordinate transformation $r' \to \chi$ with $\cos r' = (\cosh\chi)^{-1}$ leads to

$$g = \frac{1}{4} \left[\cosh^2 \chi dt'^2 - d\chi^2 - \sin^2 \chi d\Omega^2 \right] \tag{B.3}$$

which by $t' = iT$, $\chi = i\bar{\chi}$, $g \to -4g$ goes over into (1.6)[32]

$$g = \cos^2 \bar{\chi} c^2 dT^2 - d\bar{\chi}^2 - \sin^2 \bar{\chi} d\Omega^2 . \tag{B.4}$$

Introducing the radial coordinate ρ in place of the angular coordinate χ through $\rho = R \sin \bar{\chi}$ and rescaling T into $\bar{T} = RT$ results in

$$g = \frac{1}{R^2} \left\{ \left(1 - \left(\frac{\rho}{R}\right)^2\right) c^2 d\bar{T}^2 - \frac{d\rho^2}{1 - (\rho/R)^2} - \rho^2 d\Omega^2 \right\} \tag{B.5}$$

2. Forms of the DeSitter-metric depending on the time coordinate.

In (B.1) we introduce new coordinates τ, ω by

$$\left. \begin{array}{rcl} ct & = & \tau \cosh \omega \\ r & = & \tau \sinh \omega \end{array} \right\} \tag{B.6}$$

and obtain

$$g = \left(1 + \frac{K}{4} \tau^2\right)^{-2} \left\{ d\tau^2 - \tau^2 [d\omega^2 + \sinh^2 \omega d\Omega^2] \right\} . \tag{B.7}$$

If a new time coordinate $\hat{\tau}$ replaces τ such that $\tau = \frac{2}{\sqrt{K}} \sinh \hat{\tau}$ we arrive at

$$g = \frac{4}{K} \cosh^{-2} \hat{\tau} \{ d\hat{\tau}^2 - \mathrm{tgh}^2 \hat{\tau} [d\omega^2 + \sinh^2 \omega d\Omega^2] \} . \tag{B.8}$$

If we, alternatively, set $\tau = \frac{2}{\sqrt{K}} \mathrm{tg}(\tilde{\tau}/2)$ we reach the form

$$g = \frac{1}{K} \{ d\tilde{\tau}^2 - \sin^2 \tilde{\tau} [d\omega^2 + \sinh^2 \omega d\Omega^2] . \tag{B.9}$$

By further substitutions $\tilde{\tau} \to \tilde{\tau} + \frac{\pi}{2} = \tilde{T}, \omega = i\omega$ the positive-definite line-element

$$g = \frac{1}{K} \left\{ d\tilde{T}^2 + \cos^2 \tilde{T} [d\omega^2 + \sin^2 \omega d\Omega^2] \right.$$

is reached which through $\tilde{T} \to i\tilde{T}, g \to -g$ finally is transformed into

$$g = \frac{1}{K} \left\{ d\tilde{T}^2 - \cosh^2 \tilde{T} [d\omega^2 + \sin^2 \omega d\Omega^2] \right\} , \tag{B.10}$$

a form given by Lanczos (Lanczos 1922b, 539). If, in (B.5) the coordinate transformation

$$\left. \begin{array}{rcl} \bar{r} & = & \rho(1 - (\rho/R)^2)^{-1/2} e^{-t/R} \\ \bar{t} & = & t + \frac{1}{2} R\ln\left(1 - \left(\frac{r}{R}\right)^2\right) \end{array} \right\} \tag{B.11}$$

is made, then the form (3.6) of the DeSitter-cosmos appears

$$g = (d\bar{t}^2) - e^{-2\bar{t}/R}(d\bar{r}^2 + \bar{r}^2 d\Omega^2) . \tag{B.12}$$

---

[32] The reader might be worried by the use of *imaginary* coordinates. However, we always will use such transformations simultaneously for one spacelike and one timelike coordinate such that the signature of space-time remains unchanged (up to a global minus-sign).

III. A class of spherically symmetric interior fluid solutions of Einstein's equations with cosmological constant.

Here, we collect some technical knowledge about spherically symmetric gravitational fields of fluid bodies which is familiar to those working in the field of exact solutions of Einstein's equations. From reading textbooks, including Weyl's, one cannot gain as much understanding as is required.

A straight forward calculation shows that the static, spherically symmetric line element

$$g = e^{2\alpha(r)}c^2dt^2 - e^{2\beta(r)}dr^2 - r^2d\Omega^2 \tag{C.1}$$

with

$$\alpha = \ln\left[a\sqrt{f(R)} + b\sqrt{f(r)}\right] - \ln 2, \tag{C.2}$$

$$\beta = -\frac{1}{2}\ln f(r) \tag{C.3}$$

where $f(r) = 1 - \frac{2M}{r} - (\frac{r}{\ell})^2$ solves

$$G_{\alpha\beta} + \Lambda g_{\alpha\beta} = -\kappa T_{\alpha\beta} \tag{C.4}$$

with $T^{\alpha\beta} = (\mu + p)u^\alpha u^\beta - pg^{\alpha\beta}$. The energy density of the fluid is constant

$$\mu = \mu_0 = \frac{3}{\kappa\ell^2} - \frac{\Lambda}{\kappa} \tag{C.5}$$

while its pressure is given by

$$p(r) = \frac{\Lambda}{\kappa} - \frac{3b}{\kappa\ell^2} \frac{\sqrt{f(r)} + \frac{a}{3b}\left(1 + \frac{2M\ell^2}{r^3}\right)\sqrt{f(R)}}{a\sqrt{f(R)} + b\sqrt{f(r)}}. \tag{C.6}$$

We define the radius R of the fluid ball by $p(R) = 0$. This leads to

$$\ell^2(a+b)\Lambda = 3b + a\left(1 + \frac{2M\ell^2}{R^3}\right). \tag{C.7}$$

A second equation for the constant coefficients a, b follows if we join (C.1 - C.3) at r = R to the metric given in (2.4), i.e.

$$a + b = 2. \tag{C.8}$$

We can thus express a and b by the free parameters ℓ^2, R, M:

$$a = \left(1 - \frac{M\ell^2}{R^3}\right)^{-1}(3 - \ell^2\Lambda)$$

$$b = \left(1 - \frac{M\ell^2}{R^3}\right)^{-1}\left[-1 + \ell^2\left(\Lambda - \frac{2M}{R^3}\right)\right]. \tag{C.9}$$

The solution given contains as special subcases, the *inner* Schwarzschild solution ($\Lambda = M = 0$, $a = 3$, $b = -1$), the *external* Schwarzschild solution ($\Lambda = 0$, $\frac{1}{\ell^2} = 0$, $a = 0$, $b = 2$), the DeSitter-cosmos ($M = 0$, $a = 0$, $b = 2$, $\Lambda = 3/\ell^2$) and the Einstein-cosmos ($M = 0$, $\Lambda = 1/\ell^2$, $b = 0$). It belongs to a class of static spherically symmetric solutions discussed by Volkoff (Volkoff 1939) and Wyman (Wyman 1949). Weyl's constants a, b do satisfy (C.8) but not (C.7). Consequently, in his solution((eq.(2.3))R is *not* the radius of the star. As Weyl writes:"The boundary condition of vanishing pressure leads to a transcendental equation..."(RZM, p.225).

IV. The isometries of a 4-dimensional Lorentzian manifold of constant curvature.

Again, the following is known to the community of cosmologists but not to the less specialized reader of literature on General Relativity. The isometry group of a space of constant curvature must contain a 4-dimensional *transitive* subgroup because the manifold itself is an orbit of the isometry grou If all generators of this subgroup were spacelike tangent vectors, they could not span the tangent space of a Lorentzian manifold.

The 10 Killing-vector fields corresponding to the metric (1.4) are found to be

$$X^{(\rho)} = \xi^{(\rho)\mu} \frac{\partial}{\partial x^\mu} \qquad (\text{D.1})$$

with $\xi^{(\rho)\mu} := \left(1 - \frac{K\sigma^2}{4}\right) \eta^{(\rho)\mu} + \frac{K}{2} x^\mu x^\rho$ where $\sigma^2 := \eta_{\mu\nu} x^\mu x^\nu$ ($\rho, \mu, \nu = 0, 1, 2, 3$) and

$$Y^{(\kappa)(\lambda)} = \xi^{(\kappa)(\lambda)\mu} \frac{\partial}{\partial x^\mu} \qquad (\text{D.2})$$

with $\xi^{(\kappa)(\lambda)\mu} := x^{(\kappa)} \eta^{(\lambda)\mu} - x^{(\lambda)} \eta^{(\kappa)\mu}$ $\kappa \neq \lambda$ $\kappa, \lambda = 0, .., 3$. The notation $\xi^{(\rho)\mu}$ signifies the μ'th component of the Killing vector field number ρ; likewise, in $\xi^{(\kappa)(\lambda)\mu}$ the bracketed indices count the number of Killing vectors. The $X^{(\rho)}$ correspond to the translations in Minkowski-space ($K = 0$), the $Y^{(\kappa)(\lambda)}$ to the 6 Lorentz-transformations. All these Killing vector fields are *conformal* Killing fields of Minkowski-space. It is easy to check that all Killing vector fields $X^{(\rho)}$, $Y^{(\kappa)(\lambda)}$ are hypersurface-orthogonal and that at most 6 of them are spacelike. They correspond to the maximal number of isometries of the space sections.

For the form (3.6) of the metric of that part of the space time of constant curvature now called DeSitter-cosmos, we obtain as components of the Killing-form $\Xi = \Xi_\alpha dx^\alpha$

$$
\begin{aligned}
\xi_0 &= d + c_\ell x^\ell \\
\xi_i &= \frac{a}{2} \left[\epsilon_{ijk} b^j x^k + (d + c_\ell x^\ell) x_i - \frac{1}{2} c_i x^\ell x_\ell + \delta_i \right] + \frac{1}{a} c_i e^{-at} \qquad (\text{D.3})
\end{aligned}
$$

where d, c_ℓ, b^j, δ_i ($i, j, k, \ell = 1, 2, 3$) are the 10 group parameters. The case

$c_i = b^j = \delta_i = 0$, $d \neq 0$ leads to

$$\xi_\alpha = d \left[\delta_\alpha^0 + \frac{a}{2} \left(x^1 \delta_\alpha^1 + x^2 \delta_\alpha^2 + x^3 \delta_\alpha^3 \right) \right] . \qquad (\text{D.5})$$

This Killing vector field is timelike for $r^2 := \delta_{ij} x^i x^j < \frac{4}{a^2} e^{-at}$ and changes its character, i.e. becomes spacelike, at the event horizon.

References

Barbour, Julian; Pfister, Herbert (eds.) 1995. Mach's principle: from Newton's bucket to quantum gravity. *Einstein Studies* Vol. **6**, 1-536, Basel etc.: Birkhäuser.

Bergia S.; Mazzoni, L. 1999. "Vom Ursprung her miteinander kausal verbunden": Genesis and evolution of Weyl's reflections on De Sitter's universe. In (Goenner e.a. 1999, 325-342).

Deppert, Wolfgang (ed.) 1988. *Exact Sciences and Their Philosophical Foundations: Exakte Wissenschaften und ihre philosophische Grundlegung.* Frankfurt/Main - Bern etc.: Peter Lang Verlag.

DeSitter, Willem 1916a. On Einstein's theory of gravitation, and its astronomical consequences. First Paper. *Monthly Notices Roy. Astron. Soc.* **76**, 699-728.

DeSitter, Willem 1916b. On Einstein's theory of gravitation, and its astronomical consequences. Second paper. *Monthly Notices Roy. Astron. Soc.* **77**, 155-184.

DeSitter, Willem 1917a. On the relativity of inertia. Remarks concerning Einstein's latest hypothesis. *Proceedings Royal Academy Amsterdam* **19**, 1217-1225.

DeSitter, Willem 1917b. On Einstein's theory of gravitation, and its astronomical consequences. Third paper. *Monthly Notices Royal Astr. Soc.* **78**, 3-28.

Eddington, Arthur S. 1924. *The Mathematical Theory of Relativity*, 2nd edition. Cambridge: University Press.

Eddington, Arthur S. 1933. *The Expanding Universe.* Cambridge: University Press.

Ehlers, Jürgen 1988a. Hermann Weyl's contributions to the General Theory of relativity. In (Deppert e.a. 1988, 83–105).

Ehlers, Jürgen 1988b. Annotations in *Hermann Weyl: Raum-Zeit-Materie*, 7. Auflage, edited by J. Ehlers. Berlin: Springer, (RZM [7]1988).

Einstein, Albert 1917. Kosmologische Betrachtungen zur allgemeinen Relativitätstheorie. *Sitzungsberichte Preuss. Akad. Wiss.*, 142-152.

134   *H. Goenner*

Einstein, Albert 1918. Kritisches zu einer von Hrn. DeSitter gegebenen Lösung der Gravitationsgleichungen. *Sitzungsberichte Preuss. Akad. Wiss.*, 270-272.

Einstein, Albert 1922. Bemerkung zu der Arbeit von A. Friedmann "Über die Krümmung des Raumes". *Zeitschrift für Physik* **11**, 326.

Einstein, Albert 1923a. Notiz zu der Arbeit von A. Friedmann "Über die Krümmmung des Raumes". *Zeitschrift für Physik* **16**, 228.

Einstein, Albert 1923 b. Bemerkung zu der Franz Seletyschen Arbeit "Beiträge zum kosmologischen System". *Annalen der Physik* **68**, 436-38.

Einstein, Albert 1931. Zum kosmologischen Problem der Allgemeinen Relativitätstheorie. *Sitzungsberichte Preuss. Akad. Wiss.*, p. 235-237.

Einstein, Albert 1946. *The Meaning of Relativity*, 3rd edition. London: Methuen.

Einstein, Albert 1960. *Grundzüge der Relativitätstheorie*, 2nd ed. mit einem Anhang Über das kosmologische Problem. Braunschweig: Vieweg.

Einstein, Albert 1970. *Über die spezielle und allgemeine Relativitätstheorie*, 21. Auflage. Braunschweig: Vieweg.

Einstein, Albert; DeSitter, Willem 1932. On the relation between the expansion and the mean density of the universe. *Proc.eedings National Academy of Science* **18**, 213-14.

Earman, John; Eisenstaedt, Jean 1999. Einstein and Singularities. *Studies in the History and Philosophy of Modern Physics* **30**, 185-235.

Ellis, George R. 1989. The expanding universe: a history of cosmology from 1917 to 1960. *Einstein Studies* **1**, 367-431.

Forman, Paul 1971. Weimar culture, causality, and quantum theory, 1918-1927: Adaption by German physicists and mathematicians to a hostile intellectual environment. *Historical Studies in the Physical Sciences* **3**, 1-115.

Friedman, Alexander 1922. Über die Krümmung des Raumes. *Zeitschrift für Physik* **20**, 377-386.

Friedman, Alexander 1924. Über die Möglichkeit einer Welt mit konstanter negativer Krümmung des Raumes. *Zeitschrift für Physik* **21**, 326-32.

Gödel, Kurt 1949. An example of a new type of cosmological solution of Einstein's field equations of gravitation. *Review of Modern Physics* **21**, 447-50.

Goenner, Hubert; Renn, Jürgen; Ritter, Jim; Sauer, Tilmann (eds.) 1999. *The Expanding Worlds of General Relativity Einstein Studies* **7**. Basel etc.: Birkhäuser,

Heckmann, Otto 1968. *Theorien der Kosmologie*. Berichtigter Nachdruck der 1. Ausgabe 1942. Berlin: Springer.

Jeans, James 1931. *The Mysterious Universe*. Cambridge: University Press.

Kerszberg, Pierre 1989. *The Invented Universe. The Einstein-DeSitter Controversy (1916-17) and the Rise of Relativistic Cosmology*. Oxford: Clarendon.

Klein, Felix 1918. Über die Integralform der Erhaltungssätze und die Theorie der räumlich-geschlossenen Welt. *Nachrichten Gesellschaft Wissenschaften Göttingen*, 395-423; *Gesammelte Mathematische Abhandlungen* 1 (1921), 586–612.

Kottler, Friedrich 1922. Gravitation und Relativitätstheorie. In *Encyklopädie der Mathematischen Wissenschaften*, Bd. 6, Teil 2, 2. Hälfte. Leipzig: Teubner.

Kragh, Helge 1987. The beginning of the world: Georges Lemaître and the expanding universe. *Centaurus* **32**, 114-39.

Kragh, Helge 1993. Big Bang Cosmology and Steady State Theory. In Hetherington, Norris S. (ed.); *Cosmology: Historical, Literary, Philosophical, Religious and Scientific Perspectives*. New York: Garland.

Kragh, Helge 1996. *Cosmology and Controversy*. The historical development of two theories of the universe. Princeton: University Press.

Kragh, Helge 1999. Steady-State Cosmology and General Relativity: Reconciliation or Conflict? In (Goenner e.a. 1999, 377-402).

Lanczos, Kornel 1922a. Ein vereinfachendes Koordinatensystem für die Einsteinschen Gravitationsgleichungen. *Physikalische Zeitschrift* **23**, 537-539.

Lanczos, Kornel 1922b. Bemerkungen zur de Sitterschen Welt. *Physikalische Zeitschrift* **23**, 539-543.

Lanczos, Kornel 1923. Über die Rotverschiebung in der de Sitterschen Welt. *Zeitschrift für Physik* **17**, 168-189.

Lanczos, Kornel 1924. Über eine stationäre Kosmologie im Sinne der Einsteinschen Gravitationstheorie. *Zeitschrift für Physik* **21**, 73-110.

Laue, Max von 1923. *Die Relativitätstheorie*. 2. Band. Die allgemeine Relativitätstheorie und Einstein's Lehre von der Schwerkraft (2. Auflage). Braunschweig: Vieweg.

Lemaître, George 1925a. Note on DeSitter's universe. *Physical Review* **25**, 903.

Lemaître, George 1925b. Note on DeSitter's universe. *Journal of Mathematics and Physics* **4**, 188-192.

Lemaître, George 1927. Un univers homogene de masse constante et de rayon croissant rendant compte de la vitesse radiale des nebuleuses extra-galactiques. *Annales de la Société Scientifique de Bruxelles, Serie A: Sciences Mathématiques* **47**, 49-59.

Lemaître, George 1931. A homogeneous universe of constant mass and Increasing radius accounting for the radial velocity of extra-galactic nebulae. *Monthly Notices of the Astronomical Society, London* **91**, 483-490.

Lemaître, George 1931. The expanding universe. *Monthly Notices of the Astronomical Society, London* **91**, 490-501.

Merleau-Ponty, Jacques 1965. *Cosmologie du XXe siècle.* Paris: Gallimard.

Mielke, Eckehard W.; Hehl, Friedrich W. 1988. Die Entwicklung der Eichtheorien: Marginalien zu deren Wissenschafts geschichte. In (Deppert e.a. 1988, 191–232).

Møller, Christian 1952. *The theory of relativity.* Oxford: Clarendon Press. 2nd edition 1972.

Narlikar, Jayant V. 1979. *General Relativity and Cosmology.* London: MacMillan.

North, John D. 1965. *The Measure of the Universe.* Oxford: Clarendon.

Norton, John 1999. The cosmological woes of Newtonian gravitation theory. In (Goenner e.a. 1999, 271-323).

Pauli, Wolfgang 1921. *Relativitätstheorie.* Sonderabdruck aus der Encyklopädie der Mathematischen Wissenschaften. Leipzig: Teubner.

Raychaudhuri, A. 1979. *Theoretical Cosmology.* Oxford: Clarendon Press.

Riemann, Bernhard 1854. Über die Hypothesen, welche der Geometrie zu Grunde liegen. Habilitationsvortrag Göttingen. *Göttinger Abhandlungen* **13** (1867). *Werke,* 272–287.

Rindler, Wolfgang 1956. Visual horizons in world models, *Monthly Notices Roy. Astron. Soc.* **116**, 662-77.

Robertson, H. 1928. On relativistic cosmology. *Philosophical Magazine* **5**, 835-848.

Robertson, H.P. 1933. Relativistic cosmology. *Reviews of Modern Physics* **5**, 62-90.

Scholz, Erhard 1994. Hermann Weyl's contribution to geometry in the years 1918 to 1923. In Dauben, Joseph; Mitsuo, S.; Sasaki, C. (eds.); *The Intersection of History and Mathematics.* Basel etc.: Birkhäuser, 203–230.

Scholz, Erhard 1999. Weyl and the theory of connections. In Gray, Jeremy J. (ed.); *The Symbolic Universe. Geometry and Physics 1890 - 1930*. Oxford: University Press, 260–284.

Schwarzschild, Karl 1916. Über das Gravitationsfeld einer Kugel aus inkompressibler Flüssigkeit nach der Einsteinschen Theorie. *Sitzungsberichte Preuss. Akad. Wiss.*, p.424-34.

Sigurdsson, Skúli 1991. Hermann Weyl, Mathematics and Physics, 1900-1927. *Ph.D. Dissertation* Harvard University.

Sigurdsson, Skúli 1994. Unification, Geometry and Ambivalence: Hilbert, Weyl and the Göttingen Community. In Gavroglu, Kostas et al., eds; *Trends in the Historiography of Science*. Dordrecht: Kluwer, p 355-367.

Sigurdsson, Skúli 1996. Physics, Life and Contingency: Born, Schrödinger and Weyl in exile. In Ash, Mitchel G. et al., eds.; *Forced Migration and Scientific Change: Emigré German Speaking Scientists and Scholars after 1933*. New York: Cambridge University Press.

Silberstein, Ludvik 1922. *The theory of general relativity and gravitation*. Toronto: University Press.

Silberstein, Ludvik 1924a. *The theory of relativity*. 2nd edition. London:Macmillan.

Silberstein, Ludvik 1924b. Determination of the curvature invariant of Space-Time. *Philosophical Magazine* 47, 907-918.

Silberstein, Ludvik 1927. The Doppler Effect in the DeSitter's Space.Time. In reply to M. Jean Chazy. *Philosophical Magazine* 3, 1085-87.

Smith, Robert W. 1982. *The Expanding Universe. Astronomy's Great Debate 1900-1931*. Cambridge: University Press.

Sommerfeld, Arnold 1922. *Atombau und Spektrallinien*, 3. Aufl. Braunschweig: Vieweg.

Tolman, Richard C. 1934. *Relativity, Thermodynamics and Cosmology*. Oxford: Clarendon Press.

Volkoff, G.M. 1939. On the equilibrium of massive spheres. *Physical Review* 55, 413.

Weinberg, Steven 1972. *Gravitation and Cosmology*. New York: John Wiley.

Wyman, Max 1949. Radially symmetric distributions of matter. *Physical Review* 75, 1930-36.

4 Ursprünge der Eichtheorien

Norbert Straumann

4.1 Vorbemerkungen

Die Bedeutung von Hermann Weyl für die theoretische Physik ist wohl erst in neuerer Zeit richtig deutlich geworden. Von den maßgebenden Theoretikern seiner Zeit ist er offenbar als Physiker nicht ganz ernst genommen worden. Dies geht etwa aus der folgenden Stelle eines Briefes von Pauli an Weyl vom 1. Juli 1929 in gewohnter Deutlichkeit hervor:[1]

> Vor mir liegt das Aprilheft der Proceedings of the National Academy. Nicht nur enthält es eine Arbeit von Ihnen in der Rubrik "Physics", sondern, wie über Ihrer Arbeit steht, sind Sie jetzt in einem "Physical Laboratory" zu Hause: wie ich höre, sollen Sie in Amerika sogar eine Professur für theoretische Physik innehaben. Ich bewundere Ihren Mut; denn die Schlussfolgerung erscheint unabweisbar, daß Sie, wenigstens eine Zeit lang, nicht nach Ihren Erfolgen auf dem Gebiet der reinen Mathematik, sondern auf Grund Ihrer treuen, aber unglücklichen Liebe zur Physik beurteilt sein wollen. Verzeihen Sie mir, wenn ich Sie auch weiterhin als Mathematiker betrachte; sonst müßte ich ja untersuchen, wie sich das Maß Ihrer Begeisterung für die Physik zu dem Umfang verhält, in dem sich Ihre Reformvorschläge in der Physik bisher bewährt haben. Das will ich lieber nicht; denn, wenn Sie auch ein Mathematiker sind, so sind Sie doch ein solcher, der weiß, was Physik ist und was in ihr jetzt vorgeht, und es handelt sich darum, aus diesem Umstand einen möglichst großen Nutzen für den Fortschritt der Physik herauszuschlagen (Pauli 1979, 505).

Zu Paulis Haltung gegenüber Weyl schrieb mir Fierz im Zusammenhang mit der 2- Komponentengleichung von Weyl (auf die ich später näher eingehen werde):

> Bei Pauli war es so: Er hatte Widerstände gegen Weyl, weil dieser sich, als Mathematiker, ungehörig in die Theoretische Physik einmischte. Daher waren ihm auch die zweikomponentigen Spinoren zuwider. (Fierz 1991)

Aus der früheren Zeit möchte ich an dieser Stelle aber auch eine positive Reaktion zitieren. Am 8. März 1918 schrieb Einstein an Weyl:

---

[1] Auslöser des Briefes von Pauli war eine Kurzfassung der grossen Weylschen Arbeit von 1929 in den Proceedings of the National Academy (Weyl 1929d).

Hochverehrter Herr Kollege!
Die Korrektur Ihres Buches, welche ich bogenweise bekomme, lese ich mit wahrer Begeisterung. Es ist wie eine Meister-Symphonie. Jedes Wörtchen hat seine Beziehung zum Ganzen und die Anlage des Werkes ist grandios. Die prachtvolle Methode der infinitesimalen Parallel-Verschiebung von Vektoren zur Ableitung des Riemann-tensors! Wie natürlich sich das alles macht. (Einstein an Weyl 8. 3. 1918)

Weyl's Beiträge zur Theoretischen Physik sind vielfältig. Wie im folgenden ausgeführt wird, geht der Ursprung der Eichtheorien, welche die heutige Elementarteilchenphysik beherrschen, auf Weyl zurück. Ich möchte aber zuerst auch auf andere seiner Errungenschaften in der Physik hinweisen.

Sein Buch "Gruppentheorie und Quantenmechanik" (Weyl 1928b) wurde zwar seinerzeit von fast allen theoretischen Physikern angeschafft, aber offenbar nur von wenigen gelesen. Die Betonung der strukturellen Aspekte der Physik ist von den meisten wenig geschätzt worden. Gerade darin liegt aber seine Stärke und Zeitlosigkeit. Es ist wirklich erstaunlich, mit welcher Klarheit Weyl die begriffliche Struktur der Quantenmechanik bereits im Jahre 1928 herausgearbeitet hat. Man findet aber darüber hinaus auch überraschende Bemerkungen *physikalischer Natur*, welche in die Zukunft weisen. Im Vorwort zur 2. Auflage wird folgendes angekündigt:

Das Problem von Proton und Elektron wird im Zusammenhang mit den Symmetrieeigenschaften der Quantengesetze gegenüber den Vertauschungen von rechts und links, Vergangenheit und Zukunft, positiver und negativer Elektrizität aufgerollt. Im Augenblick ist keine annehmbare Lösung sichtbar; ich fürchte, daß sich hier die Wolken zu einer neuen ernsten Krise der Quantenphysik zusammenballen. (Weyl 1928b/ ²1931, VII)

Näheres dazu findet man auf S. 200-201, S. 233-234, wo die Operationen P (Raumspiegelung), T (Zeitumkehr) und C (Ladungskonjugation) diskutiert werden. Im Zusammenhang mit der Ladungskonjugation bemerkt Weyl:

Das (in der Natur nicht vorkommende) Teilchen von der Elektronenmasse m, dessen Ladung nicht $-e$, sondern $+e$ ist, werde als "positives Elektron" bezeichnet. Man erkennt aus dem Gesagten, daß die Energieniveaus des positiven Elektrons $-h\nu$ sind, wenn $h\nu$ diejenigen des negativen Elektrons sind. Abgesehen vom Vorzeichen verhalten sich beide Sorten von Teilchen gleich. Das Elektron wird außer seinen positiven auch negative Energieniveaus besitzen, die aus den positiven Energieniveaus des positiven Elektrons durch Änderung des Vorzeichens entstehen. Hier ist offenbar noch etwas nicht in Ordnung; man sollte diese negativen Energieniveaus des Elektrons streichen können.

Dies ist aber unmöglich, weil z.B. unter der Einwirkung eines Strah-
lungsfeldes Übergänge zwischen den positiven und negativen Termen
vorkommen. Daß man doppelt zu viel Terme erhält, hängt offenbar damit zusammen, daß die Größe ψ nicht *vier*, sondern nur *zwei* (Differentialgleichungen 1. Ordnung genügende) Komponenten besitzen sollte. Die Lösung dieser Schwierigkeit scheint in der Richtung zu liegen,
daß unsere Feldgleichungen mit den vier Komponenten irgendwie außer dem Elektron bereits das Proton mitumfassen. (Weyl 1928b/ [2]1931,
200)

Im Abschnitt über die Quantisierung der Maxwell-Diracschen Feldgleichungen
wird dieses Thema nochmals aufgenommen. Zunächst bemerkt Weyl:

> *Dirac* hat den Vorschlag gemacht, das Vorhandensein oder Fehlen ei-
> nes *Protons* auf dem positiven Energieniveau μ als das Fehlen bzw.
> Vorhandensein eines Elektrons auf dem entsprechenden negativen Ni-
> veau $-\mu$ zu deuten; unsere Gesetze würden dann die Protonen neben
> den Elektronen mitumfassen. (Weyl 1928b/ [2]1931, 233)

Dann beschreibt er kurz Dirac's Löchertheorie und sagt:

> Die von der Diracschen Theorie unerwünschtermaßen zugelassenen Sprün-
> ge eines Elektrons zwischen einem positiven und negativen Energieni-
> veau erscheinen jetzt als der Prozeß der "Zerstrahlung", durch welchen
> ein Proton und ein Elektron gleichzeitig vernichtet werden, und der da-
> zu inverse Prozeß. In der Astrophysik ist man zu der Annahme dieser
> durch unsere irdischen Erfahrungen nicht belegten Prozesse gekom-
> men, weil man anders nicht recht hoffen kann, die von den Sternen
> ausgestrahlte Energie zu decken. (ebda. 234)

Dazu führt nun Weyl kritisch folgendes aus:

> So bestechend diese Idee im ersten Augenblick anmutet - es ist sicher
> unmöglich, damit ohne weitere tiefgreifende Modifikationen den Tat-
> sachen der Natur gerecht zu werden. Denn sie verleiht notwendig dem
> Proton die gleiche Masse wie dem Elektron; ja es kann bewiesen wer-
> den, daß sie unter allen Umständen, auch bei strenger Berücksichtigung
> der Wechselwirkung zwischen Materie und Lichtquanten, zur Wesens-
> gleichheit von positiver und negativer Elektrizität führt; und das, wie
> immer man das Wirkungsprinzip ansetzen mag, wenn dieses nur der
> Gleichartigkeit von rechts und links Rechnung trägt.

Daran anschließend wird die Invarianz gegenüber Ladungskonjugation formuliert
und Weyl schließt:

Im Lichte der eben erwähnten Diracschen Theorie der Protonen besagt sie, daß positive und negative Elektrizität gleichartig sind, daß nämlich die Gesetze invariant sind gegenüber einer gewissen Substitution, welche die auf die Elektronen und Protonen bezüglichen Quantenzahlen miteinander vertauscht. Die Ungleichartigkeit der beiden Elektrizitätsarten scheint also ein noch tieferes Naturgeheimnis zu bergen als die Ungleichartigkeit von Vergangenheit und Zukunft.

Zu diesen Betrachtungen[2] über C, T, P meinte C.N. Yang in seinem Vortrag "Hermann Weyl's Contribution to Physics" zum 100jährigen Jubiläum an der ETH in Zürich:

Nobody, to my knowledge, absolutely nobody in the year 1930, was in any way suspecting that these symmetries were related in any manner. It was only in the 1950's that the deep connection between them was discovered. (...). What had prompted Weyl in 1930 to write the above passage is a great mystery to me. (Yang 1986, 10)

4.2 Weyls Arbeit von 1929: "Elektron und Gravitation"

Über Weyls Versuch von 1918, Elektrizität und Gravitation im Rahmen einer verallgemeinerten Riemannschen Geometrie — der Weylschen Geometrie mit nichtintegrabler Längenübertragung — in einheitlicher Weise zu beschreiben, wird an anderer Stelle in diesem Band berichtet. Deshalb werde ich mich auf eine Besprechung von Weyls großer Arbeit von 1929 beschränken, in welcher die Eichinvarianz - der eigentliche Kern der Theorie - quantentheoretisch umgedeutet wird.[3]

Schon die folgende Inhaltsangabe (Weyl 1929b) zeigt, daß hier eine schwergewichtige Arbeit vorliegt:

Einleitung. Verhältnis der allgemeinen Relativitätstheorie zu den quantentheoretischen Feldgleichungen des spinnenden Elektrons: Masse, Eichinvarianz, Fernparallelismus. Zu erwartende Modifikationen der Diracschen Theorie. - 1. Zweikomponententheorie: Die Wellenfunktion ψ hat nur zwei Komponenten. - §1. Bindung der Transformation der ψ an die Lorentztransformation des normalen Achsenkreuzes in der vierdimensionalen Welt. Asymmetrie von Zukunft und Vergangenheit, von rechts und links. - §2. In der allgemeinen Relativitätstheorie wird die Metrik in einem Weltpunkt festgelegt durch ein normales Achsenkreuz. Komponenten von Vektoren relativ zu den Achsen und den Koordinaten. Kovariante Differentiation von ψ. - §3. Allgemein invariante Fassung der Diracschen Wirkungsgröße, welche für das Wellenfeld der Materie charakteristisch ist. - §4. Die

---

[2]Vergleiche auch den Beitrag von R. Coleman und H. Korte in diesem Band, Abschnitt 5.3.

[3]Die Theorie von 1929 hat Weyl vor allem in den Beiträgen (Weyl 1929b), (Weyl 1928b/ [2]1931, 187–201) und (Weyl 1931b) dargestellt. Er kam aber auch in anderen Beiträgen darauf zurück, das letztemal in (Weyl 1950c).

differentiellen Erhaltungssätze von Energie und Impuls und die Symmetrie des Impulstensors folgen aus der doppelten Invarianz: 1. gegenüber Koordinatentransformation, 2. gegenüber Drehungen des Achsenkreuzes. Impuls und Impulsmoment der Materie. - §5. Einsteins klassische Gravitationstheorie in der neuen analytischen Formulierung. Gravitationsenergie. - §6. Das elektromagnetische Feld. Aus der Unbestimmtheit des Eichfaktors in ψ ergibt sich die Notwendigkeit der Einführung der elektromagnetischen Potentiale. Eichinvarianz und Erhaltung der Elektrizität. Das Raumintegral der Ladung. Einführung der Masse. Diskussion und Zurückweisung einer anderen Möglichkeit, in welcher die Elektrizität nicht als Begleitphänomen der Materie, sondern der Gravitation erscheint.

Bereits in der Einleitung auf der zweiten Seite der Arbeit (Weyl 1929b) wird das Prinzip der Eichinvarianz in die Quantenmechanik "hinübergerettet":

> Die Diracschen Feldgleichungen für ψ zusammen mit den Maxwellschen Gleichungen für die vier Potentiale f_p des elektromagnetischen Feldes haben eine Invarianzeigenschaft, die in formaler Hinsicht derjenigen gleicht, die ich in meiner Theorie von Gravitation und Elektrizität vom Jahre 1918 als Eichinvarianz bezeichnet hatte; die Gleichungen bleiben ungeändert, wenn man gleichzeitig ψ durch $e^{i\lambda}\psi$ und f_p durch $f_p - \partial\lambda/\partial x_p$ ersetzt, unter λ eine willkürliche Ortsfunktion in der vierdimensionalen Welt verstanden. Dabei ist in f_p der Faktor e/ch aufgenommen ($-e$: Ladung des Elektrons; c: Lichtgeschwindigkeit; $h/2\pi$: Wirkungsquantum). Auch die Beziehung dieser "Eichinvarianz" zum Erhaltungssatz der Elektrizität bleibt unangetastet. Es ist aber ein wesentlicher und für den Anschluß an die Erfahrung bedeutungsvoller Unterschied, daß der Exponent des Faktors, den ψ annimmt, nicht reell, sondern rein imaginär ist. ψ übernimmt jetzt die Rolle, welche in jener alten Theorie das Einsteinsche ds spielte. Es scheint mir darum dieses nicht aus der Spekulation, sondern aus der Erfahrung stammende neue Prinzip der Eichinvarianz zwingend darauf hinzuweisen, daß das elektrische Feld ein notwendiges Begleitphänomen nicht des Gravitationsfeldes, sondern des materiellen, durch ψ dargestellten Wellenfeldes ist. Da die Eichinvarianz eine willkürliche Funktion λ einschließt, hat sie den Charakter "allgemeinerRelativität und kann natürlich nur in ihrem Rahmen verstanden werden. (Weyl 1929b, 245f.)

Auf Weyls nähere Begründung werden wir bald eingehen.

Zur Zeit als Weyl seine Arbeit verfasste, arbeitete Einstein an einer unifizierenden Theorie, welcher ein Fernparallelismus (mit Torsion) zugrunde lag. Verschiedene Autoren, z.B. E. Wigner, wiesen auf einen Zusammenhang der Einsteinschen Theorie mit der Spintheorie des Elektrons hin. Dazu bemerkt Weyl ebenfalls bereits in der Einleitung:

An den Fernparallelismus vermag ich aus mehreren Gründen nicht zu glauben. Erstens sträubt sich mein mathematisches Gefühl a priori dagegen, eine so künstliche Geometrie zu akzeptieren; es fällt mir schwer, die Macht zu begreifen, welche die lokalen Achsenkreuze in den verschiedenen Weltpunkten in ihrer verdrehten Lage zu starrer Gebundenheit aneinander hat einfrieren lassen. Es kommen, wie ich glaube, zwei gewichtige physikalische Gründe hinzu. Gerade dadurch, daß man den Zusammenhang zwischen den lokalen Achsenkreuzen löst, verwandelt sich der Eichfaktor $e^{i\lambda}$, der in der Größe ψ willkürlich bleibt, notwendig aus einer Konstante in eine willkürliche Ortsfunktion; d.h. nur durch diese Lockerung wird die tatsächlich bestehende Eichinvarianz verständlich. (ebda., 246)

Diesen Gedanken führt Weyl in der Arbeit im einzelnen durch, nachdem er zunächst seine bekannte zweikomponentige Spinorgleichung[4] in der Speziellen Relativitätstheorie aufgestellt hat. Dabei betont er, daß letztere die Symmetrie von links und rechts aufhebt und meint: Nur diese tatsächlich in der Natur bestehende Symmetrie von rechts und links wird uns zwingen, ein zweites Paar von ψ-Komponenten einzuführen.

Massenterme werden dabei noch weggelassen, denn diese erhofft er sich von der Kopplung an die Gravitation:

Masse ist aber ein Gravitationseffekt: es besteht so die Hoffnung, für dieses Glied in der Gravitationstheorie einen Ersatz zu finden, der die gewünschte Korrektur herbeiführt. (ebda., 245)

Auf diese Bemerkung werden wir zurückkommen.

Bei der Übertragung der Spinorgleichungen in die Allgemeine Relativitätstheorie entwickelt Weyl den heute viel gebrauchten Tetradenformalismus. Er beschreibt also die Metrik durch orthonormierte Basisfelder ("normales Achsenkreuz") $\{e_\mu(x); \mu = 0, 1, 2, 3\}$. Bezeichnet $\{e^\mu(x)\}$ die zugehörige duale Basis von 1-Formen, so hat also die Metrik die Form

$$
\begin{aligned}
g(x) &= \eta_{\mu\nu} e^\mu(x) \otimes e^\nu(x), \\
(\eta_{\mu\nu}) &= diag\ (1, -1, -1, -1).
\end{aligned}
\tag{4.2.1}
$$

Natürlich wird sofort hervorgehoben, daß die Metrik nur eine Klasse von solchen Basisfeldern bestimmt: das metrische Feld ändert sich nicht, wenn die Vierbeinfelder ortsabhängigen Lorentz- Transformationen

$$
e^\mu(x) \mapsto \Lambda^\mu_\nu(x) e^\nu(x)
\tag{4.2.2}
$$

unterworfen werden. Die Gesetze müssen natürlich invariant bezüglich diesen Transformationen sein. Damit wird die Lorentz-Gruppe zu einer (nichtabelschen) Eichgruppe.

---

[4]Dazu werde ich später noch einiges sagen!

Die Zusammenhangsformen[5] $\omega = (\omega_\nu^\mu)$ haben Werte in der Lie-Algebra der homogenen Lorentz-Gruppe,

$$\omega_{\mu\nu} + \omega_{\nu\mu} = 0 \tag{4.2.3}$$

und genügen der 1. Strukturgleichung

$$de^\mu = -\omega_\nu^\mu \wedge e^\nu. \tag{4.2.4}$$

Bei lokalen Lorentz-Transformationen ändern sie sich genau so wie die Eichpotentiale einer nichtabelschen Eichtheorie:

$$\omega(x) \mapsto \Lambda(x)\omega(x)\Lambda^{-1}(x) - d\Lambda(x)\Lambda^{-1}(x). \tag{4.2.5}$$

Die Krümmungsformen Ω_ν^μ ergeben sich aus der zweiten Strukturgleichung:

$$\Omega_\nu^\mu = d\omega_\nu^\mu + \omega_\lambda^\mu \wedge \omega_\nu^\lambda. \tag{4.2.6}$$

Für ein Vektorfeld V^μ lautet die kovariante Ableitung in diesem Formalismus

$$DV^\mu = dV^\mu + \omega_\nu^\mu V^\nu. \tag{4.2.7}$$

Weyl verallgemeinert die kovariante Ableitung naturgemäß auf Spinorfelder ψ:

$$D\psi = d\psi + \frac{1}{4}\omega_{\mu\nu}\sigma^{\mu\nu}\psi, \tag{4.2.8}$$

wobei $\sigma^{\mu\nu}$ die üblichen Matrizen

$$\sigma^{\mu\nu} = \frac{1}{2}[\gamma^\mu, \gamma^\nu] \tag{4.2.9}$$

der Dirac-Theorie sind, welche infinitesimale Lorentz-Transformationen beschreiben.

Damit kann nun das Wirkungsprinzip für das gekoppelte Einstein-Dirac-System aufgestellt werden[6] . Weyl diskutiert natürlich auch gleich die Auswirkungen der beiden Symmetrieprinzipien (Weyl 1929b, 256ff.):

1. lokale Lorentz-Invarianz,

2. allgemeine Koordinateninvarianz.

All dies ist eine Art Vorbereitung des letzten Abschnittes, der mit "Elektrisches Feld" überschrieben ist. Weyl sagt:

---

[5]Gegenüber Weyl benutze ich modernere Bezeichnungen und Schreibweisen.
[6]Die Lagrange-Dichte lautet:

$$\mathcal{L} = \frac{1}{16\pi G}R - i\bar\psi\gamma^\mu D_\mu\psi. \tag{4.2.10}$$

Wir kommen jetzt zu dem kritischen Teil der Theorie. Meiner Meinung nach liegt der Ursprung und die Notwendigkeit des elektromagnetischen Feldes in folgendem begründet. Die Komponenten ψ_1, ψ_2 sind in Wahrheit nicht eindeutig durch das Achsenkreuz bestimmt, sondern nur insoweit, daß sie noch mit einem beliebigen "Eichfaktor" $e^{i\lambda}$ vom absoluten Betrag 1 multipliziert werden können. Nur bis auf einen solchen Faktor ist die Transformation bestimmt, welche die ψ unter dem Einfluß einer Drehung des Achsenkreuzes erleiden. In der speziellen Relativitätstheorie muß man diesen Eichfaktor als eine Konstante ansehen, weil wir hier ein einziges, nicht an einen Punkt gebundenes Achsenkreuz haben. Anders in der allgemeinen Relativitätstheorie: jeder Punkt hat sein eigenes Achsenkreuz und darum auch seinen eigenen willkürlichen Eichfaktor; dadurch, daß man die starre Bindung der Achsenkreuze in verschiedenen Punkten aufhebt, wird der Eichfaktor notwendig zu einer willkürlichen Ortsfunktion. (Weyl 1929b, 263)

Damit gelangt er zum Eichprinzip in der heutigen Form und stellt fest:

Aus der Unbestimmtheit des Eichfaktors in ψ ergibt sich die Notwendigkeit der Einführung der elektromagnetischen Potentiale.

An die Stelle des Maßstabsfaktors in Weyls Theorie von 1918 tritt nun ein Phasenfaktor:

$$exp \left(-\int_\gamma A \right) \to exp \left(-i\int_\gamma A \right), \qquad (4.2.11)$$

was der Ersetzung der ursprünglichen Eichgruppe $\mathcal{R}$ durch die kompakte Gruppe $U(1)$ entspricht. Die sinngemäße Übertragung des ursprünglichen Einsteinschen Einwandes führt nun zum Aharonov-Bohm-Effekt.

Natürlich deckt Weyl auch gleich den engen Zusammenhang zwischen Eichinvarianz und dem Erhaltungssatz für die Elektrizität auf. Wie in seiner ursprünglichen Theorie von 1918 betont er, daß die Stromerhaltung auf doppelte Weise folgt:

1. Feldgleichungen der Materie + Eichinvarianz ⇒ Stromerhaltung

2. Feldgleichungen des elektromagnetischen Feldes + Eichinvarianz ⇒ Stromerhaltung.

Dies entspricht einer identisch erfüllten Relation zwischen den materiellen und den elektromagnetischen Gleichungen, welche aufgrund der Eichinvarianz existieren muß. All dies ist in die Lehrbücher eingedrungen und soll hier nicht weiter ausgeführt werden.

Was die Kopplung an das Gravitationsfeld betrifft, stellt Weyl schließlich fest, daß diese eine natürliche Länge, die Planck-Länge $l_{pl} = (G\hbar/c^3)^{1/2} = 1,6 \times 10^{-33} cm$ weit unter der atomaren Größenordnung bestimmt, und bemerkt dazu:

So wird auch hier die Gravitation nur für die astronomischen Probleme
von Bedeutung sein. (Weyl 1929b, 266)

Nach der Veröffentlichung von Weyls ausführlicher Arbeit wurde auch der Ton von
Pauli freundlicher:

> Im Gegensatz zu den Bosheiten ist aber der sachliche Teil meines letz-
> ten Briefes inzwischen stark überholt, vor allem auch durch Ihre eigene
> Arbeit in der Zeitschrift für Physik. Aus diesem Grunde habe ich es
> nachher sogar bedauert, den Brief an Sie abgeschickt zu haben. Nach
> dem Studium dieser Arbeit glaube ich das, was Sie wollen und anstre-
> ben, nun wirklich verstanden zu haben. (Auf Grund der kleinen Note
> in den Proceedings of the National Academy war das noch nicht der
> Fall.) Zuerst will ich diejenige Seite der Sache hervorheben, bei der ich
> voll und ganz mit Ihnen übereinstimme: Ihr Ansatz zur Einordnung
> der Gravitation in die Diracsche Theorie des Spinelektrons.
>
> Ich bin nämlich dem Fernparallelismus ebenso feindlich gesinnt wie
> Sie, und es ist eine wahre Erlösung, daß bei Ihnen die Achsenkreuze in
> verschiedenen Punkten willkürlich gegeneinander drehbar sind. (Und
> hier muß ich Ihrer Tätigkeit in der Physik Gerechtigkeit widerfahren
> lassen. Als Sie früher die Theorie mit $g'_{ik} = \lambda g_{ik}$ machten, war dies reine
> Mathematik und unphysikalisch. Einstein konnte mit Recht kritisieren
> und schimpfen. Nun ist die Stunde der Rache für Sie gekommen; jetzt
> hat Einstein den Bock des Fernparallelismus geschossen, der auch nur
> reine Mathematik ist und nichts mit Physik zu tun hat, und Sie können
> schimpfen!) (Pauli 1979, 518, Brief vom 26. 8. 1929)

Gleich anschließend betont Pauli richtigerweise:

> Auch wenn man das Massenglied nicht streicht und 4 Komponenten
> der Wellenfunktion hat, wird Ihre Methode, die Gravitation zu be-
> handeln, ohne weiteres anwendbar sein. Damit komme ich aber zur
> anderen Seite der Sache, wo ich nicht ganz Ihre Ansicht teile, nämlich
> die Frage der ungelösten Schwierigkeiten der Diracschen Theorie (zwei
> Vorzeichen von m_0) und die Frage der Zweikomponententheorie. Nach
> meiner Meinung soll man diese Probleme nicht mit dem der Berück-
> sichtigung der Gravitation zusammenmischen. Die Hoffnung, für das
> Massenglied in der Gravitationstheorie einen Ersatz finden zu können,
> scheint mir trügerisch; die Gravitationseffekte werden numerisch immer
> viel zu klein sein. (ebda.,)

Auf seine anschließenden Fragen zur Quantisierung der Felder will ich nicht
eingehen. Im Zusammenhang mit der letzten zitierten Bemerkung von Pauli ist
es aber vielleicht angemessen zu betonen, daß auch die heutige Elementarteilchenphysik nichts über den Ursprung der Teilchenmassen aussagen kann. (Bei der

Ersetzung der Massen durch unbestimmte Teilchen-Higgs-Kopplungen wird keine physikalische Einsicht gewonnen.) Weyls Arbeit hat in allen Teilen überlebt und ist Allgemeingut geworden.

Rückblickend schrieb Weyl in einem Brief an Seelig lange nach den hier geschilderten Ereignissen:

> Aus dem Jahre 1918 datiert der von mir unternommene erste Versuch, eine einheitliche Feldtheorie von Gravitation und Elektromagnetismus zu entwickeln, und zwar auf Grund des Prinzips der Eichinvarianz, das ich neben dasjenige der Koordinaten-Invarianz stellte. Ich habe diese Theorie selber längst aufgegeben, nachdem ihr richtiger Kern: die Eichinvarianz, in die Quantentheorie herübergerettet ist als ein Prinzip, das nicht die Gravitation, sondern das Wellenfeld des Elektrons mit dem elektromagnetischen verknüpft. — Einstein war von Anfang dagegen, und das gab zu mancher Diskussion Anlaß. Seinen konkreten Einwänden glaubte ich begegnen zu können. Schließlich sagte er dann: "Na, Weyl, lassen wir das! So — das heißt auf so spekulative Weise, ohne ein leitendes, anschauliches physikalisches Prinzip — macht man keine Physik!" Heute haben wir in dieser Hinsicht unsere Standpunkte wohl vertauscht. Einstein glaubt, daß auf diesem Gebiet die Kluft zwischen Idee und Erfahrung so groß ist, daß nur der Weg der mathematischen Spekulation, deren Konsequenzen natürlich entwickelt und mit den Tatsachen konfrontiert werden müssen, Aussicht auf Erfolg hat, während mein Vertrauen in die reine Spekulation gesunken ist und mir ein engerer Anschluß an die quanten-physikalischen Erfahrungen geboten scheint, zumal es nach meiner Ansicht nicht genug ist, Gravitation und Elektromagnetismus zu einer Einheit zu verschmelzen. Die Wellenfelder des Elektrons und was es sonst noch an unreduzierbaren Elementarteilchen geben mag, müssen mit eingeschlossen werden. (Seelig 1979, 274)[7]

Mit diesem prägnanten Resumé einer verschlungenen Entdeckungsgeschichte, welche schließlich zu ungeahnten Entwicklungen in unserer Zeit führte,[8] möchte ich meine sehr unvollkommenen Ausführungen zu diesem Teil der Entwicklung beschließen.

4.3 Bemerkungen zur 2-Komponententheorie

Die 2-Komponententheorie von Weyl ist in der obigen Behandlung der Arbeit "Elektron und Gravitation" zu kurz gekommen und ich will nun noch etwas näher darauf eingehen.

---

[7]Seelig gibt keine Angabe über ds Datum des Briefes.
[8]Vergleiche dazu auch (Yang 1983, insbes. 525ff.).

Da zwei nichtäquivalente irreduzible zweidimensionale projektive Darstellungen der 1-Komponente $L_+^\uparrow$ der Lorentzgruppe existieren,[9] gibt es zwei Typen von Weylspinoren $\varphi_\alpha, \chi^{\dot\beta}$, für welche die kräftefreien Weyl'schen Gleichungen folgendermaßen lauten ($c = 1$):

$$\hat\sigma^\mu \partial_\mu \varphi = 0, \qquad \sigma^\mu \partial_\mu \chi = 0. \tag{4.3.1}$$

Dabei ist $(\sigma^\mu) = (1, -\vec\sigma), (\hat\sigma^\mu) = (1, \vec\sigma)$ ($\vec\sigma$: Pauli-Matrizen).[10] Im Spinorkalkül lauten diese Gleichungen

$$\partial^{\alpha\dot\beta} \varphi_\alpha = 0, \qquad \partial_{\alpha\dot\beta} \chi^{\dot\beta} = 0. \tag{4.3.2}$$

In seinem Handbuchartikel von 1933 schrieb Pauli zu den Weyl'schen Gleichungen:

> Indessen sind diese Wellengleichungen, wie ja aus ihrer Herleitung hervorgeht, nicht invariant gegenüber Spiegelungen (Vertauschung von links und rechts) und infolgedessen sind sie auf die physikalische Wirklichkeit nicht anwendbar. (Pauli 1933/ [2]1956, 920)

Zur Spiegelinvarianz ist nun aber folgendes anzumerken. Solange man keine Wechselwirkungen betrachtet, sind beide Weyl'schen Gleichungen durchaus spiegelinvariant. Dies geht schon daraus hervor, daß beide Gleichungen in (2) äquivalent zur *Majorana − Gleichung* sind. Betrachten wir z.B. das φ-Feld und setzen wie Weyl in (1928b/ [2]1931, 132)

$$\psi = \begin{pmatrix} \varphi \\ \epsilon\varphi^* \end{pmatrix}, \qquad \epsilon = \begin{pmatrix} 0 & 1 \\ -1 & 0 \end{pmatrix}, \tag{4.3.3}$$

so ist die 1.Gl. in (2) äquivalent zur masselosen Diracgleichung

$$\gamma^\mu \partial_\mu \psi = 0 \tag{4.3.4}$$

---

[9]Nach Übergang zur einfach zusammenhängenden Überlagerungsgruppe $SL_2\mathbb{C}$ sind dies die natürliche Darstellung σ, in heutiger Notation $D^{\frac{1}{2},0}$ sowie die dazu duale $D^{0,\frac{1}{2}} = \tilde\sigma :=^t\sigma^{*-1}$, die "komplex kontragrediente" Darstellung in Weyls Sprache (Weyl 1929b, 247f.) und(Weyl 1928b/ [2]1931, 131f.). Dabei ist hier σ^* wie in der physikalischen Literatur als komplexe Konjugation zu lesen. (Anm. des Herausgebers.)

[10]Des weiteren gehen aus den Pauli-Matrizen $\sigma^1 = \begin{pmatrix} & 1 \\ 1 & \end{pmatrix}, \sigma^2 = \begin{pmatrix} & -i \\ i & \end{pmatrix}, \sigma^3 = \begin{pmatrix} 1 & \\ & -1 \end{pmatrix},$ $\sigma^0 = \begin{pmatrix} 1 & \\ & 1 \end{pmatrix}$ die hier verwendeten Dirac-Matrizen hervor durch: $\gamma_j = \begin{pmatrix} & \sigma^j \\ -\sigma^j & \end{pmatrix}$ ($1 \leq j \leq 3$), $\gamma_0 = \begin{pmatrix} & \sigma^0 \\ \sigma^0 & \end{pmatrix}$ und $\gamma_5 = \begin{pmatrix} \sigma^0 & \\ & -\sigma^0 \end{pmatrix}$. In dem an die Tensorrechnung angelehnten Kalkül gilt bezüglich Minkowskimetrik $\gamma_0 = \gamma^0$, $\gamma_5 = \gamma^5$ und $\gamma_j = -\gamma^j$ ($1 \leq j \leq 3$). Je nach Autor sind die Darstellungen ggfs. konjugiert $A\gamma_j A^{-1}$ und/oder γ_0 mit γ_5 ausgetauscht. (Anm. des Herausgebers)

und ψ ist zudem selbstkonjugiert: Allgemein transformiert sich ein Diracspinor $\psi = \begin{pmatrix} \varphi_\alpha \\ \chi^{\dot\beta} \end{pmatrix}$ unter Ladungskonjugation C gemäß

$$C : \begin{pmatrix} \varphi \\ \chi \end{pmatrix} \longrightarrow \begin{pmatrix} -\epsilon\chi^* \\ \epsilon\varphi^* \end{pmatrix} \tag{4.3.5}$$

und dies führt in unserem Fall zur Majorana-Bedingung $\psi \xrightarrow{C} \psi$.

Niemand würde sagen, die Majorana-Theorie sei nicht spiegelinvariant. Hier ist nun folgendes zu beachten: Ein Diracfeld transformiert sich unter P gemäß

$$P : \psi \to \psi'(x) = \gamma^0 \psi(Px). \tag{4.3.6}$$

Für das Majoranafeld (14) gibt dies

$$P : \varphi \to \varphi'(x) = \epsilon\varphi^*(Px). \tag{4.3.7}$$

Unter dieser Transformation bleibt die Weyl'sche Gleichungen invariant. Meistens wird diese Operation als CP interpretiert, aber solange keine Wechselwirkungen vorhanden sind, ist dies willkürlich. Man beachte: In der Majorana-Formulierung ist C trivial, und deshalb fallen in ihr P und CP zusammen. Für die nachfolgenden historischen Bemerkungen sei auch noch an die Formulierung von *Lee und Yang* (Lee/Yang 1957) erinnert. Diese Autoren setzen in der Weyldarstellung der γ-Matrizen $\psi = \begin{pmatrix} \varphi \\ o \end{pmatrix}$, womit $(1 - \gamma^5)\psi = 0$. Die Weyl-Gleichung (2) ist wieder äquivalent zur masselosen Diracgleichung, d.h. man hat

$$\gamma^\mu \partial_\mu \psi = 0, \qquad (1 - \gamma^5)\psi = 0. \tag{4.3.8}$$

Diese Gleichungen sind natürlich unabhängig von der Darstellung der γ-Algebra, aber selbstverständlich ist die Formulierung von Lee und Yang äquivalent zu denjenigen von Weyl und von Majorana.

Nach der Entdeckung der Paritätsverletzung wurde bekanntlich die 2-Komponenten-Theorie für das Neutrino von verschiedenen Autoren vorgeschlagen.[11] Ich gehe gleich noch näher darauf ein. An dieser Stelle möchte ich aber noch eine Stelle aus dem bereits erwähnten Brief von Fierz an mich zitieren:

> Wie ich die Arbeit von Yang-Lee las, störte mich die Bemerkung: Man kann die Zweikomponenten- Theorie nicht, à la Majorana, verkürzen. Diese Formulierung ist irreführend; denn die Zweikomponenten- Theorie ist mit derjenigen von Majorana identisch. Ich habe dies Pauli geschrieben - was ich genau schrieb, weiß ich nicht, und erhielt als Antwort einen Brief: "Habe noch nie so gelacht; das ist der größte Bock, den Sie je geschossen haben."

---

[11] (Lee/Yang 1957), (Landau 1957), (Salam 1957)

Diese absurde, offenbar komplexhafte Reaktion hat mich bestürzt und
aufgeregt. So habe ich, was ich sonst nie getan habe, Pauli telefo-
niert. Er sagte, er sei gerade auf der Abreise und war sehr schwierig. Offenbar konnte ich ihn aber schließlich zur Vernunft bringen,
was die Anmerkung[12] zeigt. Die Sache war mir schon lange klar ....
(Fierz 1991)

Nachdem dies Fierz ausgeführt hat[13], schreibt er weiter:

Yang-Lee haben das offenbar nicht gesehen, weil sie an Paulis Irrlehre
glaubten, die Weylsche Theorie sei nicht spiegelinvariant (ebda.).

Einer der ersten, der die 2-Komponententheorie des Neutrinos vorschlug, war
A. Salam (1957). Dieser hatte Yang anlässlich der Seattle-Konferenz im September
1956 gehört, in der Yang über die bekannte Lösung des $\Theta - \tau$-Rätsels durch ihn
und T. D. Lee berichtete. Salam erzählt:

I remember travelling back to London on an American Air Force (MATS)
transport flight. Although I had been granted, for that night, the status of a Brigadier or a Field Marshal - I don't quite remember which -
the plane was very uncomfortable, full of crying servicemen's children -
that is, the children were crying, not the servicemen. I could not sleep.
I kept reflecting on why Nature should violate left- right symmetry
in weak interactions. Now the hallmark of most weak interactions was
the involvement in radioactivity phenomena of Pauli's neutrino. While
crossing over the Atlantic came back to me a deeply perceptive question about the neutrino which Professor Rudolf Peierls had asked when
he was examining me for a Ph.D. a few years before. Peierls' question
was: "The photon mass is zero because of Maxwell's principle of a gauge symmetry for electromagnetism; tell me, why is the neutrino mass
zero?" (Salam 1980, 527)

In dieser Nacht fiel ihm die Antwort auf die Frage von Peierls anlässlich des
Doktorexamens ein: Verantwortlich war die γ_5-Symmetrie der Neutrinogleichung.[14]
Er arbeitete sofort einige Konsequenzen dieser Symmetrie aus (z.B. den Wert des
Michel-Parameters) und unterbreitete die Idee Peierls, der ja die entscheidende

---

[12]Gemeint ist eine Fußnote in Paulis späterer Arbeit, in der er schreibt: "Independently of this
paper my attention was drawn to this equivalence by Prof. M. Fierz, whom I would like to thank
very much" (Pauli 1957).

[13]Er vermerkt hier die oben ausgeführte Tatsache, daß zwischen zweikomponentigen Weylspinoren und vierkomponentigen Majoranaspinoren ein einfacher eindeutiger Zusammenhang besteht. Dieser wurde übrigens bereits 1952 von Serpe in einer weitgehend vergessenen Arbeit in
den "Physica" beschrieben (Serpe 1952).

[14]Gemeint ist damit, daß die Neutrinogleichung sowie die Wechselwirkung des Neutrinofeldes
Ψ mit anderen Elementarteilchenfeldern invariant sind unter der Substitution $\Psi \longrightarrow \gamma_5 \Psi$. Für
die Dirac-Gleichung mit nichtverschwindender Masse ist dies nicht der Fall.

Frage gestellt hatte. Dessen Antwort war kurz und bündig: "I do not believe left-right symmetry is violated in weak nuclear forces at all." Salam hoffte darauf, mehr Resonanz am CERN zu finden,

> ... with Pauli – the father of the neutrino – nearby in Zurich. At that time CERN lived in a wooden hut just outside Geneva airport. Beside my friends, Prentki and d'Espagnat, the hut contained a gas ring on which was cooked the staple diet of CERN-Entrecôte à la crème. The hut also contained Professor Villars of MIT, who was visiting Pauli the same day in Zurich. I gave him my paper. He returned the next day with a message from the Oracle: "Give my regards to my friend Salam and tell him to think of something better." (ebda.)

4.4 Ausklang

Zunächst möchte ich wieder C.N. Yang zitieren:

> The quote above from Weyl's paper[15] also contains something which is very revealing, namely, his strong association of gauge invariance with general relativity. That was, of course, natural since the idea had originated in the first place with Weyl's attempt in 1918 to unify electromagnetism with gravity. Twenty years later, when Mills and I worked on non-Abelian gauge fields, our motivation was completely divorced from general relativity and we did not appreciate that gauge fields and general relativity are somehow related. Only in the late 1960's did I recognize the structural similarity mathematically of non-Abelian gauge fields with general relativity and understand that they both were connections mathematically. (Yang 1986, 17)

Weyl war dem Prinzip der Eichinvarianz bis ans Lebensende sehr zugetan. So schrieb er noch sechs Monate vor seinem Tod für seine *Selecta* in einem Postskript zum Wiederabdruck seiner ersten Publikation zu diesem Thema (Weyl 1918b):

> Das stärkste Argument für meine Theorie schien dies zu sein, daß die Eichinvarianz dem Prinzip von der Erhaltung der elektrischen Ladung so entspricht wie die Koordinaten-Invarianz dem Erhaltungssatz von Energie-Impuls. (Weyl 1955, 192)

Daran anknüpfend zitiere ich nochmals eine längere aufschlussreiche Passage von Yang:

---

[15]Yang bezieht sich dabei auf die von mir weiter oben zitierte Stelle: "Da die Eichinvarianz eine willkürliche Funktion λ einschließt, hat sie den Charakter "allgemeiner" Relativität und kann natürlich nur in ihrem Rahmen verstanden werden."

Weyl's reason, it turns out, was also one of the melodies of gauge theory that had very much appealed to me when as a graduate student I studied field theory by reading Paulis articles. I made a number of unsuccessful attempts to generalize gauge theory beyond electromagnetism, leading finally in 1954 to a collaboration with Mills in which we developed a non-Abelian gauge theory. In [...] we stated our motivation as follows:

The conservation of isotopic spin points to the existence of a fundamental invariance law similar to the conservation of electric charge. In the latter case, the electric charge serves as a source of electromagnetic field; an important concept in this case is gauge invariance which is closely connected with (1) the equation of motion of the electro-magnetic field, (2) the existence of a current density, and (3) the possible interactions between a charged field and the electromagnetic field. We have tried to generalize this concept of gauge invariance to apply to isotopic spin conservation. It turns out that a very natural generalization is possible.

Item (2) is the melody referred to above. The other two melodies, (1) and (3), were what had become pressing in the early 1950's when so many new particles had been discovered and physicists had to understand how they interacted with each other.

I had met Weyl in 1949 when I went to the Institute for Advanced Study in Princeton as a young "member". I saw him from time to time in the next years, 1949-1955. He was very approachable, but I don't remember having discussed physics or mathematics with him at any time. His continued interest in the idea of gauge fields was not known among the physicists. Neither Oppenheimer nor Pauli ever mentioned it. I suspect they also did not tell Weyl of the 1954 papers of Mills' and mine. Had they done that, or had Weyl somehow come across our paper, I imagine he would have been pleased and excited, for we had put together two things that were very close to his heart: gauge invariance and non-Abelian Lie groups. (Yang 1986, 19f.)

Es ist tatsächlich erstaunlich, daß Pauli in diesen späten Jahren nie mit Weyl über nicht- Abel'sche Verallgemeinerungen der Eichinvarianz gesprochen hat, da er selber – noch vor Yang und Mills – darüber gearbeitet hat. Vom 22.-25. Juli 1953 stammt ein Manuskript von Pauli, welches an Pais gerichtet ist und folgenden Titel trägt "Meson-Nucleon Interaction and Differential Geometry".[16]

Darin entwickelt Pauli gleich eine nicht-Abelsche Kaluza-Klein-Theorie für die Eichgruppe SU(2). Die höherdimensionale Mannigfaltigkeit ist 6-dimensional, wobei die Extradimensionen 2-Sphären mit ortsabhängiger Metrik sind, auf denen SU(2) ortsabhängig operiert. In "lokaler Sprache" beginnt Pauli zunächst mit

---

[16](Pauli 1999, Briefe 1614 und 1682).

der Darstellung der Geometrie eines Faserbündels mit einem homogenen Raum als typische Faser (für sein Beispiel $S^2 \cong SU(2)/U(1)$). Dabei stößt er auf die Krümmung, welche später von Yang und Mills als nicht-Abelsche Feldstärke eingeführt wurde. Anschließend stellt Pauli die 6-dimensionale Dirac-Gleichung auf und schreibt sie in an die Faserung adaptierter Form aus. In einem "Mathematical Appendix" vom Dezember 1953 wird noch manches näher ausgeführt und ebenfalls das Massenspektrum für die Diracgleichung bestimmt. Der Schlußsatz lautet: "So this leads to some rather unphysical 'shadow particles'."

Vor diesem Hintergrund wird nun die folgende Geschichte verständlicher, die sich im Frühjahr 1954 abgespielt hat.

Ende Februar wurde Yang von Oppenheimer nach Princeton eingeladen, um über die gemeinsame Arbeit mit Mills zu berichten. Darüber berichtet Yang:

> Pauli was spending the year in Princeton, and he was deeply interested in symmetries and interactions. (He had written in German a rough outline of some thoughts, which he had sent to A. Pais. Years later F.J. Dyson translated this outline into English. It started with the remark, "Written down July 22-25, 1953, in order to see how it looks," and had the title "Meson-Nucleon Interaction and Differential Geometry.") Soon after my seminar began, when I had written down on the blackboard,
>
> $$(\partial_\mu - i\epsilon B_\mu)\psi,$$
>
> Pauli asked, "What is the mass of this field B_μ?" I said we did not know. Then I resumed my presentation, but soon Pauli asked the same question again. I said something to the effect that that was a very complicated problem, we had worked on it and had come to no definite conclusions. I still remember his repartee: "That is not sufficient excuse." I was so taken aback that I decided, after a few moments' hesitation, to sit down. There was general embarassment. Finally Oppenheimer said, "We should let Frank proceed." I then resumed, and Pauli did not ask any more questions during the seminar.

> I don't remember what happened at the end of the seminar. But the next day I found the following message:

> February 24, Dear Yang, I regret that you made it almost impossible for me to talk with you after the seminar. All good wishes. Sincerely yours, W. Pauli.

> I went to talk to Pauli. He said I should look up a paper by E. Schrödinger, in which there were similar mathematics[17]. After I went back to Brookhaven, I looked for the paper and finally obtained a copy. It was a discussion of space-time-dependent representations of the γ_μ

---

[17]E. Schrödinger, Sitzungsberichte der Preussischen (Akademie der Wissenschaften, 1932), p. 105.

> As a physicist-philosopher, Weyl had written extensively about
> space, time, matter, energy, force, geometry, topology, etc., key con-
> cepts that provide the basis upon which modern physics is erected
> . . . .
> I venture to say that if Weyl were to come back today, he would
> find that amidst the very exciting, complicated and detailed deve-
> lopments in both physics and mathematics, there are fundamental
> things that he would feel very much at home with. He had helped
> to create them.

Für Weyl wäre es kaum schwierig gewesen, die mathematischen Aspekte seiner Arbeit von 1929 auf nicht-Abelsche Eichgruppen zu verallgemeinern. Von physikalischer Seite gab es aber damals für Erweiterungen in diese Richtung keine Veranlassung. Es dauerte noch Jahrzehnte, bis erkannt wurde, daß Eichsymmetrien auch für eine Theorie der schwachen und der starken Wechselwirkung entscheidend sind. Da diese Kräfte nur über kurze Abstände wirken, sind die erweiterten Eichsymmetrien des Standardmodells der Elementarteilchenphysik sehr verborgen. Gerade dieser Aspekt ist immer noch ganz unzureichend verstanden. Vielleicht werden wir eines Tages auch ein viel tieferes Verständnis für den Ursprung der Eichgruppen gewinnen. Dafür gibt es im Rahmen der Stringtheorie bereits interessante Ansätze. So weist Weyls Werk weit in die Zukunft.

Literatur

Einstein, Albert an Weyl, 8. 3. 1918. *Collected Papers* **8** B. Princeton: Princeton University Press 1998, 669f.

Fierz, Markus, Brief vom 25. Feb. 1991 an N. Straumann. Besitz des Autors.

Landau, Lew D. 1957. On the conservation laws for weak interactions. *Nuclear Physics* A **3**, 127–131.

Lee, Tsung D.; Yang, Chen Ning 1957. Parity nonconservation and a twocomponent theory of the neutrino. *Physical Review* **105**, 1671–1675. *Selected Papers* (T. D. Lee) **2**, Boston etc.: Birkhäuser 1986, 245–250; ebenso in (Yang 1983, 205–209).

Pauli, Wolfgang 1933. Die allgemeinen Prinzipien der Wellenmechanik. *Handbuch der Physik*, **24**, Teil 1, Berlin: Springer, 83–272. [2]1956 *Handbuch der Physik*, **5**, 1–168. Collected Papers **1**, 771–939.

Pauli, Wolfgang 1957. On the conservation of the lepton charge, *Nuovo Cimento* **6**, 204–215. Collected Papers **2**, 1338–1349.

Pauli, Wolfgang 1979. *Wissenschaftlicher Briefwechsel,* Band I: 1919-1929. Berlin etc.: Springer.

Pauli, Wolfgang 1990. *Die allgemeinen Prinzipien der Wellenmechanik,* herausgegeben von N. Straumann. Berlin etc.: Springer.

Pauli, Wolfgang 1999. *Wissenschaftlicher Briefwechsel,* Band IV, Teil II. Berlin etc.: Springer.

Salam, Abdus 1957. On parity conservation and neutrino mass. *Nuovo Cimento* **5,** 299–301.

Salam, Abdus 1980. Gauge unification of fundamental forces. *Review of Modern Physics* **52,** 525–538.

Schrödinger, Erwin 1932. Diracsches Elektron im Schwerefeld I. *Sitzungsberichte Preußische Akademie der Wissenschaften* Phys.-math. Klasse, 105–128. Gesammelte Abhandlungen **3,** Wien 1983, 436–460.

Seelig, Carl 1960. *Albert Einstein.* Zürich: Europa Verlag.

Serpe, J. Sur la théorie abrégée des particules de spin $\frac{1}{2}$. Physica **18,** 295–306.

Yang, Cheng Nin 1983. *Selected Papers 1945-1980* with Commentary. San Francisco: Freeman and Co.

Yang, Cheng Nin 1986. Hermann Weyl's contribution to physics. In K. Chandrasekharan (ed.), *Hermann Weyl: 1885 -1985* ; Centenary Lectures Delivered by C. N. Yang, R. Penrose, and A. Borel at the ETH Zürich. Berlin etc.: Springer, 7–21.

Part II

Hermann Weyl: Mathematician, Physicist, Philosopher

Contents

II Hermann Weyl: Mathematician, Physicist, Philosopher

Robert Alan Coleman and Herbert Korté

1   Introduction

It happens rarely that an individual is capable of pioneering work in several fields. Hermann Weyl was just such an individual, a profound thinker of wide intellectual range, a giant of our times. His vision was unique and penetrating not only in mathematics, but also in mathematical physics and in philosophy of science. Humanity, compassion and a powerful sense of the beautiful were the hallmarks of his personality and characteristic of his intellectual endeavour. The sheer range of his genius and his persistent search for a harmonious, intelligible architecture of the physical universe at once links him to the last great universalist mathematicians and thinkers of the nineteenth century such as Hilbert and Poincaré, and stands as a promise and anticipation of the future development of science and mathematics.

Weyl's long and productive intellectual life began as an apprentice to the great Hilbert. In his doctoral dissertation and subsequent publications, Weyl made important contributions to the spectral theory of self-adjoint operators. In particular, he solved the eigenvalue problem for second-order, linear, self-adjoint differential equations with *singular* boundary conditions, and he initiated the analysis of the asymptotic distribution of the eigenvalues of a self-adjoint compact operator in a Hilbert space. With respect to the latter, he created the "maximinimal" method to compute *directly* the nth eigenvalue of the compact operator.

Weyl's first book, *Die Idee der Riemannschen Fläche*, constitutes an integration of function theory, geometry and topology. It exemplifies Weyl's unique gift for harmoniously uniting previously unrelated fields. It is noteworthy that this work contains the first construction of a manifold with abstract mathematical entities as its points.

In the field of number theory, Weyl established an important and fruitful criterion for the equidistribution mod 1 of a sequence of numbers that was of fundamental significance for subsequent developments in analytic number theory. Although his main mathematical memoir on the subject appeared in 1916, he published a preliminary note in 1914 and immediately applied his results to the theory of the mean motion of the planets, once again demonstrating the breadth and versatility of his intellect.

Between 1917 and 1924, Weyl devoted a great deal of his energy to the development of the mathematical and philosophical foundations of Relativity Theory. His numerous contributions include: clarification of the role of coordinates and of invariance or symmetry principles, a group theoretic proof of the uniqueness of the Pythagorean form of the metric (the Raumproblem), the generalization of Levi-Civita's concept of parallelism, which pointed the way to the geometry of paths, and a deep analysis of the nature of motion, the role of inertia and Mach's principle. In addition, he developed the first non-Riemannian geometry, on which he based a unified field theory of gravitation and electromagnetism which incorporated the principle of gauge invariance. He also discovered the possibility of the geodesic or *causal-inertial* method for measuring the spacetime metric by being the first to distinguish between two primitive substructures of the pseudo-Riemannian structure, namely, the *conformal* structure and the *projective* structure, and then showing that these structures determine the pseudo-Riemannian metric uniquely up to a *constant*, positive factor.

During the preparation of this overview, we discovered that the standard interpretation, the Received View, of Weyl's Raumproblem, based mainly on the work of Cartan, Scheibe and Laugwitz, is in our view seriously flawed. For this reason, we have presented detailed analyses both of the Received View and of our interpretation which we call the Finsler-Metric interpretation. We show that both the mathematical and textual evidence decidedly favour the Finsler-Metric interpretation.

The central role that group theoretic techniques played in Weyl's analysis of spacetime led to what he himself considered his greatest work in mathematics, namely, a general theory of the representations and invariants of the classical Lie groups. He subsequently applied this powerful theory to the analysis of the foundations of the new physical theory of Quantum Mechanics. Weyl's analysis showed that regularities in a physical theory are most fruitfully understood in terms of symmetry groups. This emphasis on the importance of symmetry groups was later combined with Weyl's notion of local (Abelian) gauge invariance. The resulting synthesis lies at the core of the non-Abelian gauge field theories that dominate theoretical physics today.

Weyl's (1928b, 2 edn 1931) book, *Gruppentheorie und Quantenmechanik*, deals not only with the theory of Quantum Mechanics but also with Relativistic Quantum Electrodynamics. Weyl also presented a very early analysis of the discrete symmetries that are now called C, P, T and CPT. The only difference between Weyl's analysis and the currently accepted treatment of these symmetries is that Weyl treats time reversal T as a linear, unitary operator rather than as an antilinear, anti-unitary operator. Despite the fact that Dirac was motivated by Weyl's analysis of the discrete symmetry C to predict the existence of the positron and the antiproton, other renowned physicists failed to recognize the significance of Weyl's work.

Finally, Weyl contributed to the foundations of mathematics. In a book, *Das Kontinuum*, published in 1918, Weyl addressed for the first time the question of how much of classical analysis could be constructed on a *strictly* predicative basis that involves quantification *only* over the natural numbers. It was remarkable that Weyl was able to construct on this basis a *countable*, complete, ordered field W and a significant portion of classical analysis that is sufficient for the purposes of mathematical physics.

The structure of Weyl's analysis was strongly influenced by his desire to avoid Russell's theory of ramified types which Weyl rejected as a method for avoiding impredicativity because it rendered impossible most of real-number analysis. It should be noted that the ontological position that Weyl adopted in *Das Kontinuum* makes the theory of simple types insufficient for the avoidance of impredicativity. We present in some detail an account of the theories of simple and ramified types and apply the theory of simple types to the construction of the positive rationals from the positive integers and to the construction of the positive reals from the positive rationals. Our presentation serves to clarify the difficulties with impredicative definitions as Weyl saw them and to highlight the severe problems that arise if the theory of ramified types is used to avoid them. Unlike the usual accounts, our presentation of the constructions of the positive rationals and of the positive reals is more formal and explicit in conceptual detail with regard to the hierarchical levels of the various types of entities that occur. Our presentation provides, therefore, a sharp contrast with Weyl's analysis which avoids the hierarchy of types altogether.

Shortly after the publication of *Das Kontinuum*, Weyl was convinced by Brouwer to adopt an 'epistemological ontology' of the real numbers in which a real number is regarded as defined by a sequence of completely nested, open intervals. The sequences are arbitrary but never completed, becoming sequences. Weyl agreed with Brouwer that the law of the excluded middle was not valid even for existential statements pertaining to the natural numbers; consequently, Weyl rejected the analysis he presented in *Das Kontinuum*. This intuitionist view of analysis was so constraining that Weyl's enthusiastic acceptance of Brouwer's intuitionism motivated Hilbert to contest the intuitionist programme and to further develop his formalist programme for the foundations of mathematics. We present a detailed discussion of Weyl's major contribution to Brouwer's intuitionism.

Our overview is not primarily aimed at experts in mathematics and mathematical physics. Rather, its purpose is to make Weyl's scientific work accessible to a wider audience, in particular to philosophers and historians of science. This objective has influenced the mode of presentation in both scope and detail.

The portrayals of Weyl's scientific work in previous overviews have emphasized his mathematical achievements. Although we have certainly not neglected this aspect of Weyl's work, we have also highlighted his achievements in mathematical physics and in the philosophy of physics and mathematics. As was the case with Hilbert and Poincaré, mathematics and physics played a symbiotic role in Weyl's thought.

Although we have made every effort to provide a complete and balanced account of Weyl's scientific achievements, some topics have been omitted. In the domain of physics, we have not discussed his contributions to cosmology[1] and to Mie's theory[2] of matter. In the domain of mathematics, the following contributions are noteworthy: Weyl's work on Riemann matrices (Weyl, 1934b, 1936, 1937a, 1937b), his treatment of Minkowski's reduction theory (Weyl, 1935, 1940, 1942a, 1942b, 1945), and his analysis (Weyl, 1916a, 1917) of the rigidity of closed, convex, Riemannian surfaces. Weyl's development of this last topic has been discussed by Chevalley and Weil (1968).

We would like to mention that the previous overviews by Newman (1957)

---

[1] See, however, the article entitled *Weyl's contribution to cosmology* in this volume by H. Goenner.

[2] See the section entitled *Einheitliche Feldtheorie und dynamistische Materieauffassung* of the article by E. Scholz in this volume.

and by Chevalley and Weil (1968) have been very helpful in the preparation of some aspects of this work.

We wish to thank Erhard Scholz for his detailed comments particularly in connection with Weyl's Raumproblem and with Weyl's contributions to the foundations of mathematics. His thoughtful remarks led us to extend, clarify and improve significant aspects of our overview. We also wish to thank Skúli Sigurdsson for reading parts of an earlier draft and for pointing out a number of historical inaccuracies. Finally, we wish to express our appreciation to Dr. Klaus Burmeister and Dr. Bruce Plouffe of the Department of Germanic Studies at the University of Regina for their kind assistance with the translation of some of the more difficult German passages. All of our own translations have been enclosed in square brackets.

2 The Young Analyst

Hermann Weyl completed his secondary education in Altona in 1904 at the age of eighteen. He chose to attend the University of Göttingen mainly because the director of his high school was Hilbert's cousin and he had provided Weyl with a letter of recommendation. Weyl was so impressed with Hilbert that he resolved to study everything Hilbert had written. The mathematical world of Göttingen was, at that time, extremely rich and varied and was governed (Weyl 1944b, p. 122) by Hilbert, Minkowski and Felix Klein.

2.1 The Flourishing Field of Integral Equations

The theory of integral equations, popularized by Hilbert, was a major field of research. In his obituary for Hilbert, Weyl (1944b, p. 126)[3] noted that Hilbert's attention had been drawn to the subject of integral equations by a lecture presented in 1901 in Hilbert's seminar by the Swedish mathematician E. Holmgren. Holmgren's lecture dealt with the then recent and now classic paper by Fredholm in which he developed the infinite dimensional analogue of linear equations in n variables.

Fredholm's generalization can be motivated as follows. Use functional notation to write the components of the vectors f and g and of the matrix A as $f(j)$, $g(j)$ and $A(j,k)$. Then,

$$\sum_{k=1}^{n} A(j,k)f(k) = g(j), \quad \text{for } 1 \leq j \leq n \qquad (2.1.1)$$

is a system of n linear equations for the n variables $f(j)$. Define the matrix K by

$$K(j,k) = I(j,k) - A(j,k), \qquad (2.1.2)$$

where I is the matrix defined by

$$I(j,k) = \begin{cases} 1 & \text{for } j = k \\ 0 & \text{for } j \neq k \end{cases}. \qquad (2.1.3)$$

Then, the system (2.1.1) may be written in the form

$$f(j) = g(j) + \sum_{k=1}^{n} K(j,k)f(k). \qquad (2.1.4)$$

If the indices are allowed to be continuous variables and the sum is replaced by an integral, one obtains the Fredholm integral equation

$$f(x) = g(x) + \int_{a}^{b} K(x,y)f(y)dy \qquad (2.1.5)$$

---

[3]See also (Weyl 1944a, pp. 163–171). Moreover, Hellinger (1965) provides a detailed review of Hilbert's work on Integral Equations and contains detailed references to Hilbert's work.

with the kernel $K(x,y)$. Weyl (1944b, p. 164) commented that the simplicity of Fredholm's results follows from the particular form of equation (2.1.5), a form motivated by the physical problems to which Fredholm applied it. Consider once again the finite dimensional case

$$(I - K)f = g, \tag{2.1.6}$$

where f and g are column matrices. Denote the transpose of K by $\tilde{K}$. Then $(I - K)$ and $(I - \tilde{K})$ have the same rank $r \leq n$, and the homogeneous equations

$$(I - K)u = 0 \tag{2.1.7}$$

and

$$(I - \tilde{K})v = 0 \tag{2.1.8}$$

have the same number $(n - r)$ of linearly independent solutions. Let $v_1, v_2, \ldots, v_{n-r}$ be linearly independent solutions of (2.1.8), then multiplication of (2.1.6) by the row matrix $\tilde{v}_i$ for any $i \in \{1, 2, \ldots, n - r\}$ followed by transposition yields the constraint

$$\tilde{f}(I - \tilde{K})v_i = 0 = \tilde{g}v_i. \tag{2.1.9}$$

Hence, the inhomogeneous equation (2.1.6) has a solution if and only if g satisfies the constraints (2.1.9). Moreover, if f_0 is one solution of (2.1.6) and if $u_1, u_2, \ldots, u_{n-r}$ are linearly independent solutions of (2.1.7), then the general solution of (2.1.6) is given by

$$f = f_0 + \alpha_1 u_1 + \alpha_2 u_2 + \cdots + \alpha_{n-r} u_{n-r}. \tag{2.1.10}$$

Fredholm showed that these results carried over to the integral equation (2.1.5). The homogeneous equations

$$u(x) - \int_a^b K(x,y)u(y)dy = 0 \tag{2.1.11}$$

and

$$v(x) - \int_a^b \tilde{K}(x,y)v(y)dy = 0, \tag{2.1.12}$$

where $\tilde{K}(x,y) = K(y,x)$, each have the same finite number of linearly independent solutions $u_i(x)$ and $v_i(x)$ and the inhomogeneous equation (2.1.5) has a solution if and only if the constraint analogous to (2.1.9), namely,

$$\int_a^b g(x)v_i(x)dx = 0, \tag{2.1.13}$$

is satisfied for each i.

Hilbert immediately realized that the central problem in the theory of oscillations of a continuous medium, namely, the determination of the normal modes of vibration and of the corresponding resonant frequencies, could be reformulated in terms of integral equations. The dynamical laws of motion for such a medium are obtained by applying Newton's laws of motion to an infinitesimal element of the medium. One obtains a system of partial differential equations for the various fields that characterize the state of the medium. For example, for the simple case of small-amplitude vibrations of a stretched membrane, one obtains the wave equation

$$\Delta H(x,t) - \frac{1}{v^2}\frac{\partial^2}{\partial t^2}H(x,t) = 0, \qquad (2.1.14)$$

where $H(x,t)$ is the deviation of the membrane from its equilibrium position at the point x on the membrane at time t. For simply periodic motion,

$$H(x,t) = u(x)\exp(i\omega t) \qquad (2.1.15)$$

and

$$\Delta u(x) + \lambda u(x) = 0. \qquad (2.1.16)$$

The function u, defined on a domain D that corresponds to the surface of the membrane (usually a disc), must also satisfy on the boundary ∂D the condition

$$u(x) = 0 \quad \text{for } x \in \partial D, \qquad (2.1.17)$$

which follows from the fact that the membrane is clamped onto the bounding hoop. Solutions of (2.1.16) that satisfy (2.1.17) exist only for special values of λ, called eigenvalues, and the solution $u(x,\lambda)$ corresponding to an eigenvalue λ is called an eigenfunction. The analysis of other problems may lead to differential operators other than the Laplacian Δ as well as to more general domains D and to a boundary condition

$$\alpha u(x) + \beta \nabla_n u(x) = 0 \quad \text{for } x \in \partial D, \qquad (2.1.18)$$

where $\nabla_n u(x)$ denotes the derivative of $u(x)$ in the outward normal direction to ∂D. Hilbert realized that this problem was equivalent to solving the homogeneous integral equation

$$u(x) - \lambda \int_D K(x,y)u(y)dy = 0, \qquad (2.1.19)$$

where the symmetric kernel $K(x,y) = K(y,x)$ is the Green function for the appropriate differential operator and boundary condition. He also saw that (2.1.19) is the analogue of the finite dimensional eigenvalue problem

$$Ku = \frac{1}{\lambda}u \qquad (2.1.20)$$

for a real symmetric matrix K. In the finite dimensional case, there exists a set of n eigenvectors u_i corresponding to the eigenvalues λ_i. The eigenvectors corresponding to different eigenvalues are necessarily orthogonal and may be chosen to be orthonormal

$$u_i \cdot u_j = \delta_{ij}. \qquad (2.1.21)$$

Moreover, the eigenvectors u_i are linearly independent and satisfy the completeness condition

$$\sum_{i=1}^{n} u_i \tilde{u}_i = I, \qquad (2.1.22)$$

where I is the identity matrix. Also, the symmetric matrix K may be diagonalized by an orthogonal similarity transformation. Hilbert proved that these results hold also for the case of the integral equation (2.1.19). Specifically, there exists an infinite sequence of eigenvalues $\lambda_1, \lambda_2, \ldots$ such that $\lambda_k \to \infty$ as $k \to \infty$ and an orthonormal set of corresponding eigenfunctions $u_k(x)$ such that

$$u_k(x) - \lambda_k \int_D K(x,y) u_k(y) dy = 0, \qquad (2.1.23)$$

$$\int_D u_i(x) u_j(x) dx = \delta_{ij}, \qquad (2.1.24)$$

and

$$\int_D \int_D K(x,y) f(x) f(y) dx dy = \sum_i \frac{f_i^2}{\lambda_i}, \qquad (2.1.25)$$

where

$$f_i = \int_D f(x) u_i(x) dx \qquad (2.1.26)$$

is the 'Fourier' component of $f(x)$ along $u_i(x)$. Also, any function of the form

$$g(x) = \int_D K(x,y) f(y) dy \qquad (2.1.27)$$

may be expanded in a 'Fourier series'

$$g(x) = \sum_i g_i u_i(x), \qquad (2.1.28)$$

where

$$g_i = \int_D g(x) u_i(x) dx, \qquad (2.1.29)$$

and the series (2.1.28) converges uniformly.

The theory of integral equations was already a flourishing and fashionable field of research in 1904. It is not surprising that, during the first part of his career up to about 1912, Weyl worked on various aspects of this theory: singular integral equations (Weyl, 1908a, 1908b), Hilbert space theory (Weyl, 1909a, 1909b), eigenvalue problems and expansion of functions in terms of the eigenfunctions associated with various differential equations (Jerosch and Weyl, 1908) and (Weyl, 1909c, 1910d, 1910e, 1911c), the Gibbs phenomenon and its analogues (Weyl, 1910a, 1910c, 1911a) and the distribution of eigenvalues of completely continuous operators that occur in mathematical physics (Weyl, 1911b, 1912a, 1912b, 1912c, 1913b, 1915).

2.2 Singular Integral Equations

In his first publication in 1908, which served as his dissertation, Weyl considered singular integral equations with a symmetric kernel $K(x, y)$, equations of the form

$$u(x) - \lambda \int_0^\infty K(x, y)u(y)dy = 0. \qquad (2.2.1)$$

Hilbert had treated the case in which the limits of integration were both finite. Weyl gave special attention to the case in which the analogue of the Fourier integral theorem holds; so that,

$$f(x) = \int_0^\infty \int_0^\infty K(x, y)K(y, z)f(z)dzdy, \qquad (2.2.2)$$

in which case the only possible eigenvalues are $\lambda = 1$ and $\lambda = -1$. Weyl also considered the case in which the kernel has a purely continuous spectrum. This work was extended in a subsequent paper (Weyl 1908a) to the general case in which the kernel $K(x, y)$ determines a bounded quadratic functional

$$\int_0^\infty \int_0^\infty K(x, y)f(x)f(y)dxdy. \qquad (2.2.3)$$

In this case, there is both a discrete and a continuous spectrum. His work made use of Hilbert's results on the diagonalization of bounded quadratic forms

$$\sum_{i=1}^\infty \sum_{j=1}^\infty K_{ij}x_i x_j \qquad (2.2.4)$$

in infinitely many variables by means of an orthogonal transformation that preserves the metric

$$(x, x) = \sum_{i=1}^\infty x_i^2 < \infty. \qquad (2.2.5)$$

Detailed results for a number of specific cases were also given.

2.3 Ordinary Differential Equations with Singular Points

In the series of papers (Weyl, 1909c, 1910d, 1910e, 1911c) of which (Weyl, 1909c, 1910d) constituted his *Habilitationsschrift*, Weyl applied his theory of singular integral equations to the study of ordinary differential equations of the Sturm-Liouville type

$$L[u](x) \equiv D_x\big(p(x)D_x u(x)\big) - q(x)u(x) + \lambda r(x)u(x) = 0. \qquad (2.3.1)$$

In the nonsingular case, the domain is a finite closed interval $[a, b]$; $p(x)$ is continuously differentiable on $[a, b]$; $q(x)$ and $r(x)$ are continuous on $[a, b]$; and $p(x) > 0$

and $r(x) > 0$ on $[a, b]$. In addition, the solutions are required to satisfy the linear boundary conditions

$$\alpha_1 p(a) D_x u(a) + \alpha_2 u(a) = 0 \tag{2.3.2}$$

and

$$\beta_1 p(b) D_x u(b) + \beta_2 u(b) = 0, \tag{2.3.3}$$

where α_1, α_2, β_1, and β_2 are fixed real numbers. The operator L is self-adjoint; that is,

$$\int_a^b L[u](x)v(x)dx = \int_a^b u(x)L[v](x)dx. \tag{2.3.4}$$

The eigenvalues $\lambda_1, \lambda_2, \ldots$ form an increasing sequence with $\lambda_k \to \infty$ as $k \to \infty$, and the eigenfunctions $u_k(x)$ satisfy the orthonormality conditions

$$\int_a^b r(x)u_j(x)u_k(x)dx = \delta_{jk}. \tag{2.3.5}$$

Weyl considered the singular case for which the conditions stated above fail at one or both of the endpoints or for which one or both of the endpoints is infinite.

That a continuous spectrum was a possibility in the singular case was not a surprise in view of his prior work on singular integral equations; however, Weyl discovered novel phenomena that were not anticipated. Among these was the puzzling fact that at a singular end a boundary condition is required for some equations but not for others. Weyl resolved this problem in his *Habilitationsschrift* (Weyl 1910d, second paper). Consider the equation (2.3.1) with $r(x) \equiv 1$ for the finite interval $[0, R]$ where the end $x = R$ is (possibly) singular. Suppose $u_1(x)$ and $u_2(x)$ are solutions corresponding to the eigenvalues λ_1 and λ_2. Then Green's formula,

$$(\lambda_1 - \lambda_2) \int_0^R u_1(x)u_2(x)dx =$$
$$[p(x)u_1(x)D_x u_2(x) - p(x)D_x u_1(x)u_2(x)]_0^R, \tag{2.3.6}$$

shows that $u_1(x)$ and $u_2(x)$ are necessarily orthogonal if $\lambda_1 \neq \lambda_2$ since the right side of (2.3.6) vanishes as a consequence of the boundary conditions. Moreover, suppose that $u(x)$ is a complex valued eigenfunction with a complex (not real) eigenvalue λ and that $u(x)$ satisfies a real linear boundary condition at $x = 0$; then, (2.3.6) gives

$$(\lambda - \bar{\lambda}) \int_0^R u(x)\bar{u}(x)dx = p(R)[u(R)D_x \bar{u}(R) - D_x u(R)\bar{u}(R)]. \tag{2.3.7}$$

Since $\lambda \neq \bar{\lambda}$ and the integral is positive definite, $u(x)$ cannot satisfy a real linear boundary condition at $x = R$ unless $u(x)$ vanishes identically. Let $\phi(x)$ and $\theta(x)$ for $x \geq 0$ be the solutions to (2.3.1) (with $r(x) \equiv 1$) for a fixed λ in the upper half plane ($\Im\lambda > 0$) corresponding to the boundary conditions

$$\phi(0) = 1, \quad p(0)D_x \phi(0) = 0 \tag{2.3.8}$$

and

$$\theta(0) = 0, \quad p(0)D_x\theta(0) = 1, \tag{2.3.9}$$

respectively. For complex $l = l_1 + il_2$, consider the solution

$$\psi(x) = l\phi(x) - \theta(x). \tag{2.3.10}$$

The condition that $\psi(x)$ satisfy a real linear boundary condition at $x = R$ is

$$p(R)D_x\psi(R) = \gamma\psi(R), \tag{2.3.11}$$

where $\gamma \in \mathbf{R}$. From (2.3.10) and (2.3.11), one obtains

$$l = \frac{p(R)D_x\phi(R) - \gamma\phi(R)}{p(R)D_x\theta(R) - \gamma\theta(R)}. \tag{2.3.12}$$

Since $\phi(x)$ and $\theta(x)$ are linearly independent solutions of (2.3.1) for $x \geq 0$, their Wronskian cannot vanish; consequently, (2.3.12) is an invertible Möbius transformation that maps $\mathbf{R} \ni \gamma$ into a circle in the complex l-plane. Using Green's formula (2.3.6), the boundary conditions (2.3.8) and (2.3.9), and the fact that $\psi(x)$ satisfies a real linear boundary condition at $x = R$, one obtains

$$(\lambda - \bar{\lambda}) \int_0^R \psi(x)\bar{\psi}(x)dx = l - \bar{l}, \tag{2.3.13}$$

which is the constraint form of the equation of the circle (Note the l dependence of ψ given by (2.3.10)). Since the integral is a positive definite quadratic in the components of l, the interior of the circle is characterized by

$$\int_0^R \psi(x)\bar{\psi}(x)dx \leq \frac{\Im l}{\Im \lambda}. \tag{2.3.14}$$

Since $\Im\lambda > 0$, it follows that $\Im l > 0$. For fixed l, the integral on the left side of (2.3.14) is an increasing function of R. Thus, if l is on the circle corresponding to R, l is outside the circle corresponding to $R + \Delta R > R$. Therefore, for increasing R the circles are nested and have decreasing radii, and the circles must converge as $R \to \infty$ either to a *limit point* or to a *limit circle*. Weyl also showed that the kind of limit that occurs is not dependent on the value of λ chosen in the upper half plane. This independence is a result of the bijective character of the Möbius transformation (2.3.12) corresponding to λ from which it follows that for any given R, the circles corresponding to any two distinct values of λ are in bijective correspondence.

Weyl also studied the expansion of an arbitrary function $f(x)$, square integrable on $[0, \infty)$, in terms of the eigenfunctions of (2.3.1). In the nonsingular case, the expansion takes the form of a series involving a sum over the discrete sequence of eigenvalues $\lambda_1, \lambda_2, \ldots$. Weyl reformulated this sum as a Stieltjes integral, a form that can accommodate both a point spectrum and a continuous spectrum. He then proceeded to establish the correctness of the resulting formula by a direct method rather than by passing to the limit $R \to \infty$. He obtained the following results. Any eigenfunction corresponding to an eigenvalue λ with $\Im\lambda \neq 0$ is square integrable on $[0, \infty)$. In the limit point case, there is one and only one eigenfunction for each such λ so that no boundary condition at infinity can be imposed. In the limit circle case, there exists for each such λ an eigenfunction corresponding to each point on the limit circle, so that a boundary condition at infinity is required. In the limit point case, the real spectrum has in general both a discrete and a continuous part, and the eigenfunction expansion takes the form of a Stieltjes integral over the spectrum. Only those eigenfunctions corresponding to eigenvalues in the discrete spectrum are square integrable on $[0, \infty)$. In the limit circle case, the real spectrum is purely discrete, the eigenfunctions are square integrable on $[0, \infty)$ and the eigenfunction expansion has the form of a series.

The result concerning the number of boundary conditions that are required in the singular case is now formulated in terms of deficiency indices; it has been extended to operators of order m by Kodaira (1950) and by Dunford and Schwartz (1963). The results concerning the eigenfunction expansions were subsequently refined most notably by Stone (1932), Titchmarsh (1958) and Kodaira (1950). Titchmarsh and Kodaira provided the formula needed to compute the spectral measure $d\rho(\lambda)$ and thus completed the theory. This theory permits a unified treatment of the various classical expansion theorems associated with the names Fourier, Hermite, Laguerre and Bessel. In his Gibbs lecture, Weyl (1950b) concluded his discussion of the history of this problem with the comment, "It is remarkable that forty years had to pass before such a thoroughly satisfactory direct treatment emerged; this fact is a reflection on the degree to which mathematicians during this period got absorbed in abstract generalizations and lost sight of their task of finishing up some of the more concrete problems of undeniable importance."

2.4 Asymptotic Distribution of the Eigenvalues of Continuous Media

In a series of papers (Weyl, 1911b, 1912a, 1912b, 1912c, 1913b, 1915), Weyl analysed a problem of great importance to mathematical physics. The problem was to determine the asymptotic distribution of eigenvalues, that is the number $N(\lambda)$ of eigenvalues less than λ as $\lambda \to \infty$, of an elliptic partial differential operator (such as the Laplacian) for a given domain D and for given linear boundary conditions on ∂D. Given the interest in the theory of radiation and in the theory of continuous media at the time, it is not surprising that the physicists Arnold Sommerfeld and H. A. Lorentz emphasized the importance of this problem as noted by Weyl (1911b). As pointed out above, Hilbert had realized that such problems could be reformulated in terms of the integral equation (2.1.19), and it was natural for Weyl, a student of Hilbert, to attack the problem from this point of view.

In the first paper (Weyl 1911b), the method used and the results obtained for the two dimensional case are discussed briefly. The series of papers (Weyl, 1912a, 1912b, 1912c, 1913b) present a detailed analysis of the equation (2.1.16) in two and three dimensions. For the boundary conditions (2.1.17), he shows that the asymptotic form of $N(\lambda)$ is

$$N(\lambda) \sim \frac{\lambda A}{4\pi} \qquad (2.4.1)$$

in two dimensions, where A is the area of the domain D, and is

$$N(\lambda) \sim \frac{\lambda^{3/2} V}{6\pi^2} \qquad (2.4.2)$$

in three dimensions, where V is the volume of the domain D. He also shows that these asymptotic formulas are independent of both the shape of the region D and the type of linear boundary condition imposed on ∂D, and gives error bounds for these formulas. In addition, the results are generalized to the case of a self-adjoint elliptic differential operator with variable coefficients. His analysis of the much more important and also much more complicated case of electromagnetic radiation in a cavity with perfectly reflecting walls leads again to the result (2.4.2) except for a factor which allows for additional polarization degrees of freedom of the electromagnetic field. In the last paper, Weyl obtains similar results for the equally complicated case of a vibrating elastic solid.

The general strategy employed by Weyl is as follows. Consider the case of a two dimensional membrane. Denote by $G(P, Q)$ the Green function kernel for the domain D for the given boundary conditions. Divide D into two subdomains D_1 and D_2 by a line and denote by $G_1(P, Q)$ and $G_2(P, Q)$ the Green functions for these subdomains. Extend the kernels G_1 and G_2 to all of D by setting $G_i(P, Q) = 0$ if either P or Q is not in D_i. Then,

$$G = G_1 + G_2 + (G - (G_1 + G_2)), \qquad (2.4.3)$$

where $G - (G_1 + G_2)$ is positive definite. Weyl proves that the nth eigenvalue λ_n of G and the nth eigenvalue Λ_n of $G_1 + G_2$ satisfy

$$\lambda_n \leq \Lambda_n \qquad (2.4.4)$$

and

$$\lim_{n \to \infty} \frac{\lambda_n}{\Lambda_n} = 1. \qquad (2.4.5)$$

Note that λ is proportional to ν^2 where ν is the frequency; consequently, the relation (2.4.4) is physically reasonable because the additional constraint imposed on the eigenfunctions of $G_1 + G_2$ effectively stiffens the membrane with the result that the eigenfrequencies will be higher. Also, at higher frequencies the eigenfunctions oscillate more on the domain D; consequently, a progressively smaller adjustment is required to satisfy the additional constraint, with the result, expressed by (2.4.5), that the asymptotic distribution of eigenvalues remains unchanged. Suppose that the domain D of area A is covered by a net of squares D_i with sides of length a. The number of squares is A/a^2. The eigenvalue problem for a square may be solved exactly without difficulty. The eigenvalues are given by

$$\lambda_{mn} = \frac{\pi^2}{a^2}(m^2 + n^2), \qquad (2.4.6)$$

where m and n are positive integers. For each such square, the number of eigenvalues $\lambda_{mn} \leq \lambda$ is the number of lattice points (m, n) that satisfy

$$m^2 + n^2 \leq \frac{\lambda a^2}{\pi^2}, \quad m > 0, \ n > 0. \qquad (2.4.7)$$

Since each point is shared equally by the four unit squares of which it is a vertex, this number is approximately the area of the quarter disc with radius $(\lambda a^2/\pi)^{1/2}$. Thus the total number $N(\lambda)$ of eigenvalues less than λ is given by

$$N(\lambda) = \frac{A}{a^2}\frac{1}{4}\pi\frac{\lambda a^2}{\pi^2}, \qquad (2.4.8)$$

which is the result (2.4.1). In three dimensions, the result (2.4.2) is obtained in a similar way by dividing the domain D into cubic subdomains D_i by slicing D with planes.

A lemma that Weyl used in the proof of (2.4.4) was later applied by Courant to a wide range of problems, as noted by Chevalley and Weil (1968). Weyl considered the general case of a quadratic form on the space of functions that are square integrable on a domain D, determined by a square integrable, symmetric kernel according to (2.1.25). He allowed for the possibility of both positive and negative eigenvalues. Only the simple case, however, of a positive definite quadratic form defined on a finite dimensional vector space with inner product $f \cdot g$ will be discussed here. Denote by $u_1, u_2, \ldots, u_n$ the eigenvectors of the operator K that satisfy (2.1.20) and (2.1.21). The eigenvalues λ_i are ordered so that $\lambda_1 \leq \lambda_2 \leq \lambda_3 \leq \ldots$. It is convenient to set $\mu_i = 1/\lambda_i$. If the arbitrary vector f is expanded in terms of the basis u_i according to

$$f = \sum_{i=1}^{n} f_i u_i, \qquad (2.4.9)$$

then the value of the quadratic form on f is

$$K[f] = \sum_{i=1}^{n} \mu_i f_i^2. \qquad (2.4.10)$$

Since $\mu_1 \geq \mu_2 \geq \mu_3 \geq \ldots$, it is clear that μ_1 is the maximum of (2.4.10) subject to the constraint

$$\|f\|^2 = f \cdot f = \sum_{i=1}^{n} f_i^2 = 1. \qquad (2.4.11)$$

Also, μ_r is the maximum of $K[f]$ subject to the constraint (2.4.11) and the additional constraints

$$f \cdot u_1 = f \cdot u_2 = \cdots = f \cdot u_{r-1} = 1. \tag{2.4.12}$$

Of course, if some of the μ_i are the same, there will be some arbitrariness in the choice of the corresponding eigenvectors. To characterize μ_r without prior determination of the μ_i and u_i for $1 \le i \le r-1$, Weyl used the following principle. For any set of $r-1$ vectors $g_1, g_2, \ldots, g_{r-1}$, the maximum value of $K[f]$ subject to the constraint (2.4.11) and to the constraints

$$f \cdot g_1 = f \cdot g_2 = \cdots = f \cdot g_{r-1} = 0 \tag{2.4.13}$$

is greater than or equal to μ_r. The subspace spanned by the g_i is at most $r-1$ dimensional. If this subspace is identical to the subspace spanned by the first $r-1$ eigenvectors, then the maximum value obtained is μ_r itself. If the subspace spanned by the g_i is not identical to the subspace spanned by the first $r-1$ eigenvectors, then there exists a nonzero vector

$$f = \sum_{i=1}^{r-1} f_i u_i \tag{2.4.14}$$

which satisfies the constraints (2.4.13) and which may be normalized so that

$$\sum_{i=1}^{r-1} f_i^2 = 1. \tag{2.4.15}$$

For this vector,

$$K[f] = \sum_{i=1}^{r-1} \mu_i f_i^2 \ge \mu_r \sum_{i=1}^{r-1} f_i^2 = \mu_r. \tag{2.4.16}$$

Thus μ_r is the minimum of the set of maximum values of $K[f]$ on all subspaces of codimension less than r.

Let $G[f]$ and $K[f]$ be quadratic forms, chosen here to be positive definite for simplicity, and suppose that

$$G[f] = K[f] + H[f], \tag{2.4.17}$$

where $H[f]$ is positive definite (see (2.4.3)). Let the eigenvalues of G be λ_i and those of K be Λ_i. Consider any subspace of codimension less than r. The maximum of $G[f]$ on this subspace is greater than or equal to the maximum of $K[f]$ on this subspace since $H[f]$ is positive definite. Thus, the minimum of the set of all such maxima for $G[f]$ is greater than or equal to the minimum of the set of all such maxima for $K[f]$. Thus,

$$\frac{1}{\lambda_r} \ge \frac{1}{\Lambda_r}, \tag{2.4.18}$$

which indicates the way in which the principle was used to prove (2.4.4).

2.5   The Gibbs Phenomenon, The Equidistribution of Numbers mod 1 and the Problem of Mean Motion

In three papers (Weyl, 1910a, 1910c, 1911a), Weyl applied his considerable experience with various eigenfunction expansions to several examples of the Gibbs phenomenon. In its original form, the Gibbs phenomenon was concerned with the way in which the Fourier series of the function

$$H(x) = \begin{cases} 0 & -\pi < x < 0 \\ \frac{1}{2} & x = 0 \\ 1 & 0 < x < \pi \end{cases} \tag{2.5.1}$$

converges. Denote by $H_n(x)$ the nth partial sum of the Fourier series of $H(x)$. The limit of the graphs of the partial sums $H_n(x)$ includes not only the two horizontal lines but also a vertical line segment that extends from slightly below 0 to slightly above 1. Gibbs pointed out this 'overshoot' phenomenon in a letter to the editor of *Nature* (27 April 1899). On any closed interval containing the point of discontinuity such as $[-\frac{\pi}{2}, \frac{\pi}{2}]$, the sequence of functions

$$H(x) - (H_n(x) - \mathrm{Si}(nx)), \tag{2.5.2}$$

where

$$\mathrm{Si}(x) = \frac{1}{\pi} \int_{-\infty}^{x} \frac{\sin(\xi)}{\xi} d\xi, \tag{2.5.3}$$

converges uniformly to zero.

In the first paper, Weyl (1910a) considered the following case. Suppose the surface of a sphere is divided into two open regions, designated A and B, by a closed curve C which has neither points of self-intersection nor points of self-tangency and which has a continuously varying tangent. Let f_A and f_B be continuously differentiable on the closed sets $A \cup C$ and $B \cup C$ respectively and define f by

$$f(p) = \begin{cases} f_A(p) & p \in A \\ \dfrac{f_A(p) + f_B(p)}{2} & p \in C \\ f_B(p) & p \in B \end{cases} \tag{2.5.4}$$

If $f_n(p)$ is the nth order spherical harmonic expansion of $f(p)$, then at any point of the curve C the nature of the convergence for a small closed segment of a great circle that is normal to the curve C at the point is the same as that given by (2.5.2); that is, nothing new occurs. However, in the second paper, Weyl (1910c) analysed the case in which the tangent does not vary continuously along C. Weyl showed that at a point at which C has a cusp, the nature of the convergence can be described by a superposition of two Si functions. He found that the behaviour is much more dramatic if the curve C has distinct one sided tangents (a corner). Two contour graphs illustrate the mountainous terrain that the graph of $f(p) - f_n(p)$ can develop.

In this same paper, he proved a significant number-theoretic result in connection with the analysis of the Gibbs phenomenon in a one-dimensional heat problem for a circular ring consisting of two semicircles with different heat conductivities α and β which satisfy, as a result of a normalization convention, the constraint $\alpha + \beta = 1$. In his discussion of the distributions of eigenvalues for the problem, he first assumed that α (and hence β) is rational, and then he extended the result by approximating an irrational α by a sequence of rationals. He proved that for any real number α, there exists a sequence of rational numbers a_n/b_n for $(n = 1, 2, 3, \ldots)$ such that

$$\lim_{n \to \infty} n \left(\alpha - \frac{a_n}{b_n} \right) = 0, \tag{2.5.5}$$

and

$$\lim_{n \to \infty} \frac{b_n}{n} = 0. \tag{2.5.6}$$

This lemma also established that the sequence of multiples $n\alpha \bmod 1$ of an irrational number α is uniformly distributed on the interval [0,1) (identified with the circle). This result was also proved independently and almost simultaneously by Bohl and by Sierpinski (see (Weyl 1914b)). Bohl used this result to prove several important consequences for perturbations in a three-planet system.

A few years later, Weyl (1914b) re-analyzed the problem of the uniform distribution of numbers mod 1 in a much more general context. A sequence of real numbers α_i determines a sequence of real numbers β_i such that $0 \leq \beta_i < 1$. If $[\alpha_i]$ denotes the largest integer such that $[\alpha_i] \leq \alpha_i$, then

$$\beta_i = \alpha_i - [\alpha_i]. \tag{2.5.7}$$

For $0 \leq a < b < 1$, let $N_n(a, b)$ denote the number of the β_i for $1 \leq i \leq n$ which lie between a and b. By definition, the sequence α_i is uniformly distributed mod 1 if and only if

$$\lim_{n \to \infty} \frac{N_n(a, b)}{n} = b - a. \tag{2.5.8}$$

Then, a sequence α_i is uniformly distributed mod 1 if and only if for every bounded, Riemann integrable function $f(x)$ that is periodic with period 1,

$$\lim_{n \to \infty} \frac{1}{n} \sum_{k=1}^{n} f(\alpha_k) = \int_0^1 f(x)dx. \tag{2.5.9}$$

Partition [0,1] into m uniform intervals and suppose $n = lm \gg m$. Since f is periodic, $f(\alpha_k) = f(\beta_k)$. If the β_i are uniformly distributed, then approximately l of them will be in each of the intervals $((r-1)/m, r/m)$ and

$$\frac{1}{n} \sum_{k=1}^{n} f(\beta_k) \approx \frac{1}{m} \sum_{r=1}^{l} f(\frac{r}{m}), \tag{2.5.10}$$

which is the Riemann sum for the integral (2.5.9). On the other hand, consider the square pulse functions $H_{m,r}(x)$, which are periodic with period 1 and are defined on [0,1] by

$$H_{m,r}(x) = \begin{cases} 1 & x \in ((r-1)/m, r/m) \\ 0 & x \notin ((r-1)/m, r/m) \end{cases} . \tag{2.5.11}$$

Then $\sum_{k=1}^{n} H_{m,r}(\alpha_k)$ is the number of β_i in the interval $((r-1)/m, r/m)$; consequently, (2.5.8) follows from (2.5.9). The Fourier expansion theorem asserts that the functions $\exp(2\pi i k x)$ for integral k form a basis for periodic functions with period 1. It follows that (2.5.9) need be checked only for these functions; that is, one must show that for every integer k,

$$\lim_{n \to \infty} \frac{1}{n} \sum_{j=1}^{n} \exp(2\pi i k \alpha_j) = \int_0^1 \exp(2\pi i k x) dx. \tag{2.5.12}$$

For $k = 0$, the equation (2.5.12) holds for any sequence α_j. For $k \neq 0$, the integral on the right side of equation (2.5.12) is proportional to $\sin(\pi k)$ which is zero. In this way, Weyl proved that a sequence α_i is uniformly distributed mod 1 if and only if for every integer $k \neq 0$,

$$\lim_{n \to \infty} \frac{1}{n} \sum_{j=1}^{n} \exp(2\pi i k \alpha_j) = 0. \tag{2.5.13}$$

For the special case in which $\alpha_j = j\alpha$ where α is irrational, the summation in (2.5.13) is a geometric series. It follows from the formula for the sum of the series that

$$\left| \sum_{j=1}^{n} \exp(2\pi i k j \alpha) \right| \leq \frac{2}{|1 - \exp(2\pi i k \alpha)|}. \tag{2.5.14}$$

Consequently, (2.5.13) holds in this case and the multiples of an irrational number are uniformly distributed mod 1. Weyl noted that the criterion extends immediately to the case of a sequence of points $\alpha_j = (\alpha_{j1}, \alpha_{j2}, \ldots, \alpha_{jn}) \in \mathbf{R}^n$; that is, the corresponding $\beta_j = (\beta_{j1}, \beta_{j2}, \ldots, \beta_{jn})$ are uniformly distributed on the torus $[0,1)^n$ if and only if for every $k = (k_1, k_2, \ldots, k_n) \in \mathbf{Z}^n$ except $k = 0 \in \mathbf{Z}^n$,

$$\lim_{N \to \infty} \frac{1}{N} \sum_{j=1}^{N} \exp\left[2\pi i \sum_{r=1}^{n} k_r \alpha_{jr} \right] = 0. \tag{2.5.15}$$

Weyl noted that he had developed these ideas somewhat earlier but did not publish them because H. Bohr had shortly thereafter informed him of a more elementary method of proving the theorems in question (Weyl 1914b, pp. 489–490). He added that he had recognized immediately that his method applied to problems for which the method of Bohr-Rosenthal was not effective, such as the distribution mod 1 of a sequence $\alpha_j = j^s$ for some fixed power $s \in \mathbf{N}$, but that he did not exploit the method more fully until his attention was drawn to a paper in which Hardy and Littlewood discussed problems of a similar sort using arguments which they themselves characterized as 'intricate' (Weyl 1914b, p. 490). Weyl showed that his method has great power; for example, he proved that for any polynomial

$$\phi(x) = \sum_{k=1}^{r} a_k x^k, \quad a_k \in \mathbf{R}, \tag{2.5.16}$$

where a_r is irrational, then

$$\lim_{n \to \infty} \frac{1}{n} \sum_{j=0}^{n} \exp(2\pi i \phi(j)) = 0 \tag{2.5.17}$$

holds and therefore the sequence $\alpha_j = \phi(j)$ is uniformly distributed mod 1. In a subsequent paper (Weyl 1916b), the results were considerably generalized to allow for cases such as the following: a sequence $\phi(j)$ mod 1 generated by a polynomial (2.5.16) for which at least one of the coefficients a_k ($1 \le k \le r$) is irrational; a sequence $\phi(j) = (\phi_1(j), \phi_2(j), \dots, \phi_n(j))$ mod 1 generated by a set of n polynomials; and a sequence in n dimensions generated by polynomials depending on several variables. Weyl also proved that the curve on the torus $[0, 1)^n$ generated by

$$x_k \equiv \alpha_k + \gamma_k t \pmod{1}, \tag{2.5.18}$$

where $\alpha \in \mathbf{R}^n$ and $\gamma \in \mathbf{R}^n$ and the coordinates of γ do not satisfy a non-trivial linear equation with integer coefficients, is ergodic; that is, the fraction of time that x_k spends in a region of the torus is, for large time intervals, proportional to the volume of the region.

At the same time, Weyl was applying these results to the theory of planetary motion (Weyl 1914a). Consider a system of n planets with masses m_a revolving around a central star with a mass M that is assumed to be much greater than any of the m_a. As a consequence of this assumption, the centre of mass is very near the centre of the star; moreover, the mutual perturbations of the planets will be very small and the orbit of each planet will very nearly be an ellipse lying in a plane that passes through the centre of mass. The conservation of angular momentum about the centre of mass determines a *fixed* plane through the centre of mass. The total angular momentum vector is taken to be the positive normal and determines the positive sense of rotation in this plane, and a direction chosen arbitrarily in this plane fixes the zero for angles. The orbit of each of the planets is then determined by the following six parameters: the longitude of the ascending node $2\pi\omega$ or the polar angle of the point at which the orbit pierces the fixed plane while moving from its negative to its positive side; the inclination j of the orbit or the angle between the orbital plane and the fixed plane; the longitude of the perihelion $2\pi\sigma$ measured from the zero direction to the node in the fixed plane and from there to the perihelion in the orbital plane; the eccentricity r and the semimajor axis of the orbital ellipse; and the period of the orbit.

An appropriate measure of the smallness of the planetary masses is the ratio $\epsilon = \mu/M$ where μ is the reduced mass of the system. Significant perturbations of the orbital parameters occur only when the time is of the order of magnitude $1/\epsilon$; consequently, it is appropriate to replace the time $\tilde{t}$ measured in years (for the planet) by the secular time $t = \tilde{t}/\epsilon$. In the limit $\epsilon \to 0$, one obtains a system of coupled differential equations for the secular perturbations of the orbital parameters of the planets.

By forming the complex combinations

$$z = r \exp(2\pi i \sigma) \tag{2.5.19}$$

and

$$u = \sin(j) \exp(2\pi i \omega), \tag{2.5.20}$$

and by discarding nonlinear terms, a procedure that presupposes that the eccentricities and inclinations are very small, one obtains a system of equations of the form

$$D_t z_a = i \sum_b a_{ab} z_b, \tag{2.5.21}$$

where (a_{ab}) is a real, positive definite, symmetric matrix. Similar equations are obtained for the parameters u_a. The equations (2.5.21) can be decoupled by means of an orthonormal transformation of the coordinates z_a to $\tilde{z}_a$ which diagonalizes the matrix (a_{ab}). Then

$$\tilde{z}_a = \tilde{A}_a \exp[2\pi i a_a t], \tag{2.5.22}$$

and the inverse transformation gives

$$z_a = \sum_b A_{ab} \exp[2\pi i a_b t]. \tag{2.5.23}$$

The complex numbers A_{ab} may be expressed in polar form as $B_{ab} \exp[2\pi i b_{ab}]$ where $B_{ab} \in \mathbf{R}$ and $B_{ab} \geq 0$, and the index a may be suppressed if the perturbations of a single planet are under consideration. Then,

$$z = \sum_b B_b \exp[2\pi i (a_b t + b_b)]. \tag{2.5.24}$$

For $\theta = (\theta_1, \ldots, \theta_n)$ in the torus $[0, 1)^n$, set

$$f(\theta) = \sum_b B_b \exp[2\pi i \theta_b]. \tag{2.5.25}$$

Then,

$$z(t) = f(\theta(t)), \tag{2.5.26}$$

where

$$\theta_b(t) = a_b t + b_b, \tag{2.5.27}$$

and the above theorem (cf. (2.5.18)) states that

$$\lim_{T \to \infty} \frac{1}{T} \int_t^{t+T} f(\theta(t))dt = \int_0^1 \ldots \int_0^1 f(\theta)d\theta_1 \ldots d\theta_n, \qquad (2.5.28)$$

the right side of which is easy to compute.

From such considerations, Weyl was able to establish the following result: the perihelion of every planet shifts in the direction in which it revolves at a rate determined by a weighted average of the a_b. Much later in (Weyl, 1938a, 1939c), he returned to the problem and provided 'a complete solution'.

3   Riemann Surfaces

While still a *Privatdozent* at the University of Göttingen during the winter of 1911–1912, Weyl lectured on Riemann's theory of algebraic functions. A year later, he published the essential content of these lectures in his first book, *Die Idee der Riemannschen Fläche* (Weyl 1913a). Prior to this work, only the approach due to Weierstrass based on the analytic continuation of power series (function elements) was held to be rigorous; however, the view of a function as a collection of power series leaves much to be desired. As Weyl (1913a, 2 edn, 1951, p. 5) noted, "When one talks of $\sqrt{z}$ or $\log z$, one hardly envisages the totality of power series which represent pieces of these many-valued functions." It is not readily apparent, for example, whether two given power series are or are not related by analytic continuation. Riemann had provided a global perspective, but his treatment lacked rigour. Weyl succeeded in providing a rigorous foundation for Riemann's ideas. The result was a paradigm for future mathematical thought.

There were three editions of the work. The second edition was a reprint of the first and appeared in 1923 and another reprint of the second edition appeared in 1951. The third edition appeared in 1955 and involved substantial changes. Chevalley and Weil (1968, pp. 668–670) nicely characterize the differences between the first and third editions; moreover, their commentary provides useful information concerning the sources of inspiration and ideas for various aspects of Weyl's work[4].

C'est en élève de Hilbert encore, et en analyste, que Weyl dut aborder le sujet d'un des premiers cours qu'il professa à Göttingen comme jeune privatdozent, la théorie des fonctions selon Riemann. Le cours terminé et rédigé, il se retrouva géomètre, et auteur d'un livre qui devait exercer une profonde influence sur la pensée mathématique de son siècle. Peut-être s'était-il proposé seulement de remettre au goût du jour, en faisant usage des idées de Hilbert sur le principe de Dirichlet, les exposés traditionnels dont l'ouvrage classique de C. Neumann fournissait le modèle. Mais il dut lui apparaître bientôt que, pour substituer aux constants appels à l'intuition de ses prédécesseurs des raisonnements corrects et, comme on disait alors, "rigoureux" (et dans l'entourage de Hilbert on n'admettait pas qu'on trichât là-dessus), c'étaient avant tout les fondements topologiques qu'il fallait renouveler.

---

[4]Their remarks are, however, misleading in two respects. First, they say that Weyl did not make use of Brouwer's work; however, our discussion before and after remark 1 below clearly shows that Weyl did employ Brouwer's results in important ways. Secondly, Chevalley and Weil claim that Weyl's topological axioms served as a model for Hausdorff's work. However, Hausdorff (1914, reprint 1965, endnote for p. 210 on p. 456) states that he had lectured on the *Grundzüge* (basic ideas) of his theory of neighbourhoods already in the summer semester of 1912, which was before the publication of Weyl's (1913a) book. On the other hand, Weyl's lectures on Riemann's theory of algebraic functions during the winter semester of 1911–1912 presumably included a discussion of his topological axioms. For a discussion of the historical intricacies concerning Hausdorff's dependence on Weyl's work, see (Scholz 1996, p. 124) and (Scholz 1999a).

Weyl n'y semblait guère préparé par ses travaux antérieurs. Il pouvait, dans cette tâche, s'appuyer sur l'œuvre de Poincaré, mais il en parle à peine. Il mentionne, comme l'ayant profondément influencé, les recherches de Brouwer, alors dans leur première nouveauté; en réalité il n'en fait aucun usage. De fréquents contacts avec Koebe, qui dés lors s'était consacré tout entier à l'uniformisation des fonctions d'une variable complexe, durent lui être d'une grande utilité, particulièrement dans la mise au point de ses propres idées. La première édition du livre est dédiée à Félix Klein, qui bien entendu, comme Weyl le dit dans sa préface, ne pouvait manquer de s'intéresser à un travail si voisin des préoccupations de sa jeunesse ni de donner à l'auteur des conseils inspirés de son tempérament intuitif et de sa profonde connaissance de l'œuvre de Riemann. Bien qu'il n'eût jamais connu celui-ci, c'était Klein qui, à Göttingen, incarnait la tradition riemannienne. Enfin, dans l'un de ses mémoires sur les fondements de la géométrie, Hilbert avait formulé un système d'axiomes fondé sur la notion de voisinage, en soulignant qu'on trouverait là le meilleur point de départ pour "un traitement axiomatique rigoureux de l'analysis situs". De tous ces éléments si divers que lui fournissaient la tradition et le milieu, Weyl tira un livre profondément original et qui devait faire époque.

Le livre est divisé en deux chapitres, dont le premier contient la partie qualitative de la théorie. Les notions de "surface" (variété topologique de dimension 2 à base dénombrable) et de "surface de Riemann" (variété analytique complexe à base dénombrable, de dimension complexe 1) y sont définies au moyen de sytèmes d'axiomes, inspirés naturellement de celui de Hilbert, mais qui cette fois (sauf une légère omission dans la première édition) étaient destinés à subsister sans retouches, et devaient servir de modèle à Hausdorff pour son axiomatisation de la topologie générale. Dans la première et la deuxième édition, la condition de base dénombrable apparaît sous forme de condition de triangulabilité; et la triangulation joue un grand rôle dans la suite du volume; elle devait être éliminée entièrement de la troisième édition. Les questions touchant au groupe fondamental, au revêtement universel, à l'orientation, sont élucidées avec soin dans un esprit tout moderne, ainsi que les rapports entre propriétés homologiques et périodes des intégrales simples sur la surface. Dans la première et la deuxième édition, l'auteur va jusqu'à la construction, pour les surfaces orientables compactes, d'un système de "rétrosections", c'est-à-dire essentiellement d'une base privilégiée pour le premier groupe d'homologie; comme il le dit lui-même, il aurait pu, au prix d'un léger effort supplémentaire, aller jusqu'à la représentation de la surface au moyen d'un "polygone canonique" à $4g$ côtés (g désignant le genre), et à la détermination explicite du groupe fondamental, et on peut regretter qu'il ne l'ait pas fait. Mais la construction même des rétrosections, nécessairement basée sur la triangulation, disparaît dans la troisième édition, au profit d'un traitement plus purement homologique où n'interviennent que des recouvrements. En tout cas, pour tout l'essentiel, ce chapitre constitue une mise au point à peu prés définitive des questions qu'il traite.

[As a student of Hilbert and as an analyst, Weyl had to address the subject of one of the first courses that he taught at Göttingen as a young Privatdozent, the theory of functions according to Riemann. When the course was over and he had put his course notes in order, he had become a geometer and was the author of a book that would have profound influence on the mathematical thought of his century. Perhaps he had only intended to update in the fashion of the day, by making use of Hilbert's ideas on Dirichlet's principle, the traditional expositions of which the classic work of C. Neumann provided the model. But, it must soon have become evident to him that, in order to replace the constant appeals of his predecsessors to intuition with reasoning that was correct and, as one then said, "rigorous" (and in Hilbert's entourage, it was not acceptable to cheat in that regard), it was above all the topological foundations that had to be renewed. Weyl's previous work hardly prepared him at all for this task. For this project, he could have drawn from the work of Poincaré, but he hardly mentions it. He does mention, as having profoundly influenced him, Brouwer's research, which had recently appeared; however, in actual fact, he makes no use of it. Frequent contacts with Koebe, who was, from that time on, completely focussed on the uniformization of functions of a complex variable, must have been extremely useful to him, particularly in the development of his own ideas. The first edition of the book is dedicated to Felix Klein, who of course, as Weyl said in his preface, could not fail to be interested in a work so near to his own preoccupations as a young man or to give its author advice inspired by his own intuitive temperament and by his deep knowledge of Riemann's work. Although he had not known Riemann, it was Klein who, in Göttingen, incarnated the Riemannian tradition. Finally, in one of his papers on the foundations of geometry, Hilbert had formulated a system of axioms based on the notion of neighbourhood, emphasizing that one would find there the best point of departure for "a rigorous, axiomatic treatment of analysis situs". From all these diverse elements that were provided to him by tradition and by his environment, Weyl created a book that was profoundly original and groundbreaking.

This book is divided into two chapters, the first of which contains the qualitative aspects of the theory. The notions of "surface" (topological manifold of dimension 2 with a denumerable base) and of "Riemann surface" (complex analytic manifold with denumerable base and complex dimension 1) are defined by means of systems of axioms, naturally inspired by Hilbert's system, which this time (except for a minor omission in the first edition) were destined to stand without further modification, and would serve as a model for Hausdorff in his axiomatization of general topology. In the first and second editions, the condition of denumerable base appears in the form of a condition of triangulizability — and triangulation plays an important role in the rest of the volume — it was completely eliminated in the third edition. Questions pertaining to the fundamental group, the universal covering and orientation are elucidated carefully in a totally modern spirit, as well as the relationships between homological properties and periods of simple integrals on the surface. In the first and second editions, the author goes as far as the construction, for compact, orientable surfaces, of a system of "retrosections", that is, essentially a priviliged basis for the first homology group. As he says himself, he could have, with just a little more effort, gone as far as the representation of the surface by means of a "canonical polygon" with $4g$ sides (g designating the genus), and the explicit determination of the fundamental group, and one may regret that he did not do so. But, the very construction of the retrosections, necessarily based on triangulation, disappears in the third edition, in favour of a more purely homological treatment in which only coverings are used. In any case, as far as the essentials are concerned, this chapter constitutes an almost definitive development of the topics that it deals with.]

The discussion presented below is based on Weyl (1913a, 2 edn, 1951).

3.1 Analytic Function Elements

The starting point for Weyl's analysis was the method of analytic continuation of function elements developed by Weierstrass. If the complex valued function $w = u + iv = f(z)$ of the complex variable $z = x + iy$ is regular at $a \in \mathbf{C}$, that is, if $D_z f(a)$ exists, then f may be represented locally by a Taylor series

$$f(z) = A_0 + A_1(z-a) + A_2(z-a)^2 + \cdots \qquad (3.1.1)$$

which has a positive radius of convergence. This series either converges for all $z \in \mathbf{C}$, or there exists a disc $|z-a| < r$ of radius $r > 0$ centred on a such that the series converges for z inside the disc, diverges for z outside the disc and may converge or diverge for z on the rim of the disc. Weierstrass called such a power series a function element. The radius of convergence of such a series is determined by the location of the singularity nearest to a. With reference to Figure 1, consider the complex function

$$\frac{1}{1+z^2} = 1 - z^2 + z^4 - z^6 + \cdots. \qquad (3.1.2)$$

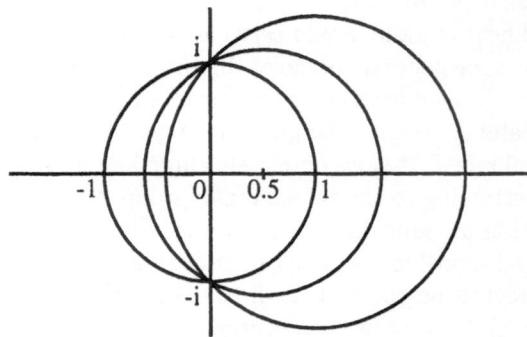

Figure 1: Analytic Continuation

The centre of the function element is $a = 0$, and the nearest singular points are at $z = \pm i$; hence, $r = 1$. It turns out that for algebraic functions, the isolated nature of the singular points is typical. By substituting

$$z = 1/2 + (z - 1/2) \tag{3.1.3}$$

in the function element (3.1.2), one obtains another function element centred on $a = 1/2$. The radius of convergence of the new series must be at least $1/2$ since this is the minimum distance from the new centre to the boundary of the old convergence disc; however, it is actually larger since the distance from the new centre to the nearest singular point ($+i$ or $-i$) is $r = \sqrt{5}/2$. Thus, the point $z = 1$ is inside the disc of convergence of the new function element and yet another function element can be obtained by means of the substitution

$$z - 1/2 = 1/2 + (z - 1), \tag{3.1.4}$$

and this function element will have a radius of convergence $r = \sqrt{2}$. Each such step is called an immediate analytic continuation, and the general process involving an arbitrary finite number of such steps is called analytic continuation. The set of function elements generated from a specific function element by analytic continuation has the property that the same set is generated by any one of its members.

An algebraic function $w = f(z)$ satisfies an equation of the form

$$w^n + p_1(z)w^{n-1} + p_2(z)w^{n-2} + \cdots + p_{n-1}(z)w + p_n(z) = 0, \tag{3.1.5}$$

where the $p_i(z)$ are polynomials in z. For most values of z, there correspond n distinct values of w and the function is multiple-valued. The multi-valued character of the function can be handled by the introduction of so-called uniformization parameters[5]. Consider the simple case

$$w^n - z^m = 0, \tag{3.1.6}$$

where m and n are mutually prime. For each value of $z \neq 0$, w takes on n distinct values. The multiple valued character of the function can be handled by considering pairs of series such as

$$z = t^n, \quad w = t^m, \quad |t| < r \tag{3.1.7}$$

and

$$z = t^{-n}, \quad w = t^{-m}, \quad |t| < r. \tag{3.1.8}$$

Note that (3.1.7) is valid in a neighbourhood of $z = 0$ while (3.1.8) is valid in a neighbourhood of ∞. Moreover, the relationship between z and w would not be altered if the uniformizing parameter t were replaced by

$$t = c_1 \hat{t} + c_2 \hat{t}^2 + c_3 \hat{t}^3 + \cdots, \tag{3.1.9}$$

where $c_1 \neq 0$ and the series is invertible and convergent in an open disc centred on $\hat{t} = 0$.

These considerations lead to the following generalization of the notion of function element. A pair of series $(P(t), Q(t))$ in integral powers of t each having at most finitely many negative powers of t is a representative of a function element if and only if[6]

1. $P(t)$ is not a constant.

2. Both $P(t)$ and $Q(t)$ converge for $|t| < r$ (with the possible exception of $t = 0$) where $r > 0$.

3. If $t_1 \neq t_2$ and $|t_1| < r$ and $|t_2| < r$, then $(P(t_1), Q(t_1)) \neq (P(t_2), Q(t_2))$.

A function element is an equivalence class $[P(t), Q(t)]$ of such pairs $(P(t), Q(t))$ any two of which are related by a transformation of the form (3.1.9). Analytic continuation of a function element is defined by the simultaneous analytic continuation of both members of a representative pair.

It is convenient to designate a standard representative for each equivalence class. Consider the three cases

$$z = P(t) = a + a_1 t + a_2 t^2 + \cdots \tag{3.1.10}$$

where $a_1 \neq 0$,

$$z = P(t) = a + a_m t^m + a_{m+1} t^{m+1} + \cdots \tag{3.1.11}$$

---

[5]The concept of a uniformization parameter was introduced prior to Weyl's work. The article on *Uniformization* in (Vinogradov 1993, Vol. 9, p. 321) mentions the work of F. Klein (1883), II. Poincaré (1907) and P. Koebe (1907).

[6]See (Weyl 1913a, 2 edn, 1951, pp. 5–6).

where $m > 1$ and $a_m \neq 0$, and

$$z = P(t) = a_{-n}t^{-n} + a_{-n+1}t^{-n+1} + \cdots \tag{3.1.12}$$

where $n > 0$ and $a_{-n} \neq 0$. The respective forms of the standard representatives are[7]

$$(a + \hat{t}, \hat{Q}(\hat{t})), \tag{3.1.13}$$

$$(a + \hat{t}^m, \hat{Q}(\hat{t})) \tag{3.1.14}$$

and

$$(\hat{t}^{-n}, \hat{Q}(\hat{t})). \tag{3.1.15}$$

Note that in the second and third cases, the standard representative is not unique; for example, $(a+\hat{t}^m, \hat{Q}(\zeta\hat{t}))$, where ζ is an mth root of unity, is as good as (3.1.14) since $\hat{t} \to \zeta\hat{t}$ is a transformation of the form (3.1.9).

The set of all possible function elements is partitioned into an infinite number of disjoint subsets called analytic forms, each of which contains all of the function elements that are mutually related by analytic continuation. Weyl showed that it was legitimate to regard these analytic forms as two-dimensional analytic manifolds, the Riemann surfaces. He was fully aware that it was far from obvious that the function elements $[P(t), Q(t)]$, equivalence classes of series, should be regarded as points on a surface. The task of justifying his point of view was all the more difficult because there did not then exist a rigorous axiomatic definition of surface that was sufficiently general to allow arbitrary elements to play the role of points.[8]

Roughly speaking, to each multiple-valued function $w = f(z)$ characterized by an equation of the form (3.1.5), there corresponds a unique Riemann surface, which Weyl showed was a two-dimensional, complex-analytic manifold. Locally, the multiple-valued function f is represented in parametric form by a pair of single valued functions

$$z = P(t) \quad \text{and} \quad w = Q(t), \tag{3.1.16}$$

where t is a local coordinate for the Riemann surface. There are, however, infinitely many analytic functions that give rise to the same Riemann surface, a fact which gives these surfaces an importance that is independent of any of the particular functions with which they are associated.

To see how function elements can be points, consider the simple case $w = \sqrt{z}$. The corresponding Riemann surface is a sphere. Suppose this sphere is mapped onto the complex t plane by stereographic projection from the south pole which corresponds to the point at infinity. At the north pole, there are two standard representatives of the function element

$$(t^2, t) \text{ and } (t^2, -t), \quad |t| < \infty, \tag{3.1.17}$$

---

[7]See equations 3, 3* and 3** in (Weyl 1913a, 2 edn, 1951, pp. 8–9).

[8]Prior to Weyl's work, surfaces had been realized essentially as submanifolds embedded in $\mathbf{R}^N$ for some $N > 2$.

which are equivalent under the transformation $t \to -t$. To describe the function element at the south pole, it is necessary to employ stereographic projection from the north pole. The orientation-preserving coordinate transformation is $s = t^{-1}$. The function element at the south pole has the two equivalent standard representatives

$$(s^{-2}, s^{-1}) \text{ and } (s^{-2}, -s^{-1}), \quad 0 < |s| < \infty. \tag{3.1.18}$$

From the standard representatives (3.1.17), one obtains by analytic continuation the function elements

$$[a^2 + 2at_1 + t_1^2, a + t_1] \tag{3.1.19}$$

and

$$[a^2 + 2at_1 + t_1^2, -a - t_1], \tag{3.1.20}$$

where $t = a + t_1$. These function elements are not equivalent. The function element (3.1.19) may be associated with the point on the sphere with coordinate $t = a$. Then the function element (3.1.20) corresponds to the point with $t = -a$. Thus each point on the sphere uniquely determines and is uniquely determined by a function element.

3.2 The General Concept of a Surface

Weyl based his definition of a general two-dimensional manifold or surface on an extension of a system of axioms that Hilbert (1902) had published. Hilbert's paper was published as an appendix in later additions of Hilbert (1899). In particular, see (Hilbert 1899, English transl., Appendix IV, 1980, pp. 150–151). Emphasizing that the objects that are the 'points of the manifold M' may be quite arbitrary, Weyl (1913a, 2 edn, 1951, pp. 17–18) formulated the axioms as follows. For each point p of the manifold M, there are certain subsets of points of the manifold that are defined to be neighbourhoods of p in M. Each neighbourhood U_0 of a point $p_0 \in M$ must contain p_0 itself. Moreover, there must exist an invertible, single valued mapping x of U_0 onto the interior points of a circular Euclidean disc K_0 such that p_0 is mapped into the centre of K_0 and such that:

1. If p is any point of U_0 and if U is a neighbourhood of p which is a subset of U_0, then the image $x(p)$ of p is in the interior of the image set $x_\vdash(U)$; that is, there exists a circular Euclidean disc K centred on $x(p)$ such that $K \subset x_\vdash(U)$.

2. If K is the interior of any circular disc with middle point[9] $x(p)$ such that $K \subset K_0$, then there exists a neighbourhood of $p \in M$ the image of which under x is a subset of K.

---

[9] Weyl used a gothic p for points in the manifold and a roman p for points in $\mathbf{R}^2$.

Weyl showed how the usual topological notions could be defined in terms of these axioms, in particular, the notions of continuity, of a continuous curve on M and of the topological equivalence of two manifolds.

For the case of Riemann surfaces, the objects that play the role of points are the function elements $[P(t), Q(t)]$ that belong to a particular analytic form. If $(P(t), Q(t))$ is a representative of a function element valid for $|t| < r$, then the substitution $t = t_0 + t_1$, where $|t_0| < r$, defines another function element. Weyl (1913a, 2 edn, 1951, p. 9) called the set of all function elements that could be obtained by such immediate analytic continuations (including the element $[P(t), Q(t)]$ itself) an analytic neighbourhood of $[P(t), Q(t)]$. The system of analytic neighbourhoods of a given function element is the set of all analytic neighbourhoods that can be obtained by using every representative and by considering every value of r compatible with the representative.

In the first paragraph of the preface and introduction of his book, Weyl (1913a, 2 edn 1951) stressed that he was greatly influenced by the then recent work of Brouwer. Weyl (1913a, 2 edn, 1951, p. 19) cites three of Brouwer's (1911a, 1911b, 1912) papers which deal with the invariance of dimension under homeomorphisms. Weyl (1913a, 2 edn, 1951, p. 20) notes that Brouwer's result can be used to replace the two conditions that Weyl used in the definition of a two-dimensional manifold by the requirement that each neighbourhood of a point $p \in M$ can be homeomorphically mapped into the interior of a circle such that the image of p is the centre of the circle. Weyl's definition of a two-dimenstional surface appears on pages 17–18. We have discussed this definition above in this subsection.

Remark 1 (Invariance of Dimension) In 1854, Riemann defined what is now called an n dimensional manifold. He did not consider the topological invariance of the dimension. The discovery by Cantor in 1877, that there exist bijective maps between $\mathbf{R}^m$ and $\mathbf{R}^n$ for $m \neq n$, showed the necessity of proving the topological invariance of dimension. The situation became more serious with Peano's discovery in 1890 of continuous mappings from $\mathbf{R}^m$ onto $\mathbf{R}^n$ for $m < n$ (the continuous space filling curve). It was only in 1911 that Brouwer (1911a, 1911c) succeeded in proving that homeomorphic manifolds were necessarily of the same dimension.

Weyl (1913a, 2 edn 1951, p. 21) imposed a further restriction on the notion of a Riemann surface, namely, that a surface admit a triangulation, that is, a covering by a countable number of regions that are smooth images of plane triangles and that overlap only along a shared edge or at a shared vertex. Weyl (1913a, 2 edn 1951, p. 22) states that his formulation of what it means for a manifold to admit a triangulation follows very closely Brouwer's (1911c) fundamental work. The method of triangulation plays an important role in the first and second editions of Weyl's book[10].

---

[10]It is clear from this and from the paragraph just before remark 1 that Weyl used Brouwer's work to a considerable extent. In view of this, it is rather odd that Chevalley and Weil, quoted above at the beginning of section 3, say, despite Weyl's claim to the contrary, that Weyl did not in fact make use of Brouwer's work.

Weyl treated a number of other topics in the first half of his book. In section 9, he presented a thorough and precise analysis of the concept of a *covering surface*, an idea first used by Poincaré. He also proved the *monodromy theorem* and discussed the special case of this theorem, the Cauchy *integral theorem*. The monodromy theorem asserts that if M is a simply connected surface, then analytic continuation of

$$z = a_{-m}t^{-m} + \cdots + a_0 + a_1 t + a_2 t^2 + \cdots, \qquad (3.2.1)$$

where t is a local uniformizing parameter (local analytic coordinate) such that $t = 0$ corresponds to a point p_0, generates on M a single valued function which is analytic except for poles provided that in the process of analytic continuation the only singular points encountered are poles. In addition, Weyl proved in section 10 that Riemann surfaces are orientable (two sided). Finally in section 11, Weyl considered the necessary and sufficient conditions for a closed one-form on M to be exact. This contribution will be discussed further below following equation (3.3.10).

3.3 Existence Theorems and the Dirichlet Principle

Weyl's principal aims in the first half of his book were to prove that corresponding to any algebraic function defined by an equation of the form (3.1.5), there exists a unique (compact) Riemann surface such that the algebraic function can be described by two single valued meromorphic functions (z, w) defined on the Riemann surface, and to discuss the most important properties of these surfaces. The main questions addressed in the second half of the book are first the existence of functions on a given compact Riemann surface with specified properties and second the problem of classifying and presenting models for the various types of compact Riemann surfaces. Weyl's analysis of the latter problem will be discussed in section 3.4.

Weyl based his treatment of these questions on the Dirichlet principle. Riemann had also employed this principle in his work. However, Weierstrass had pointed out that the existence of a minimizing function for the Dirichlet integral had not been proved. Fortunately, Hilbert had rescued the principle in 1900 and Weyl provided an elegant version of the proof in his book. The principle is used in the following way. Suppose that $w = u + iv = f(t)$ is a complex valued function on a Riemann surface for which $t = \rho + i\sigma$ is a local analytic coordinate (local uniformizing parameter). Then, the fact that f does not depend on $\bar{t}$, that is, that $D_{\bar{t}}f = 0$, yields the Cauchy-Riemann equations

$$D_\rho u = D_\sigma v, \quad D_\sigma u = -D_\rho v. \qquad (3.3.1)$$

For any not necessarily analytic function f,

$$
\begin{aligned}
df &= du + idv \\
&= (D_\rho u\, d\rho + D_\sigma u\, d\sigma) + i(D_\rho v\, d\rho + D_\sigma v\, d\sigma).
\end{aligned}
\qquad (3.3.2)
$$

Thus for an analytic function,

$$dv = -D_\sigma u \, d\rho + D_\rho u \, d\sigma = *du \tag{3.3.3}$$

and

$$df = du + i*du, \tag{3.3.4}$$

where $*$ denotes the Hodge star operator. Moreover, the equations (3.3.1) imply that both u and v are harmonic; that is, they satisfy Laplace's equation

$$\triangle u = 0 = \triangle v. \tag{3.3.5}$$

The formula (3.3.4) and the equation (3.3.5) suggest the following strategy for obtaining an analytic function on the Riemann surface. First, determine a real harmonic function u defined on the surface. Define the 1-form ω by

$$\omega = du + i*du. \tag{3.3.6}$$

This 1-form is closed since $ddu = 0$, and

$$d*du = \triangle u \, d\rho \wedge d\sigma = 0. \tag{3.3.7}$$

If the form ω is also exact, then the integral

$$\int \omega = u + i \int *du \tag{3.3.8}$$

is not only locally path independent but also globally path independent and defines an analytic function on the Riemann surface. The existence of the harmonic function u is guaranteed by the Dirichlet principle which asserts that there exists a function u that minimizes the Dirichlet integral

$$\int [(D_\rho u)^2 + (D_\sigma u)^2] \, d\rho \wedge d\sigma, \tag{3.3.9}$$

which must be harmonic.

As noted above, Weyl, at the end of the first part of his book, discussed the conditions under which a closed form is exact. The closed form $*du$ is exact if and only if for every closed curve γ on the Riemann surface

$$\oint_\gamma *du = 0. \tag{3.3.10}$$

Consider the simple case of a torus $S^1 \times S^1$ which has the shape of a doughnut. On this surface, there are two types of closed curves that cannot be continuously shrunk to zero, namely, a curve that passes once through the hole, and a curve that goes once around the hole without passing through it. If (3.3.10) holds for one curve of each type, then it holds for every closed curve γ. In his obituary notice for Hermann Weyl, Newman (1957, p. 313) evaluated this aspect of Weyl's work as follows:

Another of the new ideas which Weyl brought to his task had to wait more than twenty years to be independently rediscovered in more general form by topologists. This was the isolation of the topological part of the proof of the duality between the differentials and the 1-cycles on the surface. The 'curve-functions' introduced in §11 are 1-dimensional *co-chains* on the Riemann surface: the equation $F(\gamma) = f(p_2) - f(p_1)$ on p. 68 states precisely that F is the co-boundary of f, and shows that the symbol $F \sim 0$ has the meaning that is given to it in homology theory. The duality theorem, that the 1-dimensional connectivities derived from cycles and co-cycles are equal, is established in this section.

Weyl proved that every abstract, compact Riemann surface can be realized as the Riemann surface of some algebraic function; that is, on every abstract, compact Riemann surface there exist two globally defined meromorphic functions that are locally described by a pair of series $(z, w) = (P(t), Q(t))$ which satisfy (3.1.5) identically, and the Riemann surface determined by the algebraic function so defined is conformally equivalent to the original Riemann surface. He also provided a thorough treatment of the important theorems of Riemann-Roch and of Abel.

3.4   Classification of Compact Riemann Surfaces

Weyl completed his analysis of compact Riemann surfaces by providing a model for each of them up to a conformal equivalence. Weyl (1913a, 2 edn, 1951, p. 142) noted that the proof that such a classification is possible is based on the idea of the covering surface, and that the proof had been recently (1907) provided by Koebe and by Poincaré. The uniformization theorem states that the universal covering surface $\tilde{M}$ of any Riemann surface is conformally equivalent to the sphere, the plane, or the Lobachevskiïan plane. The Lobachevskiïan plane may be obtained by projecting the pseudosphere

$$t^2 - x^2 - y^2 = 1, \quad t > 0 \tag{3.4.1}$$

in a Minkowski spacetime with signature $(+, -, -)$ onto the plane $t = 0$ from the point $(-1, 0, 0)$. This is the analogue of stereographic projection for the sphere except that only points on the forward pseudosphere are considered.

A compact orientable surface is characterized topologically by its genus g. A sphere has genus $g = 0$. A sphere with g handles on it has genus g. Alternately, one may imagine a 'doughnut' with g holes in it.

The various Riemann surfaces are obtained as quotient manifolds $M = \tilde{M}/G$, where G is a group of conformal maps of $\tilde{M}$ onto itself such that G is discontinuous and only the identity element of G has a fixed point. If $\tilde{M}$ is the sphere, then G can consist only of the identity transformation and $M = \tilde{M}$ is a sphere ($g = 0$). If $\tilde{M}$ is the complex plane, there are three possibilities to consider, two of which lead to noncompact Riemann surfaces:

1. G consists of the identity element only. Then $M = \tilde{M}$ is the complex plane.

2. G consists of translations in some complex direction a,

$$t \to t + ma, \quad m \in \mathbf{Z}. \tag{3.4.2}$$

Then, $M = \tilde{M}/G$ is an infinitely long cylinder.

3. G consists of translations in two complex directions $a \neq 0$ and $b \neq 0$ such that $a \neq \lambda b$ for any $\lambda \in \mathbf{R}$,

$$t \to t + ma + nb, \quad (m, n) \in \mathbf{Z} \times \mathbf{Z}. \tag{3.4.3}$$

Then, $M = \tilde{M}/G$ is a torus (doughnut), a compact surface of genus 1.

For compact Riemann surfaces of genus $g > 1$, Weyl (1913a, 2 edn, 1951, p. 153) obtained the result.

> *Jeder Riemannschen Fläche entspricht demnach (von den vier vorher aufgezählten Ausnahmefällen abgesehen) eine einzige, völlig bestimmte, keine Drehungen enthaltende, diskontinuierliche Bewegungsgruppe Γ der Lobatschefskyschen Ebene. Zwei Riemannsche Flächen sind dann und nur dann konform-äquivalent (gehören, wie Riemann sich ausdrückt, derselben Klasse an oder sind Verwirklichungen einer und derselben idealen Riemannschen Fläche), wenn die zugehörigen Nicht-Euklidischen Bewegungsgruppen im Sinne der Lobatschefskyschen Geometrie kongruent sind. Umgekehrt gehört auch zu jeder diskontinuierlichen Nicht-Euklidischen Bewegungsgruppe ohne Drehungen eine bestimmte Klasse von Riemannschen Flächen.[11]*

The following translation is taken from (Weyl 1913a, 3 edn, 1955, p. 171).

> *Thus to every Riemann surface (aside from the four exceptions already listed) there corresponds a single uniquely determined discontinuous group Γ of motions of the Lobatschefskian plane; and Γ contains no rotations. Two Riemann surfaces are conformally equivalent (as Riemann expressed it, they belong to the same class or are realizations of one and the same ideal Riemann surface) if and only if the associated groups of non-Euclidean motions are congruent in the sense of Lobatschefskian geometry. Conversely, to every rotation-free discontinuous group of non-Euclidean motions there corresponds a definite class of Riemann surfaces.*

The Lobachevskiĭan plane is paved by congruent polygons with $2g$ sides, which are mapped onto each other by the action of the group G. In view of the elegance and sweeping generality of this result, it is not surprising that in the preface to the first edition of the book Weyl said,

---

[11]The emphasis is Weyl's.

Wir betreten damit den Tempel, in welchem die Gottheit (wenn ich dieses Bildes mich bedienen darf) aus der irdischen Haft ihrer Einzel-verwirklichungen sich selber zurückgegeben wird: in dem Symbol des *zweidimensionalen Nicht-Euklidischen Kristalls* wird das Urbild der Riemannschen Flächen selbst, (sowie dies möglich ist) rein und befreit von allen Verdunklungen und Zufälligkeiten, erschaubar.[12]

[We thus enter the temple in which the divine (if I may employ this metaphor) — liberated from its earthbound manifestations — is restored to itself: the very archetype of the Riemannian surface, pure and free of all obscurity and eccentricity, is (in so far as this is possible) revealed by its symbolic representation *in the two-dimensional non-Euclidean crystal.*]

3.5 Later Work on Combinatorial Topology and on the Theory of Analytic Curves

Weyl published a series of papers (Weyl, 1923a, 1923d, 1924a) written in Spanish that deal with problems in combinatorial topology. In his obituary notice, Newman (1957, p. 313) evaluated this work as follows:

It is convenient to mention here another contribution by Weyl to topology, though its connexion with continuous manifolds would have seemed remote when it was published. This was the short series of papers ... [(Weyl, 1923a, 1923d, 1924a)] written in Spanish, on 'combinatorial analysis situs', that is, the axiomatic theory of cell-complexes. A good deal of the material was of an expository character and Weyl himself seems to have attached little importance to the papers; but in fact this was the first appearance in the literature of a homology theory based on an axiomatic definition of abstract cell-complexes. Dehn and Heegaard had indeed ... developed a purely combinatorial theory, but their 'spheres' (cell-boundaries) were defined constructively, as complexes obtainable by allowed transformations from simplex-boundaries. Weyl took the opposite course of starting with an undefined boundary operation and imposing (in the form of axioms) such restrictions as were required. His postulate I_n contains, in somewhat involved form, the crucial condition $\partial\partial = 0$, but having his attention fixed from the start on invariance under subdivision, he did not bring out clearly the fact that this relation alone is a sufficient basis for a homology theory. He also considered the constructive method of defining spheres, and in this springtime of topology it was possible for him to say that, although the two methods failed to meet, it was to be hoped that

---
[12]Quoted from the forword and introduction of (Weyl 1913a, 2 edn, 1951, p. VI).

with progressive developments in analysis situs the gap would soon be closed.

Weyl also published two papers (Weyl, 1938, 1942) co-authored with his elder son Joachim Weyl and a book (Weyl 1943) in collaboration with his son Joachim Weyl on the theory of meromorphic functions and analytic curves. In the preface to the book, Weyl said:

> Five years ago my son Joachim and I discovered and brought home from the primeval forest of mathematics, a sapling which we called Meromorphic Curves (Annals of Mathematics, 1938). It looked healthy and attractive, but we did not know much about it. Soon after, a gardener from the North came along, — a skillful man of great experience, L. Ahlfors was his name, he knew; and under his care the plant, almost overnight, grew into a beautiful tree (Acta Soc. Sci. Fenn., 1941). Having learned the lesson, we set ourselves to carry out an idea dimly conceived before (Annals of Math. 1941, Proc. Nat. Acad. 1942), namely to transplant the tree Meromorph from the z-plane into the mountainous terrain of an arbitrary Riemann surface (a landscape of which I have been fond since the early days of my youth). The experiment seems to have succeeded. The leaves are out, a few buds are visible, but only the future can teach what fruits the tree will bear. In the meantime, the howling storm of war has cut us off from our wise gardener.

The analysis of algebraic functions leads to the introduction of a Riemann surface in terms of which the algebraic function is described by a pair of meromorphic functions defined on the surface and locally described by a pair of series $(z, w) = (P(t), Q(t))$ each of which has at most a finite number of negative powers of t. The negative powers of t may be dispensed with by the use of homogeneous coordinates. Then the algebraic function is described by three analytic functions locally described by

$$z_i = P_i(t), \quad i = 1, 2, 3, \tag{3.5.1}$$

where the three series

$$P_i(t) = a_{i0} + a_{i1}t + a_{i2}t^2 + \cdots \tag{3.5.2}$$

have a positive radius of convergence and at least one of the a_{i0} is not zero. Since the coordinates are homogeneous, a common positive power of t may be removed to ensure the last condition. The previous description is recovered by setting

$$z = \frac{z_1}{z_3} = \frac{P_1(t)}{P_3(t)} = P(t), \tag{3.5.3}$$

$$w = \frac{z_2}{z_3} = \frac{P_2(t)}{P_3(t)} = Q(t). \tag{3.5.4}$$

Clearly, there is no reason at all to restrict attention in (3.5.1) to just three analytic function elements. Weyl and Joachim Weyl developed the theory for the general case of n function elements. At first, they developed the theory for the case in which the domain of the curve was restricted to the complex plane, but later they generalized all of their main results to the case of curves defined *over* an arbitrary Riemann surface.

4   Spacetime

During the years 1917–1924, Weyl devoted a great deal of his energy to the development of the mathematical and philosophical foundations of Relativity Theory. Some of his most important contributions are:

1. The extension of Levi-Civita's concept of parallel transport by means of an intrinsic characterization of this notion that does not require an embedding into a flat, higher dimensional metric space. This intrinsic formulation led to the construction of significant portions of differential geometry and of dynamics on the basis of parallel transport as the fundamental notion besides and even independently of the metric ((Weyl 1918c, 4 and 5 edn) and (Weyl, 1918b, 1918d, 1919b, 1923c)).

2. The ingenious construction of a unified field theory of gravitation and electromagnetism in terms of a gauge-invariant geometry ((Weyl 1918c, 4th and 5th edition) and (Weyl, 1918b, 1919b, 1920c, 1921b, 1921e)).

3. The construction of generalized affine, projective and conformal geometries (Weyl, 1921f, 1922a, 1923c, 1929e and Robertson and Weyl (1929)), which led to subsequent developments in differential geometry such as the concept of connections on principal fiber bundles.

4. The clarification of the role and significance of invariance, symmetry and relativity principles and the clarification of the role of coordinates and the distinction between active and passive transformations (Weyl, 1927[sic]/1966, 1938b, 1939a, 1939b, 1949a, 1949b).

5. The deep group-theoretical results concerning the uniqueness of the Pythagorean form of the metric, which Weyl referred to as the 'Raumproblem'. This constitutes an important and interesting chapter in the history of the Riemann-Helmholtz-Lie-Weyl-Cartan problem of space (Weyl, 1922a, 1922b, 1923c, 1923f).

6. The discovery of the possibility of the geodesic or *causal-inertial* method for determining the spacetime metric by first distinguishing between two primitive substructures of the pseudo-Riemannian manifold structure, namely, the *conformal* structure (representing the *causal field* governing light propagation) and the *projective* structure (representing the *inertial* or *guiding field* governing all free (fall) motions) and then by showing that these structures uniquely determine the pseudo-Riemannian metric up to a constant positive factor (Weyl, 1921f, 1923c).

7. A realist field ontology of geometric structure and the analysis of the concept of motion and the role of Mach's Principle (Weyl, 1918c, 1918d, 1920b, 1921b, 1921c, 1922c, 1922a, 1923c, 1927[sic]/1966, 1931b, 1949a, 1950a).

8. The prediction and computation of the cosmological[13] red shift, based on Weyl's preferred de Sitter model, six years before the effect was empirically established by Hubble ((Weyl 1918c, 5 edn) and (Weyl, 1923b, 1930, 1934c)).

9. The invention of the idea of wormholes in connection with his analysis of mass in terms of electromagnetic field energy (Weyl, 1921c, 1924f).

Aside from his work in the foundations of mathematics, it is in his work on modern spacetime theory and related topics that Weyl's philosophical erudition manifests itself most clearly. Probably no other area of his work gives as clear a portrayal of Weyl the philosopher. It might plausibly be said in this regard that no other mathematician's work exemplifies as clearly, as concretely and as beautifully the fruitful and harmonious interplay of mathematics, physics and philosophy. His book *Raum, Zeit, Materie*, which is based on his lectures on Relativity Theory delivered in Zürich in 1917, does not merely provide an exposition of Einstein's theory. Rather, Weyl aims at an elucidation of the problem of space and time in general as this problem evolved within the history of mathematics, physics and philosophy. *Raum, Zeit, Materie* underwent a series[14] of expansions and revisions culminating in the great classical work of 1923, the fifth edition. In the preface to the fifth (German) edition, after mentioning the importance of mathematics to the work, Weyl remarks:

> Trotzdem verleugnet das Buch nicht seine philosophische Grundeinstellung: auf die *gedankliche Analyse* kommt es ihm an; die Physik liefert die Erfahrungsgrundlage, die Mathematik das scharfe Werkzeug. In der neuen Ausgabe ist diese Tendenz noch verstärkt worden; zwar das Geranke der Spekulation wurde beschnitten, aber die tragenden Grundgedanken wurden anschaulicher, sorgfältiger und vollständiger herausgearbeitet und zergliedert.

> [Despite this, the book does not disavow its basic, philosophical orientation: its central focus is *conceptual analysis*; physics provides the experiential basis, mathematics the sharp tools. In this new edition, this tendency has been further strengthened; although the growth of speculation was trimmed, the supporting foundational ideas were more intuitively, more carefully and more completely developed and analyzed.]

4.1 Weyl's Axioms for Flat Affine Geometry

In section 2 of his book *Raum, Zeit, Materie*, Weyl (1918c) provided a system of axioms for n-dimensional flat affine space. As pointed out by Rosenfeld (1988), Weyl began with Peano's axioms for an n-dimensional vector space.

---

[13] For a discussion of Weyl's contributions to cosmology see (Ehlers 1988) and the article by H. Goenner in this volume.

[14] The second edition was merely a reprint of the first edition.

Definition 2 *A vector space over the real field* **R** *is a set* $\mathcal{V}$ *together with two maps* $+ : \mathcal{V} \times \mathcal{V} \to \mathcal{V}$ *and* $\mathcal{S} : \mathbf{R} \times \mathcal{V} \to \mathcal{V}$ *called vector addition and scalar multiplication respectively. Vectors are denoted by bold upper case Roman letters and scalars are denoted by lower case Greek letters.*

The operation of vector addition, where $+(\mathbf{A}, \mathbf{B})$ *is written* $\mathbf{A} + \mathbf{B}$, *satisfies the axioms:*

VA1. *For any two vectors* **A** *and* **B**,

$$\mathbf{A} + \mathbf{B} = \mathbf{B} + \mathbf{A}.$$

VA2. *For any three vectors* **A**, **B** *and* **C**,

$$(\mathbf{A} + \mathbf{B}) + \mathbf{C} = \mathbf{A} + (\mathbf{B} + \mathbf{C}).$$

VA3. *If* **A** *and* **B** *are any two vectors, then there exists one and only one vector such that*

$$\mathbf{A} + \mathbf{X} = \mathbf{B}.$$

The vector **X** *is called the difference between* **B** *and* **A** *and it is denoted by* **B** $-$ **A**.

The operation of scalar multiplication, where $\mathcal{S}(\lambda, \mathbf{A})$ *is written* $\lambda \mathbf{A}$, *satisfies the axioms:*

SM1. *For any real numbers* λ *and* μ *and any vector* **A**,

$$(\lambda + \mu)\mathbf{A} = \lambda \mathbf{A} + \mu \mathbf{A}.$$

SM2. *For any real numbers* λ *and* μ *and any vector* **A**,

$$\lambda(\mu \mathbf{A}) = (\lambda \mu)\mathbf{A}.$$

SM3. *For any vector* **A**,

$$1\mathbf{A} = \mathbf{A}.$$

SM4. *For any real number* λ *and any vectors* **A** *and* **B**,

$$\lambda(\mathbf{A} + \mathbf{B}) = \lambda \mathbf{A} + \lambda \mathbf{B}.$$

From the axioms VA1–VA3, it follows that there exists a unique vector **0**, called the zero vector, such that for any vector **A**,

$$\mathbf{A} + \mathbf{0} = \mathbf{A}.$$

Linear dependence and linear independence of vectors are defined as follows.

Definition 3 (Linear Dependence) *A finite set of vectors* $\mathbf{A}_1, \mathbf{A}_2, \ldots, \mathbf{A}_k$ *is linearly dependent if and only if there exist* k *scalars* $\lambda_1, \lambda_2, \ldots, \lambda_k$ *at least one of which is not zero such that*

$$\lambda_1 \mathbf{A}_1 + \lambda_2 \mathbf{A}_2 + \cdots + \lambda_k \mathbf{A}_k = \mathbf{0}.$$

A finite set of vectors $\mathbf{A}_1, \mathbf{A}_2, \ldots, \mathbf{A}_k$ *is linearly independent if and only if they are not linearly dependent.*

The vector space $\mathcal{V}$ is said to be n-dimensional if and only if the following axiom is satisfied.

Axiom 4 (D: Axiom of Dimensionality) *There exist* n *vectors that are linearly independent, but every set of* $n + 1$ *vectors is linearly dependent.*

In an n-dimensional vector space $\mathcal{V}$, sets of n linearly independent vectors play a key role.

Definition 5 *A basis for the n-dimensional vector space* $\mathcal{V}$ *is a set* $\mathbf{B}_1, \mathbf{B}_2, \ldots, \mathbf{B}_n$ *of n linearly independent vectors.*

If $\mathbf{B}_1, \mathbf{B}_2, \ldots, \mathbf{B}_n$ is a basis for $\mathcal{V}$ and $\mathbf{A}$ is any vector in $\mathcal{V}$, then the $n + 1$ vectors $\mathbf{A}, \mathbf{B}_1, \mathbf{B}_2, \ldots, \mathbf{B}_n$ are necessarily linearly dependent; consequently, there exist $n + 1$ scalars $\lambda, \lambda_1, \lambda_2, \ldots, \lambda_n$ at least one of which is not zero such that

$$\lambda \mathbf{A} + \lambda_1 \mathbf{B}_1 + \lambda_2 \mathbf{B}_2 + \cdots + \lambda_n \mathbf{B}_n = \mathbf{0}. \tag{4.1.1}$$

Moreover, it must be the case that $\lambda \neq 0$. To see this, assume that $\lambda = 0$. Then,

$$\lambda_1 \mathbf{B}_1 + \lambda_2 \mathbf{B}_2 + \cdots + \lambda_n \mathbf{B}_n = \mathbf{0}. \tag{4.1.2}$$

Since the $\mathbf{B}_i$ are a basis, all of the scalars $\lambda, \lambda_1, \lambda_2, \ldots, \lambda_n$ must be zero; however, this contradicts the assumption that at least one of the scalars is not zero. One can, therefore, write

$$\mathbf{A} = A^1 \mathbf{B}_1 + A^2 \mathbf{B}_2 + \cdots + A^n \mathbf{B}_n, \tag{4.1.3}$$

where the A^i, called the components of $\mathbf{A}$ with respect to the basis $\mathbf{B}_i$, are given by

$$A^i = -\frac{\lambda_i}{\lambda}. \tag{4.1.4}$$

Remark 6 Weyl does not introduce the summation convention for tensor indices until section 6 of his book (Weyl 1918c, pp. 50–51); however, we will use it not only in this subsection but also in the rest of the section on spacetime. Thus, whenever an index appears in a term, once as an upper index and once as a lower index, it will be understood that a summation is to be carried out over the repeated index over the appropriate range; for example, equation (4.1.3) will be written

$$\mathbf{A} = A^i \mathbf{B}_i. \tag{4.1.5}$$

A flat affine space is very similar to an n-dimensional vector space except that it does not have a distinguished origin as a vector space does. Weyl defined a flat affine space as follows.

Definition 7 (Flat Affine Space) *An n-dimensional, flat affine space is a structure*

$$\mathcal{A} = \langle \mathcal{M}, \mathcal{V}, \Theta \rangle,$$

where M is a nonempty set of elements called points which are denoted by lower case Roman letters, $\mathcal{V}$ is an n-dimensional vector space, and the map $\Theta \colon M \times M \to \mathcal{V}$ satisfies the axioms:

A1. *For every $p \in M$ and every $\mathbf{A} \in \mathcal{V}$, there exists a unique $q \in M$ such that*

$$\Theta(p, q) = \mathbf{A}.$$

A2. *For any three points p, q and r in M,*

$$\Theta(p, q) + \Theta(q, r) = \Theta(p, r).$$

By instantiating both q and r to p in axiom A2, one readily sees that for every p in M, $\Theta(p, p) = 0$. Then, by instantiating r to p in A2, one sees that for every p and for every q in M

$$\Theta(p, q) = -\Theta(q, p). \tag{4.1.6}$$

Weyl introduced coordinate systems for M each of which is determined by a point $p \in M$ and a basis $\mathbf{B}_i$ for $\mathcal{V}$. Any point $q \in M$ determines a vector $\Theta(p, q)$ which can be expanded with respect to the basis $\mathbf{B}_i$ according to

$$\Theta(p, q) = x^i(q)\mathbf{B}_i; \tag{4.1.7}$$

moreover, axiom A1 guarrantees that the coordinates $x^i(q)$ uniquely distinguish the point q. Although Weyl did not do so, we shall call such coordinates *affine coordinates*.

Suppose that another system of affine coordinates $\bar{x}^i(q)$ is determined by a different point $\bar{p} \in M$ and another basis $\bar{\mathbf{B}}_i$ for $\mathcal{V}$. Then, the coordinates $\bar{x}^i(q)$ are determined by

$$\Theta(\bar{p}, q) = \bar{x}^i(q)\bar{\mathbf{B}}_i. \tag{4.1.8}$$

The relationship between the coordinates $x^i(q)$ and the coordinates $\bar{x}^i(q)$ is determined as follows. For the chosen points p and $\bar{p}$ and for any point q, it follows from axiom A2 that

$$\Theta(\bar{p}, q) = \Theta(\bar{p}, p) + \Theta(p, q). \tag{4.1.9}$$

It follows that

$$\bar{x}^i(q)\bar{\mathbf{B}}_i = \bar{x}^i(p)\bar{\mathbf{B}}_i + x^i(q)\mathbf{B}_i. \tag{4.1.10}$$

Since $\mathbf{B}_i$ and $\bar{\mathbf{B}}_i$ are both bases, each can be expanded in terms of the other, that is, there exist matrices of scalars X^i_j and $\bar{X}^i_j$ such that

$$\mathbf{B}_i = \bar{X}^j_i \bar{\mathbf{B}}_j \text{ and } \bar{\mathbf{B}}_i = X^j_i \mathbf{B}_j. \tag{4.1.11}$$

By substituting each of these equations into the other, one can readily show that

$$\bar{X}^i_k X^k_j = \delta^i_j \text{ and } X^i_k \bar{X}^k_j = \delta^i_j; \tag{4.1.12}$$

that is, X^i_j and $\bar{X}^i_j$ are inverses of each other. From (4.1.10) and (4.1.11), one obtains

$$\bar{x}^i(q)\bar{\mathbf{B}}_i = \bar{x}^i(p)\bar{\mathbf{B}}_i + x^j(q)\bar{X}^i_j\bar{\mathbf{B}}_i. \tag{4.1.13}$$

Since the $\bar{\mathbf{B}}_i$ form a basis, it follows that

$$\bar{x}^i(q) = \bar{x}^i(p) + \bar{X}^i_j x^j(q). \tag{4.1.14}$$

The components of the vector from the point q_1 to the point q_2 are $\bar{x}^i(q_2) - \bar{x}^i(q_1)$ and $x^i(q_2) - x^i(q_1)$ in the barred and unbarred systems respectively. It follows from (4.1.14) that

$$\bar{x}^i(q_2) - \bar{x}^i(q_1) = \bar{X}^i_j(x^j(q_2) - x^j(q_1)). \tag{4.1.15}$$

Weyl also discusses what would now be called an active affine transformation. In this case, a fixed coordinate system determined by a point $p \in M$ and a basis $\mathbf{B}_i$ for V is used, and a point q with coordinates $x^i(q)$ is transformed into a point $\tilde{q}$ with coordinates $x^i(\tilde{q})$ according to

$$x^i(\tilde{q}) = a^i + A^i_j x^j(q), \tag{4.1.16}$$

where the matrix A^i_j is invertible. The vector from the point q_1 to the point q_2 with coordinates $x^i(q_2) - x^i(q_1)$ is transformed into the vector from the point $\tilde{q}_1$ to the point $\tilde{q}_2$ with coordinates $x^i(\tilde{q}_2) - x^i(\tilde{q}_1)$ according to

$$x^i(\tilde{q}_2) - x^i(\tilde{q}_1) = A^i_j(x^j(q_2) - x^j(q_1)). \tag{4.1.17}$$

4.2 The Fundamental Notion of Parallel Displacement

The classical theory of physical geometry regarded the concept of 'metric congruence' as the only basic relation of geometry and developed physical geometry from this one notion alone in terms of the relative positions and displacements of physical congruence standards. However, the dynamical view of spacetime suggested by Einstein's General Theory motivated an approach to physical geometry that is very different from the metrical views of Helmholtz, Poincaré and Hilbert. It suggested that the affine geometry should be developed independently of the metric. After all, it is the affine connection that plays the essential role in the formulation of all physical laws that are expressed in terms of differential equations.

It is the affine connection that relates the state of a system at a spacetime point to the states at neighbouring spacetime events and enters into the differentials of the corresponding magnitudes. In both Newtonian physics and General Relativity, all dynamical laws presuppose the projective and affine structure and hence the Law of Inertia. In fact, the whole of tensor analysis with its covariant derivatives is based on the affine concept of infinitesimal parallel displacement and *not* on the metric.

Under the influence of the algebraic theory of invariants, Riemannian geometry had been studied as a theory of invariants of quadratic covariant tensors by E. B. Christoffel, C. G. Ricci and others. After Einstein applied Riemannian geometry to his General Theory of Relativity in 1916, Riemannian geometry became the focus of intense research. In 1917, T. Levi-Civita introduced the notion of infinitesimal transport, which helped clarify certain geometric properties of Riemannian geometry. In 1918, Weyl was able to generalize Levi-Civita's concept by introducing the affine structure on a manifold through an *intrinsic* notion of parallel displacement that does not require the use of a process of embedding the manifold into a flat higher dimensional metric space thereby freeing the concept of parallel displacement entirely from the dependence on a Riemannian metric and allowing for the possibility of founding an important part of differential geometry on this fundamental intrinsic notion alone. Weyl's generalization led to further generalizations of the geometric notion of a connection.

Weyl characterizes the notion of an *affine connection*[15] as follows:

Definition 8 (Affine Connection) *Let $T(M_p)$ denote the tangent space of M at $p \in M$. A point $p \in M$ is affinely connected with its immediate neighbourhood, if and only if for every vector $v_p \in T(M_p)$, a vector $v_{p'} \in T(M_{p'})$ is determined to which the vector $v_p \in T(M_p)$ gives rise under parallel displacement from p to the infinitesimally neighbouring point p'.*

Of the notion of parallel transport, Weyl requires that it satisfy the following condition.

Definition 9 (Parallel Displacement) *The transport of a vector $v_p \in T(M_p)$ to an infinitesimally neighbouring point $p' \in M$ constitutes a parallel displacement if and only if there exists a coordinate system $\bar{x}^i$ for the neighbourhood of $p \in M$ relative to which the transported vector $\bar{v}_{p'}$ possesses the same components as $\bar{v}_p$; that is,*

$$\bar{v}_{p'}^i - \bar{v}_p^i = d\bar{v}_p^i = 0. \tag{4.2.1}$$

Theorem 10 *If for every point p in a neighbourhood U of M, there exists a coordinate system $\bar{x}$ such that the change in the components of a vector under parallel transport to an infinitesimally near point p' is given by*

$$d\bar{v}_p^i = 0, \tag{4.2.2}$$

---

[15]Weyl used the term 'affine connection' for what is now called a symmetric, linear connection. In this overview, we use the two terms synonymously.

then locally in any other coordinate system x,

$$dv^i = -\Gamma^i_{jk}(x)v^j dx^k, \tag{4.2.3}$$

where $\Gamma^i_{jk} = \Gamma^i_{kj}$ *and conversely.*

Proof Consider an arbitrary point $p \in U$ and assume that

$$d\bar{v}^i_p = 0. \tag{4.2.4}$$

Since

$$\bar{v}^i = \bar{X}^i_j(x)v^j \tag{4.2.5}$$

locally,

$$d\bar{v}^i = \bar{X}^i_j(x)dv^j + \bar{X}^i_{jk}(x)v^j dx^k, \tag{4.2.6}$$

where the $\bar{X}^i_{jk}(x)$ are the second partial derivatives of the transformation functions $\bar{X}^i(x)$. Hence,

$$d\bar{v}^i_p = \bar{X}^i_j(x(p))dv^j_p + \bar{X}^i_{jk}(x(p))v^j_p dx^k(p). \tag{4.2.7}$$

From (4.2.4) and (4.2.7), one obtains

$$dv^i_p = -\bar{X}^{-1}{}^i_r(x(p))\bar{X}^r_{jk}(x(p))v^j_p dx^k(p). \tag{4.2.8}$$

Since p is an arbitrary point in U, it follows that locally

$$dv^i = -\Gamma^i_{jk}(x)v^j dx^k, \tag{4.2.9}$$

where $\Gamma^i_{jk}(x) = \Gamma^i_{kj}(x)$ because $\bar{X}^i_{jk}(x(p)) = \bar{X}^i_{kj}(x(p))$. Note that the structure of $\bar{X}^{-1}{}^i_r(x(p))\bar{X}^r_{jk}(x(p))$ as the sum of products of derivatives holds only on a pointwise basis; consequently, the $\Gamma^i_{jk}(x)$ do not inherit this structure.

To prove the converse, assume that (4.2.9) holds locally. Then, for an arbitrary point $p \in U$,

$$dv^i_p = -\Gamma^i_{jk}(x(p))v^j_p dx^k(p). \tag{4.2.10}$$

Consider the coordinate transformation[16]

$$\bar{x}^i = (x^i - x^i(p)) + (1/2)\Gamma^i_{jk}(x(p))(x^j - x^j(p))(x^k - x^k(p)), \tag{4.2.11}$$

which is valid in an open neighbourhood of p. Then,

$$\bar{X}^i_j(x(p)) = \delta^i_j \quad \text{and} \quad \bar{X}^i_{jk}(x(p)) = \Gamma^i_{jk}(x(p)). \tag{4.2.12}$$

From the locally valid result (4.2.6), one obtains (4.2.7). From (4.2.7), (4.2.12) and (4.2.3), it follows that at p

$$d\bar{v}^i_p = \delta^i_j dv^j_p + \Gamma^i_{jk}(x(p))v^j_p dx^k(p) = 0. \tag{4.2.13}$$

$\square$

---

[16] Weyl considered the transformation from $\bar{x}^i$ to x^i; however, the algebra is somewhat more straight-forward if one considers the transformation from x^i to $\bar{x}^i$.

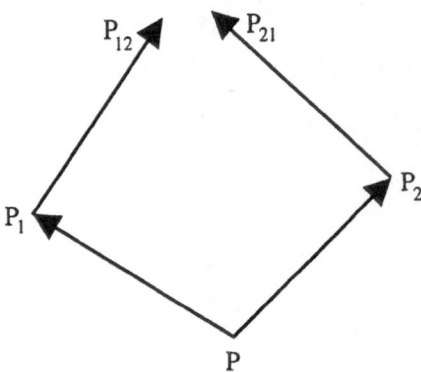

Figure 2: Symmetry of Parallel Transport

Remark 11 Weyl (1918d, §3.I.B) also provided a more synthetic argument to establish the symmetry of the affine connection. He considers two infinitesimal vectors $\vec{PP_1}$ and $\vec{PP_2}$ at a point P. The vector $\vec{PP_1}$ under parallel transport along $\vec{PP_2}$ goes into $\vec{P_2P_{21}}$. Similarly, the vector $\vec{PP_2}$ under parallel transport along $\vec{PP_1}$ goes into $\vec{P_1P_{12}}$. These relationships are illustrated in Figure 2. The condition imposed on parallel transport is that the four vectors $\vec{PP_1}$, $\vec{P_1P_{12}}$, $\vec{PP_2}$ and $\vec{P_2P_{21}}$ form a closed parallelogram; that is, the points P_{12} and P_{21} coincide. It follows that

$$\vec{PP_1} + \vec{P_1P_{12}} = \vec{PP_2} + \vec{P_2P_{21}}. \tag{4.2.14}$$

Denote the coordinates of $\vec{PP_1}$ and $\vec{PP_2}$ by dx^i and δx^i, respectively. The coordinates of $\vec{P_2P_{21}}$ and $\vec{P_1P_{12}}$ are respectively denoted by $dx^i + \delta dx^i$ and $\delta x^i + d\delta x^i$. Substitution into (4.2.14) yields

$$dx^i + \delta x^i + d\delta x^i = \delta x^i + dx^i + \delta dx^i, \tag{4.2.15}$$

or

$$d\delta x^i = \delta dx^i. \tag{4.2.16}$$

From the assumption that the vectors transform linearly, one has

$$\delta dx^i = -\delta\gamma_r^i dx^r \quad \text{and} \quad d\delta x^i = -d\gamma_r^i \delta x^r. \tag{4.2.17}$$

From the assumption that the infinitesimal transformation coefficients $\delta\gamma_r^i$ and $d\gamma_r^i$ are of the same order as the corresponding differentials δx^i and dx^i, one obtains

$$\delta\gamma_r^i = \Gamma_{rs}^i \delta x^s \quad \text{and} \quad d\gamma_r^i = \Gamma_{rs}^i dx^s. \tag{4.2.18}$$

Substitution of (4.2.17) and (4.2.18) into (4.2.16) yields

$$\Gamma_{jk}^i = \Gamma_{kj}^i. \tag{4.2.19}$$

The condition (4.2.1) imposed on the notion of a parallel transport at $p \in M$, which says that there exists a geodesic coordinate system for the neighbourhood of p, characterizes the *nature* of an affine connection on M. Moreover, a manifold M with an affine structure is homogeneous in the sense that no difference exists among the points $p \in M$ with regard to its affine nature. Nor do manifolds exist the (symmetric) affine structure of which is of a different nature.

A manifold M is said to be endowed with an affine structure if and only if at each point $p \in M$, there exists one and only one affine connection $\Gamma^i_{jk}(x(p))$ that may be characterized as the real one among the possible ones. A Riemannian manifold is therefore endowed with a definite connection $\Gamma^i_{jk}(x(p))$ at each $p \in M$ since the metric uniquely singles out from among the possible systems $\Gamma^i_{jk}(x(p))$ of parallel displacements one and only one system which is *length preserving*. Weyl called this the fundamental principle of differential geometry.

4.3   The Projective Geometry

Weyl's different approach to infinitesimal geometry, which treats the affine structure in its own right rather than as a mere aspect of the metric, led to the development of differential projective geometries or the geometries of paths.

One way to characterize the *projective structure* Π on a manifold M with an affine structure is in terms of an equivalence class of *geodesic* curves

$$\frac{d^2 x^i}{ds^2} + \Gamma^i_{jk} \frac{dx^j}{ds} \frac{dx^k}{ds} = 0 \qquad (4.3.1)$$

under arbitrary parameter diffeomorphisms, thus eliminating all the parameter descriptions and hence all possible notions of distance along the curves satisfying (4.3.1). Such an equivalence class of geodesic curves is called a geodesic *path*. While both curves and paths are one-dimensional submanifolds of M, a path does not depend in any way on the parameters used to describe it. The interest in projective geometry is, therefore, in the paths, that is, in the set of points of the curves that are solutions of (4.3.1) rather than in the possible parametric descriptions of the paths.

The approach that characterizes the projective structure in terms of an equivalence class of geodesic curves under parameter diffeomorphisms was taken mainly in the United States under the leadership of O. Veblen and T. Y. Thomas (Thomas 1925, Veblen and Thomas 1926). Essentially, an arbitrary parameter diffeomorphism of a solution curve of the equations (4.3.1) can be compensated for by a projective transformation of Γ given by

$$\Gamma^i_{jk} \rightarrow \bar{\Gamma}^i_{jk} = \Gamma^i_{jk} + \delta^i_j \phi_k + \delta^i_k \phi_j. \qquad (4.3.2)$$

Another characterization of the projective structure Π on an affine manifold takes the process of *autoparallelism of directions* as fundamental. This is the approach Weyl adopted in his lectures delivered in Barcelona and Madrid in the spring of 1922 ((Weyl 1923c); see also (Weyl 1921f)). Weyl began with the following necessary and sufficient condition for the invariance of the projective structure under a transformation $\Gamma \to \bar{\Gamma}$ of the affine structure:

Definition 12 (Projective Transformation) *A transformation $\Gamma \to \bar{\Gamma}$ preserves the projective structure of a manifold M with an affine structure if and only if*

$$(\bar{\Gamma} - \Gamma)^i_{jk} v^j v^k \propto v^i, \tag{4.3.3}$$

where v^i is an arbitrary vector.

The definition says that a change in the affine structure of M preserves the projective structure of M if and only if the vectors $v^i_{p'} \in T(M_{p'})$ and $\bar{v}^i_{p'} \in T(M_{p'})$ that result from $v^i_p \in T(M_p)$ by parallel transport under Γ and $\bar{\Gamma}$ respectively, differ at most in length but not in direction. Weyl then proved the following theorem:

Theorem 13 *A transformation $\Gamma \to \bar{\Gamma}$ is a projective transformation if and only if (4.3.2) holds, that is, if and only if*

$$(\bar{\Gamma} - \Gamma)^i_{jk} = \delta^i_j \phi_k + \delta^i_k \phi_j, \tag{4.3.4}$$

where ϕ_k is some covariant vector field.

If we choose[17] ϕ_k such that $\bar{\Gamma}_k = \bar{\Gamma}^r_{rk} = 0$ then $-\Gamma_k = (n+1)\phi_k$. Then, substitution for ϕ_k into (4.3.2), yields

$$\Pi^i_{jk} = \Gamma^i_{jk} - \frac{1}{(n+1)}(\delta^i_j \Gamma_k + \delta^i_k \Gamma_j), \tag{4.3.5}$$

where the projective coefficients Π^i_{jk} satisfy

$$\Pi^j_{jk} = \bar{\Gamma}_k = 0; \quad \Pi^i_{jk} = \Pi^i_{kj}. \tag{4.3.6}$$

These coefficients characterize the equivalence class $[\Gamma]$ of projectively equivalent connections, that is, connections equivalent under the *projective transformation* (4.3.2).

4.4 The Conformal and Weyl Geometries and Weyl's Unified Field Theory

Weyl regarded affine geometry as fundamental and developed projective geometry by an abstraction from affine geometry. He developed conformal geometry in a similar fashion by an abstraction from Riemannian, metric geometry. Riemannian geometry is characterized by a metric tensor field which is locally described by

$$ds^2 = g_{ij}(x)dx^i dx^j. \tag{4.4.1}$$

---

[17]This choice is coordinate dependent.

Such a metric field determines a unique affine connection Γ such that the metric field (4.4.1) is invariant under parallel transport. The coefficients of this connection are given by

$$\Gamma^i_{jk} = \frac{1}{2}g^{ir}(g_{rj,k} + g_{rk,j} - g_{jk,r}). \tag{4.4.2}$$

If v_p is a vector at $p \in M$, its length is given by

$$|v_p|^2 = g_{ij}(x(p))v^i_p v^j_p. \tag{4.4.3}$$

The angle between two such vectors v_p and w_p is given by

$$\cos\theta = \frac{g_{ij}(x(p))v^i_p w^j_p}{|v_p||w_p|}. \tag{4.4.4}$$

In Riemannian geometry, the parallel transport of length is path independent; consequently, it is possible to compare the lengths of any two vectors even if they are located at two different points. Under parallel transport, however, a vector suffers a path-dependent change in direction. Thus, it is not possible to define the angle between two vectors located at different points in a path-independent manner; moreover, the angle between two vectors at a given point is invariant under parallel transport if and only if both vectors are transported along the same path. Weyl desired a theory of *purely* infinitesimal geometry; consequently, he regarded the feature of Riemannian geometry that permits the comparison of length at a finite distance as unacceptable. He was therefore led to introduce the concept of gauge transformation which we now discuss.

For a conformal structure, the relative lengths of any two vectors (at the same point) $|v_p|/|w_p|$ and the angle between them (4.4.4) are determined, but the length of any one vector (4.4.3) at that point is arbitrary. Weyl expressed these facts by saying that the gauge could be chosen at each point in a smooth but otherwise arbitrary manner. Hence, the metric (4.4.1) is conventional to the extent that the metric

$$d\bar{s}^2 = \lambda(x)g_{ij}(x)dx^i dx^j \tag{4.4.5}$$

is equally valid where $\lambda(x)$, the gauge function, is positive and smooth but otherwise arbitrary. Weyl showed that for such a geometric structure, the connection is not uniquely determined. Rather, there exists an equivalence class of connections which preserve the conformal structure under parallel transport. The difference between any two connections in the equivalence class is given by

$$\bar{\Gamma}^i_{jk} - \Gamma^i_{jk} = \frac{1}{2}(\delta^i_j\theta_k + \delta^i_k\theta_j - g_{jk}g^{ir}\theta_r), \tag{4.4.6}$$

where

$$\theta_j(x)dx^j \tag{4.4.7}$$

is an arbitrary one-form field. In this type of geometry, the parallel transport of vectors is not generally well defined and the ratio of the lengths of two vectors located at different points is not determined even in a path-dependent way.

In the third edition (1920) of *Raum, Zeit, Materie* and in (Weyl, 1918b, 1919b), Weyl presented his ingenious attempt to unify gravitation and electromagnetism by constructing a gauge invariant geometry or what he called a *purely infinitesimal 'metric' geometry*. Although Weyl had to abandon his unified field theory with the advent of quantum theory, his gauge invariant differential geometry had great theoretical and aesthetic appeal, and a suitably modified notion of gauge invariance was soon applied to the quantum theory of the electron. The idea of gauge invariance was generalized by Yang and Mills (1954), and today the concept dominates the theories of all fundamental interactions.

In Einstein's geometrization of gravity, the Newtonian potential is replaced by the components of the metric tensor $g_{ij}(x)$ and the gravitational force is replaced by the components of the affine connection Γ^i_{jk}. Unfortunately, there is no place in this elegant formalism for the other long range force field, the electromagnetic field. The fact that a conformal structure does not uniquely determine a connection was used by Weyl to geometrize electromagnetism by assigning to the degrees of freedom represented by the field (4.4.7) the role of the electromagnetic potentials. The resulting geometry, called a *Weyl structure*, is intermediate between the Riemannian and conformal structures. The Weyl geometry shares with the conformal geometry the feature that at any point $p \in M$, the ratio of the lengths of two vectors and the angle between them are determined but the length of any one vector remains arbitrary. In a Weyl geometry, however, one of the connections in the equivalence class of connections compatible with the conformal geometry is chosen to play a preferred role. If for some choice of metric $g_{ij}(x)$, the connection coefficients (4.4.2) are denoted by the Christoffel symbol $\{_j{}^i{}_k\}$ and if the electromagnetic potentials are denoted by $A_j(x)$, then the *Weyl connection* is given by

$$\Gamma^i_{jk} = \{_j{}^i{}_k\} + \frac{1}{2}(\delta^i_j A_k + \delta^i_k A_j - g_{jk}g^{ir}A_r), \qquad (4.4.8)$$

which is analogous to (4.4.6). This connection is invariant under the gauge transformation[18]

$$\begin{aligned} \bar{g}_{ij}(x) &= e^{\theta(x)}g_{ij}(x) \\ \bar{A}_j(x) &= A_j(x) - D_j\theta(x), \end{aligned} \qquad (4.4.9)$$

where the gauge function is $\lambda(x) = e^{\theta(x)}$.

Weyl provided a conceptual analysis of the notion of an infinitesimal metric manifold in order to motivate the Weyl geometry on which his unified field theory is based. He argued as follows: if attention is restricted to a single point $p \in M$, then some standard of length or gauge must be chosen arbitrarily before the lengths of vectors can be determined. Hence, all that is intrinsic to the notion of a metric geometry at a point is the ability to determine the ratios of the lengths of any

---

[18]For some remarks concerning Weyl's discussion of "gauge-weight" (Eichgewicht), see the section entitled "Krümmungskonzepte, Eichkovarianz, Normaleichung" in the article by E. Scholz in this volume.

two vectors and the angle between any two vectors at a point. That these two kinds of determinations should be possible at each point is the first requirement of a metric manifold. Thus, a metric manifold must have at least a conformal structure. But Weyl required in addition that the metric structure at any point be connected to its infinitesimal neighbourhood. A conformal structure determines only an equivalence class of affine connections. Weyl showed that the additional structure required to determine a unique affine connection is provided by the congruent transport of lengths. He called this structure the "metric connection"; however, the term 'length connection' shall be used instead, in order to avoid confusion with the *modern* usage of term 'metric connection' which today denotes the symmetric, linear connection that is uniquely determined by a Riemannian metric tensor according to (4.4.2).

Definition 14 (Length Connection) *A point $p \in M$ is metrically connected with its infinitesimal neighbourhood if and only if for every length at p, there is determined at every point p' infinitesimally close to p a length to which the length at p gives rise when it is congruently displaced from p to p'.*

The only condition imposed on the concept of congruent displacement is the following:

Definition 15 (Congruent Displacement) *With respect to a choice of gauge for a neighbourhood of $p \in M$, the transport of a length l_p at $p \in M$ to an infinitesimally neighbouring point p' constitutes a congruent displacement if and only if there exists a choice of gauge for the neighbourhood of $p \in M$ relative to which the transported length $\bar{l}_{p'}$ has the same value as $\bar{l}_p$; that is*

$$\bar{l}_{p'} - \bar{l}_p = d\bar{l}_p = 0. \qquad (4.4.10)$$

Weyl called such a gauge *geodesic* at p. Note the structural similarity between this analysis and Weyl's analysis of the notion of an affine connection discussed in subsection 4.2. The proof of the following theorem closely parallels the proof of theorem 10.

Theorem 16 *If for every point p in a neighbourhood U of M, there exists a choice of gauge such that the change in an arbitrary length at p under congruent transport to an infinitesimally near point p' is given by*

$$d\bar{l}_p = 0, \qquad (4.4.11)$$

then locally with respect to any other choice of gauge,

$$dl = -lA_j(x)dx^j, \qquad (4.4.12)$$

and conversely.

Proof Consider an arbitrary point $p \in U$, and assume that there exists a choice of gauge such that

$$d\bar{l}_p = 0. \tag{4.4.13}$$

Since

$$\bar{l} = e^{\theta(x)}l \tag{4.4.14}$$

locally,

$$d\bar{l} = e^{\theta(x)}dl + le^{\theta(x)}D_j\theta(x)dx^j. \tag{4.4.15}$$

Hence,

$$d\bar{l}_p = e^{\theta(x(p))}dl_p + l_p e^{\theta(x(p))}D_j\theta(x(p))dx^j(p). \tag{4.4.16}$$

From (4.4.11) and (4.4.16), one obtains

$$dl_p = -l_p D_j\theta(x(p))dx^j(p). \tag{4.4.17}$$

Since p is an arbitrary point in U, it follows that locally

$$dl = -lA_j(x)dx^j. \tag{4.4.18}$$

Note that the structure of $D_j\theta(x(p))$ as a derivative holds only on a pointwise basis; consequently, the $A_j(x)$ do not inherit this structure.

To prove the converse, assume that (4.4.18) holds locally. Then, for an arbitrary point $p \in U$,

$$dl_p = -l_p A_j(x(p))dx^j(p). \tag{4.4.19}$$

Consider the gauge transformation

$$\bar{l} = l\exp\left[A_j(x(p))(x^j - x^j(p))\right], \tag{4.4.20}$$

which is valid in an open neighbourhood of p. Then,

$$\begin{aligned}
d\bar{l} &= \exp\left[A_j(x(p))(x^j - x^j(p))\right]dl \\
&\quad + l\exp\left[A_j(x(p))(x^j - x^j(p))\right]A_j(x(p))dx^j;
\end{aligned} \tag{4.4.21}$$

consequently, using (4.4.21) and (4.4.19), one obtains

$$d\bar{l}_p = dl_p + l_p A_j(x(p))dx^j(p) = 0. \tag{4.4.22}$$

$\square$

In Weyl's physical interpretation of the theory, the gauge field $A_j(x)$ is identified with the electromagnetic four potential, and the electromagnetic field tensor is given by

$$F_{jk}(x) = D_j A_k(x) - D_k A_j(x). \tag{4.4.23}$$

Consider two identical clocks at rest with respect to each other at some event p. A possible model for their common unit of time is a timelike vector of given length l. Suppose that the clocks separate and move along different world line paths C_1 and

C_2 which come together again at some future event q. The effect of the physical geometry on the rates of the clocks is modelled by the congruent transport of the length l along the two paths. The lengths at q will be

$$l_a = l \exp \left[- \int_{C_a} A_j(x) dx^j \right], \quad a = 1, 2. \tag{4.4.24}$$

Thus $l_1 = l_2$ if and only if

$$\oint_{C_1 - C_2} A_j(x) dx^j = \int_{\Sigma} F_{jk}(x) dx^j \wedge dx^k = 0. \tag{4.4.25}$$

Hence, in the presence of a nonvanishing electromagnetic field $F_{jk}(x)$, the clock rates l_1 and l_2 will not in general be the same; that is, there will be a second clock effect.

As Einstein pointed out, however, it is precisely this situation that suggests that Weyl's geometry conflicts with experience[19]. In Weyl's geometry, the frequency of the spectral lines of atomic clocks would depend on the location and past histories of the atoms. But experience indicates otherwise. The spectral lines are well-defined and sharp. Atomic clocks define units of time, and experience shows they are integrably transported. Thus, if we assume that the atomic and gravitational standard time, where the latter is determined by the Weyl geometry, are identical, then $F_{ij}(x) = 0$. But if $F_{ij}(x) = 0$, then a Weyl geometry is reduced to the standard Riemannian geometry that underlies General Relativity; that is, the vanishing of Weyl's *Streckenkrümmung* F is necessary and sufficient for the existence of a Riemannian metric g_{ij}.

In defending his theory against such challenges, Weyl made use of the following arguments. First, he noted that atoms, clocks and meter sticks are complicated objects for which no detailed and reliable dynamical models were then available, and that there was, therefore, no reason to assume that the clock rates are correctly modelled by the length of a timelike vector. He said (Weyl 1919b, p. 67):

> Stutzig machen könnte zunächst dies: daß nach der reinen Nahegeometrie die Streckenübertragung nicht integrabel sein soll, wenn ein elektromagnetisches Feld vorhanden ist. Steht das nicht zu dem Verhalten der starren Körper und Uhren in eklatantem Widerspruch? Das Funktionieren dieser Meßinstrumente ist aber ein physikalischer Vorgang, dessen Verlauf durch die Naturgesetze bestimmt ist, und hat als solcher nichts zu tun mit dem ideellen Prozeß der "kongruenten Verpflanzung von Weltstrecken", dessen wir uns zum mathematischen Aufbau der Weltgeometrie bedienen. Schon in der speziellen Relativitätstheorie ist der Zusammenhang zwischen dem metrischen Felde und dem Verhalten der Maßstäbe und Uhren ganz undurchsichtig, sobald man sich nicht auf quasistationäre Bewegung beschränkt. Spielen somit diese

---

[19]See Einstein's *Nachtrag* to (Weyl 1918b) which is followed by Weyl's reply.

Instrumente auch eine praktisch unentbehrliche Rolle als Indikatoren des metrischen Feldes (theoretisch wären zu diesem Zweck einfachere Vorgänge, z. B. die Lichtausbreitung, vorzuziehen), so ist es doch offenbar verkehrt, durch die ihnen direkt entnommenen Angaben das metrische Feld zu *definieren*.

[At first glance it might be surprising that according to the purely close-action geometry length transfer is non-integrable in the presence of an electromagnetic field. Does this not clearly contradict the behaviour of rigid bodies and clocks? The behaviour of these measurement instruments, however, is a physical process whose course is determined by natural laws and as such has nothing to do with the ideal process of 'congruent displacement of spacetime distance' that we employ in the mathematical construction of the spacetime geometry. The connection between the metric field and the behaviour of rigid rods and clocks is already very unclear in the theory of Special Relativity if one does not restrict oneself to quasi-stationary motion. Although these instruments play an indispensable role in praxis as indicators of the metric field, (for this purpose, simpler processes would be preferable, for example, the propagation of light waves), it is clearly incorrect to *define* the metric field through the data that are directly obtained from these instruments.]

He elaborated this idea by suggesting that the dynamical nature of such time keeping systems was such that they continually adapt to the world structure in such a way that their rates remain constant. He also distinguished between quantities that remain constant as a consequence of such dynamical adjustment and quantities that remain constant by persistence because they are isolated and undisturbed. He argued that all quantities that maintain a perfect constancy probably do so as a result of dynamical adjustment. He expressed himself as follows (Weyl 1921b, p. 261):

What is the cause of this discrepancy between the idea of congruent transfer and the behaviour of measuring-rods and clocks? I differentiate between the determination of a magnitude in Nature by "persistence" (*Beharrung*) and by "adjustment" (*Einstellung*). I shall make the difference clear by the following illustration: We can give to the axis of a rotating top any arbitrary direction in space. This arbitrary original direction then determines for all time the direction of the axis of the top when left to itself, by means of a *tendency of persistence* which operates from moment to moment; the axis experiences at every instant a parallel displacement. The exact opposite is the case for a magnetic needle in a magnetic field. Its direction is determined at each instant independently of the condition of the system at other instants by the fact that, in virtue of its constitution, the system *adjusts* itself in an unequivocally determined manner to the field in which it is situated. *A priori* we have no ground for assuming as integrable a transfer which results purely from the tendency of persistence. ... Thus, although, for example, Maxwell's equations demand the conservational equation $de/dt = 0$ for the charge e of an electron, we are unable to understand from this fact why an electron, even after an indefinitely long time, always possesses an unaltered charge, and why the same charge e is associated with all electrons. This circumstance shows that the charge is not determined by persistence, but by adjustment, and that there can exist only *one* state of equilibrium of the negative electricity, to which the corpuscle adjusts itself afresh at every instant. For the same reason we can conclude the same thing for the spectral lines of atoms. The one thing common to atoms emitting the same frequency is their constitution, and not the agreement of their frequencies on the occasion of an encounter in the distant past. Similarly, the length of a measuring-rod is obviously determined by adjustment, for I could not give *this* measuring-rod in *this* field-position any other length arbitrarily (say double or treble length) in place of the length which it now possesses, in the manner in which I can at will pre-determine its direction. The theoretical possibility of a determination of length by adjustment is given as a consequence of the *world-curvature*, which arises from the metrical field according to a complicated mathematical law. As a result of its constitution, the measuring-rod assumes a length which possesses this or that value, *in relation to the radius of curvature of the field*.

Weyl (1929c, p. 233) abandoned his theory only with the advent of the Quantum Theory of the electron. He did so because in that theory a different kind of gauge invariance associated with Dirac's theory of the electron was discovered which more adequately accounted for the conservation of electric charge. Weyl's contributions to this new theory are discussed below in subsections 5.2 and 5.3. In 1973, however, Weyl's theory was revived by Dirac (1973) in a slightly modified form, which incorporated a real scalar field $\beta(x)$. Dirac also argued that the time intervals measured by atomic clocks need not be identified with the lengths of timelike vectors in the Weyl geometry.

4.5 The Riemann-Helmholtz-Lie-Weyl Problem of Space: The Finsler-Metric Interpretation of Weyl's Raumproblem

In his lectures on the mathematical analysis of the problem of space delivered in 1922 at Barcelona and Madrid, Weyl sketched a proof demonstrating that among all the possible infinitesimal metrics that can be put on a differentiable manifold, the Pythagorean-Riemannian metric is the only type of metric that uniquely determines a symmetric linear connection. The lectures were published in 1923 together with a long appendix in which Weyl gave the details of an improved proof.

In this subsection, we present an interpretation of the conceptual analysis that underlies Weyl's solution to the Raumproblem, which we shall refer to as the Finsler-Metric interpretation. This interpretation differs radically from the currently accepted interpretation which is based on the work of (Cartan, 1922, 1923b), (Laugwitz 1958) and (Scheibe, 1957, 1988). We shall refer to this currently accepted interpretation as the Received View[20]. According to the Received View, a geometry on a manifold is roughly speaking (especially in the work of Scheibe) characterized (defined or described) by a G-field[21] together with a G-metric connection[22]. At a given point p, the group action determined by the G-field is not merely descriptive of the symmetry that the metric field has at that point, rather the group action *defines* the very notion of congruence at that point; that is, the field of group actions is *ontologically prior* and serves to characterize (define or describe) the geometry. We discuss the Received View below in subsection 4.6. In contrast, on our interpretation, Weyl's ontological position is that each geometry in the class of geometries with which he is concerned is determined by two structural fields, a possibly indefinite Finsler metric of possibly variable gauge and a length connection that is determined by a 1-form field. While group transformations play an important role in the analysis, these transformations are *not* ontologically prior to the structural fields but are ontologically dependent on them. A detailed account of the G-field interpretation together with an account of the relative merits of the two interpretations of Weyl's Raumproblem is presented in subsection 4.6. The argument in favour of the view that Weyl's Raumproblem was concerned with Finsler metrics and that Cartan invented a different Raumproblem which is concerned with G-structures, albeit because he misinterpreted what Weyl was doing, is summarized in subsection 4.7.

Remark 17 (Importance of Finsler Metrics) In a recent article, Chern (1996) points out the importance of Finsler geometry. Chern states that virtually all of the important theorems pertaining to Riemannian geometry have been extended to the setting of general Finsler metrics. Chern says that Riemannian geometry, therefore, should be regarded merely as a special case of Finsler geometry. Chern also mentions a number of fields in which Finsler metrics play an important role both in mathematics and various applied areas. It is clear then that Weyl's Raumproblem, namely, the problem of determining those general properties that are characteristic of spacetime metrics and of proving that a Finsler metric with these properties is necessarily (semi)-Riemannian, is even more relevant today than it was when Weyl first proposed it.

What is the significance and historical background of Weyl's deep group-theoretical result concerning the uniqueness of the Pythagorean-Riemannian met-

---

[20] For a sympathetic overview of the Received View, see Erhard Scholz's contribution in this volume and (Scholz 1999b).

[21] See definition 28 below in subsection 4.6.

[22] See definition 33 below in subsection 4.6.

ric? Recall that the only condition (see, definition 9) that Weyl imposed on the concept of parallel transport at $p \in M$ is the condition that says that there exists a geodesic coordinate system for the neighbourhood of $p \in M$ in which the components of the affine[23] connection vanish. For Weyl, this condition characterizes the nature of an affine connection on M. Such a connection is necessarily symmetric. An affine structure on M is homogeneous in the sense that its nature is the same everywhere. The situation is more complex in the case of the metric since there are conceivable metric structures with quite different natures. The meaning of the 'nature' of a metric will become clear after the following brief discussion concerning the conceptual-historical context of Weyl's problem.

A metric manifold is a manifold on which a distance function $f: M \times M \to \mathbf{R}$ is defined. Such a distance function must satisfy the following minimal conditions: for all $p, q, r \in M$,

(i) $f(p, q) \geq 0$ and if $f(p, q) = 0$ then $p = q$,

(ii) $f(p, q) = f(q, p)$,

(iii) $f(p, q) + f(q, r) \geq f(p, r)$ (triangle inequality).

In his famous inaugural lecture at Göttingen in 1854, Riemann examined how metrical relations can be determined on a continuous manifold, that is, what specific form should $f: M \times M \to \mathbf{R}$ have. Consider the coordinates $x^i(p)$ and $x^i(p) + dx^i(p)$ of two neighbouring points $p, p' \in M$. The measure of the distance $ds = f(p, p')$ must be some function[24] F_p at p of the differential increments $dx^i(p)$; that is,

$$ds = F_p(dx^1(p), \ldots, dx^n(p)). \tag{4.5.3}$$

Riemann (1854) states that F_p should satisfy the following requirements:

(a) **Functional Homogeneity:** If $\lambda > 0$ and $ds = F_p(dx(p))$, then

$$\lambda ds = \lambda F_p(dx(p)) = F_p(\lambda dx(p)). \tag{4.5.4}$$

(b) **Sign Invariance:** A change in sign of the differentials should leave the value of ds invariant.

---

[23] Recall that Weyl used the term 'affine connection' for what is now called a symmetric, linear connection. In this overview, we use the two terms synonymously.

[24] For the case of positive-definite Finsler metrics, it is customary to work with the function F_p that determines the infinitesimal length interval ds; however, for some purposes, it is more natural to work with the function F_p^2 that determines ds^2, particularly in the case of indefinite metrics. Note also that it is the 2-form

$$\frac{\partial^2 F_p^2(\xi)}{\partial \xi^i \partial \xi^j} d\xi^i d\xi^j \tag{4.5.1}$$

that induces by restriction Riemannian or semi-Riemannian metrics on the indicatrices defined by

$$F_p(\xi) = \pm 1. \tag{4.5.2}$$

Clearly condition (b) is satisfied by every positive homogeneous function of degree $2m$ $(m = 1, 2, 3, \ldots)$. In the simplest case, $m = 1$, and the length element ds is the square root of a homogeneous function of second degree and can be expressed in the standard form

$$ds = \left[\sum_{i=1}^{n} (dx^i(p))^2 \right]^{1/2} ; \qquad (4.5.5)$$

that is, at each point of M, there exists a coordinate system (defined up to an element of the orthogonal group $O(n)$) in which the square root of the homogeneous function of second degree can be expressed in the above standard form. Riemann's well-known general expression for the measure of length at $p \in M$ with respect to an arbitrary coordinate system is

$$ds^2 = g_{ij}(x(p))dx^i(p)dx^j(p), \qquad (4.5.6)$$

where the components of the metric tensor satisfy the symmetry condition $g_{ij} = g_{ji}$.

The assumption that $ds^2 = F_p^2$ is a quadratic differential form is not only the simplest one, but also the preferred one for other important reasons that are considered below. Riemann himself was well aware of other possibilities; for example, the possibility that ds could be the 4th root of a homogeneous polynomial of 4th order in the differentials. But Riemann restricted himself to the special case $m = 1$ because he was pressed for time and because he wanted to give specific geometric interpretations of his results.

According to Weyl (1919c), Riemann analytically motivated the restriction to the Pythagorean case in essentially the following way. Let S_p be a hypersphere around some point $p \in M$ which consists of the geometric positions of all points p' measured along the *shortest* lines; that is, $S_p = \{p'|f(p, p') = D\}$. Then, the equation of such a hypersphere around $p \in M$ may be expressed analytically as $F_p^2(dx^1, \ldots, dx^n) = D^2$. Then, the Taylor expansion of F_p^2 at p begins with quadratic terms. Riemann disregarded the higher terms and considered only the quadratic ones, which together constitute a quadratic form. If this form does not vanish and is positive in all directions, then one obtains the Pythagorean form for ds. In Weyl (1919c, 3 edn) and Weyl (1923c), it is pointed out, however, that Riemann's arguments for disregarding the higher order terms in the Taylor expansion are not compelling and Weyl suggests another justification which we now discuss.

Weyl's first requirement is essentially based on the notion of 'measurability'. The nature of the metric at a given point should be coordinate independent. In the neighbourhood of any given point $p \in M$, the choice of coordinates is essentially arbitrary. Under a change of coordinates, the differentials $dx^i(p)$ transform linearly and homogeneously; consequently, if ds is given by an expression $F_p(dx^1, \ldots, dx^n)$ with respect to a given system of coordinates, then with respect to another system of coordinates, ds is given by a function related to $F_p(dx^1, \ldots, dx^n)$ by a linear, homogeneous transformation of its arguments dx^i. In addition, it is reasonable to require that the *nature* of the metric is the same everywhere in the sense that at every point with respect to every coordinate system for a neighbourhood of the point in question, ds is represented by an element of the equivalence class $[F]$ of functions generated by any one such function, say $F_p(dx^1, \ldots, dx^n)$, by all linear, homogeneous transformations of its arguments dx^i.

For the case in which F_p is Pythagorean in form, the square root of a positive-definite quadratic form, there is just one possible equivalence class $[F]$ because every function that is the square root of a positive-definite quadratic form can be transformed to the standard expression

$$F = \left[(dx^1)^2 + \cdots + (dx^n)^2\right]^{1/2} \tag{4.5.7}$$

by means of a linear, homogeneous transformation. However, for the cases in which higher order forms are used for ds such as the fourth root of a positive-definite quartic that is functionally homogeneous, more than one equivalence class may arise. For example, the equivalence classes generated by the forms

$$ds^4 = (dx^1)^4 + (dx^2)^4 + (dx^3)^4 \tag{4.5.8}$$

and

$$
\begin{aligned}
ds^4 &= (dx^1)^4 + (dx^2)^4 + (dx^3)^4 \\
&\quad + (dx^1 dx^2)^2 + (dx^1 dx^3)^2 + (dx^2 dx^3)^2 \\
&= \left[(dx^1)^2 + (dx^2)^2 + (dx^3)^2\right]^2 \\
&\quad - \left[(dx^1 dx^2)^2 + (dx^1 dx^3)^2 + (dx^2 dx^3)^2\right]
\end{aligned}
\tag{4.5.9}
$$

are distinct.

To every possible equivalence class $[F]$ of homogeneous functions, there corresponds a *type* of metrical space. The Pythagorean-Riemannian space, for which $F_p^2 = (dx^1)^2 + \cdots + (dx^n)^2$, is one among several types of possible metrical spaces. The problem, therefore, is to single out the equivalence class $[F]$, where F corresponds to $F_p^2 = (dx^1)^2 + \cdots + (dx^n)^2$, from the other possibilities and to provide arguments for this preference.

The first satisfactory justification of the Pythagorean-Riemannian form, although limited in scope because it presupposes the full homogeneity of Euclidean space, was provided by the investigations of Hermann von Helmholtz. Helmholtz (1866, 1868) diverged from Riemann's analytic approach and made use merely of the fundamental concept of geometry, namely, the concept of the *congruent mapping*, and characterized the structure of space by requiring of space the full homogeneity of Euclidean space; however, his analysis was thereby restricted to the cases of constant curvature, positive, zero or negative. Abstracting from our experience of the movement of rigid bodies, Helmholtz was able to mathematically derive Riemann's distance formula from a number of axioms about rigid body motion in space. Helmholtz's justification was later developed and improved from a group-theoretical viewpoint by Lie ((1886/1935) and (1893, chapters 21–23)). What Weyl did was to state and prove an analogous group-theoretical result for *infinitesimal* geometry.

For the purpose of the following discussion of Weyl's proof, the terms 'metric' or 'metric structure' shall mean *any* infinitesimal *distance function* on a differentiable manifold M that is of a given *type*; that is, for any $p \in M$, F_p is a member of a particular equivalence class $[F]$. Any such metric structure has a microinvariance group $\mathbf{G}_p$ at each $p \in M$. By a microinvariance group $\mathbf{G}_p$ at $p \in M$ of a metric structure, we mean an invertible, linear map of the tangent space $T(M_p)$ onto itself that preserves the infinitesimal distance function at $p \in M$. For every $p \in M$, $\mathbf{G}_p$ is isomorphic to one and the same abstract group. Weyl called such a group the rotation group (*Drehungsgruppe*). It would be misleading, however, to adopt Weyl's terminology, since the term 'rotation group' is now used exclusively to refer to the orthogonal group $O(n)$; however, Weyl's usage allowed for the possibility that the microinvariance group may differ from $O(n)$. In the positive-definite case, the microinvariance group would be some subgroup of $O(n)$.

One must *not* think, however, that the microinvariance group *defines*[25] the geometry. Here, the focus is on the infinitesimal *distance* function. Some examples will serve to emphasize the point. The *microinvariance group* for

$$ds^4 = (dx^1)^4 + (dx^2)^4 + (dx^3)^4, \tag{4.5.10}$$

$$ds^6 = (dx^1)^6 + (dx^2)^6 + (dx^3)^6 \tag{4.5.11}$$

or in general

$$ds^\alpha = |dx^1|^\alpha + |dx^2|^\alpha + |dx^3|^\alpha \quad \text{for real } \alpha > 0 \tag{4.5.12}$$

consists of only the identity except for the case $\alpha = 2$. Other examples readily come to mind. Consider

$$ds^4 = [(dx^1)^2 + (dx^2)^2]^2 + [(dx^3)^2 + (dx^4)^2]^2, \tag{4.5.13}$$

$$ds^6 = [(dx^1)^2 + (dx^2)^2]^3 + [(dx^3)^2 + (dx^4)^2]^3, \tag{4.5.14}$$

or more generally

$$ds^{2\alpha} = [(dx^1)^2 + (dx^2)^2]^\alpha + [(dx^3)^2 + (dx^4)^2]^\alpha \quad \text{for real } \alpha > 0 \tag{4.5.15}$$

for all of which, except for the case $\alpha = 1$, the microinvariance group is isomorphic to $O(2) \times O(2)$, provided that discrete transformations are ignored. It follows from the above examples that quite different geometries may have the same microinvariance group. *It is clear, therefore, that it is a mistake to interpret Weyl's group-theoretical analysis of the uniqueness of the Pythagorean-Riemannian Form of the metric in terms of G-metrics as do the other interpretations discussed in subsection 4.6.*

---

[25] In contrast, the other interpretation discussed in subsection 4.6 defines congruence in terms of group actions.

Remark 18 Wherever Weyl discusses the Raumproblem, he always[26] discusses it within the context of Riemann's desire to single out the Pythagorean-Riemannian metric from the general class of (positive-definite) Finsler metrics. It is clear from his remarks in the eighth lecture of (Weyl 1923c) and elsewhere, that Weyl wanted to solve the problem in a more general context that would encompass the General Theory of Relativity.

Weyl did not analyze the problem of characterizing Finsler metrics which are not positive-definite nor did he consider any particular examples, other than the semi-Riemannian ones, because the group-theoretic proof presented in the seventh lecture of (Weyl 1923c) depends only on the idea that such a metric determines a microinvariance group and not on the detailed structure of the metric. Nevertheless, it is worth considering a few possibilities both for concreteness and in order to be sure that the class of indefinite metrics that are eliminated by Weyl's theorem is not empty. For any integer $n \geq 0$, the metric

$$ds^2 = \left[(dx^1)^{4n+2} - (dx^2)^{4n+2} - (dx^3)^{4n+2} - (dx^4)^{4n+2} \right]^{\frac{1}{2n+1}} \qquad (4.5.16)$$

is clearly indefinite. For $n > 0$, the 'light cones' are 'lumpy', and the microinvariance group consists of only the identity. As another example, consider the metric

$$ds^2 = \left[((dx^1)^2 - (dx^2)^2)^{2n+1} - ((dx^3)^2 + (dx^4)^2)^{2n+1} \right]^{\frac{1}{2n+1}} \qquad (4.5.17)$$

for any integer $n \geq 0$. For $n > 0$, the microinvariance group for these cases is $O(1,1) \times O(2)$. It is interesting to note that for these cases the 'light cones' are identical to the usual light cone obtained for the Minkowski metric corresponding to $n = 0$. These examples and the positive-definite cases discussed above suggest that the microinvariance group of a Finsler metric is isomorphic to a subgroup of $O(p,q)$ for some signature (p,q).

Before discussing Weyl's analysis we give explicit definitions of some concepts that play an important role.

Definition 19 (Congruence at a given point) *Any two vectors $\tilde{v}_p \in T(M_p)$ and $v_p \in T(M_p)$ are congruent if and only if*

$$F_p(\tilde{v}_p) = F_p(v_p). \qquad (4.5.18)$$

Note that this congruence relation is independent of the choice of gauge at $p \in M$.

Definition 20 (Fixed-Gauge Infinitesimal Congruent Transport) *If a metric function $F_p(v)$ is uniquely specified at each point $p \in M$, that is, if the gauge is everywhere determined, then a congruent transformation of $T(M_p)$ to $T(M_{p'})$, where p' is infinitely near p, is any infinitesimal, linear map*

$$v^i_{p'} = (\delta^i_j - \Lambda^i_{jk} dx^k) v^j_p \qquad (4.5.19)$$

---

[26] With the possible exception of (Weyl 1922b); however, even here, Weyl refers to section 18 of (Weyl 1918c) in which Finsler metrics are briefly mentioned.

which satisfies

$$F_{p'}(v_{p'}) = F_p(v_p). \tag{4.5.20}$$

Such a notion of congruent transformation can be extended by iteration, that is, by integration to permit comparison of length at distant points in a path independent way. Weyl thought that such a distant comparison of lengths was unnatural. In this respect, Riemannian geometry was not a pure near-geometry but retained a distant-geometric element. For Weyl, the distance function $F_p(v)$ determined this length of vectors at p only up to a choice of gauge. The ratios of the lengths of vectors at a given point are, therefore, determined. He also thought that it should be possible to determine the ratio of the lengths of two vectors $v_{p'}$ and v_p at *infinitesimally near* points. For such a comparison to be possible, it was necessary to equip the manifold with an additional structure, the *length connection*[27], which has been discussed above in subsection 4.4. For a given choice of gauge for M, that is, for a given choice of F_p at each $p \in M$, the length connection is determined by stating that a length l_p under congruent transport from p to p' becomes the length $l_p + dl_p$ where

$$dl_p = -l_p \omega_i(x(p)) dx^i(p) \tag{4.5.21}$$

in which the $\omega_i(x(p))$ are the components of a 1-form.

Definition 21 (Variable-Gauge Infinitesimal Congruent Transport)   *A congruent transformation from $T(M_p)$ to $T(M_{p'})$ is a linear map (4.5.19) which satisfies*

$$F_{p'}(v_{p'}) = F_p(v_p) \exp\left(-\omega_i(x(p)) dx^i(p)\right). \tag{4.5.22}$$

Note that Weyl did not make this notion of 'congruent' explicit in the proof he presented in the seventh lecture of (Weyl 1923c), but that he must have intended this notion of congruence is clear from the remarks he made at the beginning of the eighth lecture and elsewhere.

Remark 22 (Purely Infinitesimal Geometry) With the introduction of gauge invariance, a purely infinitesimal geometry according to Weyl, is determined by two structural fields:

1. a variable-gauge Finsler metric field, and

2. a length connection that is determined by a 1-form field.

Weyl used the fact that the abstract character of the microinvariance group for a specific metric (*of whatever nature*) is the same everywhere to derive additional constraints on the possible form of congruence transformations between neighbouring points. Weyl applied a group-theoretical analysis to the concept of

---

[27] As mentioned just above definition 14, Weyl used the term 'metric connection'; however, in the modern literature the term 'metric connection' is used to denote something quite different, namely, the symmetric linear connection with respect to which a symmetric Riemannian metric is invariant under parallel transport.

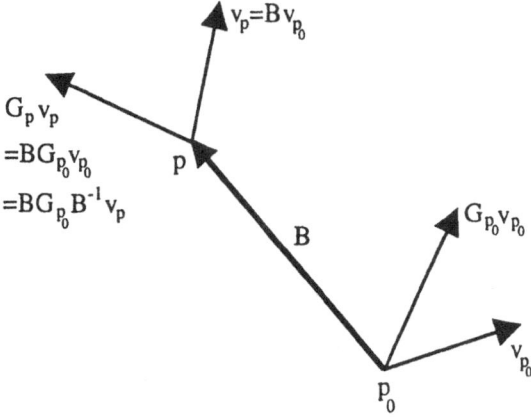

Figure 3: Similarity of the Microinvariance Groups

(length) metric connectedness by considering the congruent displacements of a tangent space $T(M_{p_0})$ at $p_0 \in M$

(a) to one *arbitrary* point $p \in M$ infinitesimally near p_0.

(b) to a sufficient number of arbitrary points $p \in M$ infinitesimally near p_0.

With respect to (a), Weyl observed that all congruent displacements A from $p_0 \in M$ to $p \in M$ consist of the composition of one congruent displacement B and an arbitrary microinvariance transformation G_{p_0} at $p_0 \in M$ which precedes B; that is, $A = B \circ G_{p_0}$, where $G_{p_0} \in \mathbf{G}_{p_0}$. To see this, consider two congruent displacements A and B from p_0 to p. Then B^{-1} is a congruent displacement from p to p_0. The composition $B^{-1} \circ A$ is, therefore, a congruent map that takes p_0 into itself; that is, $B^{-1} \circ A = G_{p_0}$ is a microinvariance at p_0.

In addition, the microinvariance groups at different points are *similar* to each other. Figure 3 illustrates the following discussion. Recall that G_{p_0} is a microinvariance at p_0, that is, $G_{p_0} \in \mathbf{G}_{p_0}$, if and only if

$$\forall v_{p_0} \left(F_{p_0}(G_{p_0} v_{p_0}) = F_{p_0}(v_{p_0}) \right), \tag{4.5.23}$$

that is, if and only if for every $v_{p_0} \in T(M_{p_0})$, the vectors $G_{p_0} v_{p_0}$ and v_{p_0} have the same length. Let $B: T(M_{p_0}) \to T(M_p)$ be a congruent map, where 'congruent' is defined by (4.5.22). Then, for every $v_{p_0} \in T(M_{p_0})$, the vectors $B v_{p_0}$ and $B G_{p_0} v_{p_0}$ in $T(M_p)$ have the same length, that is,

$$\forall v_{p_0}(F_p(BG_{p_0} v_{p_0}) = F_p(B v_{p_0})). \tag{4.5.24}$$

Letting $v_p = B v_{p_0}$ from which it follows that $v_{p_0} = B^{-1} v_p$, one obtains

$$\forall v_p(F_p(BG_{p_0}B^{-1} v_p) = F_p(v_p)); \tag{4.5.25}$$

that is, $BG_{p_0}B^{-1} = G_p \in \mathbf{G}_p$. Since this holds for any $G_{p_0} \in \mathbf{G}_{p_0}$, it follows that $\mathbf{G}_p = B\mathbf{G}_{p_0}B^{-1}$; that is, the microinvariance group at p is similar to the microinvariance group at p_0.

Thus, from the notion of congruent transport, it follows that every element of the microinvariance group $\mathbf{G}_p$ is the similarity transform of an element of the microinvariance group $\mathbf{G}_{p_0}$ at $p_0 \in M$. Weyl expressed this relationship by saying that the microinvariance groups $\mathbf{G}_{p_0}$ and $\mathbf{G}_p$ are of the same type and differ only with respect to *orientation*. Since one can continuously go from the point $p_0 \in M$ to any other point $p \in M$, the microinvariance groups on M are of the same type everywhere; that is, they differ only with respect to *orientation* and are all isomorphic to the abstract microinvariance group $\mathbf{G}$.

With respect to **(b)**, Weyl observed that if one considers two successive infinitesimal, congruent displacements of tangent vectors, one through dx^i from p_0 to p and a similar one through δx^i from p to p', then the resultant congruent infinitesimal transport arises from the resultant displacement $dx^i + \delta x^i$. A congruent transport is infinitesimal, if the change dv^i of the components v^i of an arbitrary vector are infinitesimal and of the same order as the components dx^i of the displacement. If

$$dv^i = -\epsilon \Lambda^i_{k1} v^k \qquad (4.5.26)$$

is an arbitrary infinitesimal congruent transport in the direction of the first coordinate axis to the point $(x_0^1 + \epsilon, x_0^2, \dots, x_0^n)$, and the coefficients $\Lambda^i_{k2}, \dots, \Lambda^i_{kn}$ have the corresponding meaning for the infinitesimal congruent transport in the directions of the other coordinate axes (ϵ is an infinitesimal constant), then an infinitesimal congruent transport from $p_0 \in M$ to an arbitrary point p in the neighbourhood of p_0 has the form

$$dv^i = -\Lambda^i_{kr} v^k dx^r. \qquad (4.5.27)$$

The reason Weyl emphasizes in **(b)** that one must consider a sufficient number of arbitrary points p infinitesimally near p_0 is that only in this way can one obtain the linear dependence of the congruent transport on the components dx^i of the displacement. Note that in the argument related to **(a)** this linear dependence was not an issue.

As Weyl said, everything that has been done so far constitutes merely a conceptual analysis, that is, an explication of what lies in the concepts *metric*, *congruent transport*, *parallel connection* and *length connection*. He also noted that it is necessary to make some claim that goes beyond conceptual analysis in order to single out the Pythagorean-Riemannian form from among the various kinds of possible metric structures that can be put on a differentiable manifold representing physical space.

Postulate 23 (Postulate of Freedom) *Given a choice of the metric F_{p_0} (of whatever type) at a given point $p_0 \in M$ and a choice of the length connection at $p_0 \in M$, for any choice of Λ^i_{jk} (n^3 numbers), there exists a choice of the metric F_p for every point $p \in M$ in the infinitesimal neighbourhood of $p_0 \in M$ such that* *the map*

$$v_p^i = (\delta_j^i - \Lambda_{jk}^i dx^k) v_{p_0}^j \tag{4.5.28}$$

is a congruent map for all infinitesimal dx^i.

This postulate requires that the deformability of the metric be maximal (the congruent map, the Λ_{jk}^i, may be chosen arbitrarily) in order that at this stage of analysis the possible dynamics of the world metric (of whatever type) be left as unconstrained as possible.

The *Postulate of Freedom* provides the framework for a general and concise formulation of the hypothesis of *dynamical geometry*: *Whatever the nature or type of the metric may be* (provided it is the same everywhere), *the variations from point to point in the mutual orientations of the concrete microinvariance groups are causally determined by the material content that fills space.* In contrast with Helmholtz's analysis which presupposes the homogeneity of space, the *Postulate of Freedom* allows for the possibility of replacing Helmholtz's homogeneity requirement with the possibility of subjecting the metric field to arbitrary, infinitesimal change. To assert this dynamical possibility does not require that the *nature* of the metric be specified.

Remark 24 Let us illustrate the *Postulate of Freedom* for the case of variable-gauge, possibly indefinite Riemannian geometries. Let g_{ij} be the metric at some point p_0. Then, the metric at a point p infinitesimally near p_0 is given to first order in the dx^i by the Taylor expansion

$$g_{ij} + g_{ij,k} dx^k. \tag{4.5.29}$$

The congruent transport v_p^i of a vector $v_{p_0}^i$ is given by (4.5.28). The vectors $v_{p_0}^i$ and v_p^i have the same length; therefore, according to definition 21

$$g_{ij}(p) v_p^i v_p^j = g_{ij}(p_0) v_{p_0}^i v_{p_0}^j (1 - \omega_k(p_0) dx^k), \tag{4.5.30}$$

that is,

$$\begin{gathered} (g_{ij} + g_{ij,k} dx^k)(\delta_r^i - \Lambda_{ra}^i dx^a) v_{p_0}^r (\delta_s^j - \Lambda_{sb}^j dx^b) v_{p_0}^s \\ = g_{ij} v_{p_0}^i v_{p_0}^j (1 - \omega_k(p_0) dx^k). \end{gathered} \tag{4.5.31}$$

Keeping only terms that are first order in dx^i, one obtains

$$g_{ij,k} = g_{ir} \Lambda_{jk}^r + g_{jr} \Lambda_{ik}^r - g_{ij} \omega_k. \tag{4.5.32}$$

Thus, given the geometry at p_0, that is, the metric g_{ij} and the length connection ω_i at p_0, and an arbitrary choice of congruent transport Λ_{jk}^i, the metric is determined in an infinitesimal neighbourhood of p_0 by the Taylor expansion (4.5.29) where the $g_{ij,k}$ at p_0 are given by (4.5.32).

The following postulate is motivated by the central role played by the affine (symmetric linear) connection in covariant differentiation and hence in the formulation of physical laws in the context of the General Theory of Relativity.

Postulate 25 (Uniqueness of the Affine Connection)   *The variable-gauge, Finsler, metric field (of whatever type) and the length connection field together uniquely determine an affine (symmetric linear) connection that is length preserving.*

Of course, this postulate holds in the case of Weylian geometry in which case the variable-gauge Finsler metric is a variable-gauge (semi)-Riemannian metric. The postulate also holds in the case of (semi)-Riemannian geometry for which the length connection is integrable.

Theorem 26 (Uniqueness of the Pythagorean-Riemannian Form)   *If a purely infinitesimal geometry is such that both the Postulate 23 of Freedom and the Postulate 25 of the Unique Determination of the Affine Connection are satisfied, then the variable-gauge Finsler component of the purely infinitesimal geometry must be a variable-gauge metric of the Pythagorean-Riemannian form (of some signature).*

Proof   The argument proceeds as follows. Under an infinitesimal congruent transport from p_0 to p, the components of a vector $v_{p_0} \in T(M_{p_0})$ transform according to

$$v_p^i = (\delta_k^i - \Lambda_{kr}^i dx^r)v_{p_0}^k = v_{p_0}^i - (\Lambda_{kr}^i dx^r)v_{p_0}^k.   \qquad (4.5.33)$$

Similarly, under parallel transport from p_0 to p, the components transform according to

$$\hat{v}_p^i = (\delta_k^i - \Gamma_{kr}^i dx^r)v_{p_0}^k = v_{p_0}^i - (\Gamma_{kr}^i dx^r)v_{p_0}^k.   \qquad (4.5.34)$$

Let $\bar{v}_{p_0}^i$ denote the components of v_p^i transported parallelly from p back to p_0. Then,

$$\bar{v}_{p_0}^i = (\delta_j^i + \Gamma_{kr}^i dx^r)v_p^k.   \qquad (4.5.35)$$

The composition of (4.5.33) and (4.5.35) is

$$\bar{v}_{p_0}^i = [\delta_j^i + (\Gamma_{kr}^i - \Lambda_{kr}^i)dx^r]v_{p_0}^k.   \qquad (4.5.36)$$

By hypothesis, the parallel transport is a congruent transport; consequently, the map (4.5.36) is a congruent map that leaves p_0 fixed. Thus, this map is an element of $\mathbf{G}_{p_0}$; therefore, the quantities

$$A_{kr}^i = \Gamma_{kr}^i - \Lambda_{kr}^i   \qquad (4.5.37)$$

are such that for arbitrary real coefficients λ^r, the matrix $A_k^i = A_{kr}^i \lambda^r$ is an element of the Lie algebra of the microinvariance group $\mathbf{G}_{p_0}$.

The relation

$$\Lambda_{kr}^i = \Gamma_{kr}^i - A_{kr}^i   \qquad (4.5.38)$$

requires (Postulate 25) only that there exists a symmetric, linear connection Γ that defines a congruent map (4.5.35). The additional requirement that Γ be unique (Postulate 25) entails that the decomposition (4.5.38) be unique. It follows that

$$A^i_{kr} = -A^i_{rk}. \tag{4.5.39}$$

The proof is by *reductio ad absurdum*. Suppose that there do exist non-vanishing $\tilde{A}^i_{kr} = \tilde{A}^i_{rk}$, then

$$\Lambda^i_{kr} = (\Gamma^i_{kr} + \mu\tilde{A}^i_{kr}) - (A^i_{kr} + \mu\tilde{A}^i_{kr}) \tag{4.5.40}$$

is a different decomposition for every real number μ, which contradicts the requirement that the decomposition (4.5.38) is unique. From (4.5.39), it follows that

$$\Gamma^i_{jk} = \frac{1}{2}(\Lambda^i_{jk} + \Lambda^i_{kj}); \tag{4.5.41}$$

consequently, if all possible Λ^i_{jk} are considered, then all possible Γ^i_{jk} are generated. Next, from

$$\Lambda^i_{kr} = \Gamma^i_{kr} - A^i_{kr}, \tag{4.5.42}$$

the dimension N of the Lie algebra of the microinvariance group $\mathbf{G}_{p_0}$ can be deduced from the following considerations. The dimension of the space spanned by the A^i_{kr} is nN. By the *Postulate of Freedom* 23, if only the *type* of metric is specified, then for any choice of the n^3 numbers Λ^i_{kr}, it is possible to choose a metric of the given type in an infinitesimal neighbourhood of the point in question such that the Λ^i_{kr} determine a congruent transport (Definition 21). By running through all possible metrics of the given type, one must generate all possible symmetric linear connections Γ^i_{jk}. Hence

$$n^3 = n\frac{n(n+1)}{2} + nN \tag{4.5.43}$$

from which it follows that

$$N = \frac{n(n-1)}{2}. \tag{4.5.44}$$

From the constraints (4.5.39) and (4.5.44), Weyl (1923c, Appendix 12) was able to show, by means of a complicated technical analysis, that the only possible microinvariance groups are the generalized orthogonal groups $O(p, q)$ of some signature (p, q), where $n = p + q$. □

Remark 27 The trace condition

$$A^k_{kr} = 0 \tag{4.5.45}$$

which arises from the requirement that the microinvariance groups $\mathbf{G}_p$ preserve volume, is somewhat unnatural. Nonetheless, in the Spanish lectures Weyl (1923c, p. 51) does make use of the trace condition. This condition, however, does not arise from either the Postulate of Freedom or from the requirement that the symmetric linear connection is uniquely determined by congruent transport (Postulate 25). Moreover, in the case of non-Riemannian Finsler metrics, the definition of volume is frame dependent although the volume is the same for all frames that are related by an element $G_p \in \mathbf{G}_p$.

It is noteworthy that in Appendix 12 of (Weyl 1923c), Weyl presented an improved proof that did not make use of the trace condition. It is somewhat odd that in the beginning of this appendix Weyl refers to a proof provided by Cartan (1922) but did not note that Cartan's proof required this unnatural trace condition. It is also interesting to note that Weyl emphasized that his proof was more *direct* than Cartan's proof. Weyl's remarks are in character with his constructivist attitudes even though in (Weyl 1922b, GA II, p. 295), Weyl remarks that his proof is not as intuitionistic as he would like it to be.

Thus the *Postulate of Freedom* and the requirement that the metric and the length connection uniquely determine the affine connection together entail the existence at each $p \in M$ of a non-degenerate quadratic form that is unique up to a choice of gauge at $p \in M$ and is invariant under the action of the microinvariance group $\mathbf{G}_p$ which is isomorphic to an orthogonal group of some signature.

This formulation suggests, according to Weyl, an intuitive contrast between Euclidean 'distance-geometry' and the '*near*-geometry' or '*field*-geometry' of Riemann. Weyl compared the former to a crystal 'built up from uniform unchangeable atoms in the rigid and unchangeable arrangement of a lattice', and the latter to a liquid, 'consisting of the same indiscernible unchangeable atoms, whose arrangement and orientation, however, are mobile and yielding to forces acting upon them' (Weyl 1949a, 1 edn, p. 88).

The *nature* of the metric field, that is the *nature* of the metric everywhere, is *the same* and is, therefore, absolutely determined. It reflects according to Weyl, the *a priori* structure of space or spacetime. In contrast, what is *a posteriori*, that is, accidental and capable of continuous change being causally dependent on the material content that fills space, is the mutual *orientation* of the metrics at different points. Hence, the demarcation between the *a priori* and the *a posteriori* has shifted. Euclidean geometry is still preserved for the infinitesimal neighbourhood of any given point, but the coordinate system in which the metrical law assumes the standard form $ds^2 = \sum_{i=1}^{n}(dx^i)^2$ (which coordinate system is characteristic for the orientation of the metric) is in general different from place to place.

In the context of his group-theoretical analysis, Weyl (1922b, p. 266) makes the following interesting and important statement:

> Ich bemerke dazu in erkenntnistheoretischer Hinsicht: es ist nicht richtig, zu sagen, daß der Raum oder die Welt an sich, vor aller materiellen Erfüllung, lediglich eine formlose stetige Mannigfaltigkeit im Sinne der Analysis situs ist; die *Natur* der Metrik ist ihm an sich eigentümlich, nur die gegenseitige Orientierung der Metriken in den verschiedenen

Punkten ist zufällig, a posteriori und abhängig von der materiellen Erfüllung.

[I remark from an epistemological point of view: it is not correct to say that space or the world [spacetime] is in itself, prior to any material content, merely a formless continuous manifold in the sense of analysis situs; the *nature* of the metric [its infinitesimal Pythagorean-Riemannian character] is characteristic of space in itself, only the mutual orientation of the metrics at the various points is contingent, a posteriori and dependent on the material content.]

Weyl's statement expresses the view that within the context of General Relativity, the metric field cannot be removed such that empty spacetime remains. That is, Weyl expresses the impossibility of empty spacetime in the sense of understanding 'empty' to mean not merely empty of all *matter* but also empty of all *fields*. At another place, Weyl (1949a, Engl. edn, p. 172) says:

Geometry unites organically with the field theory; space is not opposed to things (as it is in substance theory) like an empty vessel into which they are placed and which endows them with far-geometrical relationships. No empty space exists here; the assumption that the field omits a portion of the space is absurd.

The metric field does not cease to exist in a world devoid of matter but is in a state of rest: As a *rest* or *zero* field it would possess the property of *metric homogeneity*; the mutual *orientations* of the *orthogonal* groups characterizing the Pythagorean-Riemannian nature of the metric everywhere would not differ from point to point. *This means that in a matter-empty universe the metric is fixed. Consequently, the set of congruence relations on spacetime is uniquely determined.* Since the metric uniquely determines the affine connection, the homogeneous *metric* field (rest field) determines an *integrable* affine structure. Therefore, a flat Minkowski spacetime consistent with the complete absence of matter is endowed with an integrable connection and thus determines all (hypothetical) free motions. According to Weyl then, there exists in the absence of matter a homogeneous metric field, a structural field (*Strukturfeld*), which has the character of a *rest* field, and which constitutes an all pervasive background that cannot be eliminated. The structure of this *rest* field determines the extension or reference of the spacetime congruence relations and determines Lorentz invariance. The field possesses no net energy and makes no contribution to curvature.

The contrast with Helmholtz and Lie is this: both of them require homogeneity and isotropy for physical space. From a general Riemannian standpoint, the latter characteristics are valid only for a matter-empty universe. Such a universe is flat and Euclidean, whereas a universe that contains matter is *inhomogeneous, anisotropic and of variable curvature*.

4.6   The G-Metric Interpretation of Weyl's Raumproblem: The Received View

The purpose of this subsection is to analyze and to critique the G-metric interpretation of Weyl's Raumproblem due to Cartan (1922, 1923b), Scheibe (1957, 1988) and Laugwitz (1958). It is important to note that Weyl's analysis of his Raumproblem consists of two quite distinct components:

1. a *conceptual* analysis, involving the Postulate of Freedom and the unique determination of the symmetric, linear connection by congruent transport, that leads to a set of purely algebraic constraints on the Lie algebra of the microinvariance group, and

2. a *technical* proof that these algebraic constraints entail that the microinvariance group is $O(p,q)$ for some signature (p,q).

It is crucial to keep in mind that the disputed issues regarding Weyl's Raumproblem concern only the interpretation of the conceptual component.

We first summarize the Finsler-metric interpretation of Weyl's Raumproblem as presented in detail in subsection 4.5. Weyl's starting point was a problem posed by Riemann, namely, the problem of singling out the Pythagorean-Riemannian metric from the class of positive-definite Finsler metrics. Weyl was unsatisfied with Riemann's *ad hoc* solution to the problem. Moreover, in Riemannian geometry, it is possible to compare lengths at distant points in a path-independent manner. Weyl objected to this aspect of Riemannian geometry and regarded it as an enduring remnant of Euclidean geometry which should have no place in a purely infinitesimal geometry. He was, therefore, led to make his first generalization by introducing the notions of gauge and of the infinitesimal length connection. Weyl also wanted his solution of the problem to encompass the General Theory of Relativity; consequently, he allowed for the possibility that the metrics under consideration might be indefinite. For Weyl, therefore, a purely infinitesimal geometry was determined by two fields: for a given choice of local gauge,

1. a field $F_p(dx)$ that determines the lengths[28] of vectors for every $p \in M$, which satisfies

$$F_p(\lambda dx) = |\lambda| F_p(dx), \qquad (4.6.2)$$

and

---

[28]For the case of an indefinite metric, one must work with the function $F_p^2(dx)$ which is homogeneous of degree 2. It should be noted that the 2-form

$$\frac{\partial^2 F_p^2(\xi)}{\partial \xi^i \partial \xi^j} d\xi^i d\xi^j \qquad (4.6.1)$$

induces by restriction (semi)-Riemannian metrics on the indicatrices defined by $F_p^2(\xi) = \pm 1$. One can determine the signature of $F_p^2(\xi)$ from the signatures of these (semi)-Riemannian metrics.

2. a 1-form field $\omega_i dx^i$, called the length connection, that determines, for two infinitesimally near points p' and p, a length $l_{p'}$ that is congruent to a length l_p according to

$$l_{p'} = l_p - l_p \omega_i dx^i, \qquad (4.6.3)$$

where $dx^i = x^i(p') - x^i(p)$.

The notions of a congruent map of $T(M_p)$ onto itself (and hence of the microinvariance group $\mathbf{G}_p$) and of a congruent map of $T(M_p)$ onto $T(M_{p'})$ are defined in terms of the structures in 1 and 2 according to (4.5.18) and to (4.5.22), respectively. Specifically, it is **not** in general the case on the Finsler-metric interpretation of Weyl's Raumproblem that the microinvariance group $\mathbf{G}_p$ characterizes (defines or describes) the metric at p; that is, the metrics in question are **not**, on our interpretation, the so-called G-metrics[29] used in the interpretations of Weyl's Raumproblem given by Cartan, Laugwitz and Scheibe. On the basis of a group-theoretic analysis, Weyl proves that if the Postulate of Freedom holds and if congruent transport determines a unique symmetric, linear connection, then the field $F_p(dx)$ must be of the Pythagorean-Riemannian type of some signature.

To facilitate the discussion of the two basic interpretations of Weyl's theorem, we indicate how the generalized (not necessarily positive-definite) Finsler metrics, and Weyl's length connection, both of which on our view were the focus of Weyl's interest, may be constructed from the the perspective of the theory of fiber bundles. Associated with the first component of the geometry, a Finsler metric, is a function $F \colon \mathbf{R}^n \to \mathbf{R}$ that is homogeneous, $F(\lambda \xi) = |\lambda| F(\xi)$. If F is not positive-definite, it may have to satisfy other conditions which need not concern us here. Denote by $\mathbf{G}$ the set of linear maps $G \colon \mathbf{R}^n \to \mathbf{R}^n$ which satisfy $G(\mathbf{0}) = \mathbf{0}$ and

$$F(G\xi) = F(\xi). \qquad (4.6.4)$$

Then, $\mathbf{G}$ is a closed subgroup of $GL(n)$. If $L^*(M)$ denotes the bundle of linear coframes over M, then a cross section $\sigma \colon M \to \mathbf{G} \backslash L^*(M)$ determines[30] a $\mathbf{G}$-structure on M which may be described locally by a field $\Sigma_j^i(x) dx^j$ of 1-forms. Given one such description, another equally good description is given by

$$G_r^i(x) \Sigma_j^r(x) dx^j, \qquad (4.6.5)$$

where $G_j^i(x)$ is an element of $\mathbf{G}$ for each x. A generalized Finsler metric is defined on M by

$$F_p(dx^i) = F(\Sigma_j^i(x) dx^j), \qquad (4.6.6)$$

where the x^i are the coordinates of $p \in M$. If $G_p \in \mathbf{G}_p$, then

$$F_p(G_{pj}^{\ \ i} dx^j) = F_p(dx^i) \qquad (4.6.7)$$

---

[29]See definition 28.

[30]The associated fiber bundle $\mathbf{G} \backslash L^*(M)$ has the typical fiber $\mathbf{G} \backslash GL(n)$. The cross section σ determines at each $p \in M$ an equivalence class of $\mathbf{G}$-related coframes of which the members are given by (4.6.5).

or

$$F(\Sigma_r^i G_{pj}{}^r dx^j) = F(\Sigma_j^i dx^j).$$ (4.6.8)

Hence, for some $G_j^i \in \mathbf{G}$,

$$\Sigma_r^i G_{pj}{}^r = G_r^i \Sigma_j^r;$$ (4.6.9)

that is,

$$G_{pj}{}^i = \Sigma^{-1i}{}_r G_s^r \Sigma_j^s.$$ (4.6.10)

Since this equation holds for every $G_p \in \mathbf{G}_p$, it follows that $\mathbf{G}_p$ is similar to $\mathbf{G}$ and hence that all of the microinvariance groups are similar to each other. It is important to note that there exist, in general, an *infinite* number of functions $F: \mathbf{R}^n \to \mathbf{R}$ which have the same microinvariance group; consequently, corresponding to each cross section $\sigma: M \to \mathbf{G} \backslash L^*(M)$, there are in general an infinite number of Finsler metrics[31]. As for the other aspect of Weyl's purely infinitesimal geometry, the length connection, it is a connection on the principal bundle $M \times \mathbf{R}^*$ over M, where $\mathbf{R}^*$ is the group of positive reals under the operation of multiplication $*$.

We now focus on the interpretation of Weyl's Raumproblem due to Cartan, Scheibe and Laugwitz which is based on the idea that congruence is defined by means of a group.

Definition 28 (*G-Metric Congruence*) *A G-metric is determined on a manifold M if and only if for every $p \in M$, there is given a group $\mathbf{G}_p$ that acts on $T(M_p)$, where the groups $\mathbf{G}_p$ are all isomorphic to a given closed subgroup $\mathbf{G}$ of $GL(n)$ and the dependence of the group action on p is smooth. Then, any two vectors v_p and $\tilde{v}_p$ in $T(M_p)$ are* **defined** *to be congruent just in case there exists an element $G_p \in \mathbf{G}_p$ such that $\tilde{v}_p = G_p v_p$.*[32]

The principal reason that Cartan understands Weyl's view of congruence in *G*-metric terms is the following passage (Weyl 1918c, 4 edn 1922):

> ... Riemann avait déjà montré qu'on peut prendre comme forme mètrique fondamentale, et avec le même droit, une fonction homogène du 4e degré par rapport aux différentielles ou même aussi n'importe quelle fonction construite autrement, qui ne fût pas nécessairement rationelle par rapport aux différentielles. Ce qui détermine originellement et généralement la métrique en un point P, c'est le *groupe des rotations*; les propriétés métriques de la multiplicité au point P sont connues, quand on sait quelles sont parmi les transformations linéaires de l'ensemble

---

[31] Note that the field $F_p(dx)$ is determined by the pair (σ, F), that is, by a **G**-structure *together with* a function F which has **G** as its symmetry group. For the **G**-metric interpretation of Weyl's Raumproblem, however, the first component of the geometry is determined by just an **N**-structure where **N** is a closed subgroup of $GL(n)$.

[32] Precisely how the action of the group $\mathbf{G}_p$ is to be specified varies; for example, one may use a **G**-structure. In order to accomodate a generalized notion of gauge, Scheibe uses an **N**-structure where **N** is the normalizer of **G** in $\mathbf{GL(n)}$. See below.

des vecteures en P, celles qui sont des représentations congruentes de cet emsemble.

Il y a autant de sortes differentes de déterminations métriques qu'il y a de groupes essentiellement différents de transformations linéaires (il faut entendre par groupes essentiellement différents ceux qui ne diffèrent pas seulement par le choix des coordonnées). Pour la métrique pythagoricienne qui est la seule que nous ayons considérée jusqu'ici, le groupe des rotations est formé de toutes les transformations linéaires qui transforment la forme quadratique différentielle en elle-même.

[... Riemann had already shown that one could take for the fundamental quadratic form, and with equal right, a function that is homogeneous of degree 4 with respect to the differentials or even as well any such function that is differently constructed, which is not necessarily rational with respect to the differentials. That which determines fundamentally and generally the metric at a point P is the *group of rotations*; the metrical properties of the manifold at a point P are known if one knows which among the linear transformations of the set of vectors at P are congruent mappings of this set.

There are as many different kinds of metrical determinations as there are groups of linear transformations that are essentially different (it is necessary to understand by groups that are essentially different those which do not differ only by a choice of coordinates). For the Pythagorean metric which is the only one that we have considered up until now, the group of rotations consists of all linear transformations that transform the quadratic differential form into itself.]

Other passages that presumably led Scheibe and Laugwitz to conclude that for Weyl there were two Raumprobleme and that Weyl intended his second Raumproblem to be understood in G-metric terms are the following:

Schon Riemann wies darauf hin, daß als metrische Fundamentalform, zunächst mit demselben Recht, eine homogene Funktion 4. Ordnung der Differentiale oder auch irgendeine anders gebaute Funktion erwartet werden könnte, die nicht einmal rational von den Differentialen abzuhängen brauchte. Aber selbst da dürfen wir noch nicht Halt machen. Das, was ursprünglich and allgemein die Metrik in einem Punkte P bestimmt, ist die *Gruppe der Drehungen*; die metrische Beschaffenheit der Mannigfaltigkeit im Punkte P ist bekannt, wenn man weiß, welche unter den linearen Abbildungen des Vektorkörpers (d. i. der Gesamtheit aller Vektoren) im Punkte P auf sich selbst *kongruente* Abbildungen sind. Es gibt so viele verschiedene Arten von Maßbestimmungen, als es wesentlich verschiedene Gruppen linearer Transformationen gibt (wobei wesentlich verschieden solche Gruppen sind, die sich nicht bloß durch die Wahl des Koordinatensystems voneinander unterscheiden). Für die bisher allein untersuchte *Pythagoreische Metrik* besteht die Gruppe der Drehungen aus allen linearen Transformationen, welche die quadratische Fundamentalform in sich überführen.[33]

[Riemann had already pointed out that the metrical fundamental form might with equal justification be taken to be a homogeneous function of 4th order in the differentials or even any other constructible [homogeneous] function which need not depend even rationally on the differentials. But, we must not stop at this point. That which fundamentally and generally determines the metric at a point P is the *group of rotations*; the metrical structure at a point P of a manifold is known, if one knows which of the linear mappings of the tangent space (that is the totality of all vectors) at the point P onto itself are *congruent* maps. There are as many different types of metrical structures as there are essentially different groups of linear transformations (whereby such groups are essentially different which differ from each other not merely by a choice of coordinate system). For the case of the *Pythagorean* metric which is the only case that has been investigated so far, the group of rotations consists of all linear transformations which leaves the quadratic fundamental form invariant.]

...Die Metrik hängt am Begriffe der *Kongruenz*, der jedoch rein infinitesimal gefaßt werden muß. Sollen wir also die Metrik der Mannigfaltigkeit an der beliebigen Stelle P_0 und den metrischen Zusammenhang dieses Punktes mit den Punkten seiner Umgebung vollständig beschreiben, so müssen wir angeben, *welche unter den linearen Abbildungen des Vektorkörpers in P_0 auf sich selber und welche seiner linearen Abbildungen auf die Vektorkörper in den zu P_0 unendlich benachbarten Punkten P "kongruente" Abbildungen sind.*[34]

[...The metric depends on the concept of *congruence* which must however, be understood purely infinitesimally. If we want to completely describe the metric of the manifold at an arbitrary point P_0 and the metric connection of that point with the points in its neighbourhood, then we must specify, *which among the linear maps of the tangent space at P_0 onto itself and which of its linear maps onto the tangent spaces at points infinitesimally close to P_0 are "congruent" maps.*]

Essentially the same phrases appear in a number of places; for example, (Weyl 1922a, GA II, p. 341) and (Weyl 1922b, GA II, p. 263).

---

[33]See Weyl (1918c, 4 edn, pp.124–125 and 5 edn, p. 137). The content of this quote is essentially the same as the French citation above.

[34]See Weyl (1923c, p. 47). The emphasis is Weyl's.

Remark 29 Note that the above quote from the French translation of (Weyl 1918c, 4 edn) differs from the German text, also quoted above, in two respects. First, a paragraph in the German text appears as two paragraphs in the French text. Second, the sentence "Aber selbst da dürfen wir noch nicht Halt machen" is omitted in the French version.

The characterization of geometry presented in these quotations is *incompatible* with the notion of a non-Riemannian or non-semi-Riemannian Finsler metric because for the case of such metrics, there exist distinct vectors v_p and $\bar{v}_p$ in $T(M_p)$ that have the same length according to the Finsler metric but are such that there does not exist any linear mapping of $T(M_p)$ onto itself that preserves the Finsler metric at the given point p and also maps v_p into $\bar{v}_p$. It is entirely possible that Cartan seized on the words "Il y a autant de sortes differentes de déterminations métriques ..." and concluded that Weyl was characterizing metrical structure in terms of group actions, that is, in terms of what we would now call a G-structure. Despite the fact that in the immediately preceding paragraph Weyl refers to Riemann's concern about the possibility of Finsler metrics ("...une fonction homogène de 4e degré ..."), Cartan did not, for whatever reason, remark on the apparent inconsistency in Weyl's discussion. Likewise, neither Laugwitz nor Scheibe remark on this particular inconsistency. In view of the difficulties that arise if a G-metric interpretation of Weyl's Raumproblem is adopted, which will be discussed below, it makes more sense to assume that Weyl used inappropriate language because of the traditions of the Erlanger Programme in which there is a one-to-one correspondence between congruence defined in terms of structure and congruence defined in terms of a group and because such a correspondence also exists in the cases of Riemannian and semi-Riemannian geometries,[35] the geometries that were his main focus.

We first consider Cartan's analysis of Weyl's Raumproblem. As we have discussed above, Cartan obtained his understanding of Weyl's Raumproblem from the French translation of Weyl's (1918c, 4 edn 1922) book. Soon after, Cartan (1922, 1923b) wrote two papers on the subject. It is clear from these papers that Cartan adopted a G-metric view of Weyl's analysis. Indeed, Cartan (1923b) says "Ce groupe définit la *métrique* au point P." Cartan formulated his theorem for any volume-preserving subgroup of $GL(n)$. Somewhat later, the theorem, without the restriction regarding volume preservation, was reformulated[36] in terms of G-structures as follows:

---

[35] Scheibe (1988, p. 68) correctly notes that Weyl repeatedly emphasized that in his view the basic concept of geometry is congruence. It appears that Scheibe infers from this that Weyl held the view that geometric structure was *ontologically* determined by the group actions that describe the congruence relations; that is, Scheibe concludes that Weyl held a G-metric view of geometric structure. This conclusion, however, is a *non sequitur* because even if one holds that geometry is determined by a structural field of some sort such as, for example, a Riemannian metric field, one may nevertheless hold that congruence is a fundamental concept.

[36] It is unclear to the authors both who first formulated the concept of a G-structure and who first applied it to Cartan's version of Weyl's Raumproblem.

Theorem 30 (Cartan's Raumproblem) *Let M be an n-dimensional manifold and let* **G** *be a closed subgroup of* $GL(n)$[37]. *Then, if every* **G**-*structure on M admits one and only one symmetric, linear connection, the group* **G** *is an orthogonal group* $O(p,q)$ *for some signature* (p,q).

It seems probable that Cartan formulated his theorem for arbitrary subgroups of $GL(n)$ because he interpreted Weyl's analysis in G-metric terms. Although Cartan's theorem is interesting in its own right, we think that it has little to do with Weyl's Raumproblem; however, see remark 31 below. As a partial explanation of what we regard as Cartan's misinterpretation of Weyl's Raumproblem, we note that with the exception of a brief mention of a particular Finsler metric in a single sentence, Weyl did not discuss Finsler metrics in the fourth edition of his book *Raum, Zeit, Materie*, from the translation of which Cartan obtained his understanding of Weyl's Raumproblem.

In our view, the use of the terms[38] 'Weyl-Cartan theorem' and 'Weyl-Cartan Raumproblem' is inappropriate because the spirit of Cartan's analysis is quite different from our interpretation of of Weyl's analysis in the form that we have presented it in subsection 4.5. In addition, it is interesting to note that Cartan's theorem does not make use of Weyl's Postulate of Freedom; however, this postulate is essential to Weyl's proof if our view of Weyl's Raumproblem is correct.

Remark 31 It is worth noting that despite the fundamental philosophical differences, as seen by us, between Cartan and Weyl in the context of the Raumproblem, Cartan's result may also be applied[39] to Weyl's (*Finsler-metric*) Raumproblem for the special case of a *fixed* gauge as follows.

Suppose a Finsler metric, that need not be positive-definite, is given. If the microinvariance group of the Finsler metric is **G**, then the metric is determined by the function $F : \mathbf{R}^n \to \mathbf{R}$ that determines the type of the metric and by a **G**-structure that is locally described by a coframe field $\Sigma_j^i(x^i)dx^j$ according to

$$F_p(dx^i) = F(\Sigma_j^i(x^i)dx^j), \qquad (4.6.11)$$

where p is the point with coordinates x^i. Clearly, the problem of transporting the Finsler metric parallelly is equivalent to that of transporting the equivalence class of coframe fields. Thus, there is exactly one symmetric linear connection that is compatible with the Finsler metric if and only if there is exactly one symmetric linear connection that is compatible with the **G**-structure that is associated with the Finsler metric. Now, **G** is some subgroup of $O(p,q)$ for some signature (p,q); consequently, **G** is a closed subgroup of $GL(n)$. Hence, Cartan's theorem in the

---

[37] Note that $GL(n)$ means $GL(n, \mathbf{R})$.

[38] See for example (Klingenberg 1959) and (Freudenthal 1960).

[39] As far as we know, no one else has applied Cartan's theorem to Weyl's Raumproblem in the way described here. Certainly, neither Laugwitz nor Scheibe thought of applying Cartan's theorem in this way as is evident from Scheibe's (1988) comments in footnote seven on page 65 of the cited work.

form quoted above applies, and **G** must be the group $O(p, q)$ for some signature (p, q). In this way, all of the Finsler metrics except for the semi-Riemannian ones are eliminated. We emphasize that Cartan's argument does not apply to Weyl's problem in the context of Finsler metrics with variable gauge. Indeed, for a Weylian geometry the group **G** would have to be $\mathbf{R}^* \times O(p, q)$, but such a **G**-structure determines only an equivalence class of symmetric linear connections. The additional structure needed to guarantee uniqueness of the symmetric linear connection in the variable gauge case is Weyl's length connection which does not play a role in Cartan's theorem.

Remark 32 How did Weyl view Cartan's theorem? From a microinvariance or Finslerian point of view, it is clear that the only subgroups of $GL(n)$ that are relevant to Weyl's problem are subgroups of $O(p, q)$ for some signature (p, q). Although motivated by G-metric ideas, Cartan's theorem may be used, as noted above, without reference to G-metric concepts. It is in *this* purely technical sense that we have used Cartan's theorem in remark 31 to solve Weyl's Raumproblem in the fixed-gauge case. Weyl may have had in mind just such an application of Cartan's theorem restricted to subgroups (note our emphasis in the quotation) of $O(p, q)$ when Weyl (1923c, Appendix 12, p. 88) said:

> Ein ganz anderer Beweis ist von E. Cartan angegeben worden ...;
> er stützt sich auf die umfassenden und tiefgehenden älteren Unter-
> suchungen von Cartan zur Theorie der kontinuierlichen Gruppen, in
> denen es ihm gelungen war, das Problem der Aufstellung aller abstrak-
> ten Gruppen ... und ihrer Realisierung durch inf. lineare Operationen
> weitgehend zu lösen. ... *Er braucht jetzt unter den damals von ihm*
> *aufgestellten Gruppen nur diejenigen herauszusuchen, welche meinen*
> *Forderungen genügen.*[40]

> [A quite different proof has been given by E. Cartan ...; it rests on
> the earlier, general and deep investigations by Cartan on the theory
> of continuous groups in which he succeeded in solving quite generally
> the problem of the presentation of all abstract groups ... and of their
> realization by means of infinitesimal, linear operations. ... *He needs*
> *only to select from the groups presented by him those which satisfy my*
> *requirements.*]

The paper that Weyl refers to (Cartan 1922) considers all subgroups of $IGL(n)$ that preserve volumes. Note that the translations enter merely because Cartan prefers to formulate his analysis in terms of the affine tangent hyperplane rather than in terms of the tangent vector space used by Weyl. Thus, all subgroups of $SL(n)$ are considered. According to the Received View of Weyl's Raumproblem,

---

[40]The emphasis is ours.

Weyl is here concerned with the second Weyl Raumproblem. In this G-structure-context, *all* of the groups considered by Cartan are relevant; however, the emphasized text in the above quote indicates that Weyl thinks that only a restricted subclass of these groups are relevant to Weyl's problem, which would be the case if Weyl's conception of the Raumproblem was the one that we argue for, namely the Finsler-metric version.

Moreover, it should be noted that Cartan's discussion of the conceptual aspect of the Raumproblem was, to say the least, very brief. Cartan's (1922, 1923b) original formulation employs language that is quite different from that used in theorem 30. In addition, the G-metric ideas that probably motivated Cartan do not stand out in the papers cited. There is nothing overt in Cartan's presentation that would have alerted Weyl to what we think are the deep differences in their philosophical positions. It is entirely likely, therefore, that Weyl regarded Cartan's work as applying to Weyl's (Finsler-metric) Raumproblem only in the technical manner indicated above in remark 31.

Although Scheibe's (1957) published dissertation predates Laugwitz's (1958) paper, we nevertheless discuss Laugwitz's contribution first because in a later article, Scheibe (1988) embraces the distinction introduced by Laugwitz concerning the 'first' and 'second' Raumproblem discussed below. The argument presented by Laugwitz (1958) clearly illustrates how Weyl's analysis, understood as an attempt to eliminate non-Riemannian Finsler metrics, is undermined by a G-metric interpretation. Referring to (Weyl 1922b) and (Weyl 1923c), Laugwitz remarks[41]

> Das zweite Weylsche Raumproblem ... geht von einer ganz anderen Definition der Metrik aus: Dabei ist eine Gruppe von linearen homogenen Transformationen in $T(P)$ vorgegeben, und zwei Vektoren heißen genau dann gleich lang, wenn es eine Transformation aus der Gruppe gibt, welche den ersten in den zweiten überführt.

> [The second Weyl Raumproblem ... is based on a quite different definition of the metric. In this approach, a group of linear, homogeneous transformations in $T(P)$ is given by stipulation, and two vectors are defined to be of the same length just in case there exists a transformation of the group which maps the first into the second.]

Earlier in the paper, Laugwitz defines the notion of an *Eichhyperfläche*[42] (unit hypersurface) E_p at a point $p \in M$ for an arbitrary positive-definite Finsler metric.

---

[41] Laugwitz asserts that Weyl formulated two distinct Raumproblems, one presented in Weyl's commentary to Riemann's inaugural address (Weyl 1919c) (labelled the first Weyl Raumproblem) and the other (labelled the second Weyl Raumproblem) presented in (Weyl, 1922a, 1922b, 1923c). It should be noted that Laugwitz does not provide a positive, clear statement as to what the 'second Weyl Raumproblem' is. Laugwitz's main goal is to show that Weyl did not prove the 'first Weyl Raumproblem'.

[42] The English term for the hypersurface defined by (4.6.12) is *indicatrix*.

It is the set of $v_p \in T(M_p)$ which satisfy

$$F_p(v_p) = 1. \tag{4.6.12}$$

For any two directions at $p \in M$, there exist corresponding unit vectors v_p and $\tilde{v}_p$. Since these vectors have the same length, there necessarily exists, on a G-metric interpretation, a group element $G_p \in \mathbf{G}_p$ such that $\tilde{v}_p = G_p v_p$. It follows that the group $\mathbf{G}_p$ acts *transitively* on the Eichhyperfläche E_p. Laugwitz then points out on the basis of a previous theorem that the hypersurface must be a hyper-ellipsoid; that is, the metric must be Riemannian. The point that Laugwitz is making is that Weyl's assumptions (on a G-metric interpretation) eliminate *from the outset* all positive-definite Finsler metrics other than the Riemannian ones; that is, in order to eliminate these metrics, it is *not* necessary to use either the Postulate of Freedom or the unique determination of the symmetric linear connection by congruent transport. The problem of eliminating the Finsler metrics on the basis of these latter assumptions, Laugwitz calls Weyl's 'first Raumproblem'. Since, according to Laugwitz, Weyl provided a proof only in the context of the 'second Raumproblem' and since this proof could not possibly (on a G-metric interpretation) have been a proof of the 'first Raumproblem', the 'first Raumproblem' remained a conjecture (Vermutung) according to Laugwitz.

From our point of view, however, it is quite clear that for Weyl, there was only *one* Raumproblem because, in all[43] of Weyl's discussions pertaining to the Raumproblem, the elimination of the Finsler metrics, other than the Riemannian ones, is always regarded by Weyl as part of the problem. As we have shown in subsection 4.5, Weyl's proof, on a microinvariance-group interpretation, does eliminate the other-than-Riemannian Finsler metrics.

We finally turn to the work of Scheibe (1957, 1988) who also adopts an explicit G-metric view of Weyl's Raumproblem. Recall that on this view, there exists, for each $p \in M$, a group $\mathbf{G}_p$ that acts on $T(M_p)$; moreover, any two of them $\mathbf{G}_p$ and $\mathbf{G}_{p'}$ are isomorphic to each other. In addition, any two vectors v_p and $\tilde{v}_p$ in $T(M_p)$ are G-congruent just in case there exists an element $G_p \in \mathbf{G}_p$ such that $\tilde{v}_p = G_p v_p$. In the fixed-gauge case, that is, in the case that the gauge is everywhere determined once and for all, a G-congruence is determined by a $\mathbf{G}$-structure on M. The semi-Riemannian G-metrics are then singled out according to Cartan's theorem 30 by the requirement that the $\mathbf{G}$-structure admit a unique, symmetric, linear connection. Scheibe recognized that this account is not sufficiently general to encompass the case of greatest importance to Weyl, namely, that of purely infinitesimal geometry for which the choice of gauge varies as a function of position. Scheibe generalizes Cartan's approach[44] as follows.

---

[43] With the exception of (Weyl 1922b).

[44] With regard to Weyl's proof, Scheibe (1957, p. 162) says:

> Was zweitens den Weylschen Beweis des erwähnten algebraischen Satzes angeht, so ist dieser wegen seiner elementar gehaltenen Durchführung recht verwickelt.

> [Secondly, as far as Weyl's proof of the above mentioned algebraic theorem is con-

In Cartan's approach, the **G**-structure determines an action of the group on each tangent space $T(M_p)$ given by

$$G_{pj}{}^i v_p^j = \Sigma^{-1}{}_r^i(x(p)) G_s^r \Sigma_j^s(x(p)) v_p^j, \tag{4.6.13}$$

where the coframe field $\Sigma_j^i(x) dx^j$ describes the **G**-structure locally. The coframe $\Sigma(x(p))$ determines the orientation of the action of the group **G** at the point $p \in M$. Scheibe adopts the view that only the group action is important and speaks of a **G**-field, that is, of a **G**-action on each tangent space $T(M_p)$. He points out that a weaker structure suffices to describe the **G**-field, namely, an **N**-structure where **N** is the *normalizer* of **G** in $GL(n)$. If $A \in \mathbf{N}$ and $G \in \mathbf{G}$, then $\tilde{G} \in \mathbf{G}$, where

$$\tilde{G}_j^i = A^{-1}{}_r^i G_s^r A_j^s. \tag{4.6.14}$$

Thus, the **G**-action (4.6.13) is described equally well by

$$\tilde{G}_{pj}{}^i v_p^j = [A(x(p))\Sigma(x(p))]^{-1}{}_r^i G_s^r [A(x(p))\Sigma(x(p))]_j^s v_p^j. \tag{4.6.15}$$

It follows that the group action is correctly described if the coframes $\Sigma(x)$ locally describe only an **N**-structure on M. For the particular case of the group $O(p,q)$, the normalizer is the corresponding conformal group $CO(p,q)$ which means that the gauge may be varied independently at each point. The **G**-action partitions each tangent space $T(M_p)$ into equivalence classes[45], the orbits under the **G**-action. One is then faced with the problem of connecting the equivalence classes of $T(M_p)$ with the equivalence classes of $T(M_{p'})$ where $p' \in M$ is infinitesimally near $p \in M$. Scheibe (1988) proposes to do this by means of a connection on the principal fiber bundle

$$\langle M \times \mathbf{N}/\mathbf{G}, \pi, M, \mathbf{N}/\mathbf{G} \rangle. \tag{4.6.16}$$

---

cerned, it is quite complicated because of the restriction to elementary methods.]

He adds that Cartan has provided another proof that is more satisfactory. It should be noted, however, that Weyl's proof is stronger than Cartan's proof because Weyl did not need to make use of the trace condition (4.5.45). Weyl (1923c, Appendix 12) also notes that his proof is to be preferred because it is more direct.

> Im Gegensatz zu dem Cartanschen Beweis nimmt der meinige nicht den Umweg über die Untersuchung der Konstitution abstrakter Gruppen. Er stützt sich auf die klassische, von Weierstrass herrührende Theorie der einzelnen linearen Abbildung.

> [In contrast to Cartan's proof my proof does not involve the detour into the investigation of the constitution of abstract groups. It depends on the classical theory of individual linear maps that originated with Weierstrass.]

[45]The notion of gauge as a choice of length standard does not apply to G-fields except in the case of an $O(p,q)$-field. Scheibe is, therefore, forced to replace equivalence classes of elements of $T(M_p)$ defined in terms of length by the orbits under the action of $\mathbf{G}_p$. In addition, Weyl's simple notion of a length connection has to be replaced either by a connection on the principal fiber bundle (4.6.16) or by a single connection (Scheibe 1957, p. 163). To say the least, Scheibe's interpretation of Weyl's intuitive and physically motivated idea of gauge appear rather contrived. From the Finsler-metric point of view, on the other hand, the term 'gauge' *always* means 'length'.

Since an element of $\mathbf{N}$ preserves equivalence classes, such a connection will map every equivalence class at $p \in M$ into an equivalence class at $p' \in M$, where p' is infinitesimally near to p. In the particular case that $\mathbf{G} = O(p,q)$, the normalizer $\mathbf{N} = CO(p,q)$ and the factor group $\mathbf{N}/\mathbf{G} = CO(p,q)/O(p,q) = \mathbf{R}^*$, the group of positive real numbers under the operation of multiplication $*$. Then the connection[46] in question is Weyl's length connection (4.6.3). In the general case, Scheibe makes the following definitions.

Definition 33 (G-metric) *A* **G**-*metric consists of a* **G**-*field, that is, an* **N**-*structure together with a connection on the principal fiber bundle (4.6.16) which we shall call the* **G**-*metric connection.*

Definition 34 *A (not necessarily symmetric) linear connection is compatible with a* **G**-*metric if and only if the connection is reducible to a connection in the principal bundle of* **N**-*related coframes and the linear connection induces the* **G**-*metric connection.*

Scheibe (1988, See F'_3, p. 69) then states the following result:

Theorem 35 (Scheibe 1988) *For a given* **G**, *if for every* **G**-*metric, that is, for every choice of* **N**-*structure and every choice of* **G**-*metric connection, there exists one and only one symmetric linear connection that is compatible with the* **G**-*metric, then* **G** *is one of the groups* $O(p,q)$ *for some signature* (p,q).

Scheibe does not provide a proof of this 'result', nor does he give a reference to a publication in which the proof may be found.

Remark 36 Scheibe (1988, p. 72) states that the interpretation of Weyl's Raumproblem just outlined in the definitions 33 and 34 and in theorem 35 is the one that appears to him to be the correct one. Scheibe points out that in (Scheibe 1957, Published Dissertation 1957) he adopted a different interpretation in which what we have called the *G*-metric connection is replaced by a single (not necessarily symmetric) linear connection.

Having outlined Scheibe's main position, we shall now provide a critique of his work. The main difficulty with Scheibe's view, indeed with any *G*-metric interpretation of Weyl's Raumproblem, is that it rules out *ab initio* any consideration of the Finsler metrics; however, in the context of his Raumproblem, Weyl repeatedly included the elimination of the Finsler metrics, which are definitely not *G*-metrics, as part of the problem. On this point, the textual evidence is quite unambiguous. The last of the three quotes which immediately follow definition 28 comes from

---

[46]Note that in general the connection is much more complicated. Moreover, there are also degenerate cases such as that for which $\mathbf{G}$ is the group of upper triangular matrices. The normalizer $\mathbf{N}$ of this group is the group of upper triangular matrices; that is, the group is self-normalizing. In this case, the factor group $\mathbf{N}/\mathbf{G}$ consists of the identity alone. There are, moreover, n distinct equivalence classes, one for each row. The equivalence classes have different dimensions from one to n. There is, therefore, only one way to map the equivalence classes at p onto those at p'.

Weyl's (1923c) *Mathematische Analyse Des Raumproblems* in which he presents his most thorough treatment of the Raumproblem, yet this brief statement is the only one in the *Mathematische Analyse Des Raumproblems* that supports a G-metric interpretation of the Raumproblem. Regarded as a statement of an important, new ontology for metrical relations, this statement is remarkably terse[47]. Even Scheibe (1957, p. 162) remarks that it is "allzu skizzenhaft".

On the other hand, Weyl discusses Finsler metrics at great length in the fifth lecture which begins with the statement of what the Raumproblem consists of. Weyl (1923c, p. 29) starts his fifth lecture by saying:

> Wir hatten das Raumproblem in drei Teile zerlegt:
> I. Begründung der allgemeinen metrischen oder *Riemannschen* Infinitesimalgeometrie;
> II. durch die Forderung der Homogenität werden daraus die Kugelräume ausgeschieden;
> III. Frage: Durch welche Eigenschaften grundsätzlicher Natur ist unter den Kugeln diejenige von der Krümmung $\lambda = 0$ ausgezeichnet?
> Nachdem wir jetzt II. bewiesen haben, wenden wir uns der unter I. gestellten Aufgabe zu. Von zwei Annahmen ging die *Riemannsche* Geometrie aus: a) Meßbarkeit der Linienelemente, b) Gültigkeit des *Pythagoras* im Unendlichkleinen. Riemann selbst streift die Möglichkeit, das *ds* als die 4. Wurzel aus einem homogenen Polynom 4. Ordung der Differentiale anzusetzen mit Koeffizienten, welche im allgemeinen von Stelle zu Stelle veränderlich sind, oder als die 6. Wurzel aus einer ganzen rationalen Form 6. Grades, usf.
>
> [We have divided the Raumproblem into three parts:
> I. Justification of the general metrical or Riemannian infinitesimal geometry;
> II. The spaces of constant curvature are singled out by the requirement of homogeneity;
> III. Question: Which properties of a fundamental character single out from among the spaces of constant curvature the one with curvature $\lambda = 0$?
> Now that we have proven II., we turn to the problem presented in I. Riemannian geometry is based on two assumptions: a) the measurability of the line elements, b) the validity of Pythagoras in the infinitely small. Riemann himself touches upon the possibility that *ds* may be given by the 4th root of a homogeneous polynomial of 4th order in the differentials with coefficients that in general vary from place to place or by the 6th root of a completely rational form of the 6th degree, and so forth.]

[47]If the passage is interpreted in microinvariance terms, however, then it is merely a brief statement of the group-theoretic properties of Weylian geometries.

After discussing and dismissing Riemann's justification for the Pythagorean nature of the metric, Weyl (1923c, pp. 29–30) suggests an alternate approach:

> ... Man wird doch wohl verlangen müssen, daß die Natur der Metrik an jeder Stelle des Raumes die gleiche ist; d. h. man wird fordern, daß wenn an einer beliebigen Raumstelle P das ds durch einen Ausdruck $f_P(dx_1, dx_2, \ldots, dx_n)$ gegeben ist, wo f_P eine homogene (aber keineswegs notwendig rationale) Funktion 1. Ordnung seiner Argumente ist, die den verschiedenen Stellen P zugehörigen Funktionen f_P alle einer einzigen Klasse (f) angehören in dem Sinne, daß sie alle aus einer solchen Funktion f durch homogene lineare Transformation der Argumente hervorgehen; denn an den Differentialen dx_i äußert sich ja der Übergang zu einem beliebigen anderen Koordinatensystem als willkürliche lineare Transformation. Im *Pythagoreischen* Falle ist diese Forderung erfüllt, weil jede positive-definite quadratische Form durch lineare Transformation aus der Einheitsform gewonnen werden kann. Jeder solchen Klasse (f) homogener Funktionen entspricht eine Raumklasse mit bestimmt gearteter Metrik; unter diesen Raumklassen ist die *Pythagoreisch-Riemannsche*, welche der Annahme
>
> $$f^2 = (\xi^1)^2 + (\xi^2)^2 + \cdots + (\xi^n)^2$$
>
> entspricht, *eine einzige;* und es gilt diese Klasse durch innere einfache Eigenschaften aus allen anderen herauszuheben.

[... No doubt, one must require that the nature of the metric is the same at every point of the space; that is to say, one requires that if at an arbitrary point of space P, ds is given by $f_P(dx_1, dx_2, \ldots, dx_n)$ where f_p is a homogeneous (but certainly not necessarily rational) function of 1st order in its arguments, then the functions f_P associated with various points P all belong to a single class (f) in the sense that they all are generated from one such function f by means of homogeneous, linear transformations of the arguments because for the differentials dx_i, the transformation to some other arbitrary coordinate system is expressed by an arbitrary linear transformation. In the Pythagorean case, this requirement is fulfilled because each positive-definite quadratic form can be obtained from the standard one by means of a linear transformation. To each such class (f) of homogeneous functions, there corresponds a class of spaces with a definite type of metric. Among these classes of spaces, the Pythagorean-Riemannian, which corresponds to the assumption

$$f^2 = (\xi^1)^2 + (\xi^2)^2 + \cdots + (\xi^n)^2 \qquad (4.6.17)$$

is unique, and the objective is to single out this class from all the others by means of simple intrinsic properties.]

It would appear from these two passages and the context in which they occur that for Weyl the Raumproblem (I. in the first of the above two quotes) is inextricably bound up with Finsler metrics. Moreover, in the third edition of Weyl's (1919c, 3 edn 1923, pp. 26–27) commentary on Riemann's Habilitationsvortrag — the problem of selecting the Riemannian metrics out of the class of Finsler metrics has its origin in Riemann's work — Weyl refers to his (Weyl's) solution of the problem that takes into account Relativity Theory and cites in particular the Spanish lectures:

> Eine ganz andere Lösung des Raumproblems, welche der neuen durch die Relativitätstheorie geschaffenen Situation voll Rechnung trägt, rührt von Weyl her. Vgl. darüber den Vortrag "Das Raumproblem", Jahresbericht der Dtsch. Math.-Vereinig. 1923 [sic][48], ferner: Mathem. Zeitschr. Bd. 12 (1922), S. 114, und die demnächst bei Julius Springer (Berlin) erscheinenden Vorlesungen über die "Mathematische Analyse des Raumproblems".

> Geometrische Untersuchungen in Räumen, die in jedem Punkte eine beliebige Maßbestimmung tragen im Sinne der Gleichung (1) [$ds = f_P(dx_1, dx_2, \ldots, dx_n)$], sind neuerdings von P. Finsler angestellt worden (Über Kurven und Flächen in allgemeinen Räumen, Göttinger Dissertation 1918).

> [A quite different solution of the Raumproblem, which takes into account fully the new situation created by the Theory of Relativity is due to Weyl. In this regard, see the lecture "Das Raumproblem", Jahresbericht der Dtsch. Math.-Vereinig. 1923 [sic], and in addition: Mathem. Zeitschr. vol. 12 (1922), p. 114, and the lectures on the "Mathematische Analyse des Raumproblems" to be published by Julius Springer (Berlin).

> Geometric investigations of spaces which at each point are equipped with an arbitrary distance function in the sense of equation (1) [$ds = f_P(dx_1, dx_2, \ldots, dx_n)$] have been recently carried out by P. Finsler (Über Kurven und Flächen in allgemeinen Räumen, Göttinger Dissertation 1918).]

It is important to note that just before this passage, Weyl presents a detailed characterization of the Raumproblem in terms of Finsler metrics and discusses the 'first satisfying answer to this question' due to Helmholtz and Lie. It is clear, therefore, that when Weyl refers to "Eine ganz andere Lösung des Raumproblems", he is refering to the *same* problem; moreover, *immediately* after Weyl refers to his solution to the Raumproblem, Weyl cites none other than Finsler himself! Weyl's reference to his solution of the Raumproblem, which Laugwitz and Scheibe insist

---

[48]The date is 1922; see (Weyl 1922a).

can only be applicable to the so-called second Raumproblem, is immediately surrounded, before and after, by references to Finsler metrics! We think that the textual evidence (here and elsewhere) is quite clear: *Contrary to the claim of Laugwitz and Scheibe, Weyl considered only one Raumproblem, namely, the Raumproblem having to do with the elimination of other-than-(semi)-Riemannian Finsler metrics. Weyl was not even aware of the so-called second Raumproblem which was invented by Cartan albeit because Cartan misinterpreted what Weyl was doing.*

It is noteworthy that in the published version of his thesis, Scheibe (1957) does not even mention Finsler metrics. In view of the fact that the published version of Scheibe's dissertation is entitled *Über das Weylsche Raumproblem*, this omission cannot be excused as merely a minor oversight.

In a more recent work, Scheibe (1988) proposes that there are four versions of Weyl's Raumproblem. In one of these, Finsler metrics do play a role. We briefly characterize Scheibe's classification as follows:

S_1 In the fixed-gauge, positive-definite case, the class of Riemannian geometries is to be selected out of the class of positive-definite Finsler metrics.

S_2 In the fixed-gauge, indefinite case, the class of $O(p,q)$-structures (**G**-structures with $\mathbf{G} = O(p,q)$) is to be selected out of the class of **G**-structures, where **G** is any closed subgroup of $GL(n)$.

S_3 In the variable-gauge, indefinite case, the class of Weylian geometries is to be selected out of the class of G-metrics, where a **G**-metric consists of a **G**-field together with a **G**-metric connection and **G** is any[49] closed subgroup of $GL(n)$.

S_4 In the variable-gauge, indefinite case, the class of Weylian geometries is to be selected out of the class of G-metrics, where a **G**-metric consists of a **G**-field (as in S_3) and a single (not necessarily symmetric) linear connection that is compatible with the **G**-field.

Scheibe characterizes version S_1 as a conjecture due to Weyl that was proved much later by Laugwitz (1958). Version S_2 is characterized as a modern reformulation of Cartan's (1922, 1923b) work on Weyl's Raumproblem. Scheibe's versions S_3 and S_4 are due to Scheibe.

Note that there is a *striking* lack of parallelism[50] between the first version and the other three. The first version S_1 is founded on a structural (metric) field F_p which determines the lengths of vectors in each tangent space $T(M_p)$; moreover, the microinvariance groups $\mathbf{G}_p$ associated with that structure are defined in terms

---

[49]There is a restriction on **G** which Scheibe (1957) states in condition (a) at the top of page 192.

[50]It would seem that Scheibe (1988, footnote 7 on page 65) accepts Laugwitz's notions of a 'first Weyl Raumproblem' and of a 'second Weyl Raumproblem'. In view of this, Scheibe's (1988) classification is even more peculiar because there should be four cases for each of these problems corresponding to the choices of fixed or variable gauge and of positive-definite or indefinite signature.

of that structure and are, therefore, properties of that structure. In contrast, the other versions S_2, S_3 and S_4 are based on a **G**-field, that is, an action of a group **G** on each tangent space $T(M_p)$; moreover, a generalized notion of 'congruence' which replaces 'sameness of length' is defined in terms of the group actions. Ontologically speaking, these two approaches are fundamentally different from each other. It is inconceivable that Weyl, both as a philosopher and as a mathematician, would not have distinguished, both conceptually and terminologically, between such radically different ontological frameworks.

Laugwitz and Scheibe hold that the proof presented by Weyl in his Spanish lectures pertains to what they call the second Raumproblem. Of the two, only Scheibe (1988, pp. 70–72) has discussed the logic of Weyl's proof regarded as a proof of the second Raumproblem, and Scheibe's discussion is very brief and sketchy. In order to make Weyl's proof fit the G-metric conception, Scheibe interprets Weyl's analysis in a manner that completely distorts the meaning and significance of Weyl's Postulate of Freedom 23. Scheibe proposes that for Weyl, each member of the general class of geometries is determined by a **G**-field, where **G** is some closed subgroup of $GL(n)$, and by a 'full class' $\{\Lambda^i_{jk}(x)\}$ of (not necessarily symmetric) linear connections that are compatible with the **G**-field. He immediately notes that Weyl did not use this approach in characterizing Weylian geometry. Instead, Weyl used a length connection $\omega_i dx^i$ as the second component of the geometry. Scheibe asserts, however, that there is no inconsistency because the two approaches are equivalent. In order to explain what Scheibe means, we discuss the situation for the case of Weylian geometry the first component of which is an $O(p,q)$-field. An $O(p,q)$-field can equivalently be described by a conformal structure which is locally determined by a field $\hat{g}_{ij}(x)$ that satisfies some gauge-fixing condition such as $\det(\hat{g}_{ij}(x)) = -1$. A not necessarily symmetric, linear connection $\Lambda^i_{jk}(x)$ is compatible with the conformal structure $\hat{g}_{ij}(x)$ if and only if there exists a field $\omega_k(x)$ such that

$$\hat{g}_{is}\Lambda^s_{jk} + \hat{g}_{js}\Lambda^s_{ik} = \hat{g}_{ij,k} + \hat{g}_{ij}\omega_k. \tag{4.6.18}$$

The general solution of (4.6.18) for $\Lambda^i_{jk}(x)$ is given by

$$\Lambda^i_{jk}(x) = \Gamma^i_{jk}(x) + T^i_{jk}(x), \tag{4.6.19}$$

where $\Gamma^i_{jk}(x) = \Gamma^i_{kj}(x)$, $T^i_{jk}(x) = -T^i_{kj}(x)$ and

$$\Gamma^i_{jk} = K^i_{jk} + \frac{1}{2}(\delta^i_j\omega_k + \delta^i_k\omega_j - \hat{g}_{jk}\hat{g}^{ir}\omega_r) + \hat{g}^{ir}(\hat{g}_{js}T^s_{kr} + \hat{g}_{ks}T^s_{jr}). \tag{4.6.20}$$

The $K^i_{jk}(x)$ are the conformal connection coefficients given by

$$K^i_{jk} = \frac{1}{2}\hat{g}^{ir}(\hat{g}_{rj,k} + \hat{g}_{rk,j} - \hat{g}_{jk,r}), \tag{4.6.21}$$

provided that the gauge-fixing condition $\det(\hat{g}_{ij}(x)) = -1$ is used. The class $\{\Lambda^i_{jk}(x)\}$ of linear connections that are compatible with the conformal structure

$\hat{g}_{ij}(x)$ is given by (4.6.19), (4.6.20) and (4.6.21), where the fields $\omega_k(x)$ and $T^i_{jk}(x)$ may be chosen arbitrarily. Clearly, if a non-empty class $\{\Lambda^i_{jk}(x)\}$ of compatible, linear connections exists, then there exists at least one compatible linear connection $\Lambda^i_{jk}(x)$ which determines a length connection $\omega_k(x)$ according to (4.6.18); moreover, given a length connection $\omega_k(x)$, any choice of $T^i_{jk}(x)$ determines a compatible, linear connection. It is also clear that given a choice of length connection or a choice of compatible, linear connection, there is exactly one compatible, linear connection that is symmetric, namely, the linear connection for which $T^i_{jk}(x) = 0$. Scheibe's claim that a Weylian geometry can be described in different, equivalent ways is therefore correct.

How then does Scheibe's conception conflict with Weyl's Postulate of Freedom? According to Scheibe, the first component of the geometry is given as a *field*, an $O(p, q)$-field or equivalently a conformal structure $\hat{g}_{ij}(x)$. For the determination of a compatible, linear connection $\Lambda^i_{jk}(x)$, only the fields $\omega_k(x)$ and $T^i_{jk}(x)$ may be freely chosen because $\hat{g}_{ij}(x)$ and $\hat{g}_{ij,k}(x)$ are already determined. It follows that at a given point p_0, the number of free choices one can make in determining $\Lambda^i_{jk}(x(p_0))$ is

$$d = n\frac{n(n - 1)}{2} + n. \tag{4.6.22}$$

Scheibe recognizes, however, that Weyl requires that it be possible to choose the numbers $\Lambda^i_{jk}(x(p_0))$ completely arbitrarily; that is, d should be n^3. Indeed, Scheibe even paraphrases a 'lemma' due to Weyl as lemma A which begins "Every system of n^3 numbers Λ^s_{ik} can be represented ...". Thus, the acceptance of Scheibe's analysis, *as a re-construction of Weyl's argument*, leads to a numerical inconsistency. As we have noted above in remark 24, however, there is another interpretation of equation (4.6.18), which is completely in accord with the sense of the word 'Freedom' in Weyl's Postulate of Freedom. The logic of this point-wise analysis is quite different from the analysis proposed by Scheibe. According to the point-wise interpretation, the geometry is specified at a *point*; that is, $g_{ij}(x(p_0))$ and $\omega_i(x(p_0))$ are given. Then, n^3 numbers, the Λ^i_{jk} are chosen *arbitrarily* in order to specify a congruent map from $T(M_{p_0})$ to $T(M_p)$, where p is any point that is infinitesimally near p_0. Equation (4.6.18) then determines the $g_{ij,k}(x(p_0))$ and hence the metric $g_{ij}(x(p))$ by means of the Taylor expansion (4.5.29). The point is that a Weylian geometry possesses sufficient flexibility, that is, freedom to accomodate at any given point p_0 any infinitesimal linear transformation as a congruent transformation.

4.7 Concluding Remarks Concerning the Relative Merit of the Finsler and G-metric Interpretations of Weyl's Raumproblem

There are two radically different conceptions of the Raumproblem. The first, which in our judgement is Weyl's Raumproblem, has its origin in Riemann's analysis of infinitesimal distance functions on a manifold. These were later called Finsler metrics. Weyl implicitly generalized the problem in two ways, by allowing for indefinite metrics in order to encompass the General Theory of Relativity and by considering metrics with variable gauge together with an associated length connection in order to obtain a purely infinitesimal geometry. Each member of the general class of geometries under consideration is locally determined relative to a choice of gauge by two fields: a possibly indefinite Finsler metric $F_p(dx)$ and a length connection $\omega_i dx^i$. The problem is to show that if the geometry satisfies the Postulate of Freedom and determines a unique, symmetric, linear connection, then the metric $F_p(dx)$ must be Riemannian of some signature.

The second conception of the Raumproblem has its origin, in our judgement, in the work of Cartan who got his understanding of what Weyl was doing from the French translation of the fourth edition of Weyl's book *Raum, Zeit, Materie*. As we have noted above, Finsler metrics are not discussed in *Raum, Zeit, Materie* except for the mention of an example in one sentence. It would seem that Cartan interpreted another sentence (just below the other one) in Weyl's discussion to mean that congruence was to be *defined* in terms of a group action. Cartan's version of the Raumproblem was later reformulated in terms of G-structures. In order to accomodate variable 'gauge', Scheibe subsequently generalized the G-structure version of the problem. Each member of the general class of geometries under consideration is determined as follows. A subgroup $\mathbf{G}$ of $GL(n)$ which characterizes the nature of congruence is specified. An $\mathbf{N}$-structure where $\mathbf{N}$ is the normalizer of $\mathbf{G}$ in $GL(n)$ determines the action of $\mathbf{G}$ on each tangent space $T(M_p)$ and hence determines congruence at each point p. Congruent mappings between infinitely near points are determined by a connection on a fiber bundle with typical fiber $\mathbf{N}/\mathbf{G}$. The problem is to show that if the geometry determines a unique, symmetric, linear connection, then the group $\mathbf{G}$ must be $O(p,q)$ for some signature (p,q). In this version of the Raumproblem, the Postulate of Freedom does not play the role of a postulate, as it did for Weyl, rather it is a consequence of other assumptions on which the proof is based.

According to Laugwitz and Scheibe, Weyl considered both conceptions of the Raumproblem. They claim that Weyl (1919c), in the first edition of his commentary to Riemann's (1854) inaugural lecture, introduced a restricted version of the first Raumproblem for the case of fixed-gauge, positive-definite Finsler metrics, but that Weyl did not provide a proof. Rather, they claim that it was Laugwitz who provided the proof for this special case and that Laugwitz also formulated the problem for the case of fixed-gauge, indefinite Finsler metrics but was unable to prove it. In addition, they claim that the proof presented by Weyl in his Spanish lectures pertains to Weyl's second Raumproblem. The *sole* justification for this Received View is a brief statement by Weyl that characterizes the metric in terms of congruent maps.[51] This statement appears in slightly different forms in a number of places. However, Scheibe regards these brief remarks as the proclamation of an important, novel concept, the 'G-field'. If Scheibe were correct, then why

---

[51] See the quotatons that preceed remark 29.

would Weyl limit his discussion of the concept of a *G*-field to a few meager statements about this alleged new and novel idea? Furthermore, there are several major difficulties that arise if one accepts the Received View. First, the Received View faces the following glaring inconsistency. In the third edition of his Commentary on Riemann's (1854) inaugural address, Weyl (1919c, 1923, 3 edn, pp. 26) makes an explicit reference to the proof presented in his Spanish lectures. *Immediately* before this reference to his proof in the Spanish lectures, Weyl discusses the problem of selecting the Riemannian metrics from the class of Finsler metrics and Helmholtz's partial solution to this problem. Then, *immediatey* after the reference to the proof in his Spanish lectures, Weyl cites Finsler! *Thus, Weyl's reference to his proof in the Spanish lectures is sandwiched between explicit references to the Finsler version of the Raumproblem.* It is also pertinent to note that Weyl always mentions Finsler metrics whenever he discusses the Raumproblem; moreover, *Weyl never distinguished two Raumproblems; he always speakes of the Raumproblem in the singular.*

Second, Scheibe's (1988) interpretation of Weyl's analysis leads to a numerical inconsistency. Weyl clearly states in his Postulate of Freedom that all n^3 numbers Λ^i_{jk} are to be freely chosen. On a *G*-metric interpretation of Weyl's analysis, however, this is not possible. See the discussion surrounding equation (4.6.22). In this connection, it is important to note that in Weyl's analysis the Postulate of Freedom plays an important and necessary role.

Third, if one accepts the Received View, then one is forced to conclude that Weyl achieved very little: (a) Weyl formulated the so-called first Raumproblem for the restricted case of fixed gauge but he did not prove it. (b) Weyl formulated the so-called second Raumproblem but he proves it only for the case of a fixed gauge[52]. (c) Moreover, Weyl does not discuss in detail the central concept of the so-called second Raumproblem, the *G*-field, and he does not succeed in characterizing the geometries that were the focus of his concerns in his analysis of the Raumproblem, namely, the Weylian Geometries. It should be noted, however, that Weyl claimed to have solved his Raumproblem in its full generality. At the beginning of the eighth Spanish lecture, Weyl explicitly takes into account both gauge and signature. In a posthumous publication Weyl (1925/1988, p. 37) states

... Das neue gruppentheoretische Raumproblem, das vom Standpunkt der Relativitätstheorie an Stelle des Helmholtz-Lie'schen tritt, glaube

---

[52]Some adherents of the Received View claim that the analysis in Scheibe's thesis provides an interpretation of Weyl's analysis according to which Weyl proved everything in the context of a gauged version of the *G*-metric version of the Raumproblem. However, whatever Scheibe is doing, Scheibe is not providing a re-construction of Weyl's analysis! Whatever Weyl was doing, the Postulate of Freedom was an essential hypothesis both conceptually and mathematically. From a coneptual point of view, the Postulate of Freedom characterizes the flexibility required by a dynamical, physical geometry. From a mathematical point of view, the Postulate of Freedom is applied in a *pointwise* manner and crucially affects the structure of Weyl's argument. However, in Scheibe's analysis, the Postulate of Freedom does not occur as a hypothesis, either conceptually or mathematically. Whatever Scheibe's proof amounts to, it cannot, in our view, be regarded as a re-construction of Weyl's proof.

ich in meiner Schrift "Mathematische Analyse des Raumproblems" (1923, Vorlesung 7 und 8) formuliert und gelöst zu haben.

[... I believe that in my work "Mathematische Analyse des Raumproblems" (1923, lectures 7 and 8), I have formulated and solved the new group-theoretical Raumproblem, which, from the standpoint of Relativity Theory replaces the Helmholtz-Lie Raumproblem.]

These are serious difficulties for the Received View. *Furthermore, if Laugwitz and Scheibe were correct, then our analysis presented above in subsection 4.5 would constitute the first formulation and proof of the first Raumproblem in its full generality* (positive-definite/indefinite and fixed/variable gauge).

In contrast, if one holds, as we do, that for Weyl there was one and only one Raumproblem, namely, the problem that is concerned with the problem of Finsler metrics, then Weyl achieved a great deal! Interpreted in this way, Weyl's Raumproblem and his analysis of it constitutes a coherent, unified, conceptual whole which deals with the class of Finsler metrics generalized to allow for signature and variable gauge. Weyl succeeds in proving everything he claims to have proved; that is, he succeeds in characterizing the Weylian geometries by selecting them out of the class of generalized Finsler metrics.[53] There is only one difficulty for a Finsler interpretation of Weyl's Raumproblem, namely, the brief passages which characterize the metric in terms of group transformations. It is true that general Finsler metrics cannot be characterized in such group theoretic terms; however, in view of the serious difficulties associated with the Received View, it makes more sense to assume that Weyl used inappropriate language because of the traditions of the Erlanger Programme in which there is a one-to-one correspondence between congruence defined in terms of structure and congruence defined in terms of a group and because such a correspondence also exists in the cases of Riemannian and semi-Riemannian geometries, the geometries that were Weyl's main focus.

4.8 Weyl's Realist Field Ontology of Spacetime Geometry

Weyl was an early and forceful proponent of a *realist field ontology* of spacetime geometry that may be described roughly as follows: the Special and General, as well as the non-relativistic spacetime theories, are *principle* theories of spacetime structure[54]. These theories all postulate various structural constraints, and events within spacetime are held to satisfy these constraints. When interpreted physically, these mathematical structures correspond to what Weyl calls physical structural fields (*Strukturfelder*). Analogous to the electromagnetic field, these structural fields act on matter and are, within the context of the General Theory of Relativity, in turn acted on by matter.

---

[53]We have described in detail how he does this in subsection 4.5.

[54]For a discussion of the distinction between constructive and principle theories which is due to Einstein, see (Coleman and Korté 1980, pp. 1348–1349) and (Korté 1981).

Riemann's recognition that the metric structure should be separated from the manifold structure, together with his adoption of the infinitesimal standpoint, were prerequisite steps for the development of differential geometry as the mathematics of differentiable *geometric fields* on manifolds. There is, however, no indication in Riemann's work on gravitation and electromagnetism that would suggest that he anticipated the conceptual revolution concerning our understanding of space and time implied by Einstein's General Theory. We may say, however, in line with Weyl's interpretation of Riemann, that Riemann's work made this revolution possible in the following sense: *By formally separating the metric structure and thereby paving the way for the subsequent separation of the other post-differential-topological structures, such as the affine, projective and conformal structures, from the manifold, so that these structures are no longer rigidly tied to it, one deprives them of their formal geometric rigidity and, on the basis of the infinitesimal geometric standpoint or near-geometry, allows for the possibility of interpreting them as mathematical representations of flexible, dynamical, physical, structural fields* (Strukturfelder) *on the manifold of spacetime, geometrical fields which reciprocally interact with matter.* An n-dimensional manifold M whose only properties are those that fall under the concept of a manifold, Weyl (1918d) physically interprets as an n-dimensional empty world, that is, a world empty of both matter and fields. On the other hand, an n-dimensional manifold M that is an affinely connected manifold, Weyl physically interprets as an n-dimensional world filled with a gravitational field, and an n-dimensional manifold M endowed with a projective structure represents an n-dimensional non-empty world filled with an inertial-gravitational field, or what Weyl calls the *guiding field*. In a similar vein, an n-dimensional manifold M that possesses a conformal structure of Lorentz type, represents a non-empty n-dimensional world filled with a causal field. Finally, an n-dimensional manifold M endowed with a metrical structure may be interpreted physically as an n-dimensional non-empty world filled with a metric field.

A class of geometric structural fields of a given type is characterized by a particular Lie group.

Definition 37 (Microsymmetry Group) *A microsymmetry of a field at a point $p \in M$ is a local diffeomorphism that takes $p \in M$ into p and preserves the field at $p \in M$. The microsymmetry group of a field at $p \in M$ is the group of its microsymmetries at $p \in M$ under the operation of composition.*

A geometric structural field belonging to a given class has a microsymmetry group at each point $p \in M$ which is isomorphic to the Lie group that is characteristic of the class. In Relativity Theory, this microsymmetry group is isomorphic to the Lorentz group and leaves invariant a pseudo-Riemannian metric of Lorentzian signature.

The different types of geometric, structural fields may be represented from a modern mathematical point of view as cross sections of appropriate fiber bundles over the manifold M; that is, the amorphous manifold M has associated with it various geometric fields in terms of a mapping of a certain kind (called a cross section) from the manifold M to the corresponding bundle space over M. For example a vector field on M is characterized by a cross section σ of the tangent space $T(M)$, that is $\sigma: M \to T(M)$ such that $\pi \circ \sigma = \mathrm{id}_M$, where $\pi: T(M) \to M$ is the projection. In particular, Einstein's General Theory of Relativity postulates a physical field, the metrical field, which, mathematically speaking, may be characterized as a cross section of the bundle of non-degenerate, second-order, symmetric, covariant tensors of Lorentz signature over M. Weyl (1931b, p. 336) says of this world structure:

> Wie immer die Struktur genau und vollständig zu beschreiben sein mag und welches auch ihr innerer Grund ist — alle Naturgesetze zeigen, daß sie von einschneidendster Wirkung auf den Gang der physischen Geschehnisse ist: das Verhalten von starren Körpern und Uhren ist fast ausschließlich durch die Maßstruktur bestimmt, ebenso der Verlauf der Bewegung eines keiner Einwirkung unterliegenden Massenpunktes und die Ausbreitung einer Lichtwelle. Und nur an diesen Wirkungen auf die konkreten Naturvorgänge können wir die Struktur erkennen.

> [However this structure is to be exactly and completely described and whatever its inner ground might be, all laws of nature show that it constitutes the most decisive influence on the evolution of physical events: the behaviour of rigid bodies and clocks is almost exclusively determined through the metric structure, as is the pattern of the motion of a force-free mass point and the propagation of a light source. And only through these effects on the concrete natural processes can we recognize this structure.]

The views of Weyl are diametrically opposed to geometrical conventionalism and some forms of relationalism. Spacetime geometry is not about rigid rods, ideal clocks, light rays or freely falling particles, except in the derivative sense of providing information about the physically real metric field which, according to Weyl, is as physically real as is the electromagnetic field and which determines and explains the metrical behaviour of congruence standards under transport. The metrical field has physical *and* metrical significance, and the metrical significance does not consist in the mere articulation of relations obtaining between, say, rigid rods or ideal clocks.

4.9   Weyl's Discovery of the Causal-Inertial Method

The four-dimensional pseudo-Riemannian manifold is the mathematical model of the physical spacetime of General Relativity. It was Weyl (1921f) who first distinguished between two primitive substructures of that model, namely, the *conformal* and *projective* structures, and he proposed the possibility of the *geodesic* or *causal-inertial* method for determining the spacetime metric using light propagation and freely falling bodies, by showing that the *conformal* structure (representing the *causal field* governing light propagation) and the *projective* structure (representing the *inertial* or *guiding field* governing all free [fall] motions) uniquely determine the metric.[55] In a letter to Felix Klein, which was published in 1921 under the title *Zur Infinitesimal Geometrie: Einordnung der projektiven und konformen Auffassung*, Weyl (1921f) proved the important theorem[56] that states that *the projective and conformal structure of a metric space determine the metric uniquely.*

The essential reasoning of Weyl's proof is this. Suppose we are given a metric g on M. Then such a metric determines an equivalence class of conformally equivalent metrics, namely,

$$[g] = \{\bar{g} | \bar{g} = e^{\theta} g\}. \tag{4.9.1}$$

A metric g uniquely determines an affine structure (symmetric linear connection) Γ on M. Under a conformal transformation

$$g \to e^{\theta} g = \bar{g}, \tag{4.9.2}$$

the change of the components of the affine connection is given by (4.4.6), that is,

$$(\bar{\Gamma} - \Gamma)^i_{jk} = \frac{1}{2} (\delta^i_j \theta_k + \delta^i_k \theta_j - g_{jk} g^{ir} \theta_r). \tag{4.9.3}$$

Weyl shows that such a *conformal transformation* preserves the projective structure of M and hence is a *projective transformation* (satisfies definition 12) if and only if $\theta_j = 0$, in which case the conformal and projective structures are compatible. Weyl reasons as follows. A conformal transformation of the metric field g induces a transformation of the symmetric linear connection given by (4.9.3); consequently, for any vector v^i,

$$(\bar{\Gamma} - \Gamma)^i_{jk} v^j v^k = v^i \theta_k v^k - \frac{1}{2} g_{jk} v^j v^k \theta^i. \tag{4.9.4}$$

If this transformation is projective, then (4.3.4) holds, and

$$(\bar{\Gamma} - \Gamma)^i_{jk} v^j v^k = 2 v^i \phi_k v^k. \tag{4.9.5}$$

From (4.9.4) and (4.9.5), it follows that

$$\frac{1}{2} g_{jk} v^j v^k \theta^i = v^i (\theta_k - 2\phi_k) v^k. \tag{4.9.6}$$

---

[55] Kretschmann (1917) discussed the causal-inertial method before Weyl; however, Kretschmann described the motions of freely falling, neutral, massive monopoles by means of affine geodesics. Somewhat later, Lorentz (1923) re-worked Kretschmann's analysis and showed that the affine parameter did not in fact play any role, that is, that only the projective structure was needed.

[56] In the same paper, Weyl also proved another important theorem (Satz 4) concerning the conformal and projective curvature tensors. See the section entitled "Bemerkungen zum Wechselspiel von konformer und projektiver Struktur" in the article by E. Scholz in this volume.

Weyl considers the expression (4.9.6) for two vectors $v^i, w^i \in T(M_p)$ which lie in different directions at $p \in M$ and which satisfy $g_{jk}v^jv^k \neq 0$ and $g_{jk}w^jw^k \neq 0$, and obtains

$$\theta^i = 2\frac{(\theta_r - 2\phi_r)v^r}{g_{jk}v^jv^k}v^i = 2\frac{(\theta_r - 2\phi_r)w^r}{g_{jk}w^jw^k}w^i. \tag{4.9.7}$$

However, v^i and w^i are linearly independent; consequently, it is necessary that $(\theta_r - 2\phi_r)v^r = 0 = (\theta_r - 2\phi_r)w^r$ and hence $\theta^i = 0$. It also follows that $\phi_r = 0$ because $\phi_r v^r = 0$ for arbitrary timelike vectors v^i. Weyl (1923c, p. 19) remarks after the proof:

> Ist es uns in der wirklichen Welt also möglich, die Wirkungsausbreitung, insbesondere die Lichtausbreitung zu verfolgen, und vermögen wir außerdem die Bewegung freier Massenpunkte, welche dem Führungsfelde folgen, als solche zu erkennen und zu beobachten, so können wir daraus allein, ohne Zuhilfenahme von Uhren und starren Maßstäben, das metrische Feld ablesen.

> [If it is possible for us, in the real world, to discern causal propagation, and in particular light propagation, and if moreover, we are able to recognize and observe as such the motion of free mass points which follow the guiding field, then we are able to read off the metric field from this alone, without reliance on clocks and rigid rods.]

Elsewhere, Weyl (1949a, Engl. edn, p. 103) says:

> As a matter of fact it can be shown that the metrical structure of the world is already fully determined by its inertial and causal structure, that therefore mensuration need not depend on clocks and rigid bodies but that light signals and mass points moving under the influence of inertia alone will suffice.

Methodological and ontological considerations decidedly favour the geodesic method for determining the metric. The use of clocks and rigid rods is, within the context of the Special and General Theories of Relativity, an undesirable makeshift. Since neither spatial nor temporal intervals are invariants of the four-dimensional spacetime of the Special Theory of Relativity and the General Theory of Relativity, the invariant spacetime interval ds cannot be directly ascertained by means of standard clocks and rigid rods. Moreover, the concepts of a rigid body and a periodic system (such as pendulums or atomic clocks) are not fundamental or theoretically self-sufficient, but involve assumptions that presuppose quantum theoretical principles for their justification and thus lie outside the present conceptual relativistic framework. The above considerations against the employment of rigid rods and standard clocks accord essentially with a requirement suggested by Bohr and Rosenfeld (1933): *Every proper or complete physical theory should provide in and of itself, its own means for defining the quantities with which it deals.*

In his construction of relativistic chronometry, Einstein dispensed with the standard clock by using a light clock instead; however, he still used rigid bodies to construct such a clock. As discussed above, Weyl showed that it is possible to *constructively* approach the Special Theory of Relativity and the General Theory of Relativity without standard clocks *and* without rigid rods and to derive the affine and metric structures exclusively on the basis of light propagation and free (fall) motion. It was only much later that Kundt and Hoffmann (1962) as well as Marzke and Wheeler (1964) were actually able to construct a gravitational standard clock (light clock). Although this construction does not require the use of rigid bodies, they still presuppose a Riemannian spacetime geometry. That it is possible to actually construct a light clock on the basis of even more primitive structures was only recently shown by Castagnino (1968) and Ehlers, Pirani and Schild (1972).

Of course Weyl's geodesic method for discovering the metric of physical space-time would not have been feasible without some of his earlier work in differential geometry. Indeed, it was in part precisely because of his different approach to geometry in which the affine, projective and conformal structures are treated in their own right rather than as mere aspects of the metric, that Weyl was led to introduce the possibility of the causal-inertial method. The old notion of a 'geodesic path' had its inception in the context of classical *metrical* geometry and 'geodesicity' was characterized in terms of *extremal* paths of curves, which presupposed a metric. It was Weyl's metric-independent characterization of the affine structure that subsequently led to the geometry of paths and the metric-independent characterization of a *geodesic* path in terms of the process of autoparallelism of its tangent direction, which allowed for the possibility of a metric-independent (and hence non-circular) empirical characterization of observable spacetime trajectories as projective.

4.10 Ehlers, Pirani and Schild's Constructive Axiomatics

Fifty years after Weyl suggested the possibility of the causal-inertial method, Ehlers et al. (1972) were actually able to construct a procedure to synthesize the metric field directly from the conformal and projective fields.

Whereas Weyl emphasized from a physical point of view the roles of light propagation and free (fall) motion in revealing the causal and projective structures of spacetime respectively, he did not use these primitive substructures directly in order to derive from them and their compatibility relation the existence of a unique affine connection (*Weyl connection*). He considered, rather, the affine and metric structures as fundamental in his mathematical analysis and saw the projective and conformal structures as arising from them by abstraction. Thus we saw that in his group-theoretical justification of the uniqueness of the Pythagorean-Riemannian metric he considered the congruent mappings of a tangent space onto itself, and the translation of a tangent space at some point to a neighbouring point, as fundamental operations. Using the conformal and projective structures directly,

Ehlers, Pirani and Schild derived a unique pseudo-Riemannian spacetime metric solely as a consequence of a set of natural, physically well-motivated, constructive, 'geometry-free' axioms concerning the incidence and differential properties of light propagation and free fall.

The 'geometry-free' axioms are propositions about a few general qualitative assumptions concerning free (fall) motion and light propagation that can be verified directly through experience in a way that does not presuppose the full blown edifice of the General Theory of Relativity. From these axioms, the theoretical basis of the theory is reconstructed step by step. Ehlers, Pirani and Schild adopt Reichenbach's term, *constructive axiomatics* (Reichenbach, 1924, 1925), to describe the nature of their approach.

For the most part, physical theories are presented from the *deductive* axiomatic point of view; that is, the exposition begins with a set of postulates concerning the existence of high-level structures and proceeds by logical deduction to lower-level phenomena, which may be directly confronted by experiment. In the case of the General Theory of Relativity, the assumed existence of a pseudo-Riemannian structure leads to the existence of affine, projective and conformal structures, which govern the behaviour of massive particles and light rays.

The deductive approach has the advantage of making clear at the outset the ontological commitments involved and of providing a logically compelling understanding of how the lower-level phenomena are explained by the theory. However, direct contact with the higher-level structure is frequently elusive. The axioms within the deductive approach, are quite removed from the level of direct experience and, while they may be adequate for the explanation of lower-level phenomena, they may not be necessary. For example, Synge's (1960) chronogeometric axioms for the General Theory of Relativity suffice for the explanation of the phenomena explained by the theory, but it is conceivable that a Weyl structure would be sufficient. Thus within the context of a deductive axiomatic approach, only indirect and probable confirmation of the axioms representing higher-level abstractions is possible through the direct confirmation of their consequences.

The *constructive* axiomatic approach was thought to be the reverse of the deductive axiomatic procedure. The constructive axioms deal with directly observable phenomena at as low a level as possible. The aim is to formulate axioms that may be directly confronted by experiment, and then to deduce from these low-level axioms the existence of higher-level structures. As Reichenbach (1924, Gesammelte Werke, vol. 3, p. 4) remarks in his introduction of his *Axiomatik der relativistischen Raum-Zeit-Lehre*:

> Aus diesem Grunde sind alle Axiome der vorliegenden Darstellung so gewählt, daß sie aus Experimenten mit Hilfe der vorrelativistischen Optik und Mechanik abgeleitet werden können; es sind also alles Tatsachen, welche ohne Benutzung der Relativitätstheorie geprüft werden können. ... *die Tatsachenbehauptungen der Relativitätstheorie sind einzeln alle mit den Mitteln des vorrelativistischen Denkens vorstellbar:*

neu ist nur ihre Kombination im Begriffssystem einer Theorie.

[For this reason all axioms of the present representation have been chosen in such a way that they can be derived from the experiment by means of pre-relativistic optics and mechanics. All are facts that can be tested without the use of the Theory of Relativity. ... *the particular factual statements of the Theory of Relativity can all be grasped by means of pre-relativistic conceptions; only their combination within the conceptual system of a theory is new.*]

The aim of a constructive axiomatic approach to a principle theory of space-time structure is to exhibit the physical basis for the particular structural constraints which the principle theory postulates certain events must satisfy. In other words, the essential idea of the geodesic method is to *discover* through the behaviour of physical systems various intrinsic primitive geometrical spacetime structures. It is in spirit analogous to Helmholtz's procedure of deducing the existence and form of the metric of physical space.

Remark 38 The above discussion regarding the roles of deductive and constructive axiomatic schemes was written some years ago. In retrospect, this account now appears to us somewhat untutored from an epistemological point of view compared to our recently developed theory of the epistemology of geometry (Coleman and Korté, 1995a, 1995b). Although our previous results[57] constitute an effective challenge to conventionalism, they require for their proper formulation a new theory of

---

[57]Our own defense of geometric realism (see in particular (Coleman and Korté 1980) and (Korté 1981)) was inspired by, and has its roots, in the works of Hermann Weyl (1918c, 1922a, 1927[sic]/1966, 1949a, 1968) and in the constructive axiomatics of Ehlers et al. (1972). Our approach differs radically from that of any other modern spacetime realist. Instead of assuming the existence of high-level, theoretical entities such as the spacetime manifold and its geometric structures from the outset, we take *observational* entities such as light rays and particles and empirically determine the geometric structures in terms of the behaviours of these observational entities. All of our analyses, criteria and procedures in this regard presuppose only the local differential topology of spacetime. Our procedures and criteria provide a proper foundation for a number of earlier axiomatic constructions proposed by Kundt and Hoffmann (1962), Castagnino (1971), Ehlers et al. (1972), Woodhouse (1972), and Pirani (1973), all of which presupposed the ready availability of neutral massive monopoles (that are governed by a *geodesic* directing field), for the identification of which one requires the procedures and criteria which we supplied. Our work shows that the constructive axiomatics of Ehlers et al. (1972), supplemented with our analysis, is not circular. Since the conventionalist accepts the local differential topology as factually definite — that is non-conventional — and since our defense of realism involves a non-circular, non-conventional, coordinate independent construction of the conformal, projective, affine and metric structure, we claim to have provided a defense of geometric realism which undercuts conventionalism decisively. First, we have shown that the projective and conformal structures of spacetime, which are post-differential-topological, geometric structures, can be empirically determined by a procedure that is purely differential-topological in character. Since the conventionalist regards the world to be factually definite in its local differential topology, these structures have been shown by us to be non-conventional. Moreover, since the affine and metric structures are mathematically uniquely determined by mathematically compatible conformal and projective structures, the affine and metric structures are also non-conventional; that is, the pseudo-Riemannian metric is determined uniquely up to a constant positive factor.

epistemology and semantics for spacetime theories. Whereas theory plays a central role in the top-down approach of modern spacetime realists,[58] observable entities, such as light signals and massive particles, played a more or less predominant role in our constructive axiomatic approach and the role of theory remained somewhat implicit or relatively unclear prior to (Coleman and Korté, 1995a, 1995b). The issue is how the constructive axioms expressing elementary empirical evidence, whose scope of vocabulary is smaller than that of theoretical claims, can provide confirmative instances of the latter. The empiricist's solution was to introduce analytic *bridgestatments* whose function was to provide a link between evidence and theory. The relation of confirmation was regarded as a relation between evidence, hypothesis and *analytic truth* and the conventional features of a theory were thought to be exclusively localizable in these analytic statements. Quine's attack on the analytic/synthetic distinction has led to radical holistic claims that the bearing of evidence on theory must be judged by holistic criteria such as coherence and simplicity, that isolated bits of evidence cannot have confirmatory bearing on isolated bits of theory and that evidence cannot confirm a theoretical hypothesis of a theory in isolation from other theoretical hypotheses.

In (Coleman and Korté 1995a), we have sketched a new theory of semantics and epistemology which spells out the proper role of theory in our constructive axiomatic approach, decisively undercuts geometric conventionalism, and avoids radical holism. Like Glymour (1980), our new account considers the basic relation of confirmation to be a three-place relation such that the evidence confirms a hypothesis *relative to theory*, and additional assumptions taken from the theory may be used in deducing instances of the hypothesis from the evidence.

Our new epistemic account provides the necessary foundation for modeling spacetime structure. In particular, it provides the foundation for modeling the physical concepts and procedures used for the empirical determination of geometric fields by introducing a number of crucial concepts and distinctions: *formal*, *theoretic* and *physical* coordinates; symmetry versus model diffeomorphisms;[59] the

---

[58]See for example (Friedman 1983, p. 32).

[59]These concepts and distinctions are also crucial in providing a solution to Einstein's 'Hole' problem (See (Coleman and Korté, 1992b, 1994)). In (Coleman and Korté 1992b) we present a solution to this problem by providing a coherent formulation and solution to the problem of the specific nature of causality in the General Theory of Relativity, an important epistemic problem that has its origins in the early stages of the development of the General Theroy of Relativity, particularly in Einstein's collaborative work with Grossmann (Einstein and Grossmann 1914, pp. 260–261) and his subsequent concern with the compatibility of general covariance and causality (the 'Hole' problem) (Einstein 1914). There is general agreement that the General Theroy of Relativity satisfies the requirements of physical causality. However, the reasons that are advanced for holding this view are varied and far from coherent. In the physics literature, the fact that the solution in the domain of dependency of the initial data is determined *only* up to a local diffeomorphism in any local neighboorhood of this domain is explained away in a variety of ways: diffeomorphically equivalent models are asserted to be physically equivalent; the active diffeomorphism is asserted to be equivalent to a passive coordinate transformation and hence the lack of uniqueness comes down to the necessity of making a coordinate choice; the lack of uniqueness is similar to the need for a choice of gauge in electromagnetism. While these explanations are not false in any straight-forward sense, they are unsatisfactory for a number of

active versus the passive use of mensuration devices; theoretic completeness, epistemic completeness, and epistemic decidability with respect to the post-differential-topological, geometric structures of a spacetime theory; the concept that a geometric field is a *G*-structure on spacetime. This theory of epistemology for spacetime theories is firmly grounded on the fundamental principle that is accepted by conventionalists; namely, that the world is factually definite with respect to its local differential topology. Consequently, a theory of spacetime must satisfy the following requirements.

First, a spacetime theory's ontology must be sufficiently rich so that physical coordinate systems (such as, for example, radar coordinates) can be modelled; that is, the spacetime theory must tell us how to set up a physical coordinate system. In other words, a spacetime theory must tell us how to gain empirical access to the differential topology of spacetime.

Second, for each field that is to be measured, the spacetime theory must tell us

(a) that there exist appropriate probative systems the motions of which are governed by the field in question;

(b) how to identify the probative systems by means of differential-topological criteria; and

(c) how, by tracking the motions of these probative systems with respect to the physical coordinate system, one can obtain enough data to determine the field by means of a coordinate independent analysis of the data.

For example, in the case of the conformal structure, the appropriate probe is the electromagnetic field, and in the case of the projective structure the appropriate probe is the neutral, massive monopole.

Third, it is essential to distinguish between the active and passive use of technology. Mensuration devices are used passively if they are used *only* for the purpose of assigning arbitrary, physical, local coordinates to scattering events; that is, one must *not* make any assumptions about how the physical coordinate system is related to the field that is being measured. For example, Reichenbach violates this principle in his epistemological analysis of geometry because he uses rigid rods in an *active* way and therefore sets up an unnecessary and inescapable circularity to support his conventionalist conclusions.

Fourth, if a spacetime theory meets the above requirements, then, if the claims that the theory makes about post-differential-topological structures that it

---

reasons. In our solution of the problem of causality in the General Theory of Relativity (the Hole problem), we emphasize the necessity of distinguishing between formal, theoretic and physical coordinates. These distinctions permit the clear characterization of the distinction between model and symmetry diffeomorphisms, clarify the equivalence between model diffeomorphisms and passive coordinate transformations, and reveal the importance and physical significance of a number of other types of transformations. It is only by means of these distinctions that one can show in a clear and unambiguous way that the Theory of General Relativity is strictly causal.

postulates are in fact false, then one can discover this fact because every aspect of the measurement procedure is grounded in the local differential topology of spacetime.

Ehlers, Pirani and Schild adopt as primitive concepts the notions of *event*, *particle path* and *light ray*. The constructive model of spacetime is based on a triple of sets $\langle M, \mathcal{P}, \mathcal{L} \rangle$ of the objects corresponding to these primitive concepts. It is assumed that the set $M = \{e, p, q, p_1, q_1, \ldots\}$ of events has a Hausdorff topology with a countable basis. The sole purpose in assuming a topology is to permit the clear statement of local axioms through the use of such terms as 'neighbourhood'. The elements of the set of actual or possible events of M represent actual or possible locations in spacetime. The members of the sets $\mathcal{P} = \{P, Q, P_1, Q_1, \ldots\}$ and $\mathcal{L} = \{L, N, L_1, N_1, \ldots\}$ are subsets of M that represent the possible or actual paths of massive particles and light rays in spacetime.

The differential structure is not presupposed. With the first few axioms, a differential-manifold structure is introduced on the set M that is sufficient for the localization of events by means of local coordinates (such as radar coordinates). Once M is given a differential-manifold structure through the introduction of local radar coordinates by means of particles and light rays (such that any two radar coordinates are smoothly related to one another), one can do calculus on M and one may speak of tangent and direction spaces.

It is important to emphasize that the members of $\mathcal{P}$ represent possible or actual paths of *arbitrary* massive particles that may have some internal structure such as higher order gravitational and electromagnetic multipole moments and that may therefore interact in complicated ways with various physical fields. In order to constructively establish the projective structure of spacetime, it is necessary to single out a subset of $\mathcal{P}$, namely $\mathcal{P}_f$, the set of possible or actual paths of spherically symmetric, electrically neutral particles (that is, the world lines of freely falling particles). However, the set $\mathcal{P}_f \subset \mathcal{P}$, can be properly characterized only after a coordinate system (differential structure) is available. Consequently, one must employ *arbitrary* particles in the statement of those axioms that lead to the local differential structure of spacetime.

With further axioms one introduces the *conformal* and *projective* structures of M. The causal propagation of light *reveals* at each point of spacetime the infinitesimal light cone. A manifold in which the null cone is singled out in the tangent space $T(M_p)$ for all $p \in M$ has a conformal structure or a causal field defined on it. Conversely, a manifold M has a first-order conformal structure if it is endowed with an equivalence class of conformally equivalent metrics of Lorentz signature, which are proportional to one another and which define the same null cones, that is, which are defined by the relation $\bar{g} = e^{\theta} g$, where θ is any real valued function on M.

The motions of freely falling particles governed by the guiding field (geodesic directing field) *reveal* the geodesics of spacetime, that is, the geodesics corresponding to a projective equivalence class Π of symmetric connections. A manifold for which a special class of paths has been singled out, namely, the geodesics, possesses a projective structure that may be described by an equivalence class of projectively related affine connections.

All that has been established so far is that there are actual or possible projective geodesics in every conformally timelike direction. Very little connection between the conformal and the projective structures has so far been shown to exist. That there is a close relation between these structures is suggested by high energy experiments, "A massive particle ($m > 0$), though always slower than a photon, can be made to chase a photon arbitrarily closely" (Ehlers 1973b, p. 31). Ehlers, Pirani and Schild therefore assume an axiom of compatibility between these two structures and this leads to a *Weyl space*. The spacetime manifold M is now endowed with a unique symmetric affine connection (Weyl connection). Finally, a reduction to the pseudo-Riemannian structure involves setting Weyl's length curvature tensor equal to zero.

Without going into detail, the constructive axiomatic approach is roughly as follows. The projective structure, which can be measured using only the local differential structure of spacetime (see Coleman and Korté, 1980, 1982, 1984b, 1987 and section 4.11), defines an equivalence class Π of projectively related affine connections. The *Law of Causality* asserts the existence of a unique (first-order) conformal structure on spacetime, that is, a field of infinitesimal light cones. Only null one-directions are determined. No special choice of parameters along light rays is determined by this structure. The (first-order) conformal structure can also be measured using only the local differential-topological structure. By a purely mathematical process involving only differentiation, the first-order conformal structure determines a second-order conformal structure, namely, an equivalence class K of conformally related affine connections.

If the projective and conformal structures are compatible, then the intersection $\Pi \cap K$ of the equivalence class Π of projectively equivalent affine connections with the equivalence class K of conformally equivalent affine connections, contains (Ehlers et al. 1972) (see also section 4.9) a unique symmetric affine connection (Weyl connection).

Thus light propagation and free (fall) motion reveal on spacetime a conformal and a projective structure, and because of the compatibility relation between these structures a *unique* symmetric linear connection Γ (Weyl connection), such that light rays are conformal/affine null geodesics, and particles in free (fall) motion are time-like projective/affine geodesics. That is, a Weyl connection determines the parallel transport of vectors, preserving their timelike, null or spacelike character, and for any pair of non-null vectors it leaves invariant the ratio of their lengths and the angle between them, provided the vectors are transported along the same path.

Since length transfer is non-integrable in a Weyl space, a Weyl geometry reduces to a pseudo-Riemannian geometry if and only if Weyl's length-curvature tensor equals zero in which case the magnitude of a vector is path-independent under parallel transport and there exists no second clock effect.

4.11   The Laws of Motion and Mach's Principle

The transition from the kinematic to the dynamic analysis of the motions of bodies in spacetime theories has traditionally been seen to consist in the determination of a particular class of *standard motions* that are the *free motions*. The concept of force is then defined in terms of acceleration relative to the standard of *no-acceleration* provided by these free motions. This procedure is meaningful if and only if for each event of spacetime and for every timelike direction at that event, there exists one and only one standard motion through that event. From a geometric viewpoint, the determination of the class of standard motions amounts to the determination of a *geodesic path* structure or projective structure for the spacetime manifold. One introduces the concept of a class of *standard free motions* by postulating the Law of Inertia. The problem of motion and the status and meaning of the Inertial Law concerns, therefore, the nature of the transition from the kinematic to the dynamic analysis of motion.

The controversy surrounding the problem of motion has nurtured and given currency to a number of dubious claims concerning the status and meaning of the laws of inertia of various spacetime theories; for example, the laws (1) are conventional in character, (2) are definitions, (3) are circular and without empirical content, and (4) postulate the existence of free particles or of inertial frames.

Those who argue for the conventional and/or definitional character of the laws mainly on *epistemic* grounds point out that the laws do not supply *independent* criteria of what is to count as *force-free* or *natural motion*. The only way of knowing when no forces act on a body is by observing that it moves as a free particle along the geodesics of spacetime. But how, without already knowing the geodesics (or the projective structure) of spacetime, is it possible to determine which particles are free and which are not? In addition, to determine the projective structure of spacetime it is necessary to use free particles. The circularity, they argue, cannot be avoided. Others have tried to define a *free particle* with respect to an *inertial frame* as a particle the motions of which satisfy the equation $d^2x^i/dt^2 = 0$ in that frame. But how is one to determine what an inertial frame is? If an inertial frame is to be characterized as a frame in which the motion of a free particle would satisfy the equation $d^2x^i/dt^2 = 0$, then the definition is obviously circular. Hence, Newton's first law is interpreted by some as the *existence* claim: there exist physical inertial frames in which the motions of free particles would be governed by the equation $d^2x^i/dt^2 = 0$. There remains, however, the lack of noncircular criteria for identifying these frames.

Those who argue for the conventional character of the Law of Inertia from *ontological* considerations concerning the nature of spacetime structure and/or for their *relationalist* character from a Leibnizian-Machian view of motion (in which relative motion must be understood as relative motion of bodies with respect to each other or with respect to material reference frames) advance the thesis that what counts as a standard of no-acceleration or free motion is *not* dictated by a physically real and causally efficacious inertial structure of spacetime.

It is clear from Weyl's writings that he was a vigorous opponent of such an ontological view and that he advocated a realist field ontology of the inertial structure of spacetime. According to Weyl, the world does not lack a physically real projective structure. The transition from the kinematic to the dynamic behaviour of bodies in spacetime theories is not arbitrary, but, as a matter of empirical fact, is dictated by a physically real, causally efficacious, unique, projective structure that Weyl called the guiding field. In physical spacetime, according to Weyl, the projective structure has an immediate, intuitive significance. The real world is a non-empty world filled with a *metrical-gravitational* field. It is an indubitable fact, Weyl tells us, that a body that is let free in a certain world direction carries out a uniquely determined natural motion from which it can be diverted only through an external force. This natural motion[60] may be characterized geometrically as an infinitesimal autoparallelism of the initial direction R at some point $p \in M$ which consists in the parallel displacement of R at $p \in M$ to a neighbouring point $p' \in M$ which lies in the direction R at $p \in M$. The world line of a freely moving body is a geodesic, and a curve in M is a geodesic curve if and only if its tangent direction R experiences infinitesimal autoparallelism when moved along all the points of the curve. The process of autoparallelism of directions appears, physically speaking, as the tendency of persistence of the world direction of a free particle whose motion is governed by the guiding field.

The projective, guiding field, which includes both the inertial and the gravitational field, causes the inertial persistence of the motion of massive bodies. If external forces exert themselves on a body, then a motion results that is determined through the conflict between the tendency of persistence due to the guiding field and the force. The tendency of persistence of the guiding field is a type of constraining guidance, which the inertial-gravitational field exerts on every body.

Weyl's position is roughly as follows. In Galilean spacetime, the guiding field constitutes a formal spacetime structure that is rigidly tied to the spacetime manifold. In the Special Theory of Relativity the guiding field is likewise a rigid formal geometric structure, characterized by a set of coordinate systems that are equivalent up to an inhomogeneous Lorentz transformation. In both cases, the guiding field acts on matter, but is not in turn affected by matter. In the General Theory of Relativity the guiding field includes the gravitational field and thus ceases to be a rigid geometrical structural field. It is a flexible structural field that is physically real, analogous to the electromagnetic field and interacts dynamically with matter, not one-sidedly, but reciprocally. Einstein's recognition that gravity is not a force field existing in addition to a rigid inertia-determining world-geometry, but is part of the metrical and affine structure of spacetime, transformed the geometrical structural field from an absolute entity, rigidly tied to the spacetime manifold, into a dynamical, structural field on M which reciprocally interacts with matter. Thus, as Weyl emphasized, in the dualism of force and inertia, gravitation belongs to the side of inertia. Formally, this means that the dynamical inertial structure is

---

[60]See (Weyl 1918c, 5 edn, p. 219) and (Weyl 1923c, p. 13).

expressed as a single non-integrable affine connection which represents *both* inertia and gravitation. The separate existence of an integrable affine structure (namely, a rigid formal geometric structure that causally affects but is not in turn affected by matter) and a vector field representing gravitation is denied. Thus, as Ehlers (1973a) observes, the General Theory of Relativity answers Mach's criticism of Newton by removing the empirically unjustified fictitious distinction between geometry and physics.

Although Mach's Principle has been defined in many ways, there are essentially two ways to understand the principle:

a) as rejecting the absolute character of the inertial structure of spacetime in the sense of Einstein, Weyl, Trautman (1965, 1966) and Anderson (1967).

b) as rejecting the inertial structure of spacetime *per se*.

Version (b) might be characterized as *Leibnizian relativity*; that is, one understands relative motion as relative motion of bodies or physical reference frames. What Einstein is objecting to in Newtonian Mechanics, however, is the *absolute* character of the inertial structure of spacetime. He is not asserting its *fictitious* character. The General Theory of Relativity incorporates Mach's Principle as expressed in version (a) by treating the inertial structure as dynamical and not as absolute.

But it seems Einstein also sought to extend and generalize the Special Theory of Relativity by incorporating version (b) of Mach's Principle into the General Theory of Relativity. After his great papers of 1916 and 1917, Einstein (1918, p. 241) explicitly drew attention to Mach's ideas in a short paper entitled *Principielles zur Allgemeinen Relativitätstheorie* (Einstein 1918). It was in that paper, which was written as a reply to Kretschmann's (1917) criticism of Einstein's understanding of the significance of the principle of covariance, that Einstein introduced the expression '*Machsche Prinzip*', and responded to Kretschmann's criticism by summarizing the basic ideas of his General Theory of Relativity as follows:

> Die Theorie, wie sie mir heute vorschwebt, beruht auf drei Hauptgesichtspunkten, die allerdings keineswegs voneinander unabhängig sind.
> ...
>
> a) *Relativitätsprinzip:* Die Naturgesetze sind nur Aussagen über zeiträumliche Koinzidenzen; sie finden deshalb ihren einzig natürlichen Ausdruck in allgemein kovarianten Gleichungen.
>
> b) *Äquivalenzprinzip:* Trägheit und Schwere sind wesensgleich. Hieraus und aus den Ergebnissen der speziellen Relativitätstheorie folgt notwendig, daß der symmetrische "Fundamentaltensor" ($g_{\mu\nu}$) die metrischen Eigenschaften des Raumes, das Trägheitsverhalten der Körper in ihm, sowie die Gravitationswirkungen bestimmt. Den durch den Fundamentaltensor geschriebenen Raumzustand wollen wir als "G-Feld" bezeichnen.

c) *Machsches Prinzip:* Das G-Feld is *restlos* durch die Massen der Körper bestimmt. Da Masse und Energie nach den Ergebnissen der speziellen Relativitätstheorie das Gleiche sind und die Energie formal durch den symmetrischen Energietensor $(T_{\mu\nu})$ beschrieben wird, so besagt dies, daß das G-Feld durch den Energietensor der Materie bedingt und bestimmt sei.

[The theory, as I understand it today, rests on three basic points of view which are however not independent of each other.
...

a) *Relativity Principle:* The laws of nature are only propositions about spatio-temporal coincidences; therefore, they find their natural expression in generally covariant equations.

b) *Equivalence Principle:* Inertia and gravitation are essentially equivalent. From this and the result of the Special Theory of Relativity, it necessarily follows that the symmetric "fundamental tensor" $(g_{\mu\nu})$ determines the metrical properties of spacetime as well as the gravitational effects. We shall call the condition of spacetime described by the fundamental tensor, the G-field.

c) *Mach's Principle:* The G-field is completely determined through the masses of bodies. Since mass and energy are the same according to the results of Special Relativity, and since the energy is formally described through the energy tensor $(T_{\mu\nu})$, this means that the G-field is conditioned and determined through the energy tensor of matter.]

Einstein remarked in a footnote with respect to (c):

Bisher habe ich die Prinzipe a) und c) nicht auseinandergehalten, was aber verwirrend wirkte. Den Namen "Machsches Prinzip" habe ich deshalb gewählt, weil dies Prinzip eine Verallgemeinerung der Machschen Forderung bedeutet, daß die Trägheit auf eine Wechselwirkung der Körper zurückgeführt werden müsse.

[I have not distinguished between a) and c) until now, which has caused confusion. I have chosen the name "Mach's Principle", because this principle is a generalization of the Machian demand that inertia must be reduced to an interaction between bodies.]

The above passages reveal Einstein's heavy debt to Mach. His debt is not, however, restricted to specific ideas of Mach, but involves more generally Mach's empiricist programme, in particular, his insistence on the primacy of observable facts of experience, that is, Mach's insistence that only observable facts of experience be invoked to account for the phenomena of motion. Einstein (1916) was quite explicit about this in his fundamental paper on General Relativity, in which he expressed his desire to bring physics into conformity with Mach's philosophy by extending the Special Theory of Relativity to remove, from both the Special Theory of Relativity and Newton's theory, an inherent epistemological defect. The latter is brought to light by Mach's paradox, namely, Einstein's example of two fluid bodies, *A* and *B*, that are in constant *relative* rotation about a common axis. With regard to the extent to which each of the spheres bulges at its equator, infinitely many different states are possible although the relative rotation of the two bodies is the same in every case. Einstein considered the case in which *A* is a sphere and *B* is an oblate spheroid. The paradox consists in the fact that there is no readily discernible reason that accounts for the fact that one of the bodies bulges and the other does not. For Einstein, an epistemologically satisfactory solution must be based on 'an observable fact of experience'. Einstein clearly wanted to implement a Leibnizian-Machian *relational* conception of motion according to which all motion is to be interpreted as the motion of some bodies in relation to other bodies, and he wished to extend the body-relative concept of uniform inertial motion to the concept of a body-relative non-inertial motion.

In contrast, to emphasize the necessity for a physically real and causally efficacious inertial structure of spacetime, the guiding field, and to show that the Leibnizian view of relative motion is self-defeating in the General Theory of Relativity, Weyl devised the following paradox ((Weyl 1924c) and (Weyl 1949a, Engl. edn, p. 105)):

> Incidentally, without a world structure the concept of relative motion of several bodies has, as the postulate of general relativity shows, no more foundation than the concept of absolute motion of a single body. Let us imagine the four-dimensional world as a mass of plasticine traversed by individual fibers, the world lines of the material particles. Except for the condition that no two world lines intersect, their pattern may be arbitrarily given. The plasticine can then be continuously deformed so that not only one but all fibers become vertical straight lines. Thus no solution of the problem is possible as long as in adherence to the tendencies of Huyghens and Mach one disregards the structure of the world. But once the inertial structure of the world is accepted as the cause for the dynamical inequivalence of motions, we recognize clearly why the situation appeared so unsatisfactory. ... Hence the solution is attained as soon as we dare to *acknowledge the inertial structure as a real thing that not only exerts effects upon matter but in turn suffers such effects.*

Let us analyze this example using the concept of the microsymmetry group (see definition 37) of a geometric structure at an event $p \in M$. Consider a spacetime manifold equipped only with a differentiable structure, the plasticine of Weyl's example. Then *all* diffeomorphisms preserve this structure; consequently, in the absence of post-differential-topological structure, the microsymmetry group at any event $p \in M$ is an infinite-parameter group isomorphic to the group of all invertible formal power series in four variables. Clearly, given an infinite number of parameters, one can straighten out an arbitrary pattern of world lines (fibers) in the infinitesimal neighbourhood of any event. In contrast, the active microsymmetry group of a projective structure Π at any event of spacetime is a 20-parameter Lie group (see (Coleman and Korté 1981)). The fact that only a finite number of parameters are available prevents an arbitrary realignment of the world lines of material bodies in the infinitesimal neighbourhood of any given event.

Weyls response to Mach's paradox (mentioned just above) was the following. The fact that a stationary, homogeneous elastic sphere will, when set in rotation, bulge at the equator and flatten at the poles, must, according to Weyl be accounted for in the following way. The complete physical system consisting of both the body and the local inertial-gravitational field is not the same in the two situations. The cause of the effect is the state of motion of the body *with respect to* the local inertial-gravitational field, the guiding field, and is not, indeed as Weyl's plasticine example shows, cannot be the state of motion of the body relative to other bodies. To attribute the effect as Einstein and Mach did to the rotation of the body with respect to the other *bodies* in the universe is to endorse a remnant of the unjustified monopoly of the older body ontology, namely, the sovereign right of material bodies to play the role of physically real and acceptable causal agents.

The epistemic problem alluded to earlier, concerning free or natural motion, has recently become a pressing issue within the particular context of the constructive axiomatics for the General Theory of Relativity. One of the constructive axioms employed by Ehlers, Pirani and Schild, the projective axiom, is a statement of the infinitesimal version of the Law of Inertia, the law of free (fall) motion which contains Newton's first law of motion as a special case in the absence of gravitation. Since Ehlers, Pirani and Schild do not provide an independent, noncircular criterion by which to characterize free (fall) motion, their approach has been charged with circularity by philosophers such as Grünbaum (1973), Salmon (1969, 1977), Sklar (1977) and Winnie (1970). The problem is how to introduce a class of preferred motions, that is, how to characterize that particular path structure that would govern the motions of free particles (neutral, spherically symmetric, non-rotating test bodies) while avoiding the circularity problem surrounding the notion of a free particle. Although, as we just saw, Weyl is unambiguous about the ontological question concerning the reality of the guiding field, he was sensitive to the epistemological question which arises particularly within the context of the geodesic method. Within that context, we recall an earlier remark of Weyl (cited in section 4.9) to the effect that if it is possible to discern light propagation and free (fall) motion, then we are able to determine the metric from the conformal-causal and projective structures of spacetime.

Coleman and Korté, (1980, 1982, 1984b, 1987, 1989, 1990, 1992a, 1995a, 1995b) solved these and related difficulties.[61] The solution provides noncircular, empirical procedures for the identification of monopoles, for the separation of monopole particles into distinct classes each of which corresponds to a particular

---

[61]See the first footnote in remark 38 above concerning the nature and the context of our work.

path structure, for the measurement of these path structures, and for the testing of a given path structure for geodesicity.

The most difficult part of the above analysis was to discover a non-conventional procedure for the empirical determination of the projective structure which we presented in (Coleman and Korté 1980). Using the jet and jetbundle formalism of Ehresmann (1951a, 1951b, 1951c, 1952a, 1952b, 1983), we first formulate quite general path structures which are not defined at the outset in terms of *geodesic* paths and which require for their description only the local differential topological structure. We then prove a number of theorems, which are a generalization and reformulation of a result proved by Ehlers and Köhler (1977), that serve as necessary and sufficient criteria for singling out free (fall) motion. The theorems in question involve only local differential topological concepts and are coordinate and frame independent. We also show that they are epistemically effective in that they can be employed as *empirical* criteria for singling out free (fall) motion at a level of testing that requires no more structure than is needed for introducing arbitrary physical local coordinates. Our results establish that the projective axioms concerning free (fall) motion in Ehlers, Pirani and Schild's constructive axiomatics is *epistemically decidable* in a non-circular, non-conventional way. We were thus able to present in (Coleman and Korté 1984a) a non-circular formulation of Newton's Laws of Motion. According to our new formulation, the Law of Inertia asserts the existence of a unique projective structure.

> **The Law of Inertia:** There exists on spacetime a unique projective
> structure Π (or equivalently, a unique geodesic directing field Π).

In view of our work dealing with the measurement of directing fields (Coleman and Korté, 1980, 1992a) this law is a non-circular, non-conventional and falsifiable empirical statement.

Free motion is defined with reference to the projective structure Π as follows:

> **Definition of Free Motion:** A possible or actual material body is in
> a state of free motion during any part of its history just in case the
> corresponding segment of its world-line path is a solution path of the
> differential equation determined by the unique projective structure of
> spacetime.

The Law of Inertia and the definition of free motion together constitute a modern reformulation of Newton's first law of motion. Newton's second law of motion may be reformulated as follows:

> **The Law of Motion:** With respect to any coordinate system, the
> world-line path of a possible or actual material body satisfies an equa-
> tion of the form,

$$m\left[\xi_2^\alpha - \Pi^\alpha(x^i(p), \xi_1^\alpha)\right] = F^\alpha(x^i(p), \xi_1^\alpha), \qquad (4.11.1)$$

where m is a scalar constant characteristic of the material body called its inertial mass, ξ_1 is its three-velocity, ξ_2 is its three-acceleration, $\Pi^\alpha(x^i, \xi_1^\alpha)$ is the directing field (4.11.2) determined by the projective structure of spacetime and $F^\alpha(x^i(p), \xi_1^\alpha)$ is the three-force acting on the body.

The directing field (equation-of-motion structure) determined by the projective structure of spacetime is given by

$$\Pi^\alpha(\xi_1^\alpha) = \xi_1^\alpha (\Pi^0_{\rho\sigma}\xi_1^\rho\xi_1^\sigma + 2\Pi^0_{0\rho}\xi_1^\rho + \Pi^0_{00})$$
$$- (\Pi^\alpha_{\rho\sigma}\xi_1^\rho\xi_1^\sigma + 2\Pi^\alpha_{0\rho}\xi_1^\rho + \Pi^\alpha_{00}), \qquad (4.11.2)$$

where the projective coefficients Π^i_{jk} are defined in (4.3.5). Our formulation of the Law of Motion, which makes explicit its ontological dependence on the unique projective structure postulated by the Law of Inertia, together with our measurement procedure for directing fields, establishes that forces are physically real, non-conventional and empirically determinable entities.[62] Coleman and Korté (1984a) have also shown that in a world in which causal relations are determined by a

---

[62] In (Coleman and Korté 1989), we prove that if a second-order directing field is a polynomial with respect to its 3-velocity variables in every coordinate chart, then it is necessarily cubic with respect to its 3-velocity variables and hence it is geodesic; that is, it is a projective structure. This establishes that forces in the relativistic context are necessarily non-polynomial with respect to the 3-velocity variables.

We have recently shown in (Coleman and Korté 1999) that there exists an essential difference between pre-relativistic and relativistic theories with regard to the distinction between geometry and forces. In the relativistic case, the projective structure (geodesic directing field) is unique and forces are necessarily nonpolynomial in the 3-velocity variables. In contrast, in the pre-relativistic case, all of the physical directing fields, that were considered historically, are cubic in the 3-velocity variables and hence correspond to projective structures. Thus, the cubic criterion, which is an epistemically effective, local, differential-topological criterion for distinguishing between geometry and forces in the relativistic case, fails in the pre-relativistic case. It is essentially because of this circumstance that it was not possible to formulate Newton's Laws of Motion in a non-circular way in pre-relativistic physics.

Although the issue of whether or not the separation between geometry and forces is conventional remains an open question in the case of pre-relativistic *spacetime*, Poincaré's argument for the conventionality of the pre-relativistic *spatial* geometry can be refuted decisively by other means. In (Coleman and Korté 1995b), we show that Poincaré's conventionalist objections to the use of rigid rods for the measurement of geometry of space can be overcome at least in the case in which the geometry actually is Euclidean and rigid rods do in fact exist. Potential rigid rods are put through a series of tests each of which involves observations that are differential-topological in character. The rods that pass all of the tests are then used to set up numerous coordinate systems, all of the same type, and the coordinate transformation functions between any two of them are measured. This data is then analyzed in a coordinate independent manner to show that space has a Euclidean Lie pseudo-group structure which can be proven to be equivalent to a flat Riemannian structure; that is, space is Euclidean.

Although the Lie pseudo-group approach to geometric structures has its origins in the work of Cartan, this approach has only recently been re-analyzed in terms of modern differential geometry by Guillemin and Sternberg (1964, 1966). What has not been appreciated in the philosophical literature is that a structure which is expressed in terms of local coordinate transformations — a Lie pseudo-group structure — is mathematically equivalent to a structure that is readily seen to be expressed in coordinate independent terms.

conformal structure, the only possible solution to the problem of motion is the solution they have given; that is, the standard of free motion is determined by a physically real projective structure and this projective structure must be unique. In particular, they show that any equation-of-motion structure (either a curve or a path structure) that has sufficient microisotropy to be compatible with the conformal-causal structure of spacetime must be both geodesic and unique. Hence, the empirically well-supported principles of conformal causality and of the universality of free fall together require the existence of a unique Weyl structure on spacetime. This, they claim, provides for a non-conventional and empirical vindication of Weyl's realist field ontology of the inertial structure of spacetime: The transition from the kinematic to the dynamic behaviour of bodies in spacetime theories is not arbitrary but is, as a matter of empirical fact, dictated by a physically real, causally efficacious, unique, *projective* structure (guiding field).

5 Group Theory and its Applications

That spacetime is equipped with a (pseudo) Riemannian metric is a fundamental postulate of the General Theory of Relativity. Since the Riemannian type of metric is only one of a rich variety of conceivable metric forms, it is reasonable to ask what the fundamental characteristics of the Riemannian form are that single it out. As discussed in subsections 4.5, 4.6 and 4.7, Weyl analyzed this problem of space, *the Raumproblem*, in a series of articles (Weyl, 1922a, 1922b, 1923f). Weyl (1949b, p. 400) himself noted that his interest in the philosophical foundations of the General Theory of Relativity motivated his analysis of the representations and invariants of the continuous groups, "I can say that the wish to understand what really is the mathematical substance behind the formal apparatus of relativity theory led me to the study of representations and invariants of groups; and my experience in this regard is probably not unique".

A motivation of quite another sort was provided by Study when he attacked Weyl specifically, as well as other unnamed individuals, by accusing them "of having neglected a rich cultural domain (namely, the theory of invariants), indeed of having completely ignored it".[63] Weyl (1924e) replied immediately providing a new foundation for the theory of invariants of the special linear groups $SL(n, \mathbf{C})$ and its most important subgroups, the special orthogonal group $SO(n, \mathbf{C})$ and the special symplectic group $SSp(\frac{n}{2}, \mathbf{C})$ (for n even) based on algebraic identities due to Capelli. In a footnote, Weyl (1924e) sarcastically informed Study that "even if he [Weyl] had been as well versed as Study in the theory of invariants, he would not have used the symbolic method in his book *Raum, Zeit, Materie* and even with the last breath of his life would not have mentioned the algebraic completeness theorem for invariant theory".[64] Although Weyl did mention the incident again much later in an article (Weyl 1949b) concerning *Relativity as a Stimulus in Mathematical Research*, he was probably not too perturbed by Study's rebuke since Study was in the habit of attacking just about everybody. Nevertheless, there is no doubt that his attention was drawn to the theory of invariants of groups, the full analysis of which requires the theory of group representations.

In two papers, Weyl (1924b, 1924g) announced the basic methods he would employ later to systematically analyze the linear representations of semisimple Lie groups (Weyl, 1925b, 1926b, 1926c, 1926d); namely, the method of integration over the manifold of the compact real form of the given complex semisimple Lie group, a method first used by Hurwitz (1897) to construct invariants of $SL(n, \mathbf{C})$ and then by Schur (1924) in the theory of representations of the rotation group, and the idea of combining results on the representations of Lie algebras with Young's use of permutation-symmetry operators in the analysis of tensor representations (Young, 1901, 1902). In addition, Weyl made use of Cartan's work on the theory of continuous groups (Cartan 1894) and on the classification (Cartan, 1913, 1914)

---

[63] As reported by Weyl (1924e) in footnote 1 of the cited work.

[64] Weyl's point is that in the context of his book *Raum, Zeit, Materie*, the kernel-index method of tensor analysis is more appropriate than the methods of the theory of algebraic invariants.

of the representations of the Lie algebras of all semisimple Lie groups.

About the same time (1924–1926), the physicists formulated the *new* theory of Quantum Mechanics. Almost every aspect of Weyl's mathematical expertise[65] could be applied to this theory: the theory of Hilbert space, singular differential equations, eigenfunction expansions, the symmetric group, and unitary representations of Lie groups. Weyl wrote numerous articles about the theory and one of the first textbooks (Weyl 1928b) on Quantum Mechanics. The book deals not only with non-relativistic quantum mechanics but also with the relativistic quantum mechanics and with the relativistic quantum field theory of the Dirac fields and the electromagnetic fields. In this book, Weyl also discussed the invariance of the Maxwell-Dirac theory under the discrete symmetries that are now called *parity*, *time reversal* and *charge conjugation*. It is little wonder that this textbook was considered rather difficult! Following up on Dirac's discovery of spinors for the Relativistic Quantum Mechanics of the electron, Weyl later developed the general theory of spinor representations of the orthogonal groups. He also recognized that the formulation of the General Theory of Relativity would have to be modified to accommodate such fields and that the electromagnetic gauge transformations would be necessary in this context.

5.1   The Representations and Invariants of Lie Groups

A group G is a set $\{g_a | a \in \text{Index}\}$ together with a product function $\phi: G \times G \to G$ and an inverse function $\mu: G \to G$ such that these functions satisfy the following conditions. The group product is associative; that is, for any three elements g_a, g_b and g_c of G,

$$\phi(g_a, \phi(g_b, g_c)) = \phi(\phi(g_a, g_b), g_c). \tag{5.1.1}$$

There exists a unique element $1_G \in G$, called the identity such that, for any $g \in G$,

$$\phi(1_G, g) = g = \phi(g, 1_G). \tag{5.1.2}$$

For every $g \in G$, the inverse function satisfies

$$\phi(\mu(g), g) = 1_G = \phi(g, \mu(g)). \tag{5.1.3}$$

For brevity, $\phi(g_a, g_b)$ and $\mu(g)$ are frequently denoted by $g_a g_b$ and g^{-1} respectively. Then, the above conditions take the form

$$g_a(g_b g_c) = (g_a g_b)g_c, \tag{5.1.4}$$

$$1_G g = g = g 1_G \tag{5.1.5}$$

and

$$g^{-1}g = 1_G = gg^{-1}. \tag{5.1.6}$$

---

[65]In particular, Weyl's work on the theory of singular differential-integral equations (1908–1911) — see subsections 2.1–2.3 of this overview — provided him with the precise tools for solving many of the concrete problems posed by the new theory.

A finite-dimensional representation of a group G is a map ρ from G into the set of linear transformations of a finite-dimensional vector space V, such that for any g_a and g_b in G,

$$\rho(g_a)\rho(g_b) = \rho(g_a g_b),$$ (5.1.7)

and such that

$$\rho(1_G) = I,$$ (5.1.8)

where I is the identity transformation on V. It follows from these conditions that the transformations $\rho(g)$ are necessarily invertible. The vector space V is assumed to be defined over the field of complex numbers $\mathbf{C}$ unless otherwise indicated.

With respect to a basis $b_1, b_2, \ldots, b_n$ of the vector space V, the transformations $\rho(g)$ are described by invertible $n \times n$ matrices. Suppose that $\hat\rho$ and ρ are two representations of G, that the representation spaces $\hat{V}$ and V are both n-dimensional, and that there exists an invertible linear transformation $A : \hat{V} \to V$ such that for every $g \in G$,

$$\hat\rho(g) = A^{-1}\rho(g)A.$$ (5.1.9)

Then, to every basis of $\hat{V}$, there corresponds a basis of V with respect to which for every $g \in G$, the matrices that describe $\hat\rho(g)$ and $\rho(g)$ are the same. The representations $\hat\rho$ and ρ are said to be equivalent.

A representation ρ of G is said to be reducible if and only if there exists a basis for the vector space V with respect to which for every $g \in G$, the matrix corresponding to $\rho(g)$ satisfies

$$\rho(g)^i_j = 0, \text{ for } m < i \leq n \text{ and } 1 \leq j \leq m,$$ (5.1.10)

where i and j are the row and column indices respectively. The subspace V_A of V spanned by the first m basis vectors is invariant under every transformation of the representation. Hence, the restriction of the transformation $\rho(g)$ to this subspace provides an m-dimensional representation ρ_A of the group G. Such a reducible representation ρ also determines an $(n-m)$-dimensional representation as follows. Define an equivalence relation on V by stipulating that two vectors $v, w \in V$ are equivalent, $v \sim w$, if and only if their difference $v - w$ lies in the invariant m-dimensional subspace V_A. The set $V/\!\sim$ of equivalence classes is an $(n - m)$-dimensional vector space. Moreover, the equivalence relation is invariant under every transformation $\rho(g)$; consequently, the representation ρ induces an $(n-m)$-dimensional representation of G in the vector space $V/\!\sim$. In many cases, it is possible to choose a basis for V such that for every $g \in G$, the matrix corresponding to the transformation $\rho(g)$ satisfies not only the condition (5.1.10) but also the condition

$$\rho(g)^i_j = 0, \text{ for } 1 \leq i \leq m \text{ and } m < j \leq n.$$ (5.1.11)

Then, the vector space V_B spanned by the last $(n - m)$ basis vectors of V is also invariant under every transformation of the representation ρ and the restriction of ρ to this subspace determines an $(n - m)$-dimensional representation ρ_B. The representation ρ is then said to be fully reducible or decomposable, and ρ and V are the direct sums $\rho = \rho_A \oplus \rho_B$ and $V = V_A \oplus V_B$.

The character function $\chi_\rho \colon G \to \mathbf{C}$ of a representation ρ is defined by

$$\chi_\rho(g) = \sum_{j=1}^{n} \rho(g)_j^j. \tag{5.1.12}$$

The function χ_ρ does not depend on the choice of basis for V since the trace of a transformation is invariant under a similarity transformation. Any two elements g_a and g_b of G are said to belong to the same class if and only if there exists an element $g \in G$ such that

$$g_a = g^{-1} g_b g. \tag{5.1.13}$$

Since $\rho(g^{-1}) = \rho^{-1}(g)$, it follows that $\chi_\rho(g_a) = \chi_\rho(g_b)$; consequently, the character χ_ρ is a class function. Also, equivalent representations have the same character. In addition,

$$\chi_{\rho_A \oplus \rho_B} = \chi_{\rho_A} + \chi_{\rho_B}; \tag{5.1.14}$$

that is, the character of a reducible representation is the sum of the characters of its irreducible components. Moreover, if an irreducible representation occurs more than once in a representation ρ, the multiplicity appears as the coefficient of the appropriate irreducible character in the expression for χ_ρ. Also, the dimension of a representation is given by $\chi_\rho(1_G)$. Hence, much useful information about a representation ρ can be obtained quickly and easily from its character χ_ρ.

The principal goal of representation theory is to determine all inequivalent, irreducible representations of a given abstract group G in terms of which any given representation is to be analyzed. During the decade centred on the year 1900, Frobenius, Burnside and Schur had in principle worked out the theory of representations for finite groups. The results and techniques required for an appreciation of Weyl's work will now be considered.

Any finite group can be realized as a subgroup of the symmetric group S_n, the group of permutations (invertible transformations) of n integers $\{1, 2, \ldots, n\}$. Any element of S_n can be expressed as a product of independent cycles. If k distinct members of $\{1, 2, \ldots, n\}$ are denoted by $i_1, i_2, \ldots, i_k$, then the cycle $(i_1 i_2 \ldots i_k)$ is the permutation that takes i_1 into i_2, i_2 into $i_3, \ldots, i_k$ into i_1 and leaves the remaining elements unchanged. It is customary to omit cycles of length 1. The sign or parity of a cycle of length k is $(-1)^{k+1}$. The sign of an arbitrary permutation $\pi \in S_n$ is the product of the signs of its cycles. If $\pi_1, \pi_2 \in S_n$,

$$\text{sign}(\pi_1 \pi_2) = \text{sign}(\pi_1)\text{sign}(\pi_2). \tag{5.1.15}$$

The number of elements of a finite group is called its order. The order of S_n is $n!$.

Consider a finite group G of order n. The elements of G may be labelled by the integers $\{1, 2, \ldots, n\}$. Once such a coordinate labelling has been chosen and fixed, the product function of the group may be described in terms of these labels in the form

$$l = \phi(j, k), \tag{5.1.16}$$

where $j, k, l \in \{1, 2, \ldots, n\}$. These relations may be displayed in an $n \times n$ multiplication table. Every element of the group G labelled by j uniquely determines and is uniquely determined by the element of S_n that is determined by the transformation (5.1.16) for fixed j. Thus, G can be regarded as a subgroup of S_n.

The regular representation of a finite group G of order n is defined as follows. Regard the elements of G as independent basis vectors of an n-dimensional vector space V_G. To every function $f: G \to \mathbf{C}$, there corresponds the vector $f \in V_G$ given by

$$f = \sum_g f(g)g. \tag{5.1.17}$$

A linear transformation of V_G into itself is completely defined by its action on the basis vectors of V_G. For each element $a \in G$, define the linear transformation $\rho(a): V_G \to V_G$ by

$$\rho(a)g = ag \tag{5.1.18}$$

for every $g \in G$. The set of linear transformations $\{\rho(a) | a \in G\}$ determines the regular representation ρ of G. Every irreducible representation of G occurs at least once in the decomposition of the regular representation.

The group algebra A_G has the same vector space structure as V_G. The algebra product is defined in terms of the group product by formally multiplying the 'vectors' (5.1.17). The algebraic structures V_G and A_G may be more directly constructed on the set of functions $f: G \to \mathbf{C}$ by defining vector addition, scalar multiplication and multiplication pointwise according to

$$\begin{aligned}
(f_1 + f_2)(g) &= f_1(g) + f_2(g) \\
(\lambda f)(g) &= \lambda f(g), \quad \lambda \in \mathbf{C}
\end{aligned} \tag{5.1.19}$$

and

$$\begin{aligned}
(f_1 f_2)(g) &= \sum_{a \in G} f_1(a) f_2(a^{-1}g) \\
&= \sum_{a \in G} f_1(ga^{-1}) f_2(a).
\end{aligned} \tag{5.1.20}$$

To every subspace of V_G that is irreducible and invariant under the regular representation defined by (5.1.18), there corresponds a minimal left ideal $A_G e$ that is generated by an idempotent[66] element $e \in A_G$, and conversely. It is more convenient to analyze the richer of the two structures, namely, A_G.

For the group S_n, there exists a distinct irreducible representation corresponding to every partition of $n = n_1 + n_2 + \cdots + n_r$ which satisfies $n_1 \geq n_2 \geq$

---

[66] An element $e \in A_G$ is idempotent if and only if $ee = e$.

$\cdots \geq n_r$. Such a partition is conveniently represented by a Young diagram consisting of *left justified* rows of squares with n_j squares in row j for $1 \leq j \leq r$. A Young tableau is obtained by inserting the integers $\{1, 2, \ldots, n\}$ into the squares of a Young diagram in any order. To each tableau there corresponds an idempotent element of A_{S_n} given by

$$k \sum_q \sum_p \text{sign}(q) qp, \tag{5.1.21}$$

where p denotes any permutation that does not interchange integers in different rows and q denotes any permutation that does not interchange the integers in different columns and k is a normalization factor. Each such idempotent yields a minimal left ideal and a corresponding irreducible representation. All representations associated with the same diagram are equivalent. The minimal left ideals associated with a particular diagram are not all linearly independent. A complete, linearly independent set is determined by the idempotents associated with the *standard tableaux* for which the integers increase to the right and downwards.

As an example, consider the group

$$S_3 = \{1, (12), (13), (23), (123), (321)\}. \tag{5.1.22}$$

The idempotent elements corresponding to the standard tableau are

$$e_{\,1\,2\,3} \;=\; \frac{1}{6}(1 + (12) + (13) + (23) + (123) + (321)), \tag{5.1.23}$$

$$e_{\substack{1\,2\\3}} \;=\; \frac{1}{3}(1 - (13))(1 + (12))$$

$$=\; \frac{1}{3}(1 + (12) - (13) - (123)), \tag{5.1.24}$$

$$e_{\substack{1\,3\\2}} \;=\; \frac{1}{3}(1 - (12))(1 + (13))$$

$$=\; \frac{1}{3}(1 + (13) - (12) - (321)), \tag{5.1.25}$$

$$e_{\substack{1\\2\\3}} \;=\; \frac{1}{6}(1 - (12) - (13) - (23) + (123) + (321)). \tag{5.1.26}$$

These idempotents provide a resolution of the identity $1 \in A_{S_3}$; that is, the product of any two distinct idempotents vanishes and their sum is $1 \in A_{S_3}$. There are three inequivalent representations. The representations corresponding to (5.1.24) and (5.1.25) are equivalent. These representations are 2-dimensional while the representations corresponding to (5.1.23) and (5.1.26) are both 1-dimensional.

At the turn of the century, Young applied the representation theory of the symmetric group to construct all of the irreducible tensor representations of the groups $GL(n, \mathbf{C})$ and $SL(n, \mathbf{C})$. A tensor T of order k is a geometric object the components of which transform according to

$$\bar{T}^{\mu_1 \mu_2 \cdots \mu_k} = \bar{X}^{\mu_1}_{\nu_1} \bar{X}^{\mu_2}_{\nu_2} \cdots \bar{X}^{\mu_k}_{\nu_k} T^{\nu_1 \nu_2 \cdots \nu_k}, \tag{5.1.27}$$

where $\bar{X} \in GL(n, \mathbf{C})$. Corresponding to every element $\pi \in S_k$, there is a linear transformation defined by

$$(\pi^{-1} T)^{\mu_1 \mu_2 \cdots \mu_k} = T^{\mu_{\pi 1} \mu_{\pi 2} \cdots \mu_{\pi k}}, \tag{5.1.28}$$

which commutes with the linear transformation (5.1.27). The transformations (5.1.28) provide a linear representation of S_k which may be decomposed by the methods described above. For example, an arbitrary third order tensor $T^{\mu_1 \mu_2 \mu_3}$ decomposes into the four tensors

$$\frac{1}{6}(T^{\mu_1 \mu_2 \mu_3} + T^{\mu_2 \mu_1 \mu_3} + T^{\mu_3 \mu_2 \mu_1} + T^{\mu_1 \mu_3 \mu_2} + T^{\mu_2 \mu_3 \mu_1} + T^{\mu_3 \mu_1 \mu_2}), \tag{5.1.29}$$

$$\frac{1}{3}(T^{\mu_1 \mu_2 \mu_3} + T^{\mu_2 \mu_1 \mu_3} - T^{\mu_3 \mu_2 \mu_1} - T^{\mu_2 \mu_3 \mu_1}), \tag{5.1.30}$$

$$\frac{1}{3}(T^{\mu_1 \mu_2 \mu_3} + T^{\mu_3 \mu_2 \mu_1} - T^{\mu_2 \mu_1 \mu_3} - T^{\mu_3 \mu_1 \mu_2}) \tag{5.1.31}$$

and

$$\frac{1}{6}(T^{\mu_1 \mu_2 \mu_3} - T^{\mu_2 \mu_1 \mu_3} - T^{\mu_3 \mu_2 \mu_1} - T^{\mu_1 \mu_3 \mu_2} + T^{\mu_2 \mu_3 \mu_1} + T^{\mu_3 \mu_1 \mu_2}), \tag{5.1.32}$$

which correspond to the idempotent operators (5.1.23–5.1.26) respectively. This decomposition of tensor spaces into symmetry classes of tensors yields the decomposition of the representation (5.1.27) into its irreducible components.

The set of elements of a Lie group G has the structure of a smooth manifold. If this manifold has r real dimensions, open subsets of G may be coordinatized by r real variables that range over open subsets of $\mathbf{R}^r$. In general, several overlapping coordinate systems (charts) may be required to map the entire manifold. The product function of such an r-parameter Lie group is specified, at least locally, by means of r real-valued functions of $2r$ real variables,

$$\gamma^i = \phi^i(\alpha, \beta), \quad 1 \leq i \leq r, \tag{5.1.33}$$

where $\alpha, \beta, \gamma \in \mathbf{R}^r$ are the coordinates of group elements. If N charts are used, then N^3 sets of functions (5.1.33) may be required to describe the product function. The analogue for a finite group is the much simpler multiplication table (5.1.16). Every r-parameter Lie group determines a unique r-dimensional algebra, its Lie algebra. Corresponding to every Lie algebra, there is a unique, simply connected Lie group, the universal covering group, and every Lie group with the same Lie algebra is obtained as a factor group of the universal covering group with respect to one of its discrete invariant subgroups.

The group $GL(n, \mathbf{C})$ of invertible $n \times n$ matrices with complex elements provides the simplest example of the relationship between a Lie group and its Lie algebra. Let A be an arbitrary $n \times n$ matrix with complex elements and set

$$a(s) = \exp[sA] = \sum_{k=0}^{\infty} \frac{s^k}{k!} A^k, \qquad (5.1.34)$$

where s is a real parameter and A^k denotes the k-fold matrix product of A with itself. Then $a(s) \in GL(n, \mathbf{C})$ for all $s \in \mathbf{R}$ and in particular for $s = 1$. If $b(t) = \exp[tB]$, then

$$(b(t)a(s))^{-1}(a(s)b(t)) = a^{-1}(s)b^{-1}(t)a(s)b(t) = I + st[A, B] + \cdots, \qquad (5.1.35)$$

where the commutator of A and B is defined by

$$[A, B] = AB - BA. \qquad (5.1.36)$$

This result follows from the expansion (5.1.34) if all terms up to total degree 2 in s and t are kept and if the relation $a^{-1}(s) = a(-s)$ is used. The vector space of $n \times n$ matrices over $\mathbf{C}$ with the product (5.1.36) is the Lie algebra of $GL(n, \mathbf{C})$. The matrix A is determined by $2n^2$ real numbers; consequently, $GL(n, \mathbf{C})$ is a $2n^2$-parameter Lie group.

An element $a = \exp(A) \in GL(n, \mathbf{C})$ has determinant 1 and hence is an element of $SL(n, \mathbf{C})$ if and only if the sum of the diagonal elements of A (the trace of A) is zero. Any element of $GL(n, \mathbf{C})$ can be expressed as the product of an element of $SL(n, \mathbf{C})$ and an element $\exp(\alpha)I \in GL(n, \mathbf{C})$ where $\alpha \in \mathbf{C}$. The Lie algebra of $SL(n, \mathbf{C})$ consists of the Lie algebra of all $n \times n$ matrices A with zero trace. Thus, $SL(n, \mathbf{C})$ is a $(2n^2 - 2)$-parameter Lie group.

Let X_i for $1 \le i \le r$ be a basis for an arbitrary Lie algebra $\mathfrak{g}$. The commutator product of the Lie algebra $\mathfrak{g}$ is completely determined by the commutation relations

$$[X_i, X_j] = \sum_{k=1}^{r} C_{ij}^k X_k, \qquad (5.1.37)$$

where the C_{ij}^k are the *structure constants* of the Lie algebra $\mathfrak{g}$. A Lie subalgebra is a linear subspace of the Lie algebra that is closed under commutation. A subalgebra is an ideal (invariant subalgebra) if the commutator of any element of the algebra with any element of the subalgebra yields an element of the subalgebra. The set of all elements of an algebra that can be expressed as the commutator of two elements of the algebra is an invariant subalgebra called the derived algebra. The process can be iterated to obtain the derived algebra of the derived algebra and so forth. At some point in the chain, an algebra that reproduces itself under derivation will be obtained and the process terminates. If the last algebra in the chain contains only the zero element, the original algebra is said to be *solvable*. A Lie algebra that has no solvable invariant subalgebra is called *semisimple*. Cartan proved that a Lie algebra is semisimple if and only if the invariant quadratic form defined by

$$\sum_{j,k} g_{jk} A^j A^k = \sum_{j,k} \sum_{p,q} \left(C_{pj}^q C_{qk}^p \right) A^j A^k \tag{5.1.38}$$

is nondegenerate where

$$A = \sum_{i=1}^{r} A^i X_i. \tag{5.1.39}$$

Cartan also introduced the canonical basis for semisimple Lie algebras over **C**. There exist maximal commuting subalgebras of dimension l called Cartan subalgebras. Let $\vec{H} = (H_1, H_2, \ldots, H_l)$ be a basis for one of these Cartan subalgebras. The restriction of the Cartan-Killing metric (5.1.38) to the Cartan subalgebra is positive definite, and the basis vectors $\vec{H}$ may be chosen to be orthonormal with respect to this metric. The regular or adjoint representation R for a Lie algebra $\mathfrak{g}$ is defined as follows. For any $A \in \mathfrak{g}$, the linear map $R(A): \mathfrak{g} \to \mathfrak{g}$ is defined by

$$R(A)B = [B, A], \tag{5.1.40}$$

where B is an arbitrary element of $\mathfrak{g}$. One can readily show that

$$R(AB) = R(A)R(B). \tag{5.1.41}$$

One can set up the simultaneous eigenvalue problem

$$R(H_i)E_{\vec{\alpha}} = \alpha_i E_{\vec{\alpha}} \quad \text{for } i \in \{1, \ldots, l\}. \tag{5.1.42}$$

It can be shown that for the case of a semisimple Lie algebra, the $E_{\vec{\alpha}}$ for $\vec{\alpha} \neq \vec{0}$ are uniquely determined up to a normalization. The l-tuples $\vec{\alpha}$ are called *root vectors*. If $\vec{\alpha}$ is a root vector, then so is $-\vec{\alpha}$.

Remark 39 Some authors use the following, somewhat different notation. For any element $H = \sum_{i=1}^{l} c^i H_i$ of the Cartan subalgebra, one has the eigenvalue problem

$$R(H)E = \alpha(H)E. \tag{5.1.43}$$

One can show that $\alpha(H)$ is linear; consequently, α is an element of the dual vector space H^*. Clearly,

$$\alpha(H) = \alpha \left(\sum_{i=1}^{l} c^i H_i \right) = \sum_{i=1}^{l} c^i \alpha(H_i) = \sum_{i=1}^{l} c^i \alpha_i. \tag{5.1.44}$$

Moreover, the Euclidean metric on H permits the identification of each element of H^* with an element of H in a canonical manner.

The vectors $\vec{H}$ together with the vectors $E_{\vec{\alpha}}$ and $E_{-\vec{\alpha}}$ corresponding to each nonzero root vector $\vec{\alpha}$ form a basis for the Lie algebra. With respect to this canonical basis[67], the commutation relations have the form

$$\left[H_i, H_j\right] = 0, \quad \text{for } 1 \leq i, j \leq l,$$

$$\left[\vec{H}, E_{\vec{\alpha}}\right] = \vec{\alpha} E_{\vec{\alpha}},$$

$$\left[E_{\vec{\alpha}}, E_{-\vec{\alpha}}\right] = \vec{\alpha} \cdot \vec{H},$$

$$\left[E_{\vec{\alpha}}, E_{\vec{\beta}}\right] = N(\vec{\alpha}, \vec{\beta}) E_{\vec{\alpha}+\vec{\beta}}, \tag{5.1.45}$$

where $N(\vec{\alpha}, \vec{\beta})$ is a numerical factor depending on $\vec{\alpha}$ and $\vec{\beta}$. Cartan (1913, 1914) also gave a complete analysis of the irreducible representations of the Lie algebras of the semisimple Lie groups. As remarked above, this work, completed in 1914, was an important input to Weyl's analysis.

In his four part memoir on the representation theory of semisimple Lie groups, Weyl (1925b, 1926b, 1926c, 1926d) set himself the task of "weaving all these threads[68] into a unified whole". The first chapter of (Weyl 1925b) is devoted to an analysis of the irreducible representations of $SL(n, \mathbf{C})$. Cartan's analysis of the irreducible representations of the Lie algebra of $SL(n, \mathbf{C})$ is presented in detail, and its relationship to Young's analysis of the irreducible representation of $SL(n, \mathbf{C})$ corresponding to the symmetry classes of tensors defined by the action (5.1.28) of the symmetric group is explained. Young had established the irreducibility of all of the representations that he had obtained. Weyl pointed out that Cartan had not proved the *full* reducibility of the representations in his exhaustive list. The bijective correspondence between the representations obtained by Cartan and by Young showed that all of the irreducible representations of $SL(n, \mathbf{C})$ were tensor representations.

Weyl, however, proved the full reducibility of any finite-dimensional representation of $SL(n, \mathbf{C})$ and hence established that *any* linear magnitude is a direct sum of tensors by employing a method of integration over the group manifold of the compact unitary subgroup $SU(n, \mathbf{C})$. Hurwitz had first used such an integration method to construct group invariants and Schur had just recently applied the technique in the special case of the rotation group. The essence of the idea can be seen by considering a reducible representation ρ of a *finite* group G over a vector space V. Choose any basis for V and define an inner product on V by

$$(v, w) = \sum_{i=1}^n \sum_g \overline{\sum_{j=1}^n \rho(g)_j^i v^j} \sum_{k=1}^n \rho(g)_k^i w^k. \tag{5.1.46}$$

---

[67]For more details see (Gilmore 1974), (Elliott and Dawber 1987), (Bäuerle and de Kerf 1990) or (Fulton and Harris 1991).

[68]Essentially the ideas of Hurwitz (1897), Schur (1924), Young (1901, 1902) and Cartan (1913, 1914).

Then, this Hermitian inner product is invariant under all group transformations; that is, the representation is unitary with respect to this inner product,

$$(\rho(g)v, \rho(g)w) = (v, w). \tag{5.1.47}$$

Thus, if the vectors $b_1, b_2, \ldots, b_m$ span a subspace of V that is invariant under the action of ρ, the invariant inner product (5.1.46) may be used to construct a subspace of V of dimension $(n-m)$ that is orthogonal to the vectors $b_1, b_2, \ldots, b_m$ and is also invariant. Hence, any reducible representation of a finite group is fully reducible or decomposable. For a compact group such as $SU(n, \mathbf{C})$, an invariant inner product for any representation space can be constructed in a manner analogous to (5.1.46) by replacing the finite sum over the group by an invariant integral over the group manifold of $SU(n, \mathbf{C})$. Since the group manifold is compact, the total 'volume' of the group is finite and convergence difficulties do not arise. Thus a reducible representation of $SU(n, \mathbf{C})$ is fully reducible. The additional argument Weyl needed to extend this result to the representations of $SL(n, \mathbf{C})$ is discussed below in its more general context.

Weyl also used the method of integration to obtain orthonormality relations for the characters of the representations of $SU(n, \mathbf{C})$. He used these relations to obtain explicit formulas for the characters (and hence the dimensions) of the irreducible representations of $SU(n, \mathbf{C})$ and of $SL(n, \mathbf{C})$. In chapter II, Weyl (1926b) extended all of these results to the subgroups of $SL(n, \mathbf{C})$ which preserve respectively a nondegenerate symmetric quadratic form, the rotation group $SO(n, \mathbf{C})$, and a nondegenerate skewsymmetric quadratic form, the symplectic group $SSp(\frac{n}{2}, \mathbf{C})$.

In chapter III, Weyl (1926c) presented an elegant derivation of Cartan's analysis of the structure of Lie algebras including his criteria for solvable and for semisimple Lie algebras and his construction of the canonical basis (5.1.45) of a Lie algebra. Weyl then went his own way. He associated a finite group, now called the Weyl group, with the root space of every semisimple Lie algebra. Recall that the root vectors $\vec{\alpha}$ are vectors in a Euclidean space with the dimension of the Cartan subalgebra. Each root vector $\vec{\alpha}$ defines a hyperplane orthogonal to it. A reflection in this hyperplane leaves the set of root vectors invariant. The finite group generated by all such reflections, consisting of these reflections and of all possible products of these reflections is the Weyl group of the Lie algebra. It is instructive to consider the simple case of the group $SU(3)$ for which the root space is 2-dimensional and the nonzero root vectors[69] are $\pm\vec{\alpha}_1, \pm\vec{\alpha}_2$ and $\pm\vec{\alpha}_3$ where

$$\vec{\alpha}_1 = \frac{1}{\sqrt{3}}(1,0),$$

$$\vec{\alpha}_2 = \frac{1}{\sqrt{3}}\left(-\frac{1}{2}, \frac{\sqrt{3}}{2}\right),$$

---

[69]See for example (Gilmore 1974, p. 272).

$$\vec{\alpha}_3 = \frac{1}{\sqrt{3}} \left(\frac{1}{2}, \frac{\sqrt{3}}{2} \right). \tag{5.1.48}$$

The six root vectors all have the same length $1/\sqrt{3}$ and their polar angles with respect to the positive x-axis are $0°, 60°, 120°, 180°, 240°$ and $300°$. In this 2-dimensional case, the 'hyperplanes' are the lines determined by the constraint equations:

$$\vec{\alpha}_1 \cdot \vec{r} = 0, \quad \vec{\alpha}_2 \cdot \vec{r} = 0, \quad \vec{\alpha}_3 \cdot \vec{r} = 0. \tag{5.1.49}$$

The transformations corresponding to reflection in these three lines are respectively:

$$
\begin{bmatrix} \bar{x} \\ \bar{y} \end{bmatrix} = \begin{bmatrix} -1 & 0 \\ 0 & 1 \end{bmatrix} \begin{bmatrix} x \\ y \end{bmatrix},
$$

$$
\begin{bmatrix} \bar{x} \\ \bar{y} \end{bmatrix} = \begin{bmatrix} 1/2 & \sqrt{3}/2 \\ \sqrt{3}/2 & -1/2 \end{bmatrix} \begin{bmatrix} x \\ y \end{bmatrix},
$$

$$
\begin{bmatrix} \bar{x} \\ \bar{y} \end{bmatrix} = \begin{bmatrix} 1/2 & -\sqrt{3}/2 \\ -\sqrt{3}/2 & -1/2 \end{bmatrix} \begin{bmatrix} x \\ y \end{bmatrix}. \tag{5.1.50}
$$

Denote the 2×2 matrices of these transformations by A, B and C respectively. Then

$$R_+ \overset{\text{def}}{=} AB = CA = BC = \begin{bmatrix} -1/2 & -\sqrt{3}/2 \\ \sqrt{3}/2 & -1/2 \end{bmatrix} \tag{5.1.51}$$

corresponds to a rotation through $120°$, and

$$R_- \overset{\text{def}}{=} BA = AC = CB = \begin{bmatrix} -1/2 & \sqrt{3}/2 \\ -\sqrt{3}/2 & -1/2 \end{bmatrix} \tag{5.1.52}$$

corresponds to a rotation through $240°$ or equivalently through $-120°$. The group of six matrices, the identity I, the reflections A, B, C, and the rotations R_+ and R_- constitute a finite group that is isomorphic to the permutation group S_3 which is displayed above in equation (5.1.22). The map defined by

$$1 \longrightarrow I, \quad (123) \longrightarrow R_+, \quad (321) \longrightarrow R_-$$
$$(12) \longrightarrow A, \quad (23) \longrightarrow B, \quad (13) \longrightarrow C \tag{5.1.53}$$

is a group isomorphism. The Weyl group plays an important role in Weyl's analysis of the representations of the semisimple Lie groups.

Weyl also showed that it was possible to refine the choice of Cartan's canonical basis in such a way that the imposition of reality conditions determines for every semisimple Lie algebra with r complex parameters a subalgebra with r real parameters such that the restriction of the Cartan-Killing metric to this subalgebra is positive definite. Thus every complex r-parameter Lie group G contains a compact real r-parameter subgroup called the compact real form of G. Cartan had established this result by considering various cases, but Weyl's ingenious proof was based on a general analysis of the structure constants. The special unitary group $SU(n, \mathbf{C})$ is, for example, the compact real form of $SL(n, \mathbf{C})$. To obtain the compact real form of a real group such as $SL(n, \mathbf{R})$ which is $(n^2 - 1)$-dimensional over $\mathbf{R}$, it is necessary to consider its complex extension $SL(n, \mathbf{C})$ which is $(n^2 - 1)$-dimensional over $\mathbf{C}$; hence, the compact real form of $SL(n, \mathbf{R})$ is $SU(n, \mathbf{C})$ which is $(n^2 - 1)$-dimensional over $\mathbf{R}$.

In chapter IV, Weyl (1926d) proved that the representations of a semisimple Lie group are fully reducible and determined explicit formulas for the characters and dimensions of the irreducible representations. He also showed how to construct all of the irreducible representations and explained the relationship of these results to the theory of invariants. In all of these developments, the Weyl unitary trick plays a central role (Varadarajan 1974, p. 349). The essence of the method is the following. The result is first established for the compact real form of the semisimple Lie group G by a method which employs the technique of integration over the group manifold of the *compact* real form. The result is then extended to the complex group G by *analytic continuation*. For example, for the group $SL(n, \mathbf{C})$, the additional constraints (5.1.11) that must be satisfied if a reducible representation ρ is fully reducible are *polynomials* in the components of the matrix of a group element; consequently, the analytic continuation is trivial and the full reducibility of the representations of $SU(n, \mathbf{C})$, discussed above following (5.1.46), entails the full reducibility of the representations of $SL(n, \mathbf{C})$. Weyl proved that this result holds for an arbitrary semisimple Lie group.

To see how the Weyl unitary trick may be used to construct invariants of a semisimple Lie group, consider first the case of a finite group G with elements labelled by $\{1, 2, \ldots, n\}$. A function on G is just an n-tuple of numbers $\{\phi(j)\}$. The polynomial in the values $\phi(j)$ defined by

$$\psi(1, 2, 3) = \phi^2(1) + \phi(2)\phi(3) \qquad (5.1.54)$$

is not invariant under the action of S_3 defined by

$$(\pi^{-1}\psi)(1, 2, 3) = \psi(\pi 1, \pi 2, \pi 3), \qquad (5.1.55)$$

where π is one of the elements (5.1.22) of S_3; however, the polynomial

$$\sum_{\sigma \in S_3} (\phi^2(\sigma 1) + \phi(\sigma 2)\phi(\sigma 3)) = \qquad (5.1.56)$$

$$(\phi^2(1) + \phi(2)\phi(3)) + (\phi^2(2) + \phi(1)\phi(3)) + (\phi^2(3) + \phi(2)\phi(1))$$
$$+ (\phi^2(1) + \phi(3)\phi(2)) + (\phi^2(2) + \phi(3)\phi(1)) + (\phi^2(3) + \phi(1)\phi(2))$$

is such an invariant.

Now, let ρ_λ denote a finite set of linear representations, labelled by λ, of a semisimple Lie group G, and let V_λ be the vector space of dimension n_λ in which ρ_λ acts. Consider a function $f(v_1, v_2, \ldots)$ where $v_\lambda = (v_\lambda^1, v_\lambda^2, \ldots, v_\lambda^{n_\lambda}) \in V_\lambda$

and suppose that f is a polynomial (and hence analytic) in its arguments and homogeneous in each set of arguments v_λ. By definition, the function f is an invariant under the action of G if and only if for every $g \in G$

$$f(\rho_1(g)v_1, \rho_2(g)v_2, \ldots) = f(v_1, v_2, \ldots). \tag{5.1.57}$$

Suppose that f is not an invariant. The function

$$\int f(\rho_1(g)v_1, \rho_2(g)v_2, \ldots)d\mu(g), \tag{5.1.58}$$

where the integration extends over the manifold of the compact real form of G and where $d\mu$ denotes the left invariant measure of this compact subgroup, is clearly invariant under the action of this compact subgroup. Analyticity ensures that it is also an invariant of the group G. In his next two articles Weyl (1926a, 1926f) developed more fully the relationship between representation theory and the theory of invariants.

It was noted above that for the case of a finite group G, the identity element of the group algebra A_G decomposes into the sum of idempotents, each of which corresponds to an irreducible representation of G, and that every irreducible representation of G occurs at least once in this decomposition. For the particular group S_3, the idempotents are given by ((5.1.23)–(5.1.26)). The analogous result for the case of *compact* Lie groups was proved by Peter and Weyl (1927). In this case, an element of A_G is again a function $f: G \to \mathbf{C}$ and vector addition and scalar multiplication are defined as they are for the case of a finite group; however, the algebra product (5.1.20) is replaced by

$$(f_1 f_2)(g) = \int_G f_1(gk^{-1})f_2(k)d\mu(k), \tag{5.1.59}$$

where $d\mu$ is the invariant[70] measure on the group manifold. Peter and Weyl applied the theory of integral equations (see section 2) to the eigenvalue problem[71]

$$\int_G K(s,t)x(t)d\mu(t) = \lambda x(s), \tag{5.1.60}$$

where the kernel $K(s,t)$ was of the form

$$K(s,t) = \int_G f(sr^{-1})\bar{f}(tr^{-1})d\mu(r). \tag{5.1.61}$$

Such a kernel is Hermitian

$$K^\dagger(s,t) \equiv \bar{K}(t,s) = K(s,t) \tag{5.1.62}$$

---

[70]Peter and Weyl state that the measure $d\mu$ is left invariant $d\mu(ag) = d\mu(g)$ and invariant under inversion $d\mu(g^{-1}) = d\mu(g)$. These conditions imply right invariance $d\mu(ga) = d\mu(g)$.

[71]At the beginning of §3 of their paper, Peter and Weyl remark that the method they use to construct the eigenvalues and eigenvectors is a slight modification of the method developed by E. Schmidt (1905) in his dissertation.

and is invariant under the action of G,

$$K(sg, tg) = K(s, t). \tag{5.1.63}$$

The eigenfunctions of (5.1.60) are all square integrable with respect to the inner product defined by

$$(x, y) = \int_G x(t)\bar{y}(t)d\mu(t). \tag{5.1.64}$$

For any square integrable function $x: G \to \mathbf{C}$, define $U(g)$ by

$$(U(g)x)(s) = x(g^{-1}s). \tag{5.1.65}$$

Then, $U(g)$ is unitary with respect to the inner product (5.1.64) and

$$U(g_1 g_2) = U(g_1)U(g_2). \tag{5.1.66}$$

Remark 40 To see the analogy with the regular representation in the case of a finite group defined by (5.1.18), consider the action of $\rho(a)$ on an arbitrary element f of the group algebra given (5.1.17).

$$
\begin{aligned}
\rho(a)f &= \sum_g f(g)\rho(a)g \\
&= \sum_g f(g)ag \\
&= \sum_{\tilde{g}} f(a^{-1}\tilde{g})\tilde{g} \\
&= \sum_g f(a^{-1}g)g. \tag{5.1.67}
\end{aligned}
$$

Thus, the regular representation could equally well be defined directly in terms of the functions $f: G \to \mathbf{C}$ by

$$(\rho(a)f)(g) = f(a^{-1}g), \tag{5.1.68}$$

which is the direct analogue of (5.1.65).

If x is an eigenvector with eigenvalue λ of (5.1.60), then so is $U(g)x$; moreover, the set of eigenfunctions corresponding to a fixed eigenvalue λ spans a *finite-dimensional* subspace of the Hilbert space H_G of functions that are square integrable with respect to (5.1.64). The restriction of the representation (5.1.65) to this subspace provides a finite-dimensional unitary representation of G.

Since the group manifold of G is continuous, the group algebra A_G does not have a proper identity element; however, one may consider a sequence of kernels $K_n(s, t)$ which vanish unless $st^{-1} \in U_n$ where U_n is a neighbourhood of the identity of G and which satisfy

$$\int K_n(s, 1_G)d\mu(s) = 1. \tag{5.1.69}$$

In the limit in which the diameter of U_n approaches zero as $n \to \infty$, the kernel more and more nearly approximates the 'identity' of the algebra. Peter and Weyl proved that in the limit, all irreducible representations of G are obtained. Analysts had previously studied such integrals under the name 'singular integrals' (Dieudonné 1981, p. 224).

The Peter-Weyl theorem shows that the irreducible representations of a compact Lie group provide a basis in terms of which functions defined on the group manifold may be expanded. Much later, Weyl (1934a) extended the theorem to the case of functions defined on a coset space G/H where H is a closed subgroup of G. The expansion of functions defined on the sphere $S^2 = O(3)/O(2)$ in terms of the spherical harmonics $Y_{lm}(\theta, \phi)$ is a well-known special case.

Theoretical physics is dominated at present by gauge field theories. Kaluza-Klein theories can be regarded as special cases of gauge field theories. In a Kaluza-Klein theory, it is assumed that the world is $(4+m)$-dimensional[72] and is topologically the product of 4-dimensional spacetime and an m-dimensional homogeneous space G/H of a compact Lie group G. The fields that represent geometric and physical quantities are functions of both the spacetime and internal coordinates. Restricted to a given spacetime location, these fields are functions on G/H and can be expanded using Weyl's generalization of the Peter-Weyl theorem. The 'Fourier coefficients' of this expansion are spacetime fields with gauge indices associated with the representations of the group G.

The following comments of Newman (1957, p. 317–319) indicate the original and general character of Weyl's contributions to Group theory.

> Weyl's results, completed in an essential point by the famous Peter-Weyl Theorem (see below), go a long way towards solving the corresponding problem [the determination of all finite-dimensional linear representations up to an equivalence] for semisimple Lie groups, and effectively the whole way for compact Lie groups. He obtained explicit formulas for the characters and degrees of the irreducible representations by a powerful combination of the infinitesimal methods of E. Cartan, the method of invariant integration over the group-manifold of a compact Lie group (first used in a special case by Hurwitz and later by Schur) and topological considerations.

---

[72] When one speaks of a Kaluza-Klein theory rather than of a gauge field theory on spacetime with fiber G/H, one implies that the internal dimensions are real, additional, spatial dimensions of the world. The reason the internal dimensions are not perceived directly is that they form a compact space of extremely small dimension. It was shown by Wigner (1939) that in the context of Quantum Mechanics, the symmetries of Minkowski spacetime (the Poincaré group) led to the quantum numbers, mass m, spin s, three momentum $\vec{p}$ and spin projection s_z, that characterize a particle. The development of elementary particle physics following the Second World War led to the introduction of a number of internal quantum numbers such as isospin I and its projection I_z, hypercharge Y and so forth; however, there was no 'space' the symmetries of which would generate these quantum numbers. The motivation for the revival of Kaluza-Klein theories in the mid-sixties was to provide explicit models for the 'missing' internal space.

Up to this time Lie groups had been considered almost exclusively from a local, and mainly from an infinitesimal, point of view. Weyl's work contains the first important contributions to the global study of Lie groups and, as such, has been the stimulus to numerous later investigations.
...

The Peter-Weyl paper [(Peter and Weyl 1927)] preceded by only a few years the construction by Haar in 1933 of a left-invariant measure on any (separable) locally compact topological group. For compact groups, the Haar measure is also right-invariant, and the methods of Peter and Weyl could be carried over with scarcely any change to this more general case. Their work marks a decisive forward step in group-analysis, and points the way to the theory of almost periodic functions on a group, due to von Neumann, and more distantly to the modern theory of representations of locally compact topological groups by unitary transformations of a Hilbert space.

Somewhat later, Weyl (1939a) wrote a book, entitled *The Classical Groups, Their Invariants and Representations*, in which he returned to the theory of invariants and representations of the semisimple Lie groups. In this work, he satisfied his ambition "to derive the decisive results for the most important of these groups by direct algebraic construction, in particular for the full group of all non-singular linear transformations and for the orthogonal group". He intentionally restricted the discussion of the general theory and devoted most of the book to the derivation of specific results for the general linear, the special linear, the orthogonal and the symplectic groups.

5.2 Quantum Mechanics and Quantum Field Theory

Shortly after the discovery of the new theory of Quantum Mechanics, Weyl (1927a) published an analysis of its foundations in which he emphasized the fundamental role played in the new theory by the theory of Lie groups.

The first part of the paper deals with the physical interpretation of the representation of physical magnitudes by Hermitian forms on Hilbert space. The interpretation of the mathematical formalism is based on the concept of the pure case, a state of maximum homogeneity which is represented not by a vector in Hilbert space but by a ray, an equivalence class of vectors generated by taking all nonzero complex multiples of any given vector in the ray. Essentially the same point of view was adopted independently by Hilbert, von Neumann and Nordheim (1928). Weyl then analyzes the statistics of ensembles or the representation of mixed states by means of weighted averages of pure states. He sharply distinguishes the role of probability in the analysis of ensembles from the role it has in the pure case. He observes that the role of probability in the case of ensembles "is relative to our knowledge and ignorance" and has to do with disturbances that at least in principle can be made as small as desired while its role in the pure case "has a fully objective meaning which has nothing to do with disturbances".

In the second part of the paper, a group-theoretic analysis of Quantum Kinematics is presented. Suppose that Q and P are canonically conjugate Hermitian operators that satisfy Heisenberg's commutation relation

$$[Q, P] = iI. \tag{5.2.1}$$

Then Q and P are the generators of unitary transformations $\exp[i\tau Q]$ and $\exp[i\sigma P]$ where τ and σ are real. It follows from (5.2.1) that the commutator of these transformations is a pure phase; consequently, as far as ray space, the space of states, is concerned, these transformations *commute*. Weyl therefore proceeds to analyze the structure of irreducible, Abelian, unitary, transformation groups on ray space. He shows that the number of generators of such a group must be even and that there exists a canonical basis Q_μ, P_μ ($\mu = 1, 2, \ldots$) such that

$$[Q_\mu, P_\nu] = i\delta_{\mu\nu}I. \tag{5.2.2}$$

The general element of the group of unitary transformations is given by

$$U(\vec{\sigma}, \vec{\tau}) = \exp[i(\vec{\sigma} \cdot \vec{P} + \vec{\tau} \cdot \vec{Q})]. \tag{5.2.3}$$

Next, Weyl proposes that the appropriate quantum mechanical operator that corresponds to a real valued function

$$f(\vec{p}, \vec{q}) = \int_{-\infty}^{\infty} \int_{-\infty}^{\infty} \exp[i(\vec{\sigma} \cdot \vec{p} + \vec{\tau} \cdot \vec{q}]\tilde{f}(\vec{\sigma}, \vec{\tau})d^3\vec{\sigma}d^3\vec{\tau} \tag{5.2.4}$$

is the operator given by

$$F(\vec{P}, \vec{Q}) = \int_{-\infty}^{\infty} \int_{-\infty}^{\infty} \exp[i(\vec{\sigma} \cdot \vec{P} + \vec{\tau} \cdot \vec{Q}]\tilde{f}(\vec{\sigma}, \vec{\tau})d^3\vec{\sigma}d^3\vec{\tau}. \tag{5.2.5}$$

Prior to this work, no general prescription existed for functions involving both members of a conjugate pair. With this prescription, the physical magnitudes of the physical system are elegantly characterized as the real elements of the algebra of the group of unitary transformations (5.2.3).

Finally, Weyl shows for the case of the group generated by one canonical pair Q, P that the irreducible ray representation is unique and is the representation used by Schrödinger for which

$$Q\psi(q) = q\psi(q) \tag{5.2.6}$$

and

$$P\psi(q) = \frac{1}{i}\frac{d}{dq}\psi(q) \tag{5.2.7}$$

and thus that these operators are a necessary consequence of Heisenberg's commutation relations. Weyl observes that the result can be extended to the case of many degrees of freedom without difficulty.

In the third and final section of the paper, the dynamical problem is treated from the viewpoint of a one parameter (t) group of unitary transformations in system space generated by the Hamiltonian operator of the system. By employing the representation (5.2.5) for the energy function, he writes the equations of motion for the canonical operators P and Q in Hamiltonian form.

Weyl then notes that the formalism of Quantum Mechanics that he has just presented is not compatible with the Special Theory of Relativity and that a general solution of this problem is not at hand. Nevertheless, he applies his new techniques to present a semi-relativistic analysis of the interaction of a massive body with a potential field with particular attention to the important case of an electron in a Coulomb field.

One of the first textbooks on the new theory of Quantum Mechanics was published by Weyl the following year (Weyl 1928b), and a revised edition appeared in 1931. In this book, Weyl succeeds in explaining the mathematics to the physicists and the physics to the mathematicians. He not only adds to the theoretical insights discussed above, but also presents a thorough overview of the experimental foundations of the theory.

Weyl still bases the interpretation of the theory on the concept of the pure state; however, his treatment of statistical ensembles is considerably enriched (Weyl 1928b, 2 edn, 1931, p. 79). He notes that a mixed state is represented by a positive, semi-definite Hermitian form with unit trace, now called a density matrix, and that the set of all such forms is convex. If ρ_a and ρ_b belong to the set then so does $\lambda_a \rho_a + \lambda_b \rho_b$ where $\lambda_a \geq 0$, $\lambda_b \geq 0$ and $\lambda_a + \lambda_b = 1$. The extreme points of a convex set are just those elements that cannot be written in the form $\lambda_a \rho_a + \lambda_b \rho_b$ with $\lambda_a \neq 0, \lambda_b \neq 0$ and $\rho_a \neq \rho_b$. The extreme points of the set of density matrices are in bijective correspondence with the pure states. In this way, Weyl elegantly characterizes the pure states as the extreme elements of the set of mixed states.

The book (Weyl 1928b, 2 edn, 1931) is remarkably complete for such an early work and covers many topics in addition to those already mentioned. The theory of atomic spectra is presented, including an account of the selection and intensity rules. After discussing the spin of the electron and its role in accounting for the anomalous Zeeman effect, Weyl presents Dirac's (1930) theory of the Relativistic Quantum Mechanics of the electron and develops in detail the theory of an electron in a spherically symmetric field, including an analysis of the fine structure of the spectrum. He then applies the Pauli exclusion principle to explicate the periodic table of the elements. Next, he develops[73] the second quantization of the Maxwell and Dirac fields required for the analysis of many body relativistic systems. It is

---

[73] He notes in the preface of the second edition that his treatment is "in accordance with the recent work of Heisenberg and Pauli".

now customary to include such a topic under the heading of Relativistic Quantum Field Theory. The treatment is quite modern except for the confusion regarding the positive electron (anti-electron) that at that time was identified with the proton rather than with the positron, which was discovered a few years later. Weyl was quite concerned about the identification of the proton with the positive electron because his analysis of the discrete symmetries **C**, **P**, **T** and **CPT** led him to conclude that the mass of the positive electron should equal the mass of the electron. In view of the importance of the discrete symmetries **C**, **P**, **T** and **CPT** in modern Relativistic Quantum Field Theory, we present Weyl's early analysis of these symmetries in some detail in subsection 5.3.

Although Weyl apologizes for the brevity of his treatment of the empirical foundations of Quantum Mechanics, his treatment is at least as extensive as those found in more recently published textbooks. Indeed, the greater part of Weyl's discussion of the physical foundations of Quantum Mechanics will be quite familiar to anyone who has learned the theory from one of these books; however, the reader will discover that Weyl's exposition of the experimental results is enlivened by his precise appreciation of their philosophical significance.

Weyl analyzed the foundations of both the General Theory of Relativity and the Theory of Quantum Mechanics. For both theories, he provided a coherent exposition of the mathematical structure of the theory, elegant characterizations of the entities and laws postulated by the theory and a lucid account of how these postulates explain the most significant, more directly observable, lower-level phenomena. In both cases, he was also concerned with the constructive aspects of the theory, that is, with the extent to which the higher-level postulates of the theory are necessary.

There is no doubt that with regard to the General Theory of Relativity, Weyl adopted a distinctly realist philosophical position. He forcefully argued that the spacetime metric was physically real and that relationalism based on a pure body ontology was not tenable. Moreover, he was not deterred by the fact that a completely satisfactory solution to the measurement problem for the metric field was not then available. He even put forth a reductio argument, the plasticine example discussed above in section 4.11, to underscore the necessity of a guiding field. At the same time, others used the lack of an adequate measurement procedure to argue that the metric field is to some extent conventional.

In contrast to Weyl's realist position with regard to spacetime theories, Weyl's philosophical position regarding the status of Quantum Mechanics is unfortunately obscure. Passages such as the following (Weyl 1928b, 2 edn, 1931, p. 44) which argues for the reality of photons show that Weyl did not entirely abandon his realist inclinations.

Die Intensität der zur Erzeugung des Photoeffektes verwendeten mono-chromatischen Strahlung hat keinen Einfluß auf die Geschwindigkeit, mit der die Elektronen aus dem Metall herausfliegen, sondern nur auf die Häufigkeit dieses Vorganges. Auch bei schwacher Intensität, wenn nach der Wellenvorstellung Stunden nötig wären, ehe sich die über ein einzelnes Atom hinstreichende Energie der elektromagnetischen Welle zu einem Betrag von der Größe des Lichtquants aufspeichern könnte, setzt der Effekt sofort ein, mit unregelmäßig diskontinuierlich über die Metallplatte verteilten Wirkungsstellen. Das ist ein nicht minder sinnfälliger Beweis für die Lichtquanten, als es die Lichtblitze auf dem Szintillationsschirm für die korpuskular-diskontinuierliche Natur der α-Strahlen sind.

[The intensity of the monochromatic radiation that is used to generate the photoelectric effect has no influence on the speed with which the electrons are ejected from the metal but affects only the frequency of this process. Even with intensities so weak that on the classical theory hours would be required before the electromagnetic energy passing through a given atom would attain to an amount equal to that of a photon, the effect begins immediately, the points at which it occurs being distributed irregularly over the entire metal plate. This fact is a proof of the existence of light quanta that is no less meaningful than the flashes of light on the scintillation screen are for the corpuscular-discontinuous nature of α-rays.]

On the other hand, Weyl's (1928b, 2 edn, 1931, pp. 58–59) discussion of the problem of 'directional quantization' in the old quantum theory and of the way that this problem is 'resolved' in the new quantum theory has a distinctly instrumentalist flavour. Nor did he ignore the problem. In a number of places, he describes the essence of the dilemma posed by Quantum Mechanics with a dispassionate precision. Consider, for example, the following (Weyl 1928b, 2 edn, 1931, p. 67):

Die Naturwissenschaft hat konstruktiven Character. Die Merkmale, von denen sie handelt, sind keine aus den Dingen herauszuschauende unselbständige [sic][74] Momente oder Qualitäten, sondern können nur durch eine indirekte Methodik, durch Eingriff und Reaktion mit andern Körpern festgestellt werden; ihre implizite Definition ist gebunden an bestimmte Naturgesetze, denen die Reaktionen unterliegen. Man denke z. B. an die Galileische Einführung des Massenbegriffs, die im Grunde auf die folgende indirekte Definition herauskommt: "Jedem Körper kommt ein Impuls, d. i. ein mit seiner Geschwindigkeit $\vec{v}$ gleichgerichteter Vektor $m\vec{v}$ zu; der skalare Faktor m heißt Masse. Es gilt das Impulsgesetz, wonach die Summe der Impulse vor einer Reaktion zwischen verschiedenen Körpern gleich der Summe ihrer Impulse nach der Reaktion ist." Indem man die beobachteten Stoßer-

---

[74]It would appear from the context that Weyl meant to say 'selbständige' instead of 'unselbständige' here. Our translation is consistent with this assumption.

scheinungen diesem Gesetz unterwirft, erhält man Data zur Bestimmung der relativen Massen. Die Meinung war aber die, daß solche *konstruktive Merkmale den Dingen doch an sich zukämen*, auch wenn die Manipulationen, die allein zu ihrer Feststellung führen können, nicht vorgenommen werden. *In den Quanten stoßen wir auf eine grundsätzliche Schranke dieser erkenntnistheoretischen Position der konstruierenden Naturwissenschaft.*

[Natural science has a constructive character. The phenomena with which it deals are not independent manifestations or qualities which can be read off from nature, but can only be determined by means of an indirect method, through interaction with other bodies. Their implicit definition is bound up with definite natural laws which underlie the interactions. Consider, for example, the introduction of the Galilean concept of mass which essentially comes down to the following indirect definition: "Each body possesses a momentum, that is, a vector $m\vec{v}$ which has the same direction as its velocity $\vec{v}$ — the scalar factor m is called its mass. The law of momentum holds, according to which the sum of the momenta before a reaction between several bodies is the same as the sum of their momenta after the reaction." By applying this law to the observed collision phenomena, one obtains data for the determination of the relative masses. The scientific consensus was, however, that such *constructive phenomena can nevertheless be attributed to the things themselves* even if the manipulations, which alone can lead to their recognition, are not being carried out. *In Quantum Theory we encounter a fundamental limitation to this epistemological position of the constructive natural science.*]

It is difficult to accept Quantum Mechanics as an ultimate theory without at the same time giving up realism and adopting an instrumentalist view of the theory. It is clear that Weyl was fully aware of this state of affairs, and yet in all of his published work, he refrained from making an unambiguous statement of his views on this fundamental question. In view of the fact that Weyl strongly defended a realist interpretation of the General Theory of Relativity, one might have expected him to join his colleagues, A. Einstein and E. Schrödinger, in attacking the Copenhagen interpretation of Quantum Mechanics and in arguing that Quantum Mechanics must one day be replaced by a more fundamental theory. It is odd that Weyl did not participate in this debate. It is also difficult to believe that he was fully satisfied with the theory. Surely, he would have held that Schrödinger's cat had to be either alive or dead and not in some quantum superposition of these two states. Perhaps an examination of Weyl's unpublished correspondence and papers will provide an answer to this puzzle[75].

---

[75] For a discussion of Weyl's views on the role of probability in quantum mechanics, see chapter V of Sigurdsson (1991).

5.3 Weyl's Early Discussion of the Discrete Symmetries C, P, T and CPT

In his book, Weyl (1928b, 2 edn, 1931) analyzed Dirac's (1928a, 1928b) relativistic theory of the electron. This theory correctly accounted for the spin of the electron. There was a problem, however, because in addition to the positive-energy levels, the theory predicted the existence of an equal number of negative-energy levels. Dirac (1930) reinterpreted the theory by assuming that all of the negative-energy levels were normally occupied. The Pauli Exclusion Principle, which asserts that it is impossible for two electrons to occupy the same quantum state, would prevent an electron with positive energy from falling into a negative-energy state. The theory also predicted that one of the negative-energy electrons could be raised to a state of positive energy thereby creating a 'hole' or unoccupied negative-energy state. Such a hole would behave like a particle with a positive energy and a positive charge, that is, like a positive electron. At the time, the only fundamental particles that were known to exist were the electron and the proton. It was then considered rather adventurous to postulate the existence of new particles that had not been observed experimentally; consequently, it was suggested that the positive electron should be identified with the proton. Weyl was quite concerned about the identification of the proton with the anti-electron. In the preface to the second German edition of his book *Gruppentheorie und Quantenmechanik*, Weyl (1928b, 2 edn, 1931, p. VII) wrote

> Das Problem von Proton und Elektron wird im Zusammenhang mit den Symmetrieeigenschaften der Quantengesetze gegenüber den Vertauschungen von rechts und links, Vergangenheit und Zukunft, positiver und negativer Elektrizität aufgerollt. Im Augenblick ist keine annehmbare Lösung sichtbar; ich fürchte, daß sich hier die Wolken zu einer neuen ernsten Krise der Quantenphysik zusammenballen.

> [The problem of the proton and the electron is discussed in connection with the symmetry properties of the quantum laws with respect to the interchange of right and left, past and future, and positive and negative electricity. At present no acceptable solution is in sight; I fear, that in the context of this problem, the clouds are rolling together to form a new, serious crisis in quantum physics.]

Weyl had good reasons for his concern. He analyzed the invariance of the Maxwell-Dirac equations under the discrete symmetries that correspond to the transformations now called **C**, **P**, **T** and **CPT** both for the case of Relativistic Quantum Mechanics and for the case of Relativistic Quantum Field Theory, and concluded in both cases that the mass of the anti-electron should be the same as the mass of the electron. That the mass of the proton was so different from the mass of the electron, therefore, appeared to Weyl to constitute a new serious crisis in physics.

Remark 41 In a lecture presented at the Centenary for Hermann Weyl held at the ETH in Zürich, Yang (1986, p. 10) quoted from the Dover edition of Weyl's (1928b, 2 edn, 1931) book the English translation of the above passage. With regard to this passage, Yang remarked

> This was a most remarkable passage in retrospect. The symmetry that he mentioned here, of physical laws with respect to the interchange of right and left, had been introduced by Weyl and Wigner independently into quantum physics. It was called parity conservation, denoted by the symbol P. The symmetry between the past and future was something that was not well understood in 1930. It was understood later by Wigner, was called time reversal invariance, and was denoted by the symbol T. The symmetry with respect to positive and negative electricity was later called charge conjugation invariance C. It is a symmetry of physical laws when you change positive and negative signs of electricity. Nobody, to my knowledge, absolutely nobody in the year 1930, was in any way suspecting that these symmetries were related in any manner. I will come back to this matter later. What had prompted Weyl in 1930 to write the above passage is a great mystery to me.

Yang's comment is extremely misleading because it suggests that Weyl did not have a good reason for his remark. In fact, Weyl's statement was soundly based on a detailed analysis of the discrete symmetries **C**, **P**, **T** and **CPT**. It is shown below in detail that Weyl's treatment of these symmetries is the *same* as that used today except for the fact that the symmetry **T** is treated by Weyl as linear and unitary rather than as antilinear and antiunitary.

In his discussion of the relativistic quantum mechanics of the electron, Weyl used two 2-component spinors

$$\psi(x) = \begin{bmatrix} \psi_1(x) \\ \psi_2(x) \end{bmatrix} \quad \text{and} \quad \psi'(x) = \begin{bmatrix} \psi_1'(x) \\ \psi_2'(x) \end{bmatrix}. \tag{5.3.1}$$

Later in his discussion of the quantized field equations, he combined these two 2-component spinors into a single 4-component spinor. Weyl (1928b, 2 edn, 1931, p. 201) explicitly states that $\psi_1'(x)$ and $\psi_2'(x)$ will be denoted by $\psi_3(x)$ and $\psi_4(x)$ respectively. Weyl uses S_i rather than σ_i for the Pauli matrices, where $i \in \{1,2,3\}$. His equation[76] (H.III.8.15) is

$$\sigma_0 = \begin{bmatrix} 1 & 0 \\ 0 & 1 \end{bmatrix}, \ \sigma_1 = \begin{bmatrix} 0 & 1 \\ 1 & 0 \end{bmatrix}, \ \sigma_2 = \begin{bmatrix} 0 & -i \\ i & 0 \end{bmatrix}, \ \sigma_3 = \begin{bmatrix} 1 & 0 \\ 0 & -1 \end{bmatrix}. \tag{5.3.2}$$

---

[76]For ease of reference, the numbers of the equations in Hermann Weyl's book will be prefixed by an 'H'. The Roman numeral refers to the chapter in which the equation appears. Similarly, the numbers of the equations in volume I of Steven Weinberg's book will be prefixed with an 'S'. References to pages in these books will be similarly prefixed with either an 'H' or an 'S'.

Weyl also sets
$$\sigma_0' = \sigma_0 \quad \text{and} \quad \sigma_i' = -\sigma_i, \quad i \in \{1,2,3\}. \tag{5.3.3}$$
Weyl defines (H.IV.5.4) the differential operators ∇ and ∇' by

$$\nabla = \sum_{\alpha=0}^{3} \sigma_\alpha \frac{\partial}{\partial x^\alpha} = I\frac{\partial}{\partial t} + \vec{\sigma}\cdot\vec{\nabla},$$

$$\nabla' = \sum_{\alpha=0}^{3} \sigma_\alpha' \frac{\partial}{\partial x^\alpha} = I\frac{\partial}{\partial t} - \vec{\sigma}\cdot\vec{\nabla}. \tag{5.3.4}$$

Weyl's equations (H.IV.5.6) for the 2-component spinors ψ and ψ' are

$$\frac{1}{i}\nabla\psi + m\psi' = 0,$$
$$\frac{1}{i}\nabla'\psi' + m\psi = 0, \tag{5.3.5}$$

which may be written

$$-i\partial_0\psi' + i\vec{\sigma}\cdot\vec{\nabla}\psi' + m\psi = 0,$$
$$-i\partial_0\psi - i\vec{\sigma}\cdot\vec{\nabla}\psi + m\psi' = 0. \tag{5.3.6}$$

Define[77]

$$\Psi_H = \begin{bmatrix} \psi \\ \psi' \end{bmatrix} = \begin{bmatrix} \psi_1 \\ \psi_2 \\ \psi_1' \\ \psi_2' \end{bmatrix} = \begin{bmatrix} \psi_1 \\ \psi_2 \\ \psi_3 \\ \psi_4 \end{bmatrix}. \tag{5.3.7}$$

Then, the equations may be written
$$-i\begin{bmatrix} 0 & I \\ I & 0 \end{bmatrix}\partial_0\begin{bmatrix} \psi \\ \psi' \end{bmatrix} + i\begin{bmatrix} 0 & \vec{\sigma} \\ -\vec{\sigma} & 0 \end{bmatrix}\cdot\vec{\nabla}\begin{bmatrix} \psi \\ \psi' \end{bmatrix} + m\begin{bmatrix} \psi \\ \psi' \end{bmatrix} = 0. \tag{5.3.8}$$

If one defines
$$\gamma_H^0 = -i\begin{bmatrix} 0 & I \\ I & 0 \end{bmatrix}, \quad \vec{\gamma}_H = i\begin{bmatrix} 0 & \vec{\sigma} \\ -\vec{\sigma} & 0 \end{bmatrix}, \tag{5.3.9}$$

then
$$\gamma_H^\mu\gamma_H^\nu\gamma_H^\nu\gamma_H^\mu = 2\eta^{\mu\nu}I, \tag{5.3.10}$$

where
$$\eta^{\mu\nu} = \begin{bmatrix} -1 & 0 & 0 & 0 \\ 0 & 1 & 0 & 0 \\ 0 & 0 & 1 & 0 \\ 0 & 0 & 0 & 1 \end{bmatrix}. \tag{5.3.11}$$

---

[77]In order to distinguish 4-spinors and γ-matrices that correspond to Hermann Weyl's notational conventions from the 4-spinors and γ-matrices that conform to the notational conventions used in Steven Weinberg's book, these quantities will be labelled by a subscript 'H' or 'S' unless the quantity is the same in both cases.

Note that Weyl also uses this metric (p. H131). The equations (5.3.8) may be written

$$\left(\gamma_H^\mu \partial_\mu + m\right) \Psi_H = 0. \tag{5.3.12}$$

Weinberg (1995, vol. I) uses the same metric $\eta_{\mu\nu}$ as Weyl, but Weinberg's γ-matrices differ slightly from Weyl's.

$$\gamma_S^0 = \beta^\dagger \gamma_H^0 \beta = \gamma_H^0, \quad \vec{\gamma}_S = \beta^\dagger \vec{\gamma}_H \beta = -\vec{\gamma}_H, \tag{5.3.13}$$

where for both Weyl and Weinberg

$$\beta = \begin{bmatrix} 0 & I \\ I & 0 \end{bmatrix} = i\gamma^0. \tag{5.3.14}$$

A Weyl 4-spinor is related to a Weinberg 4-spinor by

$$\Psi_H = \beta\Psi_S. \tag{5.3.15}$$

Also,

$$\Psi^\dagger = \left[\psi_1^\dagger, \psi_2^\dagger, \psi_3^\dagger, \psi_4^\dagger\right] \quad \text{and} \quad \bar{\Psi} = \Psi^\dagger \beta. \tag{5.3.16}$$

Remark 42 When discussing relativistic quantum mechanics, Weyl used a bar to denote complex conjugation. Thus, in (H.IV.5.14), $\bar{\psi}_1$ means ψ_1^*. Later, when discussing the quantization of the Dirac field, he uses a bar to denote Hermitian conjugation of the operator. That this is so can be seen from his equation (H.IV.12.4′) for the equal-time anticommutation relations which may be written

$$\psi_{H\rho}^\dagger(t, \vec{x})\psi_{H\sigma}(t, \vec{x}') + \psi_{H\sigma}(t, \vec{x}')\psi_{H\rho}^\dagger(t, \vec{x}) = \delta_{\rho\sigma}\delta^3(x - \vec{x}'). \tag{5.3.17}$$

If the bar in Weyl's equation meant the expression given by (5.3.16), then instead of $\delta_{\rho\sigma}$ on the right hand side of (5.3.17), there would have to be $\beta_{\rho\sigma}$, the components of β. Weyl also uses $\tilde{\Psi}$ for what is now denoted by $\Psi^\dagger$ which can be readily seen from (H.IV.5.14). He also uses the symbol 'T' (H.III.8.18) for the matrix that is denoted here by 'β'.

We take the matrix

$$\gamma_5 = \begin{bmatrix} I & 0 \\ 0 & -I \end{bmatrix} \tag{5.3.18}$$

to be the same for both the Weyl and Weinberg cases. We now state the discrete transformations for the quantized field operators as they are given in (Weinberg 1995, vol. I) as well as the corresponding transformations that apply for the choice of matrices that correspond to Weyl's spinor equations for the electron, that is, the transformations that Weyl 'should' have obtained. Note that the matrix (S5.4.36) $\mathcal{C}$ in Weinberg's formulas is

$$\mathcal{C}_S = \gamma_S^2 \beta. \tag{5.3.19}$$

Also, the Lorentz transformation for space reflection is

$$\mathcal{P}^{\mu}_{\nu} = \begin{bmatrix} 1 & 0 & 0 & 0 \\ 0 & -1 & 0 & 0 \\ 0 & 0 & -1 & 0 \\ 0 & 0 & 0 & -1 \end{bmatrix}. \tag{5.3.20}$$

Later, we translate the transformations Weyl actually gave into our Weylian γ-matrix notation and compare the results with those derived from Weinberg's formulas.

Remark 43 (Antilinear and Antiunitary Operators) In the discussion presented below, the parity operator $\mathbf{P}$ and the charge conjugation operator $\mathbf{C}$ are characterized as *linear* and *unitary* while the time reversal operator $\mathbf{T}$ is characterized as *antilinear* and *antiunitary*. Here, we remind the reader of the definitions of these terms and alert the reader with regard to some counterintuitive aspects of the formulas for $\mathbf{C}$ and $\mathbf{T}$.

Let Φ_i denote states in a Hilbert space $\mathcal{H}$ with an inner product $\langle , \rangle$. Then, a linear and unitary operator $\mathbf{L}: \mathcal{H} \to \mathcal{H}$ satisfies

$$\mathbf{L}(\alpha_1 \Phi_1 + \alpha_2 \Phi_2) = \alpha_1 \mathbf{L}\Phi_1 + \alpha_2 \mathbf{L}\Phi_2 \tag{5.3.21}$$

and

$$\langle \mathbf{L}\Phi_1, \mathbf{L}\Phi_2 \rangle = \langle \Phi_1, \Phi_2 \rangle, \tag{5.3.22}$$

where α_1 and α_2 are any complex numbers. An antilinear and antiunitary operator $\mathbf{K}: \mathcal{H} \to \mathcal{H}$ satisfies

$$\mathbf{K}(\alpha_1 \Phi_1 + \alpha_2 \Phi_2) = \alpha_1^* \mathbf{K}\Phi_1 + \alpha_2^* \mathbf{K}\Phi_2 \tag{5.3.23}$$

and

$$\langle \mathbf{K}\Phi_1, \mathbf{K}\Phi_2 \rangle = \langle \Phi_1, \Phi_2 \rangle^* = \langle \Phi_2, \Phi_1 \rangle. \tag{5.3.24}$$

As a consequence of (5.3.23), the *wavefunction* of a state with respect to a given choice of *basis* will be conjugated under the action of an antilinear operator.

Consider the application of the operators $\mathbf{P}$ and $\mathbf{T}$ to the interaction term of the Lagrangian *operator* density (5.3.52). The linear, unitary operator acts according to

$$\mathbf{P}ie\bar{\Psi}(x)\gamma^{\mu}A_{\mu}(x)\Psi(x)\mathbf{P}^{-1} \tag{5.3.25}$$

$$= ie\mathbf{P}\bar{\Psi}(x)\mathbf{P}^{-1}\mathbf{P}\gamma^{\mu}A_{\mu}(x)\mathbf{P}^{-1}\mathbf{P}\Psi(x)\mathbf{P}^{-1} \tag{5.3.26}$$

$$= ie\mathbf{P}\bar{\Psi}(x)\mathbf{P}^{-1}\gamma^{\mu}\mathbf{P}A_{\mu}(x)\mathbf{P}^{-1}\mathbf{P}\Psi(x)\mathbf{P}^{-1} \tag{5.3.27}$$

in which the appropriate expressions for $\mathbf{P}\bar{\Psi}(x)\mathbf{P}^{-1}$, $\mathbf{P}\Psi(x)\mathbf{P}^{-1}$ and $\mathbf{P}A_{\mu}(x)\mathbf{P}^{-1}$ are then substituted. The antilinear, antiunitary operator $\mathbf{T}$, however, acts according to

$$\mathbf{T}ie\bar{\Psi}(x)\gamma^{\mu}A_{\mu}(x)\Psi(x)\mathbf{T}^{-1} \tag{5.3.28}$$

$$= -ie\mathbf{T}\bar{\Psi}(x)\mathbf{T}^{-1}\mathbf{T}\gamma^{\mu}A_{\mu}(x)\mathbf{T}^{-1}\mathbf{T}\Psi(x)\mathbf{T}^{-1} \tag{5.3.29}$$

$$= -ie\mathbf{T}\bar{\Psi}(x)\mathbf{T}^{-1}\gamma^{\mu*}\mathbf{T}A_{\mu}(x)\mathbf{T}^{-1}\mathbf{T}\Psi(x)\mathbf{T}^{-1} \tag{5.3.30}$$

in which the appropriate expressions for $\mathbf{T}\bar{\Psi}(x)\mathbf{T}^{-1}$, $\mathbf{T}\Psi(x)\mathbf{T}^{-1}$ and $\mathbf{T}A_\mu(x)\mathbf{T}^{-1}$ are then substituted. For a linear operator, the procedure is a somewhat elaborate way of describing a substitution. As pointed out below, Weyl specified all of his discrete transformations as substitutions and hence as linear, unitary transformations. On the other hand, for an antilinear, antiunitary operator, the procedure generates additional changes due to the conjugation of all complex numbers that appear in the operator expression that is transformed.

Finally, given that the wavefunctions of states with repect to a given basis are not conjugated by a linear operator but are conjugated by an antilinear operator, the expressions given below for $\mathbf{C}$ (5.3.31–5.3.33) and for $\mathbf{T}$ (5.3.40–5.3.42) look strange because a conjugation appears on the right hand side of the expressions for $\mathbf{C}$ and no conjugation appears on the right hand side of the expressions for $\mathbf{T}$. The explanation of this apparent anomaly lies in the fact that the quantities concerned are not wavefunctions but are relativistic quantum field operators which are not related to states in a straight-forward way because they are linear combinations of both annihilation and creation operators. The operators $\mathbf{C}$ and $\mathbf{T}$ affect wavefunctions in the appropriate manner. It should also be noted that the matrices that appear on the right hand side of the expressions for the effect of $\mathbf{C}$ and $\mathbf{T}$ on the spinor operators ensure that both sides of these equations transform in the same way under Lorentz transformations.

Charge Conjugation C

The charge conjugation operator $\mathbf{C}$ is a linear, unitary operator that acts (S5.5.47) on the Dirac field operator according to

$$\mathbf{C}\Psi_S(x)\mathbf{C}^{-1} = \xi^*\gamma_S^2\Psi_S^{\dagger T}(x). \tag{5.3.31}$$

The corresponding Weylian transformation is

$$\mathbf{C}\Psi_H(x)\mathbf{C}^{-1} = \xi^*\gamma_H^2\Psi_H^{\dagger T}(x). \tag{5.3.32}$$

For a 4-vector field (S5.3.45),

$$\mathbf{C}V^\mu(x)\mathbf{C}^{-1} = \xi^*V^{\mu\dagger}(x). \tag{5.3.33}$$

For the electromagnetic 4-potential,

$$A^{\mu\dagger}(x) = A^\mu(x) \tag{5.3.34}$$

and $\xi = -1$ (p. S229); consequently,

$$\mathbf{C}A^\mu(x)\mathbf{C}^{-1} = -A^\mu(x). \tag{5.3.35}$$

Parity P

Parity is a linear, unitary operator that acts (S5.5.41) on the Dirac field operator according to

$$\mathbf{P}\Psi_S(x)\mathbf{P}^{-1} = \eta^*\beta\Psi_S(\mathcal{P}x) \tag{5.3.36}$$

and

$$\mathbf{P}\Psi_H(x)\mathbf{P}^{-1} = \eta^*\beta\Psi_H(\mathcal{P}x). \tag{5.3.37}$$

The action (S5.3.44) of $\mathbf{P}$ on a 4-vector is

$$\mathbf{P}V^\mu(x)\mathbf{P}^{-1} = -\eta^*\mathcal{P}^\mu_\nu V^\nu(\mathcal{P}x). \tag{5.3.38}$$

Since the photon has intrinsic parity $\eta = -1$,

$$\mathbf{P}A^\mu(x)\mathbf{P}^{-1} = \mathcal{P}^\mu_\nu A^\nu(\mathcal{P}x). \tag{5.3.39}$$

Time Reversal T

Time reversal is an antilinear, antiunitary operator that acts (S5.5.54) on the Dirac field operator according to

$$\mathbf{T}\Psi_S(x)\mathbf{T}^{-1} = -\zeta^*\beta\gamma_5\gamma_S^2\Psi_S(-\mathcal{P}x) \tag{5.3.40}$$

and

$$\mathbf{T}\Psi_H(x)\mathbf{T}^{-1} = \zeta^*\beta\gamma_5\gamma_H^2\Psi_H(-\mathcal{P}x). \tag{5.3.41}$$

The action (S5.3.46) of $\mathbf{T}$ on a 4-vector field is given by

$$\mathbf{T}V^\mu(x)\mathbf{T}^{-1} = \zeta^*\mathcal{P}^\mu_\nu V^\nu(-\mathcal{P}x). \tag{5.3.42}$$

Since the phases ξ, ζ and η can be chosen (S5.8.4) for each particle such that

$$\xi\zeta\eta = 1, \tag{5.3.43}$$

it follows that for a photon $\zeta = 1$ because $\eta = -1$ and $\xi = -1$; consequently,

$$\mathbf{T}A^\mu(x)\mathbf{T}^{-1} = \mathcal{P}^\mu_\nu A^\nu(-\mathcal{P}x). \tag{5.3.44}$$

The CPT Operator

CPT is an antilinear, antiunitary operator that acts on the Dirac field operator according to

$$(\mathbf{CPT})\Psi_S(x)(\mathbf{CPT})^{-1} = -\xi^*\zeta^*\eta^*\gamma_5\Psi_S^{\dagger T}(-x) \tag{5.3.45}$$

and

$$(\mathbf{CPT})\Psi_H(x)(\mathbf{CPT})^{-1} = \xi^*\zeta^*\eta^*\gamma_5\Psi_H^{\dagger T}(-x). \tag{5.3.46}$$

The action on the electromagnetic 4-potential is

$$(\mathbf{CPT})A^\mu(x)(\mathbf{CPT})^{-1} = -A^\mu(-x). \tag{5.3.47}$$

Weyl's Charge Conjugation T(CPT)P

Weyl considers a discrete transformation[78] that corresponds to **T(CPT)P** rather than to **C**. Its action on the Dirac field operator is given by

$$\left(\mathbf{T(CPT)P}\right)\Psi_S(x)\left(\mathbf{T(CPT)P}\right)^{-1} = -\xi\gamma_S^2\Psi_S^{\dagger T}(x) \tag{5.3.48}$$

and

$$\left(\mathbf{T(CPT)P}\right)\Psi_H(x)\left(\mathbf{T(CPT)P}\right)^{-1} = -\xi\gamma_H^2\Psi_H^{\dagger T}(x). \tag{5.3.49}$$

This transformation acts on the electromagnetic 4-potentials according to

$$\left(\mathbf{T(CPT)P}\right)A^\mu(x)\left(\mathbf{T(CPT)P}\right)^{-1} = -A^\mu(x), \tag{5.3.50}$$

which is the same transformation as that for **C**.

  We shall now consider Weyl's analysis of the discrete transformations and shall compare them to the transformations **C**, **P**, **T** and **CPT** described above. Weyl first considers the discrete transformations in the context of Relativistic Quantum Mechanics (RQM). Later, after quantizing the field equations, he re-examines these transformations in the new context of Relativistic Quantum Field Theory (RQFT).

Remark 44 The symmetries **C**, **P**, **T** and **CPT** are realized somewhat differently in RQM and RQFT; however, we shall use the same symbols to denote the corresponding operations.

In subsection III of section 6 in chapter IV (p. 200), Weyl considers the case of Relativistic Quantum Mechanics. He first notes that the Maxwell-Dirac equations are invariant under the transformation (Weyl 1928b, 2 edn, 1931, Eq. (IV.6.12))

$$\begin{aligned}
A_0(t,\vec{x}) &\rightarrow A_0(t,-\vec{x}) \\
A_j(t,\vec{x}) &\rightarrow -A_j(t,-\vec{x}) \\
\Psi_H(t,\vec{x}) &\rightarrow \beta\Psi_H(t,-\vec{x}),
\end{aligned} \tag{5.3.51}$$

which is now called the parity transformation and is denoted by **P**. In the context of RQM, Weyl's result for **P** is the same as that used today.

  The action (Weyl 1928b, 2 edn, 1931, Eq. (IV.5.18)) for the Maxwell-Dirac theory when translated into our notation is

$$\mathbf{A} = \int \bar{\Psi}_H\gamma_H^\mu(\partial_\mu + ieA_\mu)\Psi_H\,d^4x + m\int \bar{\Psi}_H\Psi_H\,d^4x + \frac{1}{4}\int F_{\mu\nu}F^{\mu\nu}d^4x, \tag{5.3.52}$$

where $F_{\mu\nu} = \partial_\mu A_\nu - \partial_\nu A_\mu$. Note that we use units for which $\hbar = 1$ und $c = 1$; moreover, if Weyl's formulas are re-written in terms of these units, then Weyl uses

---

[78]Note that this transformation is linear and unitary because both **T** and **CPT** are antilinear and antiunitary. Consequently, Weyl's charge-conjugation operator agrees with the modern formulation despite the fact that Weyl's ⌜**T**⌝ operator was linear and unitary, because the operator **T** appears twice.

$f_\mu = e\phi_\mu$ where his ϕ_μ is our A_μ. Note that Weyl uses h to denote $\hbar$ as he notes on p. H357. Weyl denotes the three terms in (5.3.52) by M, M' and F respectively. For the purpose of translating Weyl's expression (H.IV.5.8) for M into the first integral in (5.3.52), note that the S_α in (H.IV.5.9) are 4×4 matrices S^μ where

$$S^0 = \begin{bmatrix} I & 0 \\ 0 & I \end{bmatrix} \quad \text{and} \quad S^k = \begin{bmatrix} \sigma_k & 0 \\ 0 & -\sigma_k \end{bmatrix}. \tag{5.3.53}$$

Addressing the question of whether or not the theory is symmetric under an interchange of past and future, he notes that Dirac had remarked that $M \to -M$ and $M' \to -M'$ under the transformation (Weyl 1928b, 2 edn, 1931, Eq. (IV.6.13))

$$A_\mu(x) \to -A_\mu(-x), \qquad \Psi_H(x) \to \gamma_5 \Psi_H(-x). \tag{5.3.54}$$

It follows that the equation

$$\gamma_H^\mu(\partial_\mu + ieA_\mu)\Psi_H + m\Psi_H = 0 \tag{5.3.55}$$

is invariant under (5.3.54). This transformation involves time reflection, space reflection and charge reflection; that is, it is similar in spirit to what is now called the **CPT** transformation. We shall return to this transformation in the discussion below in connection with Weyl's discussion of the quantized field equations. In the context of relativistic Quantum Mechanics, Weyl points out that because of this symmetry, the anti-electron should behave in essentially the same way as the electron. Weyl (1928b, 2 edn, 1931, p. 200) says

> Wenn es sich also um die Bewegung eines Elektrons in einem äußeren elektromagnetischen Feld handelt, gewinnt man aus einer Lösung ψ, welche die Zeit in dem Faktor $e^{-i\nu t}$ enthält, durch diese Substitution [(5.3.54)] eine neue Lösung, die statt dessen den Zeitfaktor $e^{i\nu t}$ trägt. Nur muß gleichzeitig das Viererpotential f in $-f$ verkehrt werden. Dies kann man dadurch bewirken, daß man das äußere Feld mit dem Potential φ beibehält, aber die Ladung $-e$ in $+e$ verkehrt. Das (in der Natur nicht vorkommende) Teilchen von der Elektronenmasse m, dessen Ladung nicht $-e$, sondern $+e$ ist, werde als "positives Elektron" bezeichnet. Man erkennt aus dem Gesagten, daß die Energieniveaus des positiven Elektrons $-h\nu$ sind, wenn $h\nu$ diejenigen des negativen Elektrons sind. Abgesehen vom Vorzeichen verhalten sich beide Sorten von Teilchen gleich. *Das Elektron wird außer seinen positiven auch negative Energieniveaus besitzen*, die aus den positiven Energieniveaus des positiven Elektrons durch Änderung des Vorzeichens entstehen. Hier is offenbar noch etwas nicht in Ordnung; man sollte diese negativen Energieniveaus des Elektrons streichen können. Dies ist aber unmöglich, weil z. B. unter der Einwirkung eines Strahlungsfeldes Übergänge zwischen den positiven und negativen Termen vorkommen. Daß man doppelt zu viel Terme erhält, hängt offenbar

damit zusammen, daß die Größe ψ nicht *vier*, sondern nur *zwei* (Differentialgleichungen 1. Ordnung genügende) Komponenten besitzen sollte. Die Lösung dieser Schwierigkeit scheint in der Richtung zu liegen, daß unsere Feldgleichungen mit den vier Komponenten irgendwie außer dem Elektron bereits das Proton mitumfassen.

[If one obtains a solution for the motion of an electron in an external field in which the time appears in a factor $e^{-i\nu t}$, then by means of this substitution [(5.3.54)], one obtains a new solution in which the time appears instead in a factor $e^{i\nu t}$. At the same time, however, the four-potential f must be changed into $-f$. This can be achieved by changing the charge from $-e$ to e while keeping the external field with the potential φ unchanged. The particle (which is not found in nature) which has the mass of the electron but has the charge $+e$ rather than $-e$ will be called a "positive electron". From what has been said, one sees that the energy levels of the positive electron are $-h\nu$ if those [the energy levels] of the electron are $h\nu$. Aside from this difference in sign, the two types of particles behave in the same way. *The electron will possess in addition to its positive-energy levels negative ones as well*, which arise from the positive-energy levels of the positive electron by a change of sign. Evidently, something is not right here; one should be able to get rid of these negative-energy levels. This is, however, not possible because, for example, under the influence of a radiation field, transitions occur between the positive and negative terms. That there are twice as many terms as there should be is evidently due to the fact that the quantitiy ψ should possess not *four* but only *two* components (which satisfy differential equations of the first-order). The solution of the difficulty seems to lie in the direction that our field equations with the four components somehow encompass the proton in addition to the electron.]

Weyl then notes that the transformation (5.3.54) is not a symmetry of the complete action **A** because the term F is invariant but the two terms M and M' change sign. Weyl (1928b, 2 edn, 1931, p. 200) says,

Im ganzen herrscht also, wenn die Rückwirkung der Teilchen auf das elektrische Feld in Betracht gezogen wird, *keine* Invarianz gegenüber jener Substitution.

[Our field equations as a whole, that is, when we also take into account the reaction of the particle on the radiation field, are consequently *not* invariant under this substitution.]

The point is that the other field equation,

$$\partial_\mu F^{\mu\nu} = e\bar{\psi}\gamma^\nu\psi,$$

(5.3.56)

is not invariant under (5.3.54). He proceeds to show that there does exist a transformation (Weyl 1928b, 2 edn, 1931, Eq. (IV.6.14)) which interchanges past and future that leaves the complete action **A** invariant, namely,

$$
\begin{aligned}
A_0(t, \vec{x}) &\rightarrow A_0(-t, \vec{x}) \\
A_j(t, \vec{x}) &\rightarrow -A_j(-t, \vec{x}), \quad j \in \{1, 2, 3\} \\
\Psi_H(t, \vec{x}) &\rightarrow \beta \gamma_5 \gamma_H^2 \Psi_H^*(-t, \vec{x}).
\end{aligned}
\tag{5.3.57}
$$

For the case of Relativistic Quantum Mechanics, this is the transformation that is now called *time reversal* and is denoted by **T**. To summarize, Weyl has shown that in the context of Relativistic Quantum Mechanics, the parts M, M' and F of the total action **A** transform under the transformations **P** ((5.3.51)), **CPT** ((5.3.54)) and **T** ((5.3.57)) according to

$$
\begin{aligned}
M &\rightarrow M, & M' &\rightarrow M', & F &\rightarrow F, \\
M &\rightarrow -M, & M' &\rightarrow -M', & F &\rightarrow F, \\
M &\rightarrow M, & M' &\rightarrow M', & F &\rightarrow F,
\end{aligned}
\tag{5.3.58}
$$

respectively. He ends the subsection with the caution that one must suspend judgement until the quantized field equations have been examined.

Remark 45 (Weyl's Re-interpretation of the 'Hole' Theory) Weyl initially approaches the quantization of the Maxwell-Dirac field equations from the point of view of Dirac's hole theory; that is, for each energy eigenvalue μ which may be either positive or negative, Weyl considers the occupation number n_μ which by the Pauli Exclusion Principle must be either 0 or 1. The vacuum state is then characterized by the requirement that $n_\mu = 1$ for all $\mu \leq -mc^2$. What is now called pair creation corresponds to an electron in a state of negative energy μ' jumping to a state of positive energy μ so that $n_{\mu'}$ changes from 1 to 0 and n_μ changes from 0 to 1 with the consequence that the total charge $\Sigma_\mu n_\mu$ remains fixed. What is now called pair annihilation similarly corresponds to an electron with positive energy μ dropping back down into an unoccupied state with negative energy μ' in the sea of negative-energy electrons. On this account, the total number of electrons and hence the total charge $\Sigma_\mu n_\mu$ does not change. An electron merely moves from one energy level to another. Weyl (1928b, 2 edn, 1931, p. 233) observed that the $n_\mu = 0, 1$ were merely arbitrary labels for the row of a matrix (Matrizenzeilen) and that there was nothing to prevent re-labelling them by setting for $\mu > 0$

$$
n_\mu^+ = n_\mu
\tag{5.3.59}
$$

$$
n_\mu^- = 1 - n_{-\mu}.
\tag{5.3.60}
$$

Weyl then observed that this change of notation also changed the *content* of the theory. In the new notation, the vacuum is characterized by $n_\mu^+ = 0$ and $n_\mu^- = 0$ for all μ, where necessarily $\mu > 0$ because there are now no negative-energy states. A 1-electron state has $n_\mu^+ = 1$ for some $\mu > 0$ and a 1-anti-electron state (the proton for Weyl) has $n_\mu^- = 1$ for some $\mu > 0$. Pair creation is interpreted as it is today, namely, as the simultaneous change for some $\mu > 0$ of n_μ^+ from 0 to 1 and for some $\mu' > 0$ of $n_{\mu'}^-$ from 0 to 1. The total number of particles does not remain fixed but the negative of the total charge

$$\Sigma_\mu n_\mu^+ - \Sigma_\mu n_\mu^-, \quad \mu > 0 \qquad (5.3.61)$$

and hence the total charge does remain fixed. Pair annihilation is similarly interpreted as it is today as the simultaneous change for some $\mu > 0$ of n_μ from 1 to 0 and for some $\mu' > 0$ of $n_{\mu'}$ from 1 to 0. It seems, however, that Weyl's insight was lost and this re-interpretation of the theory had to be re-invented two years later. Both Pais (1986, p. 379) and Schweber (1994, p. 77) say that the re-interpretation of the theory was first presented in the work of Fock (1933), Furry and Oppenheimer (1934), and Heisenberg (1934).

The quantization of the Maxwell-Dirac field equations is carried out in section 12 of chapter IV. At the end of the section, Weyl considers how the *quantized* field equations transform under the three discrete transformations, that correspond to **P**, **CPT** and **T**, discussed above in connection with the relativistic quantum mechanics of the electron. In addition, Weyl introduces in the context of RQFT a fourth transformation that corresponds to **T(CPT)P**. On p. H234, Weyl remarks that with respect to the first two transformations everything remains the same as for the case of Relativistic Quantum Mechanics; however, he is clearly aware that the transformations now apply to the quantum field operators rather than to the wavefunctions. For the first of these transformations, (5.3.51) parity, the operator **P** that he obtained is the same as the transformation that is used today which is given by (5.3.37) and (5.3.39), where for Weyl's transformation the appropriate choice of intrinsic parity for the electrons is $\eta = 1$. The second of the two transformations differs from **CPT** because Weyl's transformation was linear and unitary rather than antilinear and antiunitary. If we denote Weyl's transformation by $\ulcorner$**CPT**$\urcorner$, then his result would be expressed today[79] by

$$\ulcorner\mathbf{CPT}\urcorner \Psi_H(x) \ulcorner\mathbf{CPT}\urcorner^{-1} = \gamma_5 \Psi_H(-x) \qquad (5.3.62)$$

and

$$\ulcorner\mathbf{CPT}\urcorner A^\mu(x) \ulcorner\mathbf{CPT}\urcorner^{-1} = -A^\mu(-x). \qquad (5.3.63)$$

The transformation (5.3.63) is identical, except that it is linear and unitary, to (5.3.47); however, the transformation (5.3.62) has $\Psi_H(-x)$ on the right hand side rather than $\Psi_H^{\dagger T}(-x)$ as in (5.3.46). Recall that for any particle, it is always possible to choose $\xi\zeta\eta = 1$. This difference is due to the fact that Weyl did not realize that the transformation should be antilinear and antiunitary. The third of Weyl's

---

[79] Weyl employed a process of substitution rather than the concept of an operator acting on the field operators; however, for linear, unitary transformations, the two approaches are equivalent. See remark 43 above.

tansformations for the RQFT case, which we will denote by $\ulcorner\mathbf{T}\urcorner$, differs from $\mathbf{T}$ in that $\ulcorner\mathbf{T}\urcorner$ is linear and unitary rather than antilinear and antiunitary. Today, Weyl's transformation would be expressed by

$$\ulcorner\mathbf{T}\urcorner\Psi_H(x)\ulcorner\mathbf{T}\urcorner^{-1} = \beta\gamma_5\gamma_H^2\Psi^{\dagger T}(-\mathcal{P}x) \tag{5.3.64}$$

and

$$\ulcorner\mathbf{T}\urcorner A^\mu(x)\ulcorner\mathbf{T}\urcorner^{-1} = \mathcal{P}^\mu_\nu A^\nu(-\mathcal{P}x). \tag{5.3.65}$$

The transformation (5.3.65) is identical, except that it is linear and unitary, to (5.3.44). The transformation (5.3.64), however, differs from (5.3.41) in that $\Psi_H^{\dagger T}(-\mathcal{P}x)$ appears on the right hand side of (5.3.64) instead of $\Psi_H(-\mathcal{P}x)$ as in (5.3.41). For comparison with Weyl's transformation, it is appropriate to choose $\zeta = 1$ for the electron. Weyl explicitly recognized that in the case of RQFT, the fact that the field operators $\Psi_H(x)$ obeyed anticommutation relations meant that the action $\mathbf{A}$ would not be invariant under the $\ulcorner\mathbf{T}\urcorner$ operation. To summarize, for the case of RQFT, Weyl found that under the transformations $\mathbf{P}$, $\ulcorner\mathbf{CPT}\urcorner$ and $\ulcorner\mathbf{T}\urcorner$, the parts M, M' and F of the action $\mathbf{A}$ transform according to

$$\begin{aligned} M \to M, & \quad M' \to M', & \quad F \to F, \\ M \to -M, & \quad M' \to -M', & \quad F \to F, \\ M \to -M, & \quad M' \to -M', & \quad F \to F, \end{aligned} \tag{5.3.66}$$

respectively.

Weyl reacted strongly to the fact that the quantized field equations were not invariant under the transformation $\ulcorner\mathbf{T}\urcorner$ that he thought was the correct time-reversal transformation. Weyl wrote (Weyl 1928b, 2 edn, 1931, p. 234)

> *In den quantisierten Feldgleichungen spielen daher Vergangenheit und Zukunft eine wesensverschiedene Rolle*; wir erhalten keine Substitution, welche diese Gleichungen invariant läßt, aber den Ablaufssinn der Zeit umkehrt. Damit scheint mir ein prinzipiell außerordentlich bedeutsames Ziel der Physik erreicht.

> [*Hence past and future play essentially different roles in the quantized field equations*; we find no substitution which leaves these equations unchanged while reversing the direction of time. It seems to me that we have thereby reached an extraordinarily important goal of physics.]

It should be noted, however, that in an earlier section of the text, Weyl (1928b, 2 edn, 1931, p. 97) attributes the arrow of time to the statistical character of Quantum Mechanics:

> Zweitens: woher kommt der Unterschied zwischen Emission und Absorption, Prozessen, die sich durch Umkehrung der Zeitrichtung ineinander verwandeln? Die zugrunde liegenden mechanischen und Feldgesetze sind doch gegenüber der Abänderung von t in $-t$ invariant!

Antwort: *Die ausgezeichnete Richtung des Zeitablaufs liegt im Sinne der Anwendung der Wahrscheinlichkeitsrechnung;....*

[Secondly, where does the difference between emission and absorption come from, processes which transform into each other under a reversal of the direction of time? The underlying mechanical and field laws are, however, invariant under the change from t to $-t$! Answer: *The distinguished direction of time is to be understood within the context of the application of the theory of probability;....*]

The fourth discrete symmetry of the quantized field equations that Weyl discusses is the combination of the previous three, namely, $\ulcorner\mathbf{T}\urcorner(\ulcorner\mathbf{CPT}\urcorner)\mathbf{P}$. This transformation turns out to be identical to the transformation $\mathbf{T}(\mathbf{CPT})\mathbf{P}$ because the antilinear and antiunitary nature of the two operators $\mathbf{T}$ and $\mathbf{CPT}$ cancel and the resulting operator is linear and unitary. Using the operator notation, Weyl's transformation is given by

$$\left(\ulcorner\mathbf{T}\urcorner(\ulcorner\mathbf{CPT}\urcorner)\mathbf{P}\right)\Psi_H(x)\left(\ulcorner\mathbf{T}\urcorner(\ulcorner\mathbf{CPT}\urcorner)\mathbf{P}\right)^{-1} = -\gamma_H^2\Psi_H^{\dagger T}(x) \qquad (5.3.67)$$

and

$$\left(\ulcorner\mathbf{T}\urcorner(\ulcorner\mathbf{CPT}\urcorner)\mathbf{P}\right)A^\mu(x)\left(\ulcorner\mathbf{T}\urcorner(\ulcorner\mathbf{CPT}\urcorner)\mathbf{P}\right)^{-1} = -A^\mu(x). \qquad (5.3.68)$$

The transformations (5.3.67) and (5.3.68) are identical to the transformations (5.3.49) and (5.3.50), respectively, provided that one chooses the phase factor $\xi = -1$. Note that the phases for the electron that are appropriate for Weyl's transformations form a consistent set; namely, $\xi = -1$, $\zeta = 1$, $\eta = -1$ and $\xi\zeta\eta = 1$. The discrete symmetry $\ulcorner\mathbf{C}\urcorner$ given by (5.3.67) and (5.3.68) differs slightly from the transformation now called $\mathbf{C}$; however, the difference between $\ulcorner\mathbf{C}\urcorner$ and $\mathbf{C}$ is merely a matter of convention. Weyl, therefore, had a correct formulation of charge-conjugation invariance for the quantized, Maxwell-Dirac equations. Weyl also realized the great importance of the invariance of the theory under $\ulcorner\mathbf{C}\urcorner$ for Dirac's theory of the electron, which at that time assigned the proton to the role of the anti-electron. Weyl (1928b, 2 edn, 1931, p. 234) wrote

> Im Lichte der eben erwähnten Diracschen Theorie der Protonen besagt sie [the invariance under the substitutions (5.3.67) and (5.3.68)], daß positive und negative Elektrizität gleichartig sind, daß nämlich die Gesetze invariant sind gegenüber einer gewissen Substitution, welche die auf die Elektronen und Protonen bezüglichen Quantenzahlen miteinander vertauscht. Die Ungleichartigkeit der beiden Elektrizitätsarten scheint also ein noch tieferes Naturgeheimnis zu bergen als die Ungleichartigkeit von Vergangenheit und Zukunft.

[In light of Dirac's theory of the protons, mentioned just above, this [the invariance under the substitutions (5.3.67) and (5.3.68)] says that the fundamental carriers of positive and negative electricity are very similar in so far as the laws are invariant under a particular transformation which interchanges the corresponding quantum numbers of the electrons and protons. The dissimilarity of the two types of electricity appears, therefore, to hide a secret of nature which is even deeper than the dissimilarity between the past and future.]

There was thus no reason for the proton to be so different from the electron. This is clearly the motivation for Weyl's remarkable[80] statement in the preface to the second edition that we quote just above remark 41 in this subsection.

As the above discussion shows, Weyl had presented in 1931 a complete analysis, in the context of the quantized Maxwell-Dirac field equations, of the discrete symmetries that are now called **C**, **P**, **T** and **CPT**. His transformations **C** and **P** are the same as those used today. His transformations ⌜**T**⌝ and ⌜**CPT**⌝ are also very close to those used today except that Weyl's transformations were linear and unitary rather than antilinear and and antiunitary. Moreover, Weyl drew two very important conclusions from his analysis of these discrete symmetries. First, Weyl announced that the important question of the arrow of time had been solved because the field equations were not invariant under his time-reversal transformation ⌜**T**⌝. Second, Weyl pointed out that the invariance of the field equations under his charge-conjugation transformation **C** implied that the mass of the 'anti-electron' is necessarily the same as that of the electron; moreover, this result is the primary reason that Dirac (1931, p. 61) abandoned the assignment of the proton to the role of the anti-electron. Many years later Dirac (1977, p. 145) recalled:

> Well, what was I to do with these holes? The best I could think of was that maybe the mass was not the same as the mass of the electron. After all, my primitive theory did ignore the Coulomb forces between the electrons. I did not know how to bring those into the picture, and it could be that in some obscure way these Coulomb forces would give rise to a difference in the masses.
>
> Of course, it is very hard to understand how this difference could be so big. We wanted the mass of the proton to be nearly 2000 times the mass of the electron, an enormous difference, and it was very hard to understand how it could be connected with just a sort of perturbation effect coming from Coulomb forces between the electrons.
>
> However, I did not want to abandon my theory altogether, and so I put it forward as a theory of electrons and protons. Of course I was very soon attacked on this question of the holes having different masses from the original electrons. I think the most definite attack came from Weyl, who pointed out that mathematically the holes would have to have the same mass as the electrons, and that came to be the accepted view.

At another place Dirac (1971, pp. 52–55) remarks:

---

[80]See the passage by Yang quoted above in remark 41.

But still, I thought there might be something in the basic idea and so I published it as a theory of electrons and protons, and left it quite unexplained how the protons could have such a different mass from the electrons.

This idea was seized upon by Herman [sic] Weyl. He said boldly that the holes had to have the same mass as the electrons. Now Weyl was a mathematician. He was not a physicist at all. He was just concerned with the mathematical consequences of an idea, working out what can be deduced from the various symmetries. And this mathematical approach led directly to the conclusion that the holes would have to have the same mass as the electrons. Weyl just published a blunt statement that the holes must have the same mass as the electrons and did not make any comments on the physical implications of this assertion. Perhaps he did not really care what the physical implications were. He was just concerned with achieving consistent mathematics.[81]

Although it seemed unlikely, Dirac nevertheless hoped that a detailed analysis of the electromagnetic interaction between the electrons would account for the difference between the mass of the proton 'hole' and the mass of the electron. It seems likely that Dirac was convinced by Weyl's argument for the equality of these masses because it was based on an *exact* result, the invariance of the quantized, Maxwell-Dirac equations under charge conjugation.[82] As an additional reason for abandoning the idea that the proton is the anti-electron, Dirac cites the work of Tamm (1930), Oppenheimer (1930b) and himself (1930) on the instability of matter that results from the annihilation of protons and electrons. It should be noted, however, that these calculations were approximate in character. Immediately after accepting the idea that the proton is not the anti-electron, Dirac embraces Oppenheimer's (1930a, p. 563) suggestion that there are two separate Dirac 'seas', one for the electron and one for the proton. Dirac then draws the logical conclusion that there must exist two new particles, namely, the anti-electron and the anti-proton.

---

[81] However, Weyl did think that the difference between the masses of the proton and the electron portended a serious crisis in physics.

[82] Kragh (1990, p. 102–103) makes the following observation:

... When, in the early part of 1931, Dirac gave up his unitary theory and proposed the anti-electron as a separate particle, he quoted both Oppenheimer's and Weyl's objections as decisive arguments. Probably Weyl's argument, which was in spirit close to Dirac's own methodological preferences, was of more importance.

In his various recollections, Dirac repeatedly pointed out that his disinclination to postulate positively charged electrons was rooted in a lack of boldness both to rely on the mathematics of his wave equation and to disregard the restrictions set by current empirical knowledge. Had he only been faithful to the power of pure mathematical reasoning and not been led astray by what was known empirically, he immediatey would have postulated the positive electron. Indeed, this was what Weyl did, although only implicitly. On several occasions, Dirac attributed this success of Weyl's to his mathematical approach to physics. This was an approach that Dirac strongly recommended.

Dirac (1971, p. 56–59) recalls:

> At this stage in the development of the theory, Oppenheimer made a contribution. Oppenheimer accepted Weyl's conclusion that the holes had to have the same mass as the electrons and faced the physical reality that the holes were not observed in practice. Oppenheimer just said that there was some reason, which we do not understand, why the holes are never observed. He agreed that the holes could not have anything to do with protons, so there had to be some mysterious reason why they did not occur in nature.
>
> Well, Oppenheimer was really very close to the mark with this hypothesis. The reason why the holes were not oberved was simply that the experimental people had not looked for them in the right place, or if they had looked, they had not recognized what they saw.
> . . .
> It needed some years of development by the experimenters before the fact of the existence of positrons was established.

The problem of the 'arrow of time' has been an important problem for a long time. One would think that Weyl's announcement that he had solved this problem would have drawn attention to his analysis of the discrete symmetries. One would also think that the fact that Weyl's analysis led Dirac to predict the existence of two new particles, one of which was discovered experimentally one year later, would have convinced everyone of the importance of Weyl's work on **C**, **P**, **T** and **CPT**. Yet it appears that no one other than Dirac[83] paid any attention at all to Weyl's analysis of the discrete transformations. At the time, the principal researchers in Quantum Mechanics were fluent speakers of the German language; consequently, as far as language is concerned, Weyl's analysis was readily accessible. In addition, Weyl states that he was using the then recent work of Heisenberg and Pauli for the quantization of the Maxwell-Dirac field equations so that Heisenberg and Pauli at least should have had no difficulty understanding Weyl's analysis. Moreover, Wigner (1931) wrote a book in German on group theory and its applications to Quantum Mechanics and a year later wrote a paper (Wigner 1932) in which he presented the first correct analysis of time reversal in terms of an antilinear, antiunitary operator in the context of *non-relativistic* Quantum Mechanics. It is rather strange that Wigner seems to have been totally unaware[84] of Weyl's work

---

[83]In his J. Robert Oppenheimer Prize Acceptance Speech, Dirac (1971, p. 56) suggests that Oppenheimer proposed a modified theory in which protons and electrons were treated as distinct particles because of Weyl's result that the electron and anti-electron had to have the same mass; however, Dirac does not provide a specific reference and we have not found any citation of Weyl by Opppenheimer with regard to Weyl's result on the equality of the masses of the electron and the anti-electron.

[84]In a later reminiscence, Wigner (1987, p. 62) says that Dirac realized that the anti-electron was not a proton only after the discovery of the positron in 1932. It would seem that at the time that Wigner made this remark, Wigner may have forgotten that Dirac predicted the anti-electron a year before its discovery!

on time reversal in the relativistic context. Clearly, anyone who had been aware both of Weyl's analysis and of Wigner's analysis of time reversal could easily[85] have corrected Weyl's failure to treat time reversal as an antilinear, antiunitary operator. In other words, a correct analysis of **C**, **P**, **T** and **CPT** for the case of the quantized Maxwell-Dirac field equations could have been achieved in the early 1930s. Finally, we note that six years after the publication of Weyl's analysis of the discrete transformations, Kramers (1937) wrote a paper on charge conjugation in Maxwell-Dirac theory. There is no mention of Weyl's earlier analysis in Kramers' paper.

In his book *Inward Bound*, Pais (1986) provides a detailed account of the history of atomic and subatomic physics. On page 525, he discusses **P**, **T** and **C** prior to 1956. Charge conjugation is also discussed in an earlier subsection on page 38. In Pais' discussion, there is not the slightest indication of an awareness of Weyl's work on **C**, **P**, **T** and **CPT**.

Schweber (1994), in his history of Quantum Field Theory, does not discuss the discrete symmetries **C**, **P**, **T** and **CPT**; however, he does mention on page 66 the "telling argument of Weyl" to the effect that an electron hole must have the same mass as the electron. In an endnote, Schweber quotes a passage from the English edition of Weyl's book that appears in the paragraph that immediately precedes Weyl's discussion of **C**-invariance in the context of the quantized, Maxwell-Dirac field equations.

Yang (1982) wrote a paper on the history of the discrete symmetries **P**, **T** and **C**. Although Yang notes that Weyl denoted the parity transformation, then called *Spiegelung*, by i, and called the eigenvalue of the corresponding unitary operator I *signature*, Yang makes no mention of Weyl's treatment of parity in the context of the quantized, Maxwell-Dirac field equations. With regard to time reversal, Yang mentions the work of (Kramers 1930), but Weyl's analysis is not mentioned. It should be noted that both Kramers and Wigner dealt only with time reversal in the context of non-relativistic Quantum Mechanics. Yang states that the first formal development of **C** invariance was due to Kramers (1937); however, the analysis presented above clearly shows that Weyl had a full analysis of **C**-invariance in the context of the quantized Maxwell-Dirac field equations in 1931. It is rather odd that Yang overlooked Weyl's analysis of **C**-invariance because Yang cites (Yang's reference 20) the very page on which Weyl presented the symmetry transformation now called charge conjugation and expressed his (Weyl's) deep concern about the dissimilar nature of positive and negative electricity, that is, the large difference in the masses of the proton and the electron. In addition, Yang (1982, p. C8-444) refers to the preface of Weyl's (1928b, 2 edn, 1931, Dover edn) book and quotes the same passage by Weyl that we quote at the beginning of this subsection. With regard to this passage, Yang says "He [Weyl] was thinking of P, T and C, but I am not sure what crisis he was referring to." As we have already

---

[85]It is worth noting, however, that the antilinear and antiunitary character of time reversal **T** was not accepted immediately. Indeed, many years later, Pauli (1941, See Eqs (76), (77)) treated all of the discrete symmetries as linear and unitary.

noted in remark 41, at the beginning of this subsection, Yang (1986, p. 10) later expressed great puzzlement as to why Weyl even wrote in his preface to his book a passage that involved the three transformations P, T and C.

It is clear that Weyl's early analysis of the transformations **C**, **P**, **T** and **CPT** was lost to subsequent researchers[86] and had to be essentially re-invented; however, Weyl's analysis did have an impact on the development of the Maxwell-Dirac theory because his proof that the theory was invariant under charge conjugation, an *exact* result, convinced Dirac to give up his unitary theory in which the proton was identified with the anti-electron and to postulate the existence of two new particles, namely, the anti-electron and the anti-proton.

5.4 G-Structures, the Electron and Gauge Theory

In his Erlanger Programme,[87] F. Klein provided a unified approach to the various 'global' geometries by showing that each of the geometries is characterized by a particular group of transformations: Euclidean geometry is characterized by the group of translations and rotations in the plane; the geometry of the sphere S^2 is characterized by the orthogonal group $O(3)$; and the geometry of the hyperbolic plane is characterized by the pseudo-orthogonal group $O(1,2)$. In this approach, two figures are regarded as congruent if and only if there exists an element of the appropriate group that transforms one of the figures into the other. As Weyl (1929e) noted, E. Cartan adapted Klein's Erlanger Programme to infinitesimal geometry by applying Klein's notions to the tangent plane rather than to the manifold itself and thereby founded the theory of group or G-structures.

A first-order G-structure, where G is a subgroup of $GL(n)$, is determined by specifying in a smooth manner an equivalence class of privileged bases for the tangent vector space $T(M_p)$ at every point p of the manifold M. At a given point p, any two equivalent bases are related by an element of the group G. For certain groups, the assumption that M is equipped with an affine (symmetric linear connection) structure does not introduce anything extraneous to the G-structure since the G-structure uniquely determines a symmetric linear connection that preserves the G-structure under parallel transport. The Riemannian and pseudo-Riemannian structures are typical of this situation. For other groups such as the conformal group $CO(1, n-1)$ and the Galilean group, the G-structure determines only an equivalence class of symmetric linear connections, all of which preserve the G-structure under parallel transport. Moreover, for the important case of the projec-

---

[86]Schwinger (1988, pp. 107–129), however, says that he was greatly influenced by Weyl's (1928b) book *Gruppentheorie und Quantenmechanik*. Schwinger makes particular reference to Weyl's work on the discrete symmetries and says that this work "...was the starting point of my own considerations concerning the connection between spin and statistics, which culminated in what is now referred to as the TCP — or some permutation thereof — theorem".

[87]It was in 1872, when Klein succeeded K. G. C. von Staudt as professor at the Philosophical Faculty of Erlangen University, that Klein presented the "Erlanger Programme" in his inaugural lecture entitled *Betrachtungen über neuere geometrische Forschungen*. A revised version of that work appeared later in (Klein 1893).

tive structure, the corresponding group is not even of first-order[88] and it contains
the affine group as a subgroup. For an elementary exposition of these G-structures,
see (Coleman and Korté 1981) and (Coleman and Mann 1982). In the two papers
(Weyl, 1929, 1929e), the first of which is co-authored by H. P. Robertson, Weyl
considers these problems for the conformal and projective structures. In doing so,
he draws together the work of Cartan (1923a, 1986) and the work of Eisenhart
(1927), Veblen (1928b, 1928a) and Thomas (1926), and emphasizes the need for a
concept of higher order contact in these cases.

In the three papers Weyl (1929b, 1929c, 1929d), Weyl adapts Dirac's special
relativistic theory of the electron to the General Theory of Relativity. Weyl (1929d,
p. 219) notes that the spinor representation of $O(1,3)$ cannot be extended to a
representation of $GL(4)$ with the consequence that it is necessary to employ the
vierbein or Lorentz-structure formulation of the General Theory of Relativity in
order to incorporate Dirac's spinor fields. A complete exposition of this formalism
is presented in (Weyl 1929b). In this paper, he also presents a particularly simple
introduction to the theory of two-component spinors. Much later in collaboration
with R. Brauer (Brauer and Weyl 1935), Weyl developed the theory of spinors for
n-dimensions.

It is also in one of these three papers that Weyl (1929d, p. 218) explicitly
abandons his earlier attempt to unify electromagnetism with the General Theory
of Relativity by associating the electromagnetic vector potential $A_\mu(x)$ with the
additional connection coefficients that arise when a conformal structure is reduced
to a Weyl structure. The important concept of gauge invariance, however, is preserved. Rather than associating gauge transformations with the scale or gauge of
the spacetime metric tensor, Weyl associates gauge transformations with the phase
of the Dirac spinor field that represents matter. Today, the concept of gauge invariance plays a central role in theoretical physics. The following remarks of Dyson
(1983) indicate the subsequent development of gauge field theories.

> A more recent example of a great discovery in mathematical physics
> was the idea of a gauge field, invented by Hermann Weyl in 1918. This
> idea has taken only 50 years to find its place as one of the basic con-
> cepts of modern particle physics. Quantum chromodynamics, the most
> fashionable theory of the particle physicists in 1981, is conceptually
> little more than a synthesis of Lie's group-algebras with Weyl's gauge
> fields.
>
> The history of Weyl's discovery is quite unlike the history of Lie
> groups and Grassmann algebras. Weyl was neither obscure nor unrec-
> ognized, and he was working in 1918 in the most fashionable area of
> physics, the newborn theory of general relativity. He invented gauge

[88]Briefly, if $a: \mathbf{R}^n \to \mathbf{R}^n$ is a diffeomorphism such that $a(\vec{0}) = \vec{0}$, then the Taylor approximation
of second-order to a at $\vec{0}$ is an element of the group G_n^2, where the group product is determined
by composition of the Taylor series. The projective group is a subgroup of G_n^2. For more details,
see (Coleman and Korté 1981).

fields as a solution of the fashionable problem of unifying gravitation with electromagnetism. For a few months gauge fields were at the height of fashion. Then it was discovered by Weyl and others that they did not do what was expected of them. Gauge fields were in fact no good for the purpose for which Weyl invented them. They quickly became unfashionable and were almost forgotten. But then, very gradually over the next fifty years, it became clear that gauge fields were important in a quite different context, in the theory of quantum electrodynamics and its extensions leading up to the recent development of quantum chromodynamics. The decisive step in the rehabilitation of gauge fields was taken by our Princeton colleague Frank Yang and his student Bob Mills in 1954, one year before Hermann Weyl's death [Yang and Mills, 1954]. There is no evidence that Weyl ever knew or cared what Yang and Mills had done with his brain-child.

So the story of gauge fields is full of ironies. A fashionable idea, invented for a purpose which turns out to be ephemeral, survives a long period of obscurity and emerges finally as a corner-stone of physics.

Much later, Weyl returned to the problem of the interaction between gravitation and the electron (Weyl 1950c). In the formulation of the theory presented in this paper, the Lagrangian is varied independently with respect to both the metric and the connection. In (Weyl 1929b), the connection was defined in terms of the vierbein fields and the Lagrangian was varied only with respect to these fields and the matter fields. In the mixed theory, the torsion is proportional to the matter spin density and hence the connection is not metric; however, Weyl proves that the mixed theory can be reformulated as a pure metric theory by suitably modifying the Lagrangian. The resulting theory has the same form as Einstein's original theory except that the source of the curvature is no longer the usual matter, energy-momentum density but is slightly modified by the matter, spin density.

In the papers (Weyl, 1929b, 1929c, 1929d), Weyl contrasted the way in which vierbeins (orthonormal basis vectors) were employed in his own work with the way they were used in a contemporary work by Einstein (1928) in which a radically different theory of gravity was proposed. In Einstein's (1928) theory, the effects of gravity and electromagnetism are associated with a specialized torsion of spacetime rather than with the curvature of spacetime. Since the curvature vanishes everywhere, distant parallelism is a feature of the theory; however, distant parallelism appears quite unnatural from the viewpoint of Riemannian geometry, and for this reason Weyl criticized Einstein's theory.[89]

Not until 1954 did Yang and Mills (1954) generalize Weyl's electromagnetic gauge concept to the case of the non-Abelian group $O(3)$. The following year Utiyama (1956) generalized their work to the case of an arbitrary semisimple Lie group and applied his results to the theory of gravitation regarded as an $O(1, 3)$

---

[89]See in particular, (Weyl 1929d, GA III, p. 219).

or Lorentz gauge theory. Some ten years after Weyl's (1950c) comment on the role of torsion in the theory of gravitation for the special case of coupling to a Dirac field, Kibble (1961) and Sciama (1962) returned to the problem. Kibble proposed a gauge theory of the Poincaré or inhomogeneous Lorentz group while Sciama presented a Lorentz-gauge theory of gravitation that is similar in spirit to Weyl's earlier vierbein[90] formulation of General Relativity. The subsequent, rather complex development of the theory ultimately led to two equivalent formulations of the theory referred to as the Einstein-Cartan and the Einstein theories of gravity. In the Einstein or tensor formulation of the theory, the metric tensor $g = g_{ij}dx^i \otimes dx^j$ and the nonsymmetric affine connection $\Gamma^i_j = \Gamma^i_{jk}dx^k$ are regarded as the fundamental variables of the theory. In this version of the theory, the torsion tensor $T^i_{jk} = \Gamma^i_{jk} - \Gamma^i_{kj}$ is coupled *algebraically* to the spin density with the consequence that the torsion can be eliminated from the theory. The resulting theory is then the same as Einstein's metric theory except that the source is no longer just the energy-momentum tensor density but has additional terms that depend on the spin density of matter. The predicted differences resulting from this modification are, however, so slight that it is not possible at present (and probably for the foreseeable future) to detect them experimentally. As noted above, this result was first pointed out by Weyl (1950c) for the simple case of coupling to a Dirac spinor field. In the Einstein-Cartan or Poincaré gauge formulation of the theory, the orthonormal covectors $\theta^\alpha = e^\alpha_i dx^i$, that are dual to the orthonormal basis vectors $e_\alpha = e^i_\alpha \partial x_i$, and the Lorentz-connection forms $\Lambda^\alpha_\beta = \Lambda^\alpha_{\beta i}dx^i$ are regarded as the fundamental variables of the theory. Nester (1977) has given a thorough analysis of the effective equivalence between the Einstein-Cartan and the Einstein theories of gravitation. References to the literature, additional details concerning the historical development of the Einstein-Cartan theory of gravity and an account of the physical arguments that underlie the two versions of the theory may be found in the work of Hehl (1985), and Hehl, von der Heyde, Kerlick and Nester (1976).

---

[90]Sciama (1961) also used Weyl's vierbein formalism to analyze the Einstein-Schrödinger unified field theory in both its real nonsymmetric and its complex Hermitian versions.

6 Foundations of Mathematics

The history of formal deduction goes back to Aristotle and Euclid, and the development of mathematical analysis can be traced back to Archimedes. It was not until the middle of the nineteenth century that these developments began to converge, due to the work of Boole and Frege. The subsequent efforts to provide a secure foundation for mathematics led to the discovery around 1900 of the antinomies which engendered a foundational crisis.

This foundational crisis in mathematics was an irresistible challenge to Weyl who provided in 1918 a constructive analysis of the real numbers in *Das Kontinuum* (Weyl 1918a). Weyl requires that all relations that are used to construct sets be explicitly definable in terms of a small number of elementary relations by the application of a few principles of construction (definition) that are used finitely many times. In an earlier paper, Weyl (1910b) presented some of these principles of construction in a rather terse form in connection with his desire to make Zermelo's vague concept of 'definite property' in Zermelo's *Aussonderungsaxiom* more precise. In *Das Kontinuum*, Weyl also restricted quantification to the natural numbers in order to avoid the complications associated with Russell's theory of types that would arise if quantification over subsets of natural numbers were allowed. Weyl succeeded in reconstructing a considerable portion of today's real analysis on a strictly predicative foundation. A few years later, Weyl was 'converted' to the intuitionistic constructivism of Brouwer in which a real number is conceived of as a 'becoming free-choice' sequence of nested open intervals. In a series of polemical lectures presented to the Mathematische Kolloquium Zürich, Weyl (1921d, GA II, pp. 143–180) described the new construction of real numbers and continuous functions. He characterized his earlier work in *Das Kontinuum* as an 'atomistic approach' to the concept of real number. In contrast, Weyl (1921d, GA II p. 173) asserted that it is contrary to the essence of the new continuum for it to be divided into separate pieces. Nevertheless, Weyl (1918a) noted in the second preface to a reprinted edition of *Das Kontinuum* that the methods used in that work might still be of fundamental value.

> The point of view adopted in this monograph continues to strike me as a natural transitional stage in the development of foundational research. However, in the period since its appearance, my work has been superseded by two trends identified by the catchwords Intuitionism and Formalism. Still, this deeper grounding of the foundation has not led to an even moderately satisfying or defensible conclusion; things remain in a state of flux. It seems not to be out of the question that the limitations prescribed in the present treatise — i.e., unrestricted application of the concepts "existence" and "universality" to the natural numbers, but not to sequences of natural numbers — can once again be of fundamental significance.[91]

---

[91] This quotation appeared originally in a preface to the 1932 reprint edition of *Das Kontinuum*.

Weyl's other related publications in the foundations of mathematics are (Weyl, 1919a, 1924e, 1925a, 1927[sic]/1966, 1928a, 1929a, 1931a, 1944a, 1946, 1948, 1985).

In subsection 6.1, we provide some historical background of the various approaches to the foundations of mathematics in order to put Weyl's work into perspective. The subsection 6.2 provides an account of Russell's theories of types that is more accessible than those that are currently available in the literature. These other accounts either are overly simplified or are presented in an excessively terse and abstract manner. It is important for the reader to have an understanding of the theories of types because Weyl's determination to avoid at all cost the complications and conceptual obstacles of the ramified theory of types *strongly* influenced the entire structure of the analysis presented in *Das Kontinuum*. The notions of 'definability' and 'impredicativity' also played a key role in *Das Kontinuum*. These concepts are explained in a concrete manner in subsection 6.3 in the context of applying the *simple* theory of types to the construction of the reals. Next, Weyl's predicative analysis of *Das Kontinuum* is presented in subsection 6.4. Finally, subsection 6.5 discusses the reasons for Weyl's 'conversion' to Brouwer's intuitionistic programme and explicates Weyl's (1921d) major contribution to intuitionism.

6.1 Historical Background: Logicism, Formalism and Intuitionism

Regardless of their philosophical orientation concerning the nature of mathematics, most mathematicians would agree that mathematical theories are about abstract objects, although some entities, such as the natural numbers, are also regarded as intuitive. The natural numbers can be intuitively apprehended, and it seems impossible to reduce our knowledge of the sequence of natural numbers to anything more basic. Although, at the most elementary level of mathematics, the quantifiers are restricted to this fundamental domain, the tendency has been to go beyond it for the most part. But while abstraction, generalization and formalization yield mathematical power, elegance and parsimony, they can also open up the possibility of meaninglessness and contradiction. The introduction of infinitesimals into analysis around the late seventeenth century by Newton, Leibniz, Euler, Jakob Bernoulli and Johann Bernoulli serves as a good historical example. The differential and integral calculus had been formulated by Newton and Leibniz in terms of quantities that were infinitely small yet nonzero. While the notion of the 'infinitesimal' certainly helped to foster mathematical physics — the leading physicists and mathematicians being the same during that period — the infinitesimal constructions and definitions were too inconsistent to be useful. Newton postulated the existence of a special number which, being infinitely small, could be multiplied by any finite number and still remain negligible. On the other hand, one had to divide by that number which meant that it had to be a nonzero quantity. Similarly, Leibniz's dx was less than any assignable quantity and yet was nonzero.

We have taken it from the English translation that appears in (Weyl 1918/1994).

The notion of the infinitesimal was attacked (e.g., Berkeley) or eyed with suspicion (e.g., d'Alembert) throughout the eighteenth century. The nineteenth century experienced a return to Euclidean rigour, which by the end of the late nineteenth century culminated under the influence of Weierstrass in the epsilon-delta method, a method similar to Archimedes' 'method of exhaustion'.[92]

The reconstruction of the calculus in terms of the notion of limit and its epsilon-delta definition constituted a reduction of the calculus to the arithmetic basis of real numbers. The demand of logical rigour, however, which motivated the arithmetization of analysis now carried over into inquiries about the logical foundation of the real number system itself. It soon became apparent that the new foundational rigour depended on generalizations and postulations of abstract entities that were as far removed from geometrical and physical intuition as were the infinitesimals of early analysis. Subsequent developments showed that the arithmetic foundation of the new edifice of analysis was no less infected with the danger of meaninglessness and paradox than was the early analysis involving infinitesimals. Dedekind had defined real numbers as boundary values of *arbitrary* Dedekind cuts without concern as to how these sets of rationals were to be defined. Functions in turn were defined as arbitrary sets of pairs of these real numbers subject only to certain obvious restrictions. This set-theoretic foundation of the Dedekind-Weierstrass construction of the real numbers and analysis presupposed Cantor's general theory of infinite sets, which spawned a realm of mathematics that was even farther removed from the physical sciences, and many mathematicians such as Poincaré, were worried about the extreme generality involved in set-theoretic reasoning, the existence of infinite sets and the hierarchies of infinities. Thinking of the real numbers in terms of a geometric continuum has at least the advantage of being able to appeal to some type of physical model. The abstract notion of an infinite set required for the Dedekind-Weierstrass construction of real numbers, however, does not lend itself to such an intuitive interpretation. Poincaré and others who did not believe in the existence of infinite sets were thus pleased when a

---

[92]Weierstrass' work was preceded by, among others, Bolzano, who proposed the epsilon-delta method fifty years earlier, and Cauchy, who provided a clear definition of limit, which he defined on the basis of infinitesimal quantities. But, Cauchy thought of the infinitesimals as independent variables rather than as infinitesimal, fixed quantities and showed how appeals to infinitesimals and geometric intuitions could ultimately be avoided.

It was only in 1961 that A. Robinson was able to provide an adequate, formal framework for infinitesimals by employing a non-standard model of the theory of real numbers; however, the concept of infinitesimals still remains somewhat controversial.

Leibniz considered the infinitesimals to be useful fictions rather than metaphysical facts. Leibniz was right. The crucial issue is not their material existence or non-existence, but whether it is legitimate to utilize them in proving theorems. Non-standard analysis legitimizes their use. To prove a theorem about standard objects of the standard model, one embeds the standard objects into a non-standard model which contains additional mathematical objects such as infinitesimals. It may then be easier to find a proof using non-standard objects. However, the objects in the enriched non-standard model which correspond to those of the standard model have the property that propositions about them are true in the non-standard model if and only if they are true in the standard model.

series of logical contradictions (particularly Russell's paradox) arose around the turn of the century in the set-theoretic foundations underlying the works of Frege and others.

It was not until Frege, who toward the end of the nineteenth century developed the first comprehensive philosophical system for a foundation of mathematics (usually referred to as logicism) that philosophers and mathematicians became cognizant of the lack of a satisfactory account of the foundations of mathematics. The two main alternatives to logicism, the formalism of Hilbert and the intuitionism of Brouwer and Weyl were subsequent developments in the foundations of mathematics that took place during the early part of this century alongside the Frege-Russell-Whitehead programme.

In 1908, the young Dutchman Luitzen Egbertus Jan Brouwer (1881-1966) dared to challenge the sacrosanct doctrine that the laws of classical logic have an absolute validity independently of the subject matter and proposed drastic measures to end the crisis in the foundations of mathematics due to the discovery of the set-theoretic antinomies around the turn of the century. Brouwer's (1976a) dissertation of 1907 already contained some of his ideas, which were worked out more fully in his article of 1908, entitled "The unreliability of the logical principles" (Brouwer 1976a, pp. 107–111). In this article, he expressed and defended the startling conclusion that the logical principle of the excluded middle does not offer a reliable guide in matters of mathematical thinking. Brouwer gave the name intuitionism to his constructivist conception of mathematics. Although his so called Amsterdam school developed an explicit philosophy of intuitionism, some of the basic intuitionist principles were appealed to and were shared by a minority of important mathematicians such as Kronecker, Poincaré, and Borel.

Brouwer's criticism of Hilbert's belief that each particular mathematical problem can be solved by being either affirmed or refuted is a direct consequence of his attitude toward the principle of the excluded middle. Whereas Frege insisted on the gap between truth and the recognition of truth, Brouwer argued against a verificationist-transcendent conception of truth for mathematical statements; that is, Brouwer rejected a platonistic semantics, according to which all mathematical statements are subject to a conception of truth for which the principle of bivalence holds. The assumption of bivalence for a given mathematical statement is permissible, according to Brouwer, only if it is assured that there exists a verificationist means to recognize it as either true or false. It is not legitimate to assume bivalence merely on the ground that at least one of the two possibilities must be the case; one must have an effective method for finding a proof or disproof to establish which possibility is in fact the case.

Brouwer argued that one must restrict oneself to the consideration of those mathematical objects that in a definite sense can be effectively constructed. Mathematical existence means mental constructibility in principle, and no clear meaning can be given to the existence of mathematical objects independently of their mental construction. To claim that a mathematical object having a certain property exists means that, and is proved when, a method is known that in principle would allow such an object to be constructed.

The admissible means of construction are not pre-determined by a set of rules. A formalization of the constructive processes is rejected. One merely requires that the constructions are intuitively unambiguous. A proof is valid when it is a construction, the individual steps of which are immediately evident.

Infinite sets or sequences must not be viewed as actually complete or as actually infinite. Classical mathematics treated infinite sets as though they could actually be given completely. Brouwer, however, admits only potential infinity. The totality of the natural numbers cannot be considered as an actually complete construction, but only as something that is constructively evolving. But if an infinite set is regarded as only potentially complete, then one cannot constructively decide whether all its elements do or do not have a determinate property. The principle of the excluded middle is therefore strictly inapplicable in the case of infinite sets.

Proponents of logicism and formalism essentially accepted the way in which classical mathematics was done, but they nonetheless saw a need for a foundation for classical mathematics because they regarded the latter to be in need of justification. The intuitionists, on the other hand, considered this alleged need for justification and hence for foundation as evidence that something was fundamentally amiss with the nature of traditional mathematics and mathematical activity. The intuitionists argued that if mathematics were correctly done, it would not require any justification. Rather, what was needed was a *reconstruction* of classical mathematics. According to the intuitionists a correct understanding of the nature of mathematics and mathematical activity would ultimately mean that this reconstruction would have to penetrate as far as propositional logic and bring about a revision of something as basic as our understanding of the propositional operators.

For logicism, this justification was to be achieved directly by means of a reduction of the basic mathematical principles to the principles of logic, that are assumed to be more fundamental and hence more secure. Logicism's main concern was the problem of the nature of mathematical truth. It undertook to solve the epistemological and ontological problems of the philosophy of mathematics by defining all mathematical notions in terms of logical ones and by defining all mathematical theorems from the analytic truths of logic, thus attempting to establish that mathematics deals only with analytic relations among concepts. By merely assuming first-order logic with '$\in$' as the only primitive binary predicate denoting membership and the two set-theoretic axioms of extensionality and abstraction, Frege (and others such as Dedekind) was able to define the sets of the natural and real numbers and was able to formally construct within the system the proofs of the known theorems of arithmetic about these sets and their members. Frege reduced mathematics to logic plus the theory of sets or properties, properties of properties, and so forth. Holding, however, that the two set-theoretic principles of abstraction and extensionality were universally valid logical principles, Frege believed that he had shown conclusively that the truths of mathematics were all universally valid logical principles. It turned out in the course of events that Frege had cause to doubt the universal validity of the principle of abstraction.

In general, logicism neither permits the postulation of special mathematical entities nor recognizes any type of mathematical intuition as a source of synthetic a priori mathematical knowledge. This position stands in sharp contrast with that of Poincaré and the later intuitionists, who accepted the principle of mathematical induction as a given primitive not reducible to logic or anything else. Even Hilbert maintained that mathematics does have an extralogical subject matter, namely expressions, and that its elementary truths are intuitive (*anschaulich*).

For a formalist such as Hilbert, a reduction of mathematics to logic does not go far enough to secure mathematics, for even if such a reduction were successful one would still be left with the question of that logic's consistency. Hilbert proposed, as did the logicists, that classical mathematics be formulated as a formal axiomatic system. Indeed, due to the Russell-Whitehead programme, the formalists inherited at least a clear formulation of what it was that had to be proved consistent. In such a system, signs are manipulated according to certain definite and explicit rules that ignore the interpretations of the signs and deal simply with sequences of marks. There are well-formed formulas (strings of marks satisfying certain conditions as to their shapes and occurrences) and proofs, that is sequences of well formed formulas, each of which is either an axiom or is derivable from previous formulas by a single application of one of the transformation rules. Statements about the formulas and proofs of the system and their relationships are meaningful and are the subject of what Hilbert called proof-theory. The main purpose of proof-theory was to establish the consistency of the formal system in question. Since the statements of metamathematics themselves constitute a deductive system, it was necessary to state the methods of inference that are to be allowed to be used in the metatheory. Hilbert specified that statements of metamathematics are to be proved by the most elementary intuitively obvious methods, called 'finitary' methods. It is not easy to give a precise definition of 'finitary' methods, but it was assumed that these methods would be no stronger than those employed in the system for which a consistency proof was to be devised. In other words, to provide a credible proof of the consistency of a formal theory, the principles used at the meta-level must be more convincing than the principles occurring in the given formal theory being examined for consistency. In sum, Hilbert proposed three separate and distinct theories (Kleene 1967, p. 65):

a) the informal theory of which the formal system constitutes a formalization;
b) the formal system or object theory; and
c) the metatheory, in which the formal system is described and studied.

Hilbert thought that while (c) was restricted to finitary methods and (a) need not be, a consistency proof for (b) whose interpretation is (a) could nevertheless be given by (c)[93].

---

[93]It should be noted that Hilbert was using the term 'consistency' to mean absolute and not

When Weyl (1918a) entered the foundational debate, neither Gödel's incompleteness results nor Hilbert's formalist programme had been established. Already in 1904, however, Hilbert proposed to save classical mathematics and to meet the crisis caused by the paradoxes by suggesting that mathematics be formulated as an axiomatic theory to establish its consistency. Thus, Hilbert's concerted effort in 1922 and thereafter to actually carry out his 1904 proposal is not merely a belated response to meet the crisis caused by the set-theoretic paradoxes, but constitutes in no small measure a direct response to his most famous pupil's challenge to classical mathematics[94]. Hilbert (1922, GA III, pp. 159–160) said:

> Was Weyl und Brouwer tun, kommt im Prinzip darauf hinaus, daß sie die einstigen Pfade von Kronecker wandeln: sie suchen die Mathematik dadurch zu begründen, daß sie alles ihnen unbequem Erscheinende über Bord werfen und eine Verbotsdiktatur à la Kronecker errichten. Dies heißt aber, unsere Wissenschaft zerstückeln und verstümmeln, und wir laufen Gefahr, einen großen Teil unserer wertvollsten Schätze zu verlieren, wenn wir solchen Reformatoren folgen. Weyl und Brouwer verfehmen [sic] die allgemeinen Begriffe der Irrationalzahl, der Funktion, ja schon der zahlentheoretischen Funktion, die Cantorschen Zahlen höherer Zahlklassen usw.; der Satz, daß es unter unendlichvielen ganzen Zahlen stets eine kleinste gibt, und sogar das logische "Tertium non datur" z. B. in der Behauptung: entweder gibt es nur eine endliche Anzahl von Primzahlen oder unendlichviele, sind Beispiele verbotener Sätze und Schlußweisen. Ich glaube, daß, so wenig es Kronecker damals gelang, die Irrationalzahl abzuschaffen ... ebensowenig werden Weyl und Brouwer heute durchdringen; nein: Brouwer ist nicht, wie Weyl meint, die Revolution, sondern nur die Wiederholung eines Putschversuches mit alten Mitteln, der seinerzeit, viel schneidiger unternommen, doch gänzlich mißlang und jetzt zumal, wo die Staatsmacht durch Frege, Dedekind and Cantor so wohl gerüstet und befestigt ist, von vornherein zur Erfolglosigkeit verurteilt ist.
>
> Zusammenfassend möchte ich sagen: Wenn man von einer mathematischen Krise spricht, so darf man jedenfalls nicht, wie es Weyl tut,

---

relative consistency. The latter involves the mapping of a theory T onto another theory T', such that if T' is consistent, then so is T. For example, Descartes' analytical geometry constitutes a consistency proof for geometry relative to the consistency of the theory of real numbers. The proof of absolute consistency for a formalized system consists in the deduction of a metamathematical proposition that makes a statement about all the proofs of the formal theory, and the question of whether contradictory proofs could be derived in the system must be considered by finitary methods which are intuitively convincing.

[94] It was not long after Hilbert had begun his formalist programme in earnest that Brouwer responded by saying "an incorrect theory even if it cannot be hampered by a contradiction is nevertheless incorrect, just as a criminal policy unchecked by a reprimanding court is nonetheless criminal"(Brouwer 1924), (van Heijenoort 1967, pp. 334–345), (Brouwer 1976a, pp. 268–274).

Hilbert answered in 1928: "To take the law of the excluded middle away from the mathematician would be like denying the astronomer the telescope or the boxer the use of his fists"(Hilbert 1926), (van Heijenoort 1967, p. 476).

von einer neuen Krise sprechen. Der Circulus vitiosus ist von Weyl
künstlich in die Analysis hineingetragen. Seine Darstellung der Un-
sicherheit der Resultate der heutigen Analysis entspricht nicht dem
wirklichen Sachverhalt. Und was die von ihm und Brouwer so stark
betonten konstruktiven Tendenzen angeht, so hat eben Weyl meiner
Meinung nach den richtigen Weg zur Realisierung dieser Tendenzen
verfehlt. Erst der hier in Verfolgung der Axiomatik eingeschlagene Weg
wird, wie ich glaube, den konstruktiven Tendenzen, soweit sie natürlich
sind, völlig gerecht.

[What Weyl and Brouwer do is essentially to walk the former paths
of Kronecker: they seek a foundation for mathematics by throwing over-
board everything that seems uncomfortable to them and by erecting a
Kronecker-like, dictatorial prohibition. This means, however, the dissolution and mutilation of our science, and we run the risk of losing a
great part of our most valuable treasures if we follow such reformers.
Weyl and Brouwer outlaw the general concepts of irrational number,
function and even the number-theoretic function, Cantor's transfinite
numbers, and so forth. The theorem that there always exists a smallest
number among the infinitely many whole numbers and even the "Tertium non datur", e.g., the assertion, either there exists only finitely
many prime numbers or there exist infinitely many, are examples of
theorems or inferences that are disallowed. I believe that just as Kronecker failed to eliminate the irrational numbers Weyl and Brouwer
will fail in their present endeavour. No! Brouwer is not the revolution as Weyl thinks, rather, he is merely a repetition of an attempted
putsch using old means. The earlier attempt was much more daringly
undertaken, yet it was completely unsuccessful. Particularly because
the power of the state [of mathematics] is now so completely armed
and fortified as a result of the works of Frege, Dedekind and Cantor,
the present attempt is condemned to failure right from the start.
To sum up I would say: If one speaks of a mathematical crisis, one
may certainly not speak, as Weyl does, of a new crisis. Weyl artificially
brings the circulus vitiosus into analysis. His characterization of the
uncertainty of the results of current analysis does not correspond to the
actual states of affairs. As far as the constructive tendencies so heavily
emphasised by him and Brouwer are concerned, Weyl, in my opinion,
has chosen the wrong path for the realization of these tendencies. Only
the path chosen here in pursuit of axiomatization will, as I believe, do
full justice to the constructive tendencies, as far as they are natural.]

Many years later in his article on Hilbert's mathematical work, Weyl (1944a)
remarked:

I regret that in his opposition to Brouwer, Hilbert never openly acknowledged the profound debt which he, as well as all other mathematicians, owe Brouwer for his revelation.

Hilbert was not willing to make the heavy sacrifices which Brouwer's standpoint demanded, and he saw, at least in outline, a way by which the cruel mutilation could be avoided. At the same time, he was alarmed by signs of wavering loyalty within the ranks of mathematicians some of whom openly sided with Brouwer. My own article on the *Grundlagenkrise* in *Math. Zeit.* vol. 10 (1921), written in the excitement of the first postwar years in Europe, is indicative of the mood. Thus Hilbert returns to the problem of foundations in earnest. He is convinced that complete certainty can be restored without "committing treason to our science." There is anger and determination in his voice when he proposes *"die Grundlagenfragen einfürallemal aus der Welt zu schaffen."*

Of the three rival systems, intuitionism stands out not only by virtue of its overall philosophical orientation and underlying motivation but also because its underlying conceptual framework is still regarded by some as a viable approach to the foundations of mathematics. Both logicism and formalism developed a philosophy of mathematics that depended for its adequacy upon the success of a specific mathematical programme. For Frege, the programme consisted in the construction of mathematical principles from logic, and for Hilbert, it consisted in the development of finitistic consistency proofs for mathematical theories. Consequently, the philosophical frameworks of both logicism and formalism were seriously undermined when it was shown that their respective mathematical programmes were inadequate.

The Fregean programme collapsed due to Russell's discovery of the set-theoretic paradoxes. Russell communicated this contradiction to Frege in 1902 just as the last volume of Frege's foundational work was going to print. Intuitively, the paradox may be stated as follows: If every condition determines a set, then consider the set y determined by the condition $x \notin x$; that is, $y = \{x | x \notin x\}$. Now, does y have itself as an element? By the rules of propositional logic, either it does or it does not. If it does, then $y \in \{x \mid x \notin x\}$ and so y must satisfy the defining condition of the set, that is, $y \notin y$. On the other hand, if $y \notin y$, then y satisfies the defining condition of the set y and hence $y \in y$. Since Frege's system permits the formal deduction of Russell's paradox, it is inconsistent and hence incapable of distinguishing between truth and falsehood. Since everything is provable in such a system, it can hardly serve as a foundation of mathematics. In order to eliminate the set-theoretic paradoxes (the logical paradoxes), Russell (1903) introduced the (simple) theory of types, a version of which is discussed in subsection 6.2. These paradoxes are eliminated because an expression $\phi \in \psi$ is well formed only if ψ is of a type that can contain things of the type ϕ. In particular, $x \in x$ is not well formed.

Hilbert's programme is thought to have suffered a similar fate due to Gödel's second incompleteness theorem[95], according to which every consistency proof of an arithmetical theory requires the use of some principles not belonging to the theory in question. Gödel showed in 1931 that the arithmetic of integers cannot be fully axiomatized within a formal system of the post-Fregean logicist programme of Whitehead and Russell, and that a proof of the meta-mathematical statement "this system is inconsistent" cannot be achieved by using methods that are no stronger than those allowed in the system. It follows that consistency cannot be proved by using weaker finitistic methods. Gödel's results showed that the features that Hilbert expected could be established by the proper use of formalization could not be realized.

6.2 The Simple and Ramified Theory of Types

Some formulations of set theory permit sets which have different types of entities as elements such as the set $\{1, \{1\}, (2,3), \{\{(2,3)\}\}\}$. If the use of the predicate '$\in$' is not restricted and if every predicate P defines a set $B = \{x|P(x)\}$, then one is led to consider the sets of entities $A = \{x|x \in x\}$ and $S = \{x|\neg x \in x\}$. The latter, however, leads to Russell's paradox or contradiction, which is the observation that $S \in S$ if and only if $\neg S \in S$. In order to avoid this paradox and others of a similar nature that we will refer to as the *set-theoretic* paradoxes, Russell (1903, Appendix B) proposed a hierarchy of logical types[96] for individuals and sets. The essential idea is that the types are assigned a level and that an expression of the form '$\phi \in \psi$' is not regarded as well formed unless the level of 'ψ' is greater[97] than the level of 'ϕ'. Thus, Russell's theory of (simple) types eliminated his contradiction by prohibiting the expressions '$x \in x$' and '$\neg x \in x$'.

The theory of (simple) types did not, however, solve all of the difficulties. There were, for example, other paradoxes such as the *liar paradox*. Poincaré (1906) suggested that all of the paradoxes, including the set-theoretic paradoxes were the result of a kind of viciously circular reasoning. Russell (1908) formulated the principle as follows:

Principle 46 (Vicious-Circle Principle) *If, provided a certain collection had a total, it would have members only definable in terms of that total, then the said collection has no total.*
[Russell added the following footnote:]

---

[95]The issue concerning the impact of Gödel's theorems on Hilbert's programme remains controversial. See, for example, (Detlefsen 1986). Also see (Peckhaus 1990) for a thorough historical discussion of Hilbert's programme.

[96]Russell regarded ordered pairs (couples), ordered triples (trios) and so forth as distinct types that could be used in the formation of higher types.

[97]Russell was undecided as to whether or not types had to be *minimal*. The set of all individuals $D^{\{1\}}$ is a set of minimal types and so is the set $D^{\{\{1\}\}}$ of all subsets of individuals. However, a variable that ranges over the union $D^{\{1\}} \cup D^{\{\{1\}\}}$, provided such a union is permitted, is not of a minimal type. Thus, a set such as $\{1, 2, \{1, 2\}\}$ is possible only if non-minimal types are allowed. If only minimal types are permitted, then for an expression '$\phi \in \psi$' to be well-formed, it is necessary but not sufficient that the level of 'ψ' be exactly one greater than the level of 'ϕ'.

When I say that a collection has no total, I mean that statements about all *its members are nonsense . . . .*

Remark 47 Bound variables can be used in the introduction of some new entity only if their range is already determined and therefore does not change with the introduction of the new object. As Church (1976) put it, "to avoid impredicativity the essential restriction is that quantification over any domain (type) must not be allowed to add new members to the domain, as it is held that adding new members changes the meaning of quantification over the domain in such a way that a vicious circle results". And Gödel (1944) remarks that if the entities in question are constructed by us then "there must clearly exist a definition (namely the description of the construction) which does not refer to a totality to which the object belongs, because the construction of a thing can certainly not be based on a totality of things to which the thing itself belongs".

Guided by this principle, Russell (1908) developed a theory of ramified types, a version of which is described below. As will be shown below, Russell's ramified theory of logical types eliminates the very possibility of impredicative definitions. However, it does so at a price; namely, it is impossible to prove many important results of standard analysis within the resulting system of logic. It is for this reason that Weyl rejected the ramified theory of types. His decision strongly influenced the way in which he constructed the reals from the natural numbers in his book *Das Kontinuum* as described below in subsection 6.4.

We shall now present versions of both the *simple* and the *ramified theory of types*. There is no unique formulation of these theories. The formulations presented here are partially inspired by the formulations of Hilbert and Ackermann (1950) and of Church (1976) and by an appreciation of the standard procedure for the construction of the real numbers from the natural numbers N (positive integers).

Remark 48 (Necessity of both Relations and Sets) As far as we know, Hilbert and Ackermann (1950) were the first to propose the 'functional' notation for the type labels of the simple theory of types. Church adopted their notation for the simple theory of types and modified it for the ramified theory of types. We extend these formulations by allowing for both relations and sets, and we apply the resulting formalism for the simple theory of types to the construction of the rationals and the reals in subsection 6.3.

It is also important to note that in order to address the issue of impredicative definition as understood by Poincaré and Weyl, it is *necessary* to employ a formalism that distinguishes between relations and sets. For someone who is a Platonist with respect to the real numbers, a relation with an impredicative *form*, that determines a real number, is not a problem because the relation is used merely to *select* a real number from a pre-existing totality of real numbers. On the other hand, for someone, such as Weyl, who is a constructivist with regard to the real numbers, the use of a relation with an impredicative *form* to determine a real number, in the absence of any known procedure for explicitly defining each and every conceivable real number, may in fact be a definition of that real number. However, such a definition would be impredicative and hence viciously circular.

We consider first the theory of simple of types. A theory of (simple) types is constructed from a base which is a set of individuals. A basic set of relations and operations for the set of individuals is given. Other kinds of entities are constructed by combining Kronecker products, logical relations and the basic relations and operations. Each kind of entity is assigned a type and the types are grouped into sets of types characterized by an integer called 'level'. We begin by defining the labels or types for the various kinds of entities and the level associated with each type.

Definition 49   *1. The type of an individual is 1. By definition, this type is a* **closed** *type. The* **level** *of of the type 1 is zero.*

    2. If α_j for $j \in \{1, 2, \ldots, n\}$ are closed types of whatever level, then

$$\alpha_1, \alpha_2, \ldots, \alpha_n \tag{6.2.1}$$

    is a **list** *type. The level of this list type is the maximum of the levels of the types that appear in the list.*

    3. From a list type (6.2.1), one forms the closed *types $\langle \alpha_1, \alpha_2, \ldots, \alpha_n \rangle$ and $\{\alpha_1, \alpha_2, \ldots, \alpha_n\}$. The level of these closed types is one more than the level of the list type (6.2.1).*

The set of all individuals is of type $\{1\}$ and is denoted by $D^{\{1\}}$. The set $D^{\{1\}}$ is of level 1. The variables that range over $D^{\{1\}}$ are x_j^1 (or simply x_j), and the names of particular individuals are c_j^1 (or simply c_j). The subscripts on variables and constants of a given type serve only as tags that are used to ensure that distinct entities of a given type have distinct names. At this stage, the only closed type is 1; consequently, for the application of provision 2 of definition 49, all of the α_j equal 1. It follows that the list types generated by provision 2 of definition 49 are

$$1 \qquad 1,1 \qquad 1,1,1 \qquad 1,1,1,1 \tag{6.2.2}$$

and so forth. The entities corresponding to these types are individuals, ordered pairs of individuals, ordered triples of individuals, ordered quadruples of individuals, and so forth. The sets of these entities are

$$D^{\{1\}}, \quad D^{\{1,1\}} = D^{\{1\}} \times D^{\{1\}}, \quad D^{\{1,1,1\}} = D^{\{1\}} \times D^{\{1\}} \times D^{\{1\}},$$
$$D^{\{1,1,1,1\}} = D^{\{1\}} \times D^{\{1\}} \times D^{\{1\}} \times D^{\{1\}}, \tag{6.2.3}$$

and so forth. The elements of all of these sets are of level zero, and the sets themselves are all of level 1. It is convenient to introduce abbreviations for the types (6.2.2) such that the type corresponding to an n-tuple of individuals is denoted by n. The variables that range over $D^{\{n\}}$ are x_j^n, and the names of particular elements of $D^{\{n\}}$ are c_j^n. Sometimes it is convenient to regard $D^{\{m+n\}}$ as a Kronecker product $D^{\{m\}} \times D^{\{n\}}$ or $D^{\{l+m+n\}}$ as a Kronecker product $D^{\{l\}} \times D^{\{m\}} \times D^{\{n\}}$, and so forth. Then, one has types such as m, n and l, m, n. The corresponding variables and constants are $x_j^{m,n}, x_j^{l,m,n}$ and $c_j^{m,n}, c_j^{l,m,n}$ respectively. In addition, instead of using a variable $x_j^{m,n}$ or a constant $c_j^{m,n}$, one may use an ordered pair of variables (x_k^m, x_l^n) or of constants (c_k^m, c_l^n) respectively. Of course, there are situations in which mixtures such as (x_k^m, c_l^n) are appropriate. An extreme application of this notation is the use of $(x_1, x_2, \ldots, x_n)$ for a variable with domain $D^{\{n\}}$.

The application of provision 3 of definition 49 yields the closed types

$$
\begin{array}{ccccc}
\langle 1 \rangle & \langle 2 \rangle & \langle 3 \rangle & \cdots & \langle n \rangle & \cdots \\
\{1\} & \{2\} & \{3\} & \cdots & \{n\} & \cdots
\end{array}
\tag{6.2.4}
$$

which are of level one.

For any type α, the set of all entities of that type is denoted by $D^{\{\alpha\}}$. If the level of the type α is greater than zero, we use X_j^α for the variables that run over $D^{\{\alpha\}}$ and R_j^α for the names of particular individuals in $D^{\{\alpha\}}$. An entity $R_j^{\langle \alpha \rangle}$ is a relation defined on $D^{\{\alpha\}}$. If desired, one may write $R_j^{\langle \alpha \rangle} : D^{\{\alpha\}}$ in order to emphasize the domain of definition of the relation; however, it is usually not important to emphasize the domain of definition of a relation unless the domain is a subdomain of $D^{\{\alpha\}}$. In particular, a relation $R_1^{\langle n \rangle}$ defined on $D^{\{n\}}$ defines a set[98]

$$
R_1^{\{n\}} \stackrel{\text{def}}{=} \exists x_1^n R_1^{\langle n \rangle}(x_1^n).
\tag{6.2.5}
$$

In some circumstances, it may be appropriate to regard $R_1^{\{n\}}$ as a subdomain of $D^{\{n\}}$ and to consider relations defined on $R_1^{\{n\}}$. It may be that a relation $R_2^{\langle n \rangle} : R_1^{\{n\}}$ only makes sense if its argument is restricted to the subdomain. The subset of $R_1^{\{n\}}$ defined by the relation $R_2^{\langle n \rangle}$ is given by

$$
R_2^{\{n\}} \stackrel{\text{def}}{=} \exists x_1^n \in R_1^{\{n\}} R_2^{\langle n \rangle}(x_1^n),
\tag{6.2.6}
$$

where the range of the quantifier '∃' has been restricted to the subdomain $R_1^{\{n\}}$. Clearly, if the relations $R_1^{\langle n \rangle}$ and $R_2^{\langle n \rangle}$ are not the universal relations, then $R_2^{\{n\}} \subset R_1^{\{n\}} \subset D^{\{n\}}$.

From the closed type 1 and the closed types listed in (6.2.4), a great many list types may be formed, far too many to permit a systematic listing of them all. We consider some examples as well as their application. The list type

$$
\langle 1 \rangle, \langle 1 \rangle, 3, \{2\}
\tag{6.2.7}
$$

---

[98]The symbol '∃x' is the set-formation quantifier; that is, '∃x' means 'the set of all x such that'. It is interesting to note that Weyl (1946, GA IV, p. 270) used '[x]' as the set-formation quantifier.

which is of level one, leads to the additional closed types

$$\langle\langle 1\rangle, \langle 1\rangle, 3, \{2\}\rangle \quad \text{and} \quad \{\langle 1\rangle, \langle 1\rangle, 3, \{2\}\}, \tag{6.2.8}$$

which are of level 2. A constant $R^{\langle 1\rangle, \langle 1\rangle, 3, \{2\}}$ of type (6.2.7) is a 4-tuple

$$(R_1^{\langle 1\rangle}, R_2^{\langle 1\rangle}, c^3, R_1^{\{2\}}) \tag{6.2.9}$$

which is an element of

$$D^{\{\langle 1\rangle\}} \times D^{\{\langle 1\rangle\}} \times D^{\{3\}} \times D^{\{\{2\}\}} = D^{\{\langle 1\rangle, \langle 1\rangle, 3, \{2\}\}}. \tag{6.2.10}$$

A variable $X_1^{\langle 1\rangle, \langle 1\rangle, 3, \{2\}}$ that ranges over this set may also be expressed as a 4-tuple $(X_1^{\langle 1\rangle}, X_2^{\langle 1\rangle}, x_1^3, X_1^{\{2\}})$. A relation $R^{\langle\langle 1\rangle, \langle 1\rangle, 3, \{2\}\rangle}$ is defined on the set $D^{\{\langle 1\rangle, \langle 1\rangle, 3, \{2\}\}}$ (or possibly a subset of this set). The subset determined by this relation is given by

$$R_1^{\{\langle 1\rangle, \langle 1\rangle, 3, \{2\}\}} \stackrel{\text{def}}{=} \exists X_1^{\langle 1\rangle, \langle 1\rangle, 3, \{2\}} R^{\langle\langle 1\rangle, \langle 1\rangle, 3, \{2\}\rangle}(X_1^{\langle 1\rangle, \langle 1\rangle, 3, \{2\}}). \tag{6.2.11}$$

As another example, consider the simple list type $\langle 1\rangle$ of level one that leads to the closed types $\langle\langle 1\rangle\rangle$ and $\{\langle 1\rangle\}$. A relation $R^{\langle 1\rangle}: D^{\{1\}}$ is an element of $D^{\{\langle 1\rangle\}}$, the set of all monadic relations or the set of all properties of individuals. A relation $R_1^{\langle\langle 1\rangle\rangle}: D^{\{\langle 1\rangle\}}$ is a property of properties, which determines the set[99] of all properties of individuals which have the property $R_1^{\langle\langle 1\rangle\rangle}$, namely,

$$R_1^{\{\langle 1\rangle\}} \stackrel{\text{def}}{=} \exists X_1^{\langle 1\rangle} R_1^{\langle\langle 1\rangle\rangle}(X_1^{\langle 1\rangle}). \tag{6.2.12}$$

One has the relations $R_1^{\{\langle 1\rangle\}} \subseteq D^{\{\langle 1\rangle\}}$, $R_1^{\{\langle 1\rangle\}} \in D^{\{\{\langle 1\rangle\}\}}$ and $R_1^{\langle\langle 1\rangle\rangle} \in D^{\{\langle\langle 1\rangle\rangle\}}$. Similarly, the simple list type $\{1\}$ of level 1 leads to the closed types $\langle\{1\}\rangle$ and $\{\{1\}\}$ of level 2. A relation $R_1^{\langle\{1\}\rangle}$ is defined on $D^{\{\{1\}\}}$ and determines the subset

$$R_1^{\{\{1\}\}} \stackrel{\text{def}}{=} \exists X_1^{\{1\}} R_1^{\langle\{1\}\rangle}(X_1^{\{1\}}) \tag{6.2.13}$$

of $D^{\{\{1\}\}}$. One has the relations $R_1^{\{\{1\}\}} \subseteq D^{\{\{1\}\}}$, $R_1^{\{\{1\}\}} \in D^{\{\{\{1\}\}\}}$ and $R_1^{\langle\{1\}\rangle} \in D^{\{\langle\{1\}\rangle\}}$. From the closed types of levels 2, 1 and 0, one forms the list types of level 2 and hence the closed types of level 3 such as $\{\{\{1\}\}\}, \langle\{\{1\}\}\rangle, \langle\{\langle 1\rangle\}\rangle$ and $\langle\langle\{1\}\rangle\rangle$.

It is useful to consider the use of types from a more general perspective. For any type α, a relation $R_1^{\langle\alpha\rangle}$ is defined on some subdomain[100] of $D^{\{\alpha\}}$. The variables that range over $D^{\{\alpha\}}$ are denoted by X_j^α, and the names of entities that belong to $D^{\{\alpha\}}$ are R_j^α. The subset of $D^{\{\alpha\}}$ determined by $R_1^{\langle\alpha\rangle}: D^{\{\alpha\}}$ is given by

$$R_1^{\{\alpha\}} \stackrel{\text{def}}{=} \exists X_1^\alpha R_1^{\langle\alpha\rangle}(X_1^\alpha). \tag{6.2.14}$$

---

[99] If the relation $R_1^{\langle\langle 1\rangle\rangle}$ is the universal relation on $D^{\{\langle 1\rangle\}}$, then $R_1^{\{\langle 1\rangle\}}$ may be $D^{\{\langle 1\rangle\}}$ itself. For the set relations, we use the sequence of symbols $\subset, \subseteq, =, \supseteq, \supset$.

[100] Of course, $D^{\{\alpha\}}$ itself is a possible subdomain.

Then, $R_1^{\{\alpha\}} \subseteq D^{\{\alpha\}}$, and $R_1^{\alpha} \in R_1^{\{\alpha\}}$. If α is a list type

$$\alpha_1, \alpha_2, \ldots, \alpha_n \qquad (6.2.15)$$

then the set of ordered n-tuples of the given types is

$$D^{\{\alpha_1,\alpha_2,\ldots,\alpha_n\}} = D^{\{\alpha_1\}} \times D^{\{\alpha_2\}} \times \cdots \times D^{\{\alpha_n\}}. \qquad (6.2.16)$$

A relation $R_1^{\langle\alpha_1,\alpha_2,\ldots,\alpha_n\rangle} : D^{\{\alpha_1,\alpha_2,\ldots,\alpha_n\}}$ determines a set

$$R_1^{\{\alpha_1,\alpha_2,\ldots,\alpha_n\}} \stackrel{\text{def}}{=} \exists X^{\alpha_1,\alpha_2,\ldots,\alpha_n} R_1^{\langle\alpha_1,\alpha_2,\ldots,\alpha_n\rangle}(X_1^{\alpha_1,\alpha_2,\ldots,\alpha_n}), \qquad (6.2.17)$$

where $R_1^{\{\alpha_1,\alpha_2,\ldots,\alpha_n\}} \subseteq D^{\{\alpha_1,\alpha_2,\ldots,\alpha_n\}}$.

It is rather cumbersome to discuss the natural numbers, the integers, the rationals and the reals without using functions. We use $\alpha|\beta$ as the type label for a function whose argument is of type β and whose value is of type α. As names for particular functions of type $\alpha|\beta$, we use $F_j^{\alpha|\beta}$. The set of all functions of type $\alpha|\beta$ is denoted by $D^{\{\alpha|\beta\}}$, and the variables that run over this set are $X_j^{\alpha|\beta}$. We now consider the procedures for defining functions in terms of relations.

First, any relation $R_1^{\langle\alpha,\beta\rangle}$ defines a set-valued function $F_1^{\{\alpha\}|\beta} : D^{\{\beta\}} \to D^{\{\{\alpha\}\}}$ according to

$$\forall X_1^{\beta} \left(F_1^{\{\alpha\}|\beta}(X_1^{\beta}) \stackrel{\text{def}}{=} \exists X_1^{\alpha} R_1^{\langle\alpha,\beta\rangle}(X_1^{\alpha}, X_1^{\beta}) \right). \qquad (6.2.18)$$

Of course, it may be desirable to restrict both the domain and range to subsets of $D^{\{\beta\}}$ and $D^{\{\{\alpha\}\}}$ respectively.

To define a function $F_2^{\alpha|\beta} : D^{\{\beta\}} \to D^{\{\alpha\}}$, one needs a relation $R_2^{\langle\alpha,\beta\rangle} : D^{\{\alpha,\beta\}}$ that satisfies the following two constraints,

$$\forall X_1^{\beta} \exists X_1^{\alpha} R_2^{\langle\alpha,\beta\rangle}(X_1^{\alpha}, X_1^{\beta}) \qquad (6.2.19)$$

and

$$\forall X_1^{\alpha} \forall X_2^{\alpha} \forall X_1^{\beta} \left\{ \left[R_2^{\langle\alpha,\beta\rangle}(X_1^{\alpha}, X_1^{\beta}) \wedge R_2^{\langle\alpha,\beta\rangle}(X_2^{\alpha}, X_1^{\beta}) \right] \right.$$
$$\left. \Longrightarrow X_1^{\alpha} = X_2^{\alpha} \right\}. \qquad (6.2.20)$$

The condition (6.2.19) says that for every input there exists an output, and the condition (6.2.20) says that the output is unique. The function $F_2^{\alpha|\beta} : D^{\{\beta\}} \to D^{\{\alpha\}}$ is defined[101] by

$$\forall X_1^{\beta} \left(F_2^{\alpha|\beta}(X_1^{\beta}) \stackrel{\text{def}}{=} \daleth X_1^{\alpha} R_2^{\langle\alpha,\beta\rangle}(X_1^{\alpha}, X_1^{\beta}) \right). \qquad (6.2.21)$$

---

[101] The symbol '$\daleth x$' which means 'the x such that' is the definite-description quantifier.

As a final example of the notation we note that a function $F^{\alpha|\beta}: D^{\{\beta\}} \to D^{\{\alpha\}}$ defines a function $F_{\vdash}^{\{\alpha\}|\{\beta\}}: D^{\{\{\beta\}\}} \to D^{\{\{\alpha\}\}}$, the forward map of subsets induced by $F^{\alpha|\beta}$, which is given by

$$\forall X_1^{\{\beta\}} \left[F_{\vdash}^{\{\alpha\}|\{\beta\}}(X_1^{\{\beta\}}) \stackrel{\text{def}}{=} \exists X_1^{\alpha} \exists X_1^{\beta} \left(X_1^{\beta} \in X_1^{\{\beta\}} \wedge X_1^{\alpha} = F^{\alpha|\beta}(X_1^{\beta}) \right) \right].$$
(6.2.22)

The theory of (simple) types will be applied below in the construction of the positive rationals from the natural numbers (positive integers) and in the construction of the positive reals from the positive rationals. In the context of these constructions, the procedures for defining functions (6.2.18) and (6.2.21) will be illustrated by concrete examples.

Remark 50 In a theory of (simple) types, the use of the membership symbol '$\in$' is restricted. An expression of the form $\phi^{\alpha} \in \psi^{\beta}$ is well formed if and only if ϕ^{α} is an object of type α, ψ^{β} is an object of type β and $\beta = \{\alpha\}$. In addition, an expression of the form $\phi^{\alpha} = \psi^{\beta}$ is well formed if and only if $\alpha = \beta$. Also, there are an infinite number of null sets, one for each type; moreover, each of the sets that is roughly characterized by 'the set of all objects of a given type' is a universal set.

We shall now discuss the so called ramified theory of types. In the definition of (simple) types, the level of a relation that is being defined depends solely on the levels of the types that appear as *independent variables* in the relation. It is clear, however, that constants and/or bound variables may appear in the definition of a relation. If in the definition of a quantity of a given type, a bound variable occurs which has a level equal to or greater than the level of the quantity that is being defined, then the definition is said to be *impredicative*. Impredicativity can also occur if a constant appears in the definition and the definition of the constant involves bound variables of too high a level. As a simple example, consider

$$\forall x_1 \left(R_1^{\langle 1 \rangle}(x_1) \stackrel{\text{def}}{=} \forall X_1^{\{1\}} R_1^{\langle 1,\{1\}\rangle}(x_1, X_1^{\{1\}}) \right).$$
(6.2.23)

The level of $R_1^{\langle 1 \rangle}$ is 1 and its definition involves quantification over $X_1^{\{1\}}$ which is also of level 1. As another example, define the set $R_2^{\{1\}}$ by

$$R_2^{\{1\}} \stackrel{\text{def}}{=} \exists x_1 \forall X_1^{\{1\}} R_2^{\langle 1,\{1\}\rangle}(x_1, X_1^{\{1\}}).$$
(6.2.24)

This definition is also impredicative. One can define a relation $R_3^{\langle 1 \rangle}$ by

$$\forall x_1 \left(R_3^{\langle 1 \rangle}(x_1) \stackrel{\text{def}}{=} R_1^{\langle 1,\{1\}\rangle}(x_1, R_2^{\{1\}}) \right),$$
(6.2.25)

which is implicitly impredicative because the definition (6.2.24) of the constant $R_2^{\{1\}}$ is impredicative.

According to Russell, Poincaré and others, such impredicative definitions involve a vicious circularity. In order to avoid impredicative definitions, Russell introduced the theory of *ramified types* ($\overset{\text{def}}{=}$ *r*-types). Each of the closed types of the simple theory is divided into an infinite number of sublevels. The level and sublevel numbers are combined to form an integer called the *order* O of the *r*-type. In the definition of a quantity of order O, quantification, whether explicit or implicit may occur only over quantities of order lower than O. The recursive definition of *r*-types and their order are as follows:

Definition 51   *1. The r-type of individuals is 1. The* order *of this type is zero. By definition, the r-type 1 is a closed r-type.*

2. *If α_j for $j \in \{1, 2, \ldots, m\}$ are closed r-types of whatever orders, then*

$$\alpha_1, \alpha_2, \ldots, \alpha_m \tag{6.2.26}$$

is a list *r-type. The order of this list r-type is the maximum N of the order of the closed r-types that appear in the list (6.2.26).*

3. *From a list r-type (6.2.26) one forms the closed r-types $\langle \alpha_1, \alpha_2, \ldots, \alpha_m \rangle / n$ and $\{\alpha_1, \alpha_2, \ldots, \alpha_m\}/n$, where $n \in \{1, 2, 3, \ldots\}$. The order of these closed r-types is $N + n$.*

Definition 52 *An r-type*

$$\langle \alpha_1, \alpha_2, \ldots, \alpha_m \rangle / k \tag{6.2.27}$$

respectively

$$\{\alpha_1, \alpha_2, \ldots, \alpha_m\}/k \tag{6.2.28}$$

is said to be directly lower *than the r-type*

$$\langle \beta_1, \beta_2, \ldots, \beta_m \rangle / n \tag{6.2.29}$$

respectively

$$\{\beta_1, \beta_2, \ldots, \beta_m\}/n \tag{6.2.30}$$

if and only if $\alpha_1 = \beta_1, \alpha_2 = \beta_2, \ldots, \alpha_m = \beta_m$ and $k < n$.

We turn now to the question of how the entities corresponding to the various *r*-types are generated. As in the case of (simple) type theory, the constants that refer to individuals are denoted by c_j, where $j \in \{1, 2, 3, \ldots\}$. Again, the range $\{1, 2, 3, \ldots\}$ for a subscript will be assumed in the following. The set of individuals is denoted by $D^{\{1\}/1}$ and the variables that range over this domain are x_j. The variables x_j and the constants c_j are of *order* zero. The set $D^{\{1\}/1}$ is of order one. For ordered *n*-tuples, the constants and variables are denoted by c_j^n and x_j^n respectively. These quantities are also of order zero. The corresponding set $D^{\{n\}/1}$ is of order 1. As in the case of (simple) type theory, one may use an ordered *n*-tuple of individual constants and variables instead of c_j^n and x_j^n; for example, instead of c_1^2 and x_1^3, one may use (c_1, c_2) and (x_1, x_2, x_3) respectively. Of course, mixtures such as (x_1, c_1, c_2, x_2) are also allowed. Next, consider relations $R_j^{\langle n \rangle/1} : D^{\{n\}/1}$ whose definition involve quantification only over individuals; that is, only the quantifiers $\forall x_j$ and $\exists x_j$ appear in the definition. These relations are of order one. The set of all such relations is denoted by $D^{\{\langle n \rangle/1\}/1}$ which is of order 2. The variables $X_j^{\langle n \rangle/1}$ range over the domain $D^{\{\langle n \rangle/1\}/1}$. A relation $R_1^{\langle n \rangle/1} : D^{\{n\}/1}$ determines a subset of $D^{\{n\}/1}$ given by

$$R_1^{\{n\}/1} \stackrel{\text{def}}{=} \exists x_1^n R^{\langle n \rangle/1}(x_1^n), \tag{6.2.31}$$

which is of order one because only quantification over individuals $(\forall x_j, \exists x_j, \exists x_j)$ is involved. In the following, we shall use 'II' as a metavariable which takes values in $\{\forall, \exists\}$. The set of all sets of type (6.2.31) is denoted by $D^{\{\{n\}/1\}/1}$, and the variables that range over this domain are $X_j^{\{n\}/1}$.

For convenience, we use a pair of brackets [ ] as a metasymbol that stands for either a pair of braces { } or a pair of angle brackets ⟨ ⟩. List types can now be formed from the closed types 1 and $[n]/1$; for example, one can have tuples such as

$$R_1^{2, \langle 1 \rangle/1, \{2\}/1, \langle 1 \rangle/1} = \left(c_1^2, R_1^{\langle 1 \rangle/1}, R_1^{\{2\}/1}, R_2^{\langle 1 \rangle/1} \right). \tag{6.2.32}$$

If one allows for repetitions, such tuples can be indicated by $R_1^{m, [n]/1}$. The set of all such tuples is

$$D^{\{m, [n]/1\}/1} = D^{\{m\}/1} \times D^{\{[n]/1\}/1}, \tag{6.2.33}$$

and the variables $X_j^{m, [n]/1}$ range over this domain. A relation

$$R_1^{\langle m, [n]/1 \rangle/1} : D^{\{m, [n]/1\}/1} \tag{6.2.34}$$

determines a set

$$R_1^{\{m, [n]/1\}/1} \stackrel{\text{def}}{=} \exists X_1^{m, [n]/1} R^{\langle m, [n]/1 \rangle/1} \left(X_1^{m, [n]/1} \right), \tag{6.2.35}$$

which is a subset of $D^{\{m, [n]/1\}/1}$ and an element of $D^{\{\{m, [n]/1\}/1\}/1}$. Both the relation $R_1^{\langle m, [n]/1 \rangle/1}$ and the set $R_1^{\{m, [n]/1\}/1}$ are of order two.

Next, define a relation on the set of m-tuples by

$$\forall x_1^m \left(R_1^{\langle m \rangle/2}(x_1^m) \stackrel{\text{def}}{=} IIX_1^{[n]/1} R_1^{\langle m, [n]/1 \rangle/1}(x_1^m, X_1^{[n]/1}) \right). \tag{6.2.36}$$

The relation $R^{\langle m \rangle/2}$ is of order two because quantification $IIX_1^{[n]/1}$ over entities of order one occurs in its definition. The set of such relations is denoted by $D^{\{\langle m \rangle/2\}/1}$. The relation $R_1^{\langle m \rangle/2}$ defines the set

$$R_1^{\{m\}/2} \stackrel{\text{def}}{=} \exists x_1^m R_1^{\langle m \rangle/2}(x_1^m), \tag{6.2.37}$$

which is a subset of $D^{\{m\}/2}$ and an element of $D^{\{\{m\}/2\}/1}$. At this stage, one can form tuples of order two such as

$$R_1^{l,[m]/1,[n]/2,[r,[s]/1]/1} \tag{6.2.38}$$

which are elements of

$$D^{\{l,[m]/1,[n]/2,[r,[s]/1]/1\}/1}, \tag{6.2.39}$$

a set of order three. A relation

$$R_1^{\langle l,[m]/1,[n]/2,[r,[s]/1]/1\rangle/1}, \tag{6.2.40}$$

which is of order three, is an element of

$$D^{\{\langle l,[m]/1,[n]/2,[r,[s]/1]/1\rangle/1\}/1}, \tag{6.2.41}$$

a set of order four. The relation (6.2.40) determines a set

$$R_1^{\{l,[m]/1,[n]/2,[r,[s]/1]/1\}/1} \tag{6.2.42}$$
$$\stackrel{\text{def}}{=} \exists X_1^{l,[m]/1,[n]/2,[r,[s]/1]/1} R_1^{\langle l,[m]/1,[n]/2,[r,[s]/1]/1\rangle/1} \left(X_1^{l,[m]/1,[n]/2,[r,[s]/1]/1} \right),$$

which is a subset of (6.2.39) and an element of $D^{\{\{l,[m]/1,[n]/2,[r,[s]/1]/1\}/1\}/1}$, a set of order four.

By suitably quantifying over the second-order and/or first-order variables of (6.2.40), one can define relations of the kind $R_1^{\langle l,[m]/1\rangle/2}$ and $R_1^{\langle l\rangle/3}$ which are elements of $D^{\{\langle l,[m]/1\rangle/2\}/1}$ and $D^{\{\langle l\rangle/3\}/1}$ respectively. The corresponding sets $R_1^{\{l,[m]/1\}/2}$ and $R_1^{\{l\}/3}$ are elements of $D^{\{\{l,[m]/1\}/2\}/1}$ and $D^{\{\{l\}/3\}/1}$ respectively. Now, one can form tuples of entities of order less than or equal to three and so forth.

The important point to note is that entities of a similar kind belong to different sets; for example, the sets $R_1^{\{l\}/1}$, $R_1^{\{l\}/2}$ and $R_1^{\{l\}/3}$ are elements of $D^{\{\{l\}/1\}/1}, D^{\{\{l\}/2\}/1}$ and $D^{\{\{l\}/3\}/1}$ respectively. Similarly, the sets $R_1^{\{l,[m]/1\}/1}$ and $R_1^{\{l,[m]/1\}/2}$ are elements of $D^{\{\{l,[m]/1\}/1\}/1}$ and $D^{\{\{l,[m]/1\}/2\}/1}$ respectively.

Remark 53 Just as in the case of (simple) types (see remark 50), the use of the symbols '$\in$' and '$=$' is restricted in the case of r-types. On a strict interpretation, the rules would be precisely the same in the context of r-type theory except that 'type' would be replaced by 'r-type'; however, such strict rules are extremely inhibiting (perhaps even prohibitive), and more relaxed systems of rules have been proposed. One such relaxation of the rules is the notion (Church 1976, p. 748) of the cumulative range of a variable. As noted above, one encounters in the theory of r-types sequences of sets of the form

$$D^{\{[\alpha]/1\}/1}, D^{\{[\alpha]/2\}/1}, D^{\{[\alpha]/3\}/1}, \ldots. \tag{6.2.43}$$

To say the least, it is rather unnatural for entities of such similar kind to belong to different sets; moreover, one cannot talk about the union of the sets in (6.2.43). In order to partially overcome this constraint, it was proposed by Church that the range of a variable be cumulative in the sense that it includes the range of every variable of directly lower type; for example, $X_1^{[\alpha]/3}$ ranges not only over $D^{\{[\alpha]/3\}/1}$ but also over $D^{\{[\alpha]/2\}/1}$ and $D^{\{[\alpha]/1\}/1}$. This relaxed rule permits one to write '$X_1^{[\alpha]/3} \in D^{\{[\alpha]/1\}/1}$' and '$X_1^{[\alpha]/3} \in D^{\{[\alpha]/2\}/1}$' as well as '$X_1^{[\alpha]/3} \in D^{\{[\alpha]/3\}/1}$'. One can in effect talk about the union of the sets (6.2.43). The cumulative rule also extends to variables that are of list r-types; for example, the variable $(X_1^{\langle\alpha\rangle/3}, X_1^{\{\beta\}/2})$ ranges over the set

$$\left(D^{\{\langle\alpha\rangle/1\}/1} \cup D^{\{\langle\alpha\rangle/2\}/1} \cup D^{\{\langle\alpha\rangle/3\}/1}\right) \times \left(D^{\{\{\beta\}/1\}/1} \cup D^{\{\{\beta\}/2\}/1}\right). \quad (6.2.44)$$

6.3   Definability, Impredicativity and the Construction of the Reals

Examples (6.2.23)–(6.2.25) of impredicative definitions in a theory of (simple) types were discussed above in subsection 6.2. In addition, it was shown how the theory of ramified types prevents impredicative definition by forcing a defined quantity to be of a higher order than the order of any quantified variable that appears either explicitly or implicitly in the definiens.

Remark 54 Impredicative specification of an entity is not necessarily viciously circular. It is permissible to *select* impredicatively a member of a totality each member of which either has already been defined or or is assumed to exist. However, if all of the members of a putative totality have not already been defined, then an impredicative specification of an element of that totality is not permissible because it may be a *definition* rather than a *selection*, and impredicative *definitions* are viciously circular.

The constructions presented below serve to illustrate the notion of a *defined totality*. In particular, the positive rationals are constructively defined in terms of the natural numbers. In contrast, the positive reals are really only characterized rather than defined in terms of the positive rationals. In this context, it is explained that the impredicative specification of the least upper bound of a set of reals is really an *impredicative definition* and is therefore viciously circular. It is important to emphasize that this particular impredicative definition cannot be avoided on the basis of the theory of simple types. Because Weyl held a conceptualistic view of sets, he was particularly concerned in *Das Kontinuum* with avoiding impredicative definitions. He also thought that the theory of ramified types was not a workable solution of the problem of impredicative definitions. These concerns shaped his construction of the reals from the natural numbers. The following constructions also serve to contrast the usual construction of the reals from the natural numbers with Weyl's construction.

Using the language of (simple) types described in subsection 6.2, we present two constructions; the construction of the positive rationals, which we denote by $\hat{Q}$, from the natural numbers (positive integers) N, and the construction of the positive reals, which we denote by $\hat{R}$, from the positive rationals $\hat{Q}$, where a real is taken to be a *Dedekind Cut*. The sets of individuals for these constructions are $D^{\{1\}} = N^{\{1\}}$ and $D^{\{1\}} = \hat{Q}^{\{1\}}$ respectively. In each case, the following conventions are used for constants and variables of the indicated levels:

	Constants	Variables	
Level 0	a, b, c	x, y, z	
Level 1	A, B, C	X, Y, Z	,
Level 2	$\mathcal{A}, \mathcal{B}, \mathcal{C}$	$\mathcal{X}, \mathcal{Y}, \mathcal{Z}$	
Level 3	Δ, Γ, Λ	Φ, Ψ, Ω	

$$(6.3.1)$$

where subscripts may also be used to distinguish additional constants and variables.

In the construction of the positive rationals, the set of natural numbers $N^{\{1\}}(= D^{\{1\}})$ and the basic operations of addition and multiplication as well as the basic relations of equality '=' and order '<' are assumed to be given for the natural numbers. A positive rational will be an equivalence class of ordered pairs, that is, an element of $N^{\{\{2\}\}}$. These equivalence classes are defined by a relation $E^{\langle 2,2 \rangle}$ defined on $N^{\{2\}} \times N^{\{2\}} = N^{\{2,2\}}$. Roughly, the equivalence class $\overline{(a_1, a_2)}$ (a positive rational) that contains the ordered pair (a_1, a_2) corresponds to the (positive) fraction $\frac{a_1}{a_2}$. The relation $E^{\langle 2,2 \rangle}$ is defined by

$$\forall(x_1, x_2)\forall(y_1, y_2) \left(E^{\langle 2,2 \rangle}((x_1, x_2), (y_1, y_2)) \stackrel{\text{def}}{=} (x_1 y_2 = x_2 y_1) \right). \qquad (6.3.2)$$

This definition of $E^{\langle 2,2 \rangle}$ may be motivated by starting with the equality of fractions,

$$\frac{x_1}{x_2} = \frac{y_1}{y_2}, \qquad (6.3.3)$$

and cross multiplying. The property of being an equivalence relation on $D^{\{2\}}$ is defined by a relation $\mathcal{E}^{\langle\langle 2,2 \rangle\rangle} : N^{\{\{2,2\}\}}$ defined by

$$\forall X^{\langle 2,2 \rangle} \left\{ \mathcal{E}^{\langle\langle 2,2 \rangle\rangle}(X^{\langle 2,2 \rangle}) \stackrel{\text{def}}{=} \left[\forall(x_1, x_2) X^{\langle 2,2 \rangle}((x_1, x_2), (x_1, x_2)) \right. \right.$$

$$\wedge \forall(x_1, x_2)\forall(y_1, y_2) \left(X^{\langle 2,2 \rangle}((x_1, x_2), (y_1, y_2)) \iff X^{\langle 2,2 \rangle}((y_1, y_2), (x_1, x_2)) \right)$$

$$\wedge \forall(x_1, x_2)\forall(y_1, y_2)\forall(z_1, z_2) \left(\left(X^{\langle 2,2 \rangle}((x_1, x_2), (y_1, y_2)) \right. \right.$$

$$\left. \left. \left. \wedge X^{\langle 2,2 \rangle}((y_1, y_2), (z_1, z_2)) \right) \implies X^{\langle 2,2 \rangle}((x_1, x_2), (z_1, z_2)) \right) \right] \right\}. \qquad (6.3.4)$$

Thus, the relation $E^{\langle 2,2 \rangle}$ given by (6.3.2) is an equivalence relation on $D^{\{2\}}$ just in case $\mathcal{E}^{\langle\langle 2,2 \rangle\rangle}(E^{\langle 2,2 \rangle})$.

Define the set-valued function $R^{\{2\}|2}\colon D^{\{2\}} \to D^{\{\{2\}\}}$ by

$$\forall (x_1,x_2)\left\{ R^{\{2\}|2}(x_1,x_2) \overset{\text{def}}{=} \exists (y_1,y_2)E^{\langle 2,2\rangle}((y_1,y_2),(x_1,x_2))\right\}. \qquad (6.3.5)$$

Then, the equivalence class of ordered pairs that contains the ordered pair (a_1,a_2) is $R^{\{2\}|2}(a_1,a_2)$, which was informally denoted above by $\overline{(a_1,a_2)}$. Note that $R^{\{2\}|2}(a_1,a_2) \in D^{\{\{2\}\}}$. Define the relation $\hat{\mathbf{Q}}^{\langle\{2\}\rangle}\colon D^{\{\{2\}\}}$ by

$$\forall X^{\{2\}}\left(\hat{\mathbf{Q}}^{\langle\{2\}\rangle}(X^{\{2\}}) \overset{\text{def}}{=} \exists(x_1,x_2)\left(X^{\{2\}} = R^{\{2\}|2}(x_1,x_2)\right)\right). \qquad (6.3.6)$$

Then, an element $A^{\{2\}} \in \mathbf{N}^{\{\{2\}\}}$ is a positive rational just in case $\hat{\mathbf{Q}}^{\langle\{2\}\rangle}(A^{\{2\}})$; moreover, the set of positive rationals $\hat{\mathbf{Q}}^{\{\{2\}\}}$ is given by

$$\hat{\mathbf{Q}}^{\{\{2\}\}} \overset{\text{def}}{=} \exists X^{\{2\}}\hat{\mathbf{Q}}^{\langle\{2\}\rangle}(X^{\{2\}}). \qquad (6.3.7)$$

Clearly, $\hat{\mathbf{Q}}^{\{\{2\}\}} \subset \mathbf{N}^{\{\{2\}\}}$, where '$\subset$' is defined in the usual way[102] in terms of '$\in$'.

It is important to note that the set $\hat{\mathbf{Q}}^{\{\{2\}\}}$ is *explicitly* defined. An explicit set of labels for its elements is provided, namely, the elements of $\mathbf{N}^{\{2\}}$; moreover,

$$R^{\{2\}|2}(a_1,a_2) = R^{\{2\}|2}(b_1,b_2) \qquad (6.3.8)$$

just in case

$$a_1 b_2 = a_2 b_1. \qquad (6.3.9)$$

In addition, the criterion for membership is equally explicit. Thus,

$$(a_1,a_2) \in R^{\{2\}|2}(b_1,b_2) \qquad (6.3.10)$$

just in case (6.3.9) holds.

The order relation on the positive rationals is defined in terms of the order relation on $\mathbf{N}^{\{1\}}$ by

$$\forall X^{\{2\}} \in \hat{\mathbf{Q}}^{\{\{2\}\}}\forall Y^{\{2\}} \in \hat{\mathbf{Q}}^{\{\{2\}\}}\left\{ \mathcal{L}^{\langle\{2\},\{2\}\rangle}\left(X^{\{2\}},Y^{\{2\}}\right)\right.$$
$$\overset{\text{def}}{=} \exists(x_1,x_2)\exists(y_1,y_2)\left[(x_1,x_2) \in X^{\{2\}} \wedge (y_1,y_2) \in Y^{\{2\}}\right.$$
$$\left.\left.\wedge (x_1 y_2 < x_2 y_1)\right]\right\}. \qquad (6.3.11)$$

Just as one allows expressions of the form '$\phi = \psi$' provided that ϕ and ψ are of the same type, one may write '$X^{\{2\}} < Y^{\{2\}}$' instead of '$\mathcal{L}^{\langle\{2\},\{2\}\rangle}\left(X^{\{2\}},Y^{\{2\}}\right)$' provided that $X^{\{2\}}$ and $Y^{\{2\}}$ are positive rationals.

We turn now to the definition of the arithmetic operations on the positive rationals. First, consider the operation of addition which is a function

$$\mathcal{A}^{\{2\}|\{2\},\{2\}}\colon \hat{\mathbf{Q}}^{\{\{2\}\}} \times \hat{\mathbf{Q}}^{\{\{2\}\}} \to \hat{\mathbf{Q}}^{\{\{2\}\}}. \qquad (6.3.12)$$

---

[102]Recall that we use '$\subset$' for 'proper subset' and '$\subseteq$' for 'subset or equal'.

Since the set-formation quantifier '∃' is used to form the output set, the function is determined by the relation $\mathcal{A}^{\langle 2, \{2\}, \{2\}\rangle}: D^{\{2, \{2\}, \{2\}\}}$ defined by

$$\forall (z_1, z_2)\forall X^{\{2\}} \in \hat{\mathbf{Q}}^{\{\{2\}\}}\forall Y^{\{2\}} \in \hat{\mathbf{Q}}^{\{\{2\}\}} \left\{ \mathcal{A}^{\langle 2, \{2\}, \{2\}\rangle} \left((z_1, z_2), X^{\{2\}}, Y^{\{2\}}\right) \right.$$

$$\stackrel{\text{def}}{=} \exists (x_1, x_2)\exists (y_1, y_2) \left[(x_1, x_2) \in X^{\{2\}} \wedge (y_1, y_2) \in Y^{\{2\}} \right.$$

$$\left.\left. \wedge\, (z_1 = x_1 y_2 + x_2 y_1) \wedge (z_2 = x_2 y_2) \right] \right\}. \quad (6.3.13)$$

The function $\mathcal{A}^{\{2\}|\{2\},\{2\}}$ is then defined by

$$\forall X^{\{2\}} \in \hat{\mathbf{Q}}^{\{\{2\}\}}\forall Y^{\{2\}} \in \hat{\mathbf{Q}}^{\{\{2\}\}} \left\{ \mathcal{A}^{\{2\}|\{2\},\{2\}} \left(X^{\{2\}}, Y^{\{2\}}\right) \right.$$

$$\left. \stackrel{\text{def}}{=} \rfloor(z_1, z_2)\mathcal{A}^{\langle 2, \{2\}, \{2\}\rangle} \left((z_1, z_2), X^{\{2\}}, Y^{\{2\}}\right) \right\}. \quad (6.3.14)$$

Of course, one has to show that the output set is indeed a rational; that is, one has to show that

$$\forall X^{\{2\}} \in \hat{\mathbf{Q}}^{\{\{2\}\}}\forall Y^{\{2\}} \in \hat{\mathbf{Q}}^{\{\{2\}\}} \left\{ \hat{\mathbf{Q}}^{\langle\{2\}\rangle} \left(\mathcal{A}^{\{2\}|\{2\},\{2\}} \left(X^{\{2\}}, Y^{\{2\}}\right)\right) \right\}. \quad (6.3.15)$$

The functions for multiplication $\mathcal{M}^{\{2\}|\{2\},\{2\}}$ and division $\mathcal{D}^{\{2\}|\{2\},\{2\}}$ are similarly defined in terms of relations $\mathcal{M}^{\langle 2, \{2\}, \{2\}\rangle}$ and $\mathcal{D}^{\langle 2, \{2\}, \{2\}\rangle}$ respectively. Moreover, in the definition of these relations, only the expressions for z_1 and z_2 change. For $\mathcal{M}^{\langle 2, \{2\}, \{2\}\rangle}$,

$$z_1 = x_1 y_1 \quad \text{and} \quad z_2 = x_2 y_2, \quad (6.3.16)$$

and for $\mathcal{D}^{\langle 2, \{2\}, \{2\}\rangle}$,

$$z_1 = x_1 y_2 \quad \text{and} \quad z_2 = x_2 y_1, \quad (6.3.17)$$

where the formulas (6.3.17) are appropriate for the case $X^{\{2\}}$ divided by $Y^{\{2\}}$.

For the construction of the positive reals from the positive rationals, we take the set of individuals to be $\hat{\mathbf{Q}}^{\{1\}} (\stackrel{\text{def}}{=} D^{\{1\}})$, and assume that the basic operations of addition, multiplication and division as well as the relations of equality '=' and order '<' on $\hat{\mathbf{Q}}^{\{1\}}$ are given. A positive real will be a *Dedekind Cut* in $\hat{\mathbf{Q}}^{\{1\}}$, which is a subset of $\hat{\mathbf{Q}}^{\{1\}}$ that satisfies the conditions stated below. In general, a subset $A^{\{1\}}$ of $\hat{\mathbf{Q}}^{\{1\}}$ is defined by a relation $A^{\langle 1\rangle}: \hat{\mathbf{Q}}^{\{1\}}$:

$$A^{\{1\}} \stackrel{\text{def}}{=} \rfloor x A^{\langle 1\rangle}(x). \quad (6.3.18)$$

The property of 'being a cut' is defined by a relation $\mathcal{C}^{\langle\{1\}\rangle}: \hat{\mathbf{Q}}^{\{\{1\}\}}$ defined by

$$\forall X^{\{1\}} \left\{ \mathcal{C}^{\langle\{1\}\rangle}(X^{\{1\}}) \stackrel{\text{def}}{=} \right.$$

$$\left[\left(\exists x(x \in X^{\{1\}}) \wedge \exists x \neg (x \in X^{\{1\}})\right) \right.$$

$$\wedge \forall x \left(x \in X^{\{1\}} \Longrightarrow \exists y(x < y \wedge y \in X^{\{1\}})\right)$$

$$\left.\left. \wedge \forall x \left(x \in X^{\{1\}} \Longrightarrow \forall y(y < x \Longrightarrow y \in X^{\{1\}})\right) \right] \right\}. \quad (6.3.19)$$

Thus, a subset $A^{\{1\}} \subset D^{\{1\}}$ is a positive real just in case $\mathcal{C}^{\langle\langle\{1\}\rangle\rangle}(A^{\{1\}})$. The set of positive reals is given by

$$\mathbf{R}^{\{\{1\}\}} \stackrel{\text{def}}{=} \exists X^{\{1\}} \mathcal{C}^{\langle\langle\{1\}\rangle\rangle}(X^{\{1\}}). \tag{6.3.20}$$

Superficially, the definitions (6.3.20) and (6.3.7) are similar; however, there is an enormous difference. The set $\hat{\mathbf{Q}}^{\{\{2\}\}}$ is defined in terms of the relation $\hat{\mathbf{Q}}^{\{\{2\}\rangle}$ by definition (6.3.6) which in turn is defined in terms of the function $R^{\{2\}|2}$ given by (6.3.5), and this function provides a *constructive means to specify* (define) every possible positive rational in the explicit sense indicated above in the paragraph containing equation (6.3.8). In contrast, the relation $\mathcal{C}^{\langle\langle\{1\}\rangle\rangle}$ provides a means of testing any given subset of $\hat{\mathbf{Q}}^{\{1\}}$ for admissibility as an element of $\hat{\mathbf{R}}^{\{\{1\}\}}$, but it does not define any of the elements of $\hat{\mathbf{R}}^{\{\{1\}\}}$; that is, it does not provide a means to determine the membership of even a single cut.

For concreteness, we give a few examples of subsets of $\hat{\mathbf{Q}}^{\{1\}}$ that are cuts. Every element of $\hat{\mathbf{Q}}^{\{1\}}$ determines a cut. Define the function $R^{\{1\}|1}: \hat{\mathbf{Q}}^{\{1\}} \to \hat{\mathbf{Q}}^{\{\{1\}\}}$ by

$$\forall x_2 \left(R^{\{1\}|1}(x_2) \stackrel{\text{def}}{=} \exists x_1 (x_1 < x_2) \right). \tag{6.3.21}$$

Then, for any a, $\mathcal{C}^{\langle\langle\{1\}\rangle\rangle}(R^{\{1\}|1}(a))$, that is, $R^{\{1\}|1}(a)$ is a cut and, therefore, an element of $\hat{\mathbf{R}}^{\{\{1\}\}}$, namely, the positive real corresponding to the positive rational a. The cuts corresponding to the positive square roots of positive rationals are determined by the function $S^{\{1\}|1}: \hat{\mathbf{Q}}^{\{1\}} \to \hat{\mathbf{Q}}^{\{\{1\}\}}$ defined by

$$\forall x_2 \left(S^{\{1\}|1}(x_2) \stackrel{\text{def}}{=} \exists x_1 (x_1 x_1 < x_2) \right); \tag{6.3.22}$$

for example $\sqrt{2}$ is represented by the cut $S^{\{1\}|1}(2)$. These cuts are well defined because in each case an explicit formula for determining membership is given. The difference between the set of rationals $\hat{\mathbf{Q}}^{\{\{2\}\}} \subset \mathbf{N}^{\{\{2\}\}}$ and the set of reals $\hat{\mathbf{R}}^{\{\{1\}\}} \subset \hat{\mathbf{Q}}^{\{\{1\}\}}$ is that *each and every* element of the set of rationals $\hat{\mathbf{Q}}^{\{\{2\}\}}$ is well defined while only some elements of the set of reals $\hat{\mathbf{R}}^{\{\{1\}\}}$ are well defined. Later, when the least upper bound of a set of positive real numbers is considered, quantification over $\hat{\mathbf{R}}^{\{\{1\}\}}$ is involved in the definition of the positive real number that is the least upper bound. If all of the elements of $\hat{\mathbf{R}}^{\{\{1\}\}}$ were *already* well defined, then the definition would merely select a positive real number from a *predefined* totality; however, if all of the members of $\hat{\mathbf{R}}^{\{\{1\}\}}$ are not predefined, the definition of the least upper bound might 'create' a new positive real number and that in turn changes the very domain over which quantification occurs. Such a definition is *impredicative* in character.

The order relation $\mathcal{L}^{\langle\langle\{1\},\{1\}\rangle\rangle}$ on $\hat{\mathbf{R}}^{\{\{1\}\}}$ is defined by

$$\forall X^{\{1\}} \in \hat{\mathbf{R}}^{\{\{1\}\}} \forall Y^{\{1\}} \in \hat{\mathbf{R}}^{\{\{1\}\}} \left\{ \mathcal{L}^{\langle\langle\{1\},\{1\}\rangle\rangle} \left(X^{\{1\}}, Y^{\{1\}} \right) \right.$$

$$\left. \stackrel{\text{def}}{=} \left[\forall x \left(x \in X^{\{1\}} \Longrightarrow x \in Y^{\{1\}} \right) \wedge \exists x \left(\neg x \in X^{\{1\}} \wedge x \in Y^{\{1\}} \right) \right] \right\}. \tag{6.3.23}$$

For convenience, one may write '$A^{\{1\}} < B^{\{1\}}$' instead of $\mathcal{L}^{\langle\{1\},\{1\}\rangle}\left(A^{\{1\}}, B^{\{1\}}\right)$ provided that $A^{\{1\}} \in \hat{\mathbf{R}}^{\{\{1\}\}}$ and $B^{\{1\}} \in \hat{\mathbf{R}}^{\{\{1\}\}}$. Intuitively, $A^{\{1\}} < B^{\{1\}}$ just in case $A^{\{1\}} \subset B^{\{1\}}$, where '$\subset$' means 'proper subset'.

We now define the operations of addition, multiplication and division for the positive reals in terms of the corresponding operations for the positive rationals. Addition is a function $\mathcal{A}^{\{1\}|\{1\},\{1\}} : \hat{\mathbf{R}}^{\{\{1\}\}} \times \hat{\mathbf{R}}^{\{\{1\}\}} \to \hat{\mathbf{R}}^{\{\{1\}\}}$. Since the quantifier '$\exists$' is used to form the output set, the function is determined by the relation $\mathcal{A}^{\langle 1,\{1\},\{1\}\rangle}$ defined by

$$\forall z \forall X^{\{1\}} \in \hat{\mathbf{R}}^{\{\{1\}\}} \forall Y^{\{1\}} \in \hat{\mathbf{R}}^{\{\{1\}\}} \left\{ \mathcal{A}^{\langle 1,\{1\},\{1\}\rangle}\left(z, X^{\{1\}}, Y^{\{1\}}\right) \right.$$
$$\left. \overset{\text{def}}{=} \exists x \exists y \left[x \in X^{\{1\}} \wedge y \in Y^{\{1\}} \wedge (z = x+y)\right]\right\}. \qquad (6.3.24)$$

Then, the function $\mathcal{A}^{\{1\}|\{1\},\{1\}}$ is defined by

$$\forall X^{\{1\}} \in \hat{\mathbf{R}}^{\{\{1\}\}} \forall Y^{\{1\}} \in \hat{\mathbf{R}}^{\{\{1\}\}} \left\{ \mathcal{A}^{\{1\}|\{1\},\{1\}}\left(X^{\{1\}}, Y^{\{1\}}\right) \right.$$
$$\left. \overset{\text{def}}{=} \exists z \mathcal{A}^{\langle 1,\{1\},\{1\}\rangle}\left(z, X^{\{1\}}, Y^{\{1\}}\right)\right\}. \qquad (6.3.25)$$

Of course, one has to show that the output set is indeed a positive real; that is, one has to show that

$$\forall X^{\{1\}} \in \hat{\mathbf{R}}^{\{\{1\}\}} \forall Y^{\{1\}} \in \hat{\mathbf{R}}^{\{\{1\}\}} \left\{ \mathcal{C}^{\langle\{1\}\rangle}\left(\mathcal{A}^{\{1\}|\{1\},\{1\}}\left(X^{\{1\}}, Y^{\{1\}}\right)\right)\right\}. \qquad (6.3.26)$$

The functions multiplication $\mathcal{M}^{\{1\}|\{1\},\{1\}}$ and division $\mathcal{D}^{\{1\}|\{1\},\{1\}}$ are similarly defined in terms of relations $\mathcal{M}^{\langle 1,\{1\},\{1\}\rangle}$ and $\mathcal{D}^{\langle 1,\{1\},\{1\}\rangle}$ respectively. Moreover, in the definition of these relations, only the expression for z changes. For $\mathcal{M}^{\langle 1,\{1\},\{1\}\rangle}$,

$$z = xy \qquad (6.3.27)$$

and for $\mathcal{D}^{\langle 1,\{1\},\{1\}\rangle}$,

$$z = x/y, \qquad (6.3.28)$$

where the formula (6.3.28) is appropriate for the division of $X^{\{1\}}$ by $Y^{\{1\}}$. Note that the expressions (6.3.27) and (6.3.28) are as simple as they are because the *positive* reals are being constructed from the *positive* rationals. For the construction of the reals from the rationals, the definitions of multiplication and division each involve the consideration of a number of cases that depend on the signs of the arguments of the functions.

Next, we consider the problem of defining the least upper bound of a set of reals. First, a set of reals is defined by a relation[103] $\mathcal{A}^{\langle\{1\}\rangle} : \hat{\mathbf{R}}^{\{\{1\}\}}$, that is, by a

---

[103]We use $\mathcal{A}^{\langle\{1\}\rangle}$ to denote the relation that determines a set of reals $\mathcal{A}^{\{\{1\}\}}$ because Weyl used $\mathcal{A}$ for the analogous quantity in his work. There should be no confusion with the relation $\mathcal{A}^{\langle 1,\{1\},\{1\}\rangle}$ that defines the addition function $\mathcal{A}^{\{1\}|\{1\},\{1\}}$ because the type labels distinguish the various entities.

relation defined on the subdomain $\hat{\mathbf{R}}^{\{\{1\}\}}$ of $\hat{\mathbf{Q}}^{\{\{1\}\}}$. The set of reals $\mathcal{A}^{\{\{1\}\}} \subseteq \hat{\mathbf{R}}^{\{\{1\}\}}$ itself is given by

$$\mathcal{A}^{\{\{1\}\}} \overset{\text{def}}{=} \exists X^{\{1\}} \in \hat{\mathbf{R}}^{\{\{1\}\}} \mathcal{A}^{\langle\langle 1\rangle\rangle}(X^{\{1\}}). \tag{6.3.29}$$

It is assumed that $\mathcal{A}^{\{\{1\}\}}$ is not empty, that is,

$$\exists X^{\{1\}} \in \hat{\mathbf{R}}^{\{\{1\}\}} \left(X^{\{1\}} \in \mathcal{A}^{\{\{1\}\}} \right); \tag{6.3.30}$$

moreover, it is assumed that $\mathcal{A}^{\{\{1\}\}}$ is bounded from above, that is,

$$\exists X^{\{1\}} \in \hat{\mathbf{R}}^{\{\{1\}\}} \forall Y^{\{1\}} \in \hat{\mathbf{R}}^{\{\{1\}\}} \left(Y^{\{1\}} \in \mathcal{A}^{\{\{1\}\}} \implies Y^{\{1\}} \le X^{\{1\}} \right). \tag{6.3.31}$$

The least upper bound is defined to be the union of all of the elements of $\mathcal{A}^{\{\{1\}\}}$. First, define the set of all subsets of reals by

$$\hat{\mathbf{R}}^{\{\{\{1\}\}\}} \overset{\text{def}}{=} \exists \mathcal{X}^{\{\{1\}\}} \forall Y^{\{1\}} \left(Y^{\{1\}} \in \mathcal{X}^{\{\{1\}\}} \implies \mathcal{C}^{\langle\langle 1\rangle\rangle}(Y^{\{1\}}) \right), \tag{6.3.32}$$

and consider the relation $\Gamma^{\langle 1, \{\{1\}\}\rangle} \colon \hat{\mathbf{Q}}^{\{1\}} \times \hat{\mathbf{R}}^{\{\{\{1\}\}\}}$ defined by

$$\forall x \forall \mathcal{X}^{\{\{1\}\}} \in \hat{\mathbf{R}}^{\{\{\{1\}\}\}} \left\{ \Gamma^{\langle 1, \{\{1\}\}\rangle} \left(x, \mathcal{X}^{\{\{1\}\}} \right) \right.$$
$$\left. \overset{\text{def}}{=} \exists Y^{\{1\}} \in \hat{\mathbf{R}}^{\{\{1\}\}} \left(x \in Y^{\{1\}} \wedge Y^{\{1\}} \in \mathcal{X}^{\{\{1\}\}} \right) \right\}. \tag{6.3.33}$$

Then, for the set of reals given by (6.3.29), one has the relation $\Gamma_{\mathcal{A}}^{\langle 1\rangle} \colon \hat{\mathbf{Q}}^{\{1\}}$ defined by

$$\forall x \left(\Gamma_{\mathcal{A}}^{\langle 1\rangle}(x) \overset{\text{def}}{=} \Gamma^{\langle 1, \{\{1\}\}\rangle}(x, \mathcal{A}^{\{\{1\}\}}) \right), \tag{6.3.34}$$

and the least upper bound for the set of reals $\mathcal{A}^{\{\{1\}\}}$ is given by

$$\Gamma_{\mathcal{A}}^{\{1\}} \overset{\text{def}}{=} \exists x \Gamma_{\mathcal{A}}^{\langle 1\rangle}(x) = \exists x \Gamma^{\langle 1, \{\{1\}\}\rangle}(x, \mathcal{A}^{\{\{1\}\}}). \tag{6.3.35}$$

If one unpacks all of the definitions, one obtains

$$\Gamma_{\mathcal{A}}^{\{1\}} \overset{\text{def}}{=} \exists x \exists Y^{\{1\}} \in \hat{\mathbf{R}}^{\{\{1\}\}} \left(x \in Y^{\{1\}} \wedge Y^{\{1\}} \in \mathcal{A}^{\{\{1\}\}} \right). \tag{6.3.36}$$

Intuitively, $\Gamma_{\mathcal{A}}^{\{1\}}$ is the union of the elements of $\mathcal{A}^{\{\{1\}\}}$. First, it is necessary to show that $\Gamma_{\mathcal{A}}^{\{1\}} \in \hat{\mathbf{R}}^{\{\{1\}\}}$, that is, that $\mathcal{C}^{\langle\langle 1\rangle\rangle}\left(\Gamma_{\mathcal{A}}^{\{1\}}\right)$ holds. Then, it is clear that $\Gamma_{\mathcal{A}}^{\{1\}}$ is an upper bound, that is,

$$\forall Y^{\{1\}} \in \hat{\mathbf{R}}^{\{\{1\}\}} \left(Y^{\{1\}} \in \mathcal{A}^{\{\{1\}\}} \implies Y^{\{1\}} \le \Gamma_{\mathcal{A}}^{\{1\}} \right) \tag{6.3.37}$$

because $\Gamma_{\mathcal{A}}^{\{1\}}$ is the union of the elements of $\mathcal{A}^{\{\{1\}\}}$. Finally, one has to show that

$$\forall X^{\{1\}} \in \hat{\mathbf{R}}^{\{\{1\}\}} \left\{ \left[\forall Y^{\{1\}} \in \hat{\mathbf{R}}^{\{\{1\}\}} \left(Y^{\{1\}} \in \mathcal{A}^{\{\{1\}\}} \implies (Y^{\{1\}} \leq X^{\{1\}}) \right) \right] \right.$$
$$\left. \implies \Gamma_{\mathcal{A}}^{\{1\}} \leq X^{\{1\}} \right\}, \quad (6.3.38)$$

that is, that $\Gamma_{\mathcal{A}}^{\{1\}}$ is the least upper bound. The rather lengthy proofs are not essential for our purposes.

The important point is that the definition of the real $\Gamma_{\mathcal{A}}^{\{1\}}$ is impredicative because its definition (6.3.36) involves quantification '$\exists Y^{\{1\}} \in \hat{\mathbf{R}}^{\{\{1\}\}}$' over a variable that has the same level as the real $\Gamma_{\mathcal{A}}^{\{1\}}$. If each and every real is somehow defined at the outset, then the definition of the least upper bound of a set of reals merely serves to select a particular real from a previously defined totality of reals, and no circularity results. On the other hand, if *only* some reals are well defined, even if the set of such reals is quite large, then the formation of the least upper bound of a set of reals might 'create' or define a previously undefined real thereby changing the domain over which the quantification '$\exists Y^{\{1\}} \in \hat{\mathbf{R}}^{\{\{1\}\}}$' ranges and hence changing the meaning of the definition. As we have noted above, there are many who hold that such impredicative definitions involve a vicious circularity.

If one tries to use the ramified theory of types to avoid the impredicativity in the definition of the least upper bound of a set of reals, then the reals are not all of the same order. There could, in principle, be an infinite sequence of sets of reals of different order

$$\hat{\mathbf{R}}^{\{\{1\}/1\}/1}, \hat{\mathbf{R}}^{\{\{1\}/2\}/1}, \hat{\mathbf{R}}^{\{\{1\}/3\}/1}, \ldots. \quad (6.3.39)$$

It is this partitioning of the reals that makes it impossible to construct anything similar to the usual analysis; that is, the difficulties are not merely due to the fact that the notation is viciously cumbersome. With regard to the problem of the existence of entities of similar kind but of different order (what we call 'order' Weyl called 'level'), Weyl (1949a, 1 edn, pp. 49–50) remarks:

> One could escape this dilemma only if, for every property of the second level [order], there existed a property of the first level [order] equal to it (not in meaning but) in extension. As long as the sequence of natural numbers is accepted as an extensionally definite aggregate, one might consider as the properties of the first level [order] those which are generated by the definitional principles ... from the one basic relation 'n follows upon m'. In this case, our wish will hardly be fulfilled. We would have the task of extending the principles of construction for the properties of the first level [order] in such a manner that every set of the second level [order] demonstrably coincides with one of the first. But there is not the slightest indication that this is possible. Russell, in order to extricate himself from the affair, causes reason to commit hara-kiri, by postulating the above assertion in spite of its lack of support by

any evidence ('axiom of reducibility'). In a little book *Das Kontinuum* ..., I have tried to draw the honest consequence and constructed a field of real numbers of the first level [order], within which the most important operations of analysis can be carried out.

In another place Weyl (1946) says:

What we can get in this way constitutes the ground level [order], or level [order] 1. One could build over it a second level [order] containing relations which are constructed by applying the quantifiers to the totality of relations of this or that type constructible on the first level [order], and proceed in the same manner from the second to a third level [order] etc. One would obtain a 'ramified hierarchy' of types and levels [orders]. But in this way, ... nothing resembling our classical Calculus will result. The temptation to pass beyond the first level [order] of construction must be resisted; instead, one should try to make the range of constructible relations as wide as possible by *enlarging the stock of basic operations*. It is *a priori* clear that iteration in some form must find a place among these irreducible principles of construction ....

The purpose of the proposal discussed in remark 53 at the end of subsection 6.2 is to overcome at least partially the partitioning of the reals just mentioned as well as other similar problems.

6.4   Weyl's Predicative Analysis

Weyl's view of the source of the problem underlying impredicativity may be described in this way[104]. First, consider the following two basic viewpoints concerning sets:

(a) For any given set, there exists the property to be a member of the set;

(b) For any property, there 'exists' a set of elements that have the property.

According to (a), the existence of a set is ontologically prior to the property of membership; that is, set existence is ontologically independent of definability, and the issue of circular impredicativity does not arise. Weyl (1946) says, "if with regard to a bag of potatoes or a curve drawn by pencil on paper, the property of a potato to be in the bag or of a point to lie on the curve is introduced, then the set ... is prior to the property."

According to (b), sets are viewed in a conceptualistic fashion. Since concepts or properties are plausibly held not to exist independently of being grasped by

---

[104]See especially (Weyl 1919a) and (Weyl 1921d).

someone and hence, it would seem, independently of their definition, and since a property such as 'being even', for example, is prior to the set of all even numbers, the conceptualistic view of sets suggests that they be regarded as entities that exist on the basis of their definability or constructibility. As Russell does, Weyl accepts the conceptualist viewpoint of sets and regards sets to be synonymous with a property or a predicate. Moreover, the rejection of 'actual infinity' would seem to require the conceptualist point of view where infinite sets are concerned. Thus, to every propositional function ϕ, there corresponds a set ϕ, and the expressions 'a has the property A', '$A(a)$', or 'a belongs to the set A' have the same meaning. If we regard 'x is a member of A' as meaning only that 'x has the property A', then the restriction to predicative definitions becomes necessary. Moreover, two propositional functions ϕ, ψ define the same set and are said to be co-extensive[105], if and only if every entity that satisfies ϕ also satisfies ψ.

Now consider the traditional characterization of a real number and of the least upper bound of an arbitrary bounded set of reals. A real number is a subset $A^{\{1\}}$ of the rationals $\mathbf{Q}^{\{1\}}$ that is determined by a property $A^{\langle 1 \rangle}$ defined on $\mathbf{Q}^{\{1\}}$ and has the property of being a Dedekind-cut. Thus, $\mathcal{C}^{\langle\langle 1 \rangle\rangle}\left(A^{\{1\}}\right)$, where $\mathcal{C}^{\langle\langle 1 \rangle\rangle}$ is defined by (6.3.19). The set of reals is denoted by $\mathbf{R}^{\{\{1\}\}}$. A set of reals $\mathcal{A}^{\{\{1\}\}}$ is determined by a relation $\mathcal{A}^{\langle\langle 1 \rangle\rangle}\colon \mathbf{R}^{\{\{1\}\}}$. If the set of reals $\mathcal{A}^{\{\{1\}\}}$ is non-empty and bounded above, the least upper bound of $\mathcal{A}^{\{\{1\}\}}$ is the real number $\Gamma_{\mathcal{A}}^{\{1\}}$ given by

$$\Gamma_{\mathcal{A}}^{\{1\}} \overset{\text{def}}{=} \exists x \exists Y^{\{1\}} \in \mathbf{R}^{\{\{1\}\}} \left(x \in Y^{\{1\}} \wedge Y^{\{1\}} \in \mathcal{A}^{\{\{1\}\}} \right). \tag{6.4.1}$$

Note that the real $\Gamma_{\mathcal{A}}^{\{1\}}$ is defined in terms of a relation $\Gamma^{\langle 1, \{\{1\}\}\rangle}$ of level three. Intuitively, the set $\Gamma_{\mathcal{A}}^{\{1\}}$ is the union of all of the sets in $\mathcal{A}^{\{\{1\}\}}$. Weyl pointed out that this definition is problematic. What it means for a subset of the rationals to be a cut is perfectly well defined. If one is presented with a well defined subset of the rationals $A^{\{1\}}$, then whether or not that subset satisfies the Cutness property, that is, whether or not $\mathcal{C}^{\langle\langle 1 \rangle\rangle}\left(A^{\{1\}}\right)$ holds can be determined. It does not follow, however, that the set of reals $\mathbf{R}^{\{\{1\}\}}$ is *extensionally definite* in the sense that there is a complete list of constructive definitions that determine every possible cut. The essence of the difficulty is clarified in the following discussion.

Weyl considered the natural numbers as extensionally definite. Since Weyl showed that it is possible to explicitly construct *all* of the rationals as equivalence classes of ordered quadruples of the natural numbers, the rationals form a set that is also extensionally definite. A real $A^{\{1\}}$, however, is determined by a property $A^{\langle 1 \rangle}$ of the rationals which satisfies $\mathcal{C}^{\langle\langle 1 \rangle\rangle}\left(A^{\langle 1 \rangle}\right)$, where $\mathcal{C}^{\langle\langle 1 \rangle\rangle}$ is defined by

$$\forall X^{\langle 1 \rangle} \left\{ \mathcal{C}^{\langle\langle 1 \rangle\rangle}(X^{\langle 1 \rangle}) \overset{\text{def}}{=} \right.$$

$$\left[\left(\exists x X^{\langle 1 \rangle}(x) \wedge \exists x \neg X^{\langle 1 \rangle}(x) \right) \right.$$

---

[105]Weyl used the term 'umfangsgleich'.

$$\wedge \forall x \left(X^{\langle 1 \rangle}(x) \implies \exists y (x < y \wedge X^{\langle 1 \rangle}(y)) \right)$$

$$\wedge \forall x \left(X^{\langle 1 \rangle}(x) \implies \forall y (y < x \implies X^{\langle 1 \rangle}(y)) \right) \Big] \Big\}. \tag{6.4.2}$$

The set version of this definition is given by (6.3.19). Weyl points out that the set

$$\mathcal{C}^{\{\langle 1 \rangle\}} \stackrel{\text{def}}{=} \exists X^{\langle 1 \rangle} \mathcal{C}^{\langle\langle 1 \rangle\rangle} \left(X^{\langle 1 \rangle} \right), \tag{6.4.3}$$

that is, the set of all cut-defining properties is not extensionally definite. Suppose the contrary, and consider any property $\mathcal{A}^{\langle\langle 1 \rangle\rangle} : \mathcal{C}^{\{\langle 1 \rangle\}}$. Then, the property $\Gamma_{\mathcal{A}}^{\langle 1 \rangle}$ defined by

$$\forall x \left\{ \Gamma_{\mathcal{A}}^{\langle 1 \rangle}(x) \stackrel{\text{def}}{=} \exists X^{\langle 1 \rangle} \in \mathcal{C}^{\{\langle 1 \rangle\}} \left(X^{\langle 1 \rangle}(x) \wedge \mathcal{A}^{\langle\langle 1 \rangle\rangle}(X^{\langle 1 \rangle}) \right) \right\} \tag{6.4.4}$$

also satisfies $\mathcal{C}^{\langle\langle 1 \rangle\rangle} \left(\Gamma_{\mathcal{A}}^{\langle 1 \rangle} \right)$. Thus, $\Gamma_{\mathcal{A}}^{\langle 1 \rangle} \in \mathcal{C}^{\{\langle 1 \rangle\}}$. But this is a contradiction because by virtue of its dependence on $\mathcal{A}^{\langle\langle 1 \rangle\rangle}$, the property $\Gamma_{\mathcal{A}}^{\langle 1 \rangle}$ has a *meaning* that differs from the meaning of the elements in $\mathcal{C}^{\{\langle 1 \rangle\}}$. It is, therefore, *different as a relation*. The assumption that leads to this contradiction, namely, that the set $\mathcal{C}^{\{\langle 1 \rangle\}}$ is extensionally definite, is therefore false.

According to Weyl, the solution to this problem, which he believed was not achievable, would be the following *completeness* result. First, one must have a well defined set of procedures for constructing relations on the set of rationals. Then, the set of properties of the rationals that are both constructive and predicative[106] is extensionally definite. The subset $\mathcal{K}^{\{\langle 1 \rangle\}}$ of these relations which are cuts is also extensionally definite. The set $\mathcal{K}^{\{\langle 1 \rangle\}}$ is the set of reals. In addition, one must prove that for any predicative, constructively defined property $\mathcal{A}^{\langle\langle 1 \rangle\rangle} : \mathcal{K}^{\{\langle 1 \rangle\}}$, which determines a non-empty set of reals that is bounded above, the constructively defined but impredicative relation $\Gamma_{\mathcal{A}}^{\langle 1 \rangle}$ defined by

$$\forall x \left\{ \Gamma_{\mathcal{A}}^{\langle 1 \rangle}(x) \stackrel{\text{def}}{=} \exists X^{\langle 1 \rangle} \in \mathcal{K}^{\{\langle 1 \rangle\}} \left(X^{\langle 1 \rangle}(x) \wedge \mathcal{A}^{\langle\langle 1 \rangle\rangle}(X^{\langle 1 \rangle}) \right) \right\} \tag{6.4.5}$$

is coextensive with one of the relations in $\mathcal{K}^{\{\langle 1 \rangle\}}$; that is,

$$\exists X^{\langle 1 \rangle} \in \mathcal{K}^{\{\langle 1 \rangle\}} \forall x \left(X^{\langle 1 \rangle}(x) \iff \Gamma_{\mathcal{A}}^{\langle 1 \rangle}(x) \right). \tag{6.4.6}$$

Although the definition (6.4.5) is impredicative, no circularity results because it is used merely to pick out a coextensive relation from a previously defined extensionally definite set $\mathcal{K}^{\{\langle 1 \rangle\}}$. Weyl (1921d, Part I, §1, p. 42, GA II, pp. 145–146) says with regard to the above:

> Um dem Satz von der Existenz der oberen Grenze einer jeden Menge reeller Zahlen einen klaren Sinn zu erteilen und seine Wahrheit sicherzustellen, wäre also dies erforderlich: es müsste ein in sich bestimmter

---

[106]Their definition involves quantification only over the rationals.

und begrenzter Inbegriff von Eigenschaften, "$\mathcal{K}$-Eigenschaften", konstruiert werden, für welchen nachweislich der Satz gilt, dass eine Eigenschaft $\Gamma_{\mathcal{A}}$, welche nach dem obigen Schema aus der Gesamtheit der $\mathcal{K}$-Eigenschaften konstruiert ist, stets mit einer bestimmten $\mathcal{K}$-Eigenschaft umfangsgleich ist. Dies ist niemals versucht worden; es liegt nicht das leiseste Anzeichen dafür vor, dass eine solche Konstruktion möglich ist; sie ist von vornherein so ungeheuer unwahrscheinlich, dass man niemandem vernünftigerweise zumuten kann, danach zu suchen.[107]

[In order to give a clear meaning to the theorem of the existence of the [least] upper bound for each set of real numbers and to ensure its truth, the following would be required: a definite and closed totality of properties, "$\mathcal{K}$-properties", would have to be constructed for which the provable theorem holds that a property $\Gamma_{\mathcal{A}}$, that is constructed according to the above schema from the totality of $\mathcal{K}$-properties, is always extensionally equivalent to a determinate $\mathcal{K}$-property. This has never been attempted and there does not seem to be the slightest indication that such a construction is possible. Such a construction is from the outset so daunting and improbable that one could not reasonably expect anyone to search for it.]

Although Weyl was not able to achieve the completeness result just stated, he was able to prove a somewhat weaker *completeness* result. He was able to constructively define the notion of a sequence of elements of $\mathcal{K}^{\{(1)\}}$ and to prove that if such a sequence is *Cauchy*, then there exists an element of $\mathcal{K}^{\{(1)\}}$ to which the sequence converges.

As we saw, it is Weyl's contention that even if a concept's meaning determines its extension, it does not follow that the concept in question is extensionally definite and that it makes sense to speak of the objects falling under the concept as a determinate closed totality. If $\mathcal{P}$ is a property of objects that belong to the extension of some concept $\mathcal{C}$, then the proposition 'x has $\mathcal{P}$' ($x \in P$) is either true or false. If the concept $\mathcal{C}$ is also extensionally-definite, then not only does it make sense to ask of an arbitrary object x whether it has the property $\mathcal{P}$, but it is then also legitimate to ask whether there exists among the objects falling under $\mathcal{C}$ one that possesses the property $\mathcal{P}$. That the concept of natural number is extensionally definite is supported by the inductive generating process of the natural numbers given to us in intuition (Anschauung). The concept 'rational number' is likewise extensionally definite. On the other hand, concepts such as 'object', 'property of rational numbers', etc., are not extensionally definite. Consequently, existential and universal propositions about properties of rational numbers are meaningful only if the extensionally indefinite concepts have been restricted to extensionally definite ones, namely, $\mathcal{K}$-properties of rational numbers. Weyl showed in *Das Kontinuum* that this restriction is possible by constructing all properties and relations from

---

[107] In this quotation we have changed the symbols to accord with our notation.

a few primary relations of the natural numbers and by restricting quantification only to the latter ground category.

Weyl's approach in *Das Kontinuum* is not as radical as Brouwer's version of intuitionism. In *Das Kontinuum*, Weyl accepts the law of the excluded middle for the natural and rational numbers but not for the real numbers or for properties of real numbers. To emphasize, Weyl assumes as the basis for construction only one category of objects, the intuitively given basic category (Grundkategorie) of natural numbers $\mathbf{N}$ (without zero) and the one basic binary relation (Grundeigenschaft) $S(x, y)$ between numbers, namely, 'x is followed by y'. All the other relations of the various types are explicitly constructed, the quantifiers $\exists x$ or $\forall x$ being applied only to natural numbers and not to arguments of higher level.[108] Moreover, no axioms are postulated. Although Weyl accepts something like the hierarchy of levels as legitimate, he does not regard type theories as practical enough to work with and bases his reconstruction of analysis solely on *arithmetically definable* sets whose definitions involve quantification[109] only over $\mathbf{N}$.

Weyl proposes eight rules for the construction of new mathematical properties. The first six of these basically define a 'restricted' calculus of propositional functions. If we let the fundamental category of individuals be the natural numbers and consider a number of basic propositional functions that correspond to certain basic mathematical relations such as 'is greater than', 'is the product of,' etc., then further propositional functions may be constructed by (a) logical operations such as negation, conjunction, disjunction and existential quantification, and (b) identification of variables and substitution of individual constants for variables:

1. One may construct from the propositional function ϕ its negation $\neg\phi$. *Example:* If $\mathcal{A}(x_1, x_2), x_1, x_2 \in \mathbf{N}$ means 'x_2 follows x_1', then $\neg\mathcal{A}(x_1, x_2)$ means that 'x_2 does not follow x_1'.

2. If ϕ is a propositional function of several distinct variables then one can derive a propositional function having a smaller number of distinct variables by identification of one or several variables. *Example:* From $\mathcal{A}(x_1, x_2, x_3)$ one may construct the relations $\mathcal{A}(x_1, x_2, x_1)$ and $\mathcal{A}(x_1, x_1, x_1)$.

3. From the propositional functions ϕ, ψ, one may construct their conjunction $(\phi \wedge \psi)$. Identification of none, some or all of the variables in ϕ and ψ will in general yield different conjunctions. *Example:* If $\mathcal{A}(x_1, x_2), \mathcal{B}(x_3, x_4)$ are two propositional functions, then one may construct the expressions $\mathcal{A}(x_1, x_2) \wedge \mathcal{B}(x_3, x_4)$, $\mathcal{A}(x_1, x_2) \wedge \mathcal{B}(x_2, x_4)$ or $\mathcal{A}(x_1, x_2) \wedge \mathcal{B}(x_1, x_2)$.

4. From the propositional functions ϕ and ψ, one may construct their

---

[108]This statement is not quite accurate. See remark 55 below equation (6.4.13).
[109]Again, see remark 55 below.

disjunction $(\phi \vee \psi)$. *Remark:* As in **3**, different disjunctions may be constructed depending on how the variables of ϕ are related to or identified with those of ψ.

5. From a propositional function ϕ, a propositional function with fewer variables may be constructed by the substitution of constants, that is, natural numbers for the variables of ϕ. *Example:* From the propositional function $\mathcal{A}(x_1, x_2, x_3)$ of three variables, one may construct a propositional function $\mathcal{A}(x_1, x_2, a)$ of two variables.

6. If $\phi(x_1, x_2, \ldots, x_n)$ is a propositional function, then one can construct propositional functions with fewer variables by existential quantification. *Example:*

$$\psi(x_1, x_2, \ldots, x_{n-1}) \stackrel{\text{def}}{=} \exists x_n \phi(x_1, x_2, \ldots, x_n), \tag{6.4.7}$$

$$\chi(x_1, x_3, \ldots, x_{n-1}) \stackrel{\text{def}}{=} \exists x_2 \exists x_n \phi(x_1, x_2, \ldots, x_n). \tag{6.4.8}$$

What has been described so far is a restricted calculus of propositional functions. To this formalism, Weyl adds a principle of generalized substitution and a principle of iteration, which are discussed below. Using this enlarged set of rules, he is able to construct from the natural numbers and the fundamental relations on these, the positive fractions **F**, the rationals **Q**, the arithmetically definable reals **W** and a considerable portion of classical analysis.

Natural numbers are entities of level 0. All relations among ordered n-tuples of natural numbers $\mathcal{A}(x_1, x_2, \ldots, x_n)$ — that can be constructed by combining the fundamental *arithmetic relations* on the natural numbers a finite number of times and by quantifying *only* over the natural numbers — are relations of level 1. The sets of ordered n-tuples[110] (several dimensional sets) that can be obtained by quantifying over the natural number variables, such as

$$A = \{\langle x_1, x_2 \rangle \mid \mathcal{A}(x_1, x_2)\}, \tag{6.4.9}$$

and the set-valued functions (functional connections) that can be obtained by quantifying over some of the natural number variables, such as

$$A(x_3, x_4, x_5) = \{\langle x_1, x_2 \rangle \mid \mathcal{A}(x_1, x_2, x_3, x_4, x_5)\}, \tag{6.4.10}$$

are also entities of level 1. Weyl also employs relations of level 2 between natural numbers and sets of natural numbers of the form

$$\mathcal{A}(x_1, x_2, \ldots, x_n, X_1, X_2, \ldots, X_m). \tag{6.4.11}$$

However, in the *formation* of these relations, quantification over entities other than level 0 is not permitted; moreover, the only set-valued functions that may be

---

[110]In this subsection, angled brackets are used to enclose ordered n-tuples.

constructed have the form

$$A(x_k, \ldots, x_n, X_1, X_2, \ldots, X_m) = \hspace{3cm} (6.4.12)$$
$$\{\langle x_1, x_2, \ldots, x_j \rangle \mid \mathcal{A}(x_1, x_2, \ldots, x_j, x_k, \ldots, x_n, X_1, X_2, \ldots, X_m)\}.$$

The following comprehension principle summarizes the above restrictions on the construction of sets.

$$\forall y_1, \forall y_2, \ldots, \forall y_m, \forall Y_1, \forall Y_2, \ldots, \forall Y_n, \exists A, \forall x_1, \forall x_2, \ldots, \forall x_k \hspace{2cm} (6.4.13)$$
$$[\langle x_1, x_2, \ldots, x_k \rangle \in A \iff \mathcal{A}(x_1, x_2, \ldots, x_k, y_1, y_2, \ldots, y_m, Y_1, Y_2, \ldots, Y_n)].$$

The relation $\mathcal{A}$ must not involve the set A itself for otherwise the definition of A would be impredicative. Also, $\mathcal{A}$ must be arithmetically definable in a way that involves quantification *only* over the natural numbers. Note that, to the right of $\exists A$, only quantifiers over the natural numbers may appear. It is these quantifiers that pertain to the *construction* of the set A. In the set notation, these variables appear to the left of the | in the definitions (6.4.9),(6.4.10) and (6.4.12). The quantifiers to the left of $\exists A$ merely permit arbitrary substitution of entities with values of the same type as the quantified variables. Thus, the comprehension principle (6.4.13) also encompasses Weyl's generalized substitution principle and, by repeated application, his iteration principle.

Remark 55 Weyl is using a restricted second-order theory in the following sense. For the *definition* of relations and the *formation* of sets, that is, for purposes of *construction*, quantification is permitted only over the natural numbers; however, he uses variables of level 1 as 'place-holders' for arbitrary entities of level 1. Weyl's rules amount to permitting the use of universal quantification over variables of level 1 for the purposes of *reference* only.

Before discussing Weyl's presentation of these principles, let us consider his general strategy for constructing the real numbers and some of his notational conventions. With the exception of the natural numbers themselves, all other numbers are sets of n-tuples of natural numbers for some n. Thus, the positive fractions **F** are equivalence classes of ordered pairs $\langle x_1, x_2 \rangle$; the rationals **Q** and the arithmetically constructed real numbers **W** are sets of ordered quadruples; and the arithmetically constructible complex numbers **C** are sets of ordered octuples. It is for this reason that the independent variables of a relation are almost always of level 1 (sets). The dependent variables of a relation always range over the natural numbers because the relation is being used to construct a number (set) by quantifying over these variables, and quantification is restricted to the domain **N**. Thus the relations considered are generally of the form $\mathcal{R}(x_1, \ldots, x_m \mid X_1, \ldots, X_n)$, where the vertical stroke separates the dependent variables x_i from the independent variables X_j.

One of the principles discussed above governing the formation of relations from other relations permits the substitution of a constant of the appropriate type for a variable. Consider the simple case of a relation of the form $\mathcal{R}(x_1, x_2 \mid X)$, where X is a variable that ranges over subsets of $\mathbf{N}^2$. Then, if $A \subset \mathbf{N}^2$ and A is a constant set, one may form the relation $\mathcal{R}(x_1, x_2 \mid A)$. The extended substitution principle permits the substitution for X of any function the values of which are sets of the appropriate type. In this case, the relation $\mathcal{R}$ itself defines such a function.

$$R(X) = \{\langle x_1, x_2 \rangle \mid \mathcal{R}(x_1, x_2 \mid X)\}. \tag{6.4.14}$$

Thus, it is permissible to form the relation $\mathcal{R}(x_1, x_2 \mid R(X))$. Of course, any other function having values of the appropriate type could also be substituted.

The principle of iteration permits the use of extended substitution an arbitrary, finite number of times. Weyl specifically draws attention to the following four patterns of iteration.

First, one may repeatedly compose a relation with itself to form the relations

$$\mathcal{R}(1; x_1, x_2 \mid X) \overset{\text{def}}{=} \mathcal{R}(x_1, x_2 \mid X), \tag{6.4.15}$$

$$\mathcal{R}(2; x_1, x_2 \mid X) \overset{\text{def}}{=} \mathcal{R}(1; x_1, x_2 \mid R(X))$$
$$\overset{\text{def}}{=} \mathcal{R}(x_1, x_2 \mid R(X)), \tag{6.4.16}$$

and in general

$$\mathcal{R}(n+1; x_1, x_2 \mid X) \overset{\text{def}}{=} \mathcal{R}(n; x_1, x_2 \mid R(X)). \tag{6.4.17}$$

Remark 56 In a paper entitled *Weyl Vindicated: "Das Kontinuum" 70 Years Later*, Feferman (1988) discusses Weyl's contributions to the foundations of mathematics. In section five of that paper, Feferman 're-examines' Weyl's *Das Kontinuum* and points out that one of Weyl's principles of iteration (the one defined above by (6.4.15)–(6.4.17)) is in conflict with Weyl's intention to use sets of the lowest level; however, Weyl's system of axioms does satisfy the requirement of predicativity. Feferman presents three reconstructions, called $K^{(\alpha)}$, W and $K^{(\beta)}$, of Weyl's system of axioms. In the system $K^{(\alpha)}$, Weyl's rules are restricted in order to maintain the strictly arithmetical interpretation that Weyl wanted. The system W is an extension of $K^{(\alpha)}$ that also admits a strictly arithmetical interpretation; however, W is much more flexible than $K^{(\alpha)}$ and one can formalize a much greater body of mathematics in W than in $K^{(\alpha)}$. The system $K^{(\beta)}$, which can be embedded in a simple extension of W, incorporates all of the principles of definition that Weyl used, in particular his iteration principles. Feferman's paper ends with an overview of the scientifically applicable mathematics that can be formalized in W and concludes "Weyl's program is now substantially vindicated for scientifically applicable mathematics by means of systems like W".

Second, it might happen that one or more independent variables are present that do not participate in the iteration; for example, for a relation $\mathcal{S}(y \mid X, Y)$ where X is a variable that ranges over subsets of $\mathbf{N}^2$ and Y is a variable that ranges over subsets of $\mathbf{N}$, one obtains

$$\mathcal{S}(1; y \mid X, Y) \overset{\text{def}}{=} \mathcal{S}(y \mid X, Y), \tag{6.4.18}$$

$$\mathcal{S}(2; y \mid X, Y) \overset{\text{def}}{=} \mathcal{S}(1; y \mid X, S(X, Y)), \tag{6.4.19}$$

$$\mathcal{S}(3; y \mid X, Y) \overset{\text{def}}{=} \mathcal{S}(2; y \mid X, S(X, Y)), \tag{6.4.20}$$

and so forth, where

$$S(X, Y) = \{y \mid \mathcal{S}(y \mid X, Y)\}. \tag{6.4.21}$$

Third, two relations may be substituted repeatedly into each other as in the following example. Let X and Y be variables that range over $\mathbf{N}^2$ and $\mathbf{N}$ respectively, and set

$$R(X, Y) = \{\langle x_1, x_2 \rangle \mid \mathcal{R}(x_1, x_2 \mid X, Y)\} \tag{6.4.22}$$

and

$$S(X, Y) = \{y \mid \mathcal{S}(y \mid X, Y)\}. \tag{6.4.23}$$

Then,

$$\mathcal{R}(1; x_1, x_2 \mid X, Y) \overset{\text{def}}{=} \mathcal{R}(x_1, x_2 \mid X, Y)$$
$$\mathcal{S}(1; y \mid X, Y) \overset{\text{def}}{=} \mathcal{S}(y \mid X, Y), \tag{6.4.24}$$
$$\mathcal{R}(2; x_1, x_2 \mid X, Y) \overset{\text{def}}{=} \mathcal{R}(1; x_1, x_2 \mid R(X, Y), S(X, Y))$$
$$\mathcal{S}(2; y \mid X, Y) \overset{\text{def}}{=} \mathcal{S}(1; y \mid R(X, Y), S(X, Y)), \tag{6.4.25}$$

and in general

$$\mathcal{R}(n+1; x_1, x_2 \mid X, Y) \overset{\text{def}}{=} \mathcal{R}(n; x_1, x_2 \mid R(X, Y), S(X, Y))$$
$$\mathcal{S}(n+1; y \mid X, Y) \overset{\text{def}}{=} \mathcal{S}(n; y \mid R(X, Y), S(X, Y)). \tag{6.4.26}$$

Fourth, a sequence of relations $\mathcal{R}(x_1, x_2 \mid X, n)$, where X ranges over subsets of $\mathbf{N}^2$, may be given that determine a sequence of set-valued functions $R(X, n)$ just as in (6.4.14). Then, the relations $\mathcal{R}^*(x_1, x_2 \mid X, n)$ may be constructed by using $R(X, n)$ to substitute at the nth step as follows:

$$\mathcal{R}^*(x_1, x_2 \mid X, 1) \overset{\text{def}}{=} \mathcal{R}(x_1, x_2 \mid X, 1), \tag{6.4.27}$$

$$\mathcal{R}^*(x_1, x_2 \mid X, 2) \overset{\text{def}}{=} \mathcal{R}^*(x_1, x_2 \mid R(X, 2), 1), \tag{6.4.28}$$

$$\mathcal{R}^*(x_1, x_2 \mid X, 3) \overset{\text{def}}{=} \mathcal{R}^*(x_1, x_2 \mid R(X, 3), 2), \tag{6.4.29}$$

and in general

$$\mathcal{R}^*(x_1, x_2 \mid X, n+1) \overset{\text{def}}{=} \mathcal{R}^*(x_1, x_2 \mid R(X, n+1), n). \tag{6.4.30}$$

Remark 57 In the following, the set-valued function corresponding to a relation of the form

$$\mathcal{R}(x_1, \ldots, x_n \mid y_1, \ldots, y_k, Y_1, \ldots, Y_l) \tag{6.4.31}$$

is denoted by

$$R(y_1, \ldots, y_k, Y_1, \ldots, Y_l); \qquad (6.4.32)$$

that is,

$$R(y_1, \ldots, y_k, Y_1, \ldots, Y_l) = \{\langle x_1, \ldots, x_n \rangle | R(x_1, \ldots, x_n | y_1, \ldots, y_k, Y_1, \ldots, Y_l)\}. \qquad (6.4.33)$$

Let us consider the construction of the positive fractions **F**. Denote the relation $x_3 = x_1 x_2$ among the natural numbers x_i by $\Pi(x_1, x_2, x_3)$. Define the relation

$$\mathcal{R}(x_1, x_2 \mid y_1, y_2) \overset{\text{def}}{=} \exists z (\Pi(x_1, y_2, z) \wedge \Pi(x_2, y_1, z)). \qquad (6.4.34)$$

This relation is an equivalence relation on the set of ordered pairs $\langle x_1, x_2 \rangle \in \mathbf{N}^2$. The equivalence class containing the ordered pair $\langle m_1, m_2 \rangle$ is $R(m_1, m_2)$, where

$$R(y_1, y_2) = \{\langle x_1, x_2 \rangle | \mathcal{R}(x_1, x_2 | y_1, y_2)\}, \qquad (6.4.35)$$

and is denoted by $\frac{m_1}{m_2}$. Each of these equivalence classes is a positive fraction.

Define a relation $\mathcal{M}(z_1, z_2 \mid X, Y)$, where $X = \frac{x_1}{x_2} \in \mathbf{F}$ and $Y = \frac{y_1}{y_2} \in \mathbf{F}$ by

$$\mathcal{M}(z_1, z_2 \mid X, Y) \overset{\text{def}}{=} \mathcal{R}(z_1, z_2 \mid x_1 y_1, x_2 y_2). \qquad (6.4.36)$$

Then, $Z = M(X, Y) \in \mathbf{F}$ is the product of X and Y. Similarly, define

$$\mathcal{A}(z_1, z_2) \overset{\text{def}}{=} \mathcal{R}(z_1, z_2 \mid x_1 y_2 + x_2 y_1, x_2 y_2), \qquad (6.4.37)$$

where $\frac{x_1}{x_2} = X$ and $\frac{y_1}{y_2} = Y$. Then, $Z = A(X, Y) \in \mathbf{F}$ is the sum of X and Y. The reciprocal of a fraction X is obtained from the relation

$$\mathcal{D}(z_1, z_2 \mid X) \overset{\text{def}}{=} \mathcal{R}(z_1, z_2 \mid x_2, x_1) \qquad (6.4.38)$$

as the function $D(X)$.

A rational $X \in \mathbf{Q}$, is an equivalence class of ordered quadruples of natural numbers $\langle x_1, x_2, x_3, x_4 \rangle$. The equivalence relation is defined by

$$\mathcal{R}(x_1, x_2, x_3, x_4 \mid y_1, y_2, y_3, y_4) \overset{\text{def}}{=}$$
$$\exists z [\Pi(x_1 y_4 + x_2 y_3, x_4 y_2, z) \wedge \Pi(x_2 y_4, x_3 y_2 + x_4 y_1, z)]. \qquad (6.4.39)$$

The equivalence class that contains $\langle m_1, m_2, m_3, m_4 \rangle$ is

$$\frac{m_1}{m_2} \ominus \frac{m_3}{m_4} \overset{\text{def}}{=} R(m_1, m_2, m_3, m_4). \qquad (6.4.40)$$

Roughly, this equivalence class corresponds to the difference of the positive fractions $\frac{m_1}{m_2}$ and $\frac{m_3}{m_4}$. Weyl used '÷' instead of '$\ominus$', but such usage would now be confusing.

Addition is defined by means of the relation $\mathcal{A}$

$$\mathcal{A}(z_1, z_2, z_3, z_4 \mid X, Y) \stackrel{\text{def}}{=}$$
$$\mathcal{R}(z_1, z_2, z_3, z_4 \mid x_1 y_2 + x_2 y_1, x_2 y_2, x_3 y_4 + x_4 y_3, x_4 y_4), \qquad (6.4.41)$$

where $\mathcal{R}$ is defined by (6.4.39). This definition may be motivated by the following:

$$
\begin{aligned}
\frac{z_1}{z_2} \ominus \frac{z_3}{z_4} &= \left(\frac{x_1}{x_2} \ominus \frac{x_3}{x_4}\right) + \left(\frac{y_1}{y_2} \ominus \frac{y_3}{y_4}\right) \\
&= \left(\frac{x_1}{x_2} + \frac{y_1}{y_2}\right) \ominus \left(\frac{x_3}{x_4} + \frac{y_3}{y_4}\right) \\
&= \left(\frac{x_1 y_2 + x_2 y_1}{x_2 y_2} \ominus \frac{x_3 y_4 + x_4 y_3}{x_4 y_4}\right). \qquad (6.4.42)
\end{aligned}
$$

The sum of X and Y is then $Z = A(X, Y)$. The relations corresponding to the other operations may be similarly constructed.

Weyl employed the method of Dedekind cuts in a predicative manner to construct the real numbers. A cut is a nonempty set S^* of ordered quadruples of natural numbers such that:

1. If $X = \frac{x_1}{x_2} \ominus \frac{x_3}{x_4}$ is a rational, then either $X \cap S^* = \emptyset$ or $X \subset S^*$.

2. If X is a rational and $X \subset S^*$, then for any rational $Y < X$, $Y \subset S^*$.

3. There exists a rational X such that $X \cap S^* = \emptyset$.

4. For any rational X such that $X \subset S^*$, there exists a rational $Y > X$ such that $Y \subset S^*$.

A real number is a cut S^*. It is important to note that Weyl recognizes the existence of *only* those sets A that are arithmetically definable according to the comprehension principle (6.4.13). Since the number of such sets is countable, the field $\mathbf{W}$ of all arithmetically definable real numbers is also countable. The usual field $\mathbf{R}$ of real numbers contains 'arbitrary' cuts not just the arithmetically definable cuts. Of course, Weyl would not have agreed that the additional real numbers exist.

For every rational number $X = \frac{x_1}{x_2} \ominus \frac{x_3}{x_4}$ there corresponds a real number X^* defined by

$$X^* = \{\langle y_1, y_2, y_3, y_4 \rangle \mid (y_1 x_4 + y_2 x_3) x_2 y_4 < (x_1 y_4 + x_2 y_3) y_2 x_4\}, \qquad (6.4.43)$$

where the condition defining this real number is simply a restatement of the relation $Y < X$. The real number $A^* = \sqrt{2}$ is defined by

$$A^* = \{\langle x_1, x_2, x_3, x_4 \rangle \mid [(x_1 x_4 \leq x_2 x_3] \vee \qquad (6.4.44)$$
$$[(x_1 x_4 > x_2 x_3) \wedge (x_1 x_4)^2 + (x_2 x_3)^2 < 2 x_1 x_2 x_3 x_4 + 2(x_2 x_4)^2]\},$$

where the defining condition is a restatement of $(X \leq 0) \vee (X > 0 \wedge X^2 < 2)$ in which X is rational.

Addition of two real numbers, $X^* \in \mathbf{W}$ and $Y^* \in \mathbf{W}$, is defined by means of the relation

$$\mathcal{A}^*(z_1, z_2, z_3, z_4 \mid X^*, Y^*) \stackrel{\text{def}}{=}$$

$$\exists x_1, \exists x_2, \exists x_3, \exists x_4, \exists y_1, \exists y_2, \exists y_3, \exists y_4 \left\{ \langle x_1, x_2, x_3, x_4 \rangle \in X^* \wedge \right.$$

$$\left. \langle y_1, y_2, y_3, y_4 \rangle \in Y^* \wedge \left[\frac{z_1}{z_2} \ominus \frac{z_3}{z_4} = A \left(\frac{x_1}{x_2} \ominus \frac{x_3}{x_4}, \frac{y_1}{y_2} \ominus \frac{y_3}{y_4} \right) \right] \right\}, (6.4.45)$$

where A is the function defined by (6.4.41). Then, $Z^* = A^*(X^*, Y^*)$ is the sum of X^* and Y^*. The definitions of the other operations are more complicated but differ from the definitions used in the standard construction of $\mathbf{R}$ only by reformulating the conditions in terms of quadruples of natural numbers instead of in terms of rationals.

Constructing all numbers as sets of tuples of natural numbers rather than permitting types of higher level as in the usual construction makes the bookkeeping more complicated in Weyl's approach. Weyl resists the temptation to define the positive fractions as equivalence classes of ordered pairs of natural numbers, the rationals as equivalence classes of ordered pairs of positive fractions (themselves equivalence classes) and the reals as sets (cuts) of rationals. He does so because such a layered approach requires that quantification over higher types (sets and sets of these sets and so forth) be permitted. In combination with the demand for strictly predicative constructions, quantification over higher types leads immediately[111] to Russell's ramified[112] theory of types. For example, different *orders* of sets of n-tuples of natural numbers can be constructed by using the comprehension principle

$$\exists A \forall x_1, \forall x_2, \ldots, \forall x_n [\langle x_1, x_2, \ldots, x_n \rangle \in A \iff \mathcal{A}(x_1, x_2, \ldots, x_n)], \qquad (6.4.46)$$

where for sets of order zero, the relation $\mathcal{A}$ can involve quantification only over the natural numbers; for sets of order one, $\mathcal{A}$ can involve quantification over the natural numbers and sets of order zero; for sets of order two, $\mathcal{A}$ can involve quantification over the natural numbers, sets of order zero and sets of order one; and so forth. Of course, sets of sets and other higher types are similarly ramified, and the full range of possibilities is exceedingly complicated. Weyl rejected this approach because it makes it impossible to state mathematical theorems in a general fashion. He justly regarded the complexities of his approach simple by comparison.

Weyl was able to prove that the class $\mathbf{W}$ of all arithmetically definable real numbers is a complete, ordered field. This statement requires some clarification in view of the customary claim that the only complete, ordered field is $\mathbf{R}$. It has already been pointed out that Weyl did not accept the existence of the additional

[111]See subsection 6.3.
[112]See subsection 6.2.

elements of **R** because they correspond to *arbitrary* cuts that are not arithmetically definable. Weyl also did not accept sequences that are not arithmetically definable. A real valued sequence is determined by a relation $\mathcal{R}(x_1, x_2, x_3, x_4 \mid y)$ such that for every y the set $\mathcal{R}(y) \in \mathbf{W}$. The crucial point is that the sequence is specified by means of an arithmetically definable relation and not by choosing in some arbitrary manner real numbers one after another. Then, the claim that **W** is complete may be more fully stated as follows: *Every arithmetically definable sequence of arithmetically definable real numbers that is a Cauchy sequence converges to an arithmetically definable real number.*

Weyl also considers a number of principles that are from the customary viewpoint equivalent to the Cauchy convergence principle and that have been used as the starting point of analysis, namely,

1. A sequence of nested intervals, the lengths of which tend to zero, determines a unique real number;

2. A monotone, increasing sequence of real numbers that is bounded above converges to a real number;

3. The Dedekind Cut Principle: Let **A** and **B** be two sets of real numbers such that each element of **A** is less than every element of **B**. Moreover, suppose that for every positive rational A there exists a real number $X^* \in \mathbf{A}$ and a real number $Y^* \in \mathbf{B}$ such that the real number A^* determined by the rational A according to (6.4.43) satisfies $A^* > Y^* - X^*$. Then, there exists a unique real number C^* that is greater than or equal to every element of **A** and less than or equal to every element of **B**;

4. A bounded set of real numbers has a least upper bound and a greatest lower bound;

5. Each bounded, infinite set of real numbers has an accumulation point.

Weyl proved that 1 and 2 hold in his system but that 3, 4 and 5 do not hold unless the sets of real numbers referred to are restricted to contain only real numbers that correspond to rational numbers. Another important result that does not hold in Weyl's system is the Dirichlet Principle, which plays an important role in Weyl's theory of Riemann surfaces. He noted that this principle does not hold even in the weaker formulation in which only the existence of a lower bound rather than the existence of a minimum is asserted. Weyl also showed that the Heine-Borel theorem holds with the usual restriction to arithmetically definable sequences.

Weyl was also able to show that essentially the entire analysis of continuous functions could be developed within his system. In particular he proved the following main theorems about continuous functions.

The Intermediate Value Theorem: Every real valued function F^* that is continuous on a closed interval $[A^*, B^*]$ takes on every value between $F^*(A^*)$ and $F^*(B^*)$ at some point C^* such that $A^* < C^* < B^*$.

The Extreme Value Theorem: If F^* is a real valued function that is continuous on a closed interval $[A^*, B^*]$, then there exist two real values C_1^* and C_2^* such that for any $X^* \in [A^*, B^*]$, $F^*(C_1^*) \leq F^*(X^*) \leq F^*(C_2^*)$

Uniform Continuity: Every real valued function that is continuous on a closed interval is uniformly continuous on that closed interval.

Weyl also noted that the first of these results could be extended to continuous functions of several variables and that the Fundamental Theorem of Algebra was also valid in his system of analysis. Of course, these theorems must be appropriately interpreted. A continuous function has an arithmetically definable value at an arithmetically definable point. Thus, the Intermediate Value Theorem asserts that for any arithmetically definable real number between $F^*(A^*)$ and $F^*(B^*)$, there exists an arithmetically definable real number at which the function F^* has that intermediate value.

Series are defined in terms of sequences (of the appropriately restricted kind!). Polynomials with arithmetically definable coefficients are arithmetically definable and so are the roots of these polynomials; consequently, **W** is algebraically closed in the sense that the subfield of **R** that consists of the algebraic real numbers is contained in **W**. The power series with arithmetically definable sequences of coefficients are arithmetically definable, and have arithmetically definable values at arithmetically definable points. Hence, all of the elementary functions including the exponential, the logarithm and the trigonometric functions are arithmetically definable. Also, at least for continuous functions, the theory of differentiation and integration including the fundamental theorem of calculus is developed in the same manner as in the more customary analysis.

Weyl showed that a significant portion of classical analysis can be constructed on a strictly predicative basis that involves quantification only over the natural numbers. He thereby provided grounds for the hope that a secure foundation for all of analysis might be found. For a discussion of subsequent work on predicative analysis, the reader is referred to the work of (Feferman, 1964, 1985, 1988), (Grzegorczyck 1954) and (Lorenzen, 1971).

6.5 The Continuum as a Medium of Free Becoming

As pointed out above, Weyl achieved a great deal in his foundational work on analysis published in 1918 in *Das Kontinuum*. It is therefore surprising that just a few years later, he essentially repudiated his 1918 programme and enthusiastically endorsed Brouwer's intuitionistic programme despite the fact that Brouwer's intuitionism would impose severe limitations on the domain of mathematics, limitations that surpassed *by far* those that were a consequence of the programme presented in *Das Kontinuum*. According to a letter (see (van Dalen 1995, p.162)) dated 28 January 1927 from Brouwer to Fraenkel, Weyl met Brouwer during the summer of 1919 in Engadin and was converted[113] to Brouwer's views as a result of private discussions with Brouwer.[114] It would appear that two aspects of Brouwer's ideas attracted Weyl. First, Weyl thought that Brouwer's conception of a real number as a free-choice, becoming sequence of nested intervals and the associated conception of the continuum as a medium of free becoming captured the essence of the continuum much more naturally than the atomistic conception of the continuum presented in *Das Kontinuum* of 1918. Second, Weyl followed Brouwer in rejecting the principle of the excluded middle not only for existential statements concerning sequences of natural numbers, the set of which is not extensionally well defined, but also for existential statements about the natural numbers themselves. Since the constructive principles employed in *Das Kontinuum* freely used existential and universal quantification over the natural numbers, Weyl's rejection of the principle of the excluded middle for existential statements about the natural numbers meant that he regarded the analysis of *Das Kontinuum* as completely undermined. Later, however, Weyl (1918a, Preface to the 1932 reprint) thought that there was still considerable value in the analysis of *Das Kontinuum* in view of the fact that the intuitionistic programme of Brouwer had not led to "an even moderately satisfying or defensible conclusion".

The structure of the argument in part II of Weyl's (1921d) paper proceeds from a strictly epistemic view of a real number as a sequence of better and better approximations each of which is determined by an open interval which contains the real number. Because such an approximation can in practice only be presented to some finite degree, the sequence must be one that is always in the process of *becoming* better and better. Having enticed the reader to accept this epistemic view of a real number, Weyl then analyses the meaning of an existential statement $\exists \alpha \phi(\alpha)$, where α is a real variable. Weyl concludes that the principle of the excluded middle is not valid for such existential statements and that such statements are in fact an "empty invention of the logicians". Weyl thereby makes it *appear* that his conclusion that the principle of the excluded middle is not valid

---

[113]For a discussion of the difference between the intuitionism of Brouwer and the intuitionism of Weyl, see the paper by Majer (1988). For a discussion of some possible cultural-historical influences underlying Weyl's adoption of Brouwer's intuitionism see (Mehrtens 1990).

[114]Dalen's paper is a useful source of information concerning the relationship between Brouwer and Weyl; however, in some respects Dalen's paper should be approached with caution since some of the issues that probably could have been clarified have been left ambiguous.

for such existential statements follows as a logical consequence of *mathematical* considerations rather than of *philosophical* considerations. In several places in this context, Weyl asserted that to adopt the contrary view was foolish (unvernünftig) and meaningless (sinnlos). It seems reasonable to conclude, therefore, that Weyl did believe that the invalidity of the principle of the excluded middle follows from mathematical considerations. Weyl's reasoning here may be successful as a piece of polemics[115] but it ultimately fails because there is nothing intrinsic to the mathematics that compels one to adopt an epistemic view of real numbers and hence to adopt a version of mathematical truth that is epistemically constrained as opposed to objective. One can adopt a view of real numbers that is consistent with a verificationist-transcendent view of truth in mathematics in which case the law of the excluded middle for existential statements about real numbers is perfectly acceptable. Ultimately, therefore, the dispute comes down to whether one accepts a verificationist-transcendent notion of mathematical truth or one that is epistemically constrained. Contrary to the impression one gets from Weyl's manner of presentation, the issue is a matter of philosophy and not one of mathematics.

Weyl begins his exposition with an account of the new concept of *real number*. The concept is firmly epistemic in character. He says that if the decimal expansion of a real number α is known up to the k-th decimal place with an error of ± 1 in the k-th place, then the real number α lies inside the open interval

$$\left(\frac{m-1}{10^k}, \frac{m+1}{10^k}\right), \tag{6.5.47}$$

where m is a certain integer. Thus, to a real number, there corresponds a sequence of nested open intervals that are characterized by two natural numbers m and k. As an example, consider the decimal expansion $2.42674\ldots$, which determines the open intervals

$$\left(\frac{23}{10}, \frac{25}{10}\right), \left(\frac{241}{10^2}, \frac{243}{10^2}\right), \left(\frac{2425}{10^3}, \frac{2427}{10^3}\right),$$
$$\left(\frac{24266}{10^4}, \frac{24268}{10^4}\right), \left(\frac{242673}{10^5}, \frac{242675}{10^5}\right), \ldots \tag{6.5.48}$$

The corresponding pairs of integers (m, k) are $(24, 1)$, $(242, 2)$, $(2426, 3)$, $(24267, 4)$ and $(242674, 5)$ respectively. For the sake of mathematical simplicity, Weyl switches to binary intervals

$$\left(\frac{m-1}{2^k}, \frac{m+1}{2^k}\right), \tag{6.5.49}$$

---

[115] In a letter (van Dalen 1995, p. 147) written to Brouwer, in May of 1920, Weyl says: "Finally I have sent what I have long promised you [A copy of his paper on the 'New Crisis' (Weyl 1921d)]. It should not be viewed as a scientific publication, but as a propaganda pamphlet; thence the size. I hope that you will find it suitable for this purpose, and moreover suited to rouse the sleepers; that is why I want to publish it".

where m and k are integers (ganze Zahlen) in binary form. Weyl notes that one must use open intervals of the form (6.5.49) rather than intervals of the form

$$\left(\frac{m}{2^k}, \frac{m+1}{2^k}\right) \tag{6.5.50}$$

for fixed k and variable m because the open intervals (6.5.49) overlap as follows

$$\tag{6.5.51}$$

whereas the intervals (6.5.50) do not; consequently, for fixed k, the intervals (6.5.49) cover the real line, but the intervals (6.5.50) do not.

Just after introducing the concept of a real number as a sequence of nested, open intervals of the form (6.5.49), Weyl (1921d, p. 49, GA II, p. 152) says:

> Da jedes der Dualintervalle durch zwei ganzzahlige Charaktere gekenn-zeichnet werden kann ... [m and h in (6.5.49)] und das Enthaltensein eines Intervalls in einem andern sich durch eine einfache Relation zwi-schen diesen ihren Charakteren ausdrückt, so bedeutet es nur eine un-wesentliche Vereinfachung unserer Überlegungen, wenn wir zunächst statt der Folgen ineinander geschachtelter Dualintervalle keiner Ein-schränkung unterworfene *Folgen natürlicher Zahlen* betrachten.

> [Since every binary interval may be characterized by means of two whole numbers ... and since the inclusion of an interval in another in-terval may be expressed through a simple relation between these whole numbers, it is therefore an inessential simplification of our considera-tions if we instead consider *sequences of natural numbers* that are not restricted rather than sequences of nested binary intervals.]

Weyl (1921d, p. 51, GA II, p. 154) also uses the relationship between real numbers and sequences of natural numbers to establish that the real numbers are uncount-able. See remark 59 below and the following paragraph. The relationship between real numbers and sequences of natural numbers is not, however, discussed until the beginning of section 4 of Weyl's paper. The following is an explanation of what Weyl says.

The open binary interval

$$\left(\frac{n-1}{2^{k+1}}, \frac{n+1}{2^{k+1}}\right) \tag{6.5.52}$$

is contained in the open interval (6.5.49) if and only if

$$2m - 1 \leq n \leq 2m + 1; \tag{6.5.53}$$

consequently, $n \in \{2m - 1, 2m, 2m + 1\}$. These values for n determine three open intervals (6.5.52) that are related to the open interval (6.5.49) as indicated in the following diagram.

(6.5.54)

Note that only one of the three smaller intervals does not share a boundary point with the larger open interval. If an open interval i_2 is contained in an open interval i_1 and if the interval i_2 does not share a boundary point with i_1, then Weyl (1921d, p. 71, GA II, p. 171) says that i_2 lies *entirely* within i_1 (ganz im Innern). The reason for considering only sequences of nested open intervals such that a given interval lies *entirely* within the preceding interval is that a sequence of the form

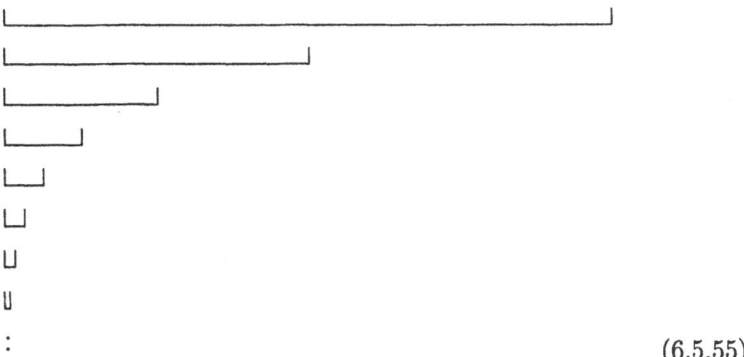

(6.5.55)

has a limit that is not *inside* any of the intervals. It follows that of the three possibilities allowed by (6.5.53), only $n = 2m$ is permissible. Therefore, if there is to be any real variability in a sequence of binary intervals, the number k *cannot* be constrained so that it increases uniformly by 1 between successive intervals of the sequence. Weyl reduces the problem of choosing a sequence of nested binary intervals to that of choosing an initial binary interval i together with a sequence of natural numbers n_j, where $j \in \mathbb{N}$. At the first level for which the length of the initial interval i is divided by 2^1 as illustrated in (6.5.54), there is only one suitable subinterval, which is assigned the number 1. At the second level, the length is divided by 2^2 and the pattern of subintervals is as follows.

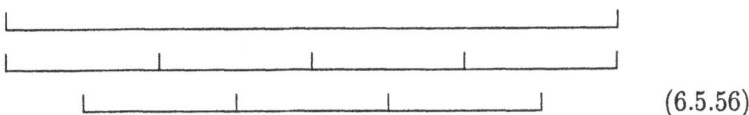

(6.5.56)

The two subintervals that share a common boundary point with i are discarded, and the other 5 are numbered $2, 3, 4, 5$ and 6 from left to right and top to bottom.

At the next level, the length of i is divided by 2^3 and one obtains the following pattern.

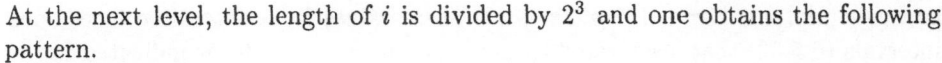

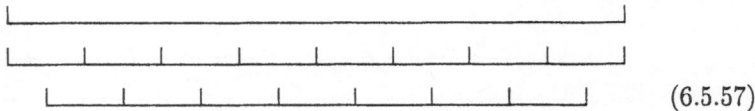

(6.5.57)

The 13 subintervals that are *entirely* contained in i are numbered from left to right and top to bottom $7, 8, 9, \ldots, 18$ and 19 respectively. This process of subdividing i may be continued indefinitely. At level n, one obtains

$$2^n + (2^n - 1) - 2 = 2^{n+1} - 3 \tag{6.5.58}$$

additional, acceptable subintervals. Thus, a binary interval i and a sequence of natural numbers n_j, where $j \in \mathbf{N}$, determines a sequence of binary intervals $i, i_1, i_2 \ldots$, where i_1 is the subinterval of i with number n_1, i_2 is the subinterval of i_1 with number n_2, i_3 is the subinterval of i_2 with number n_3, and so forth.

Remark 58 It should be noted that the sequence of open intervals of the type (6.5.49) that is determined by a binal expansion of a number differs somewhat from the sequence just described. Suppose one has a binal expansion $0.d_1 d_2 \ldots d_n \ldots$ where $d_j \in \{0, 1\}$. Then, the open interval that is centered on $0.d_1 d_2 \ldots d_n$ is given by (6.5.49) with $m = d_1 d_2 \ldots d_n$ and $k = n$. If the next digit $d_{n+1} = 0$, then the next open interval is centered on the same point but is of half the length; that is, the successive intervals are related as in the following diagram.

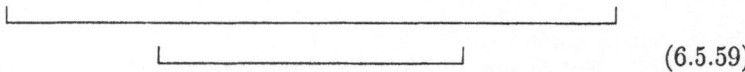

(6.5.59)

However, if the next digit $d_{n+1} = 1$, then the next interval is shifted to the right as in the following diagram.

(6.5.60)

The two successive intervals share a common boundary point, which is *not* a pattern that the sequences of nested intervals that are considered by Weyl may have. As an example, consider the successive open intervals determined by 1.10110110, where the first interval is centered on 1.1.

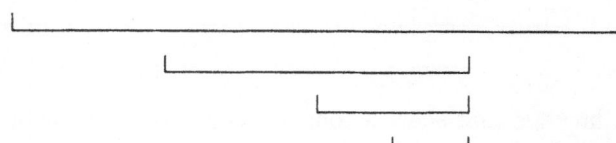

$$\begin{array}{l} \sqsubset\!\sqsupset \\ \sqcup\!\!\sqcup \\ \cup \\ \| \\ \vdots \end{array} \qquad (6.5.61)$$

Of course, one can pick out of such a sequence a subsequence that does satisfy the formation rules proposed by Weyl, namely, the subsequence of all those open intervals that correspond to adding a zero digit. On the other hand, given a sequence of open intervals that satisfy Weyl's requirements and that are characterized by (m_j, k_j) for $j \in \mathbf{N}$, the binary fractions $(m_j - 1)/2^{k_j}$ always increase and provide ever more precise binal expansions. Although at any finite order, the conversion between binal expansions and Weyl's sequences of *completely* nested, open intervals is straight-forward, Brouwer's (1921) article on the existence of reals that do not have decimal expansions indicates that there are subtleties in the limit of infinite sequences.

Not only is a real number regarded as defined by an initial interval together with a sequence of *completely* nested, open[116] intervals, but the concept of a *sequence* is quite different from the one that was and had been in general use. Following Brouwer, Weyl regards a sequence as consisting of an *unbounded*, as opposed to *actually infinite*, series of free or arbitrary choices. One can always add an additional term to the sequence, but the sequence is never completed. Weyl speaks of a *becoming* (werdende) sequence. One can only pose questions concerning such sequences that depend for their answers only on that part of the sequence that has already been specified such that the answers cannot be affected by the subsequent development of the sequence. The view of a real number as a *becoming, free-choice* sequence of completely nested, open intervals is decidedly epistemic in character. A real number is *known* only to the extent that it is bracketed, that is, to the extent that it is known to lie in a given open interval. Given an approximate specification of a real number by means of an open interval, one can always choose to specify it more precisely by freely choosing a smaller, completely nested subinterval; however, although such refinement can be extended without limit, it can never be fully completed. With regard to the concept of *sequence* that was generally accepted by the mathematical community at that time, Weyl notes that the individual terms of the sequence were also arbitrary; however, the sequence was not *becoming*. In some way, each and every one of the infinite number of terms was regarded as being fixed; moreover, statements that involved the entire sequence, not just some finite portion of it, were regarded as determinate in the sense that they were either true or false. Weyl regarded this standpoint as senseless and untenable. With regard to the *concept of sequence* Weyl (1921d, pp. 49–50, GA II, p. 152) says

---

[116]It is interesting to know that Brouwer (1921) used *closed* intervals. In addition, Brouwer seems to have required that they are merely nested rather than completely nested.

Die Schwierigkeit liegt im Begriff der *Folge*. Wenn der heutigen Analysis überhaupt ein Standpunkt zugrunde liegt, von dem aus ihre Aussagen and Beweise verständlich sind, so ist es der: die Folge entsteht dadurch, dass die einzelnen Zahlen der Reihe nach willkürlich gewählt werden; das Resultat dieser unendlich vielen Wahlakte liegt fertig vor, und im Hinblick auf die fertige unendliche Folge kann ich z. B. fragen, ob unter ihren Zahlen die 1 vorkommt. Aber dieser Standpunkt ist sinnwidrig und unhaltbar; denn die Unerschöpflichkeit liegt im Wesen des Unendlichen. Eine einzelne *bestimmte* (und ins Unendliche hinaus bestimmte) Folge kann nur durch ein *Gesetz* definiert werden. Entsteht hingegen eine Folge Schritt für Schritt durch freie Wahlakte, so will sie als eine *werdende* betrachtet sein, und nur solche Eigenschaften können sinnvollerweise von einer werdenden Wahlfolge ausgesagt werden, für welche die Entscheidung "ja oder nein" (kommt die Eigenschaft der Folge zu oder nicht) schon fällt, wenn man in der Folge bis zu einer gewissen Stelle gekommen ist, ohne dass die Weiterentwicklung der Folge über diesen Punkt des Werdens hinaus, wie sie auch ausfallen möge, die Entscheidung wieder umstossen kann. So können wir mit Bezug auf eine Wahlfolge wohl fragen, ob in ihr an vierter Stelle die Zahl 1 auftritt, aber nicht, ob in ihr die Zahl 1 überhaupt nicht auftritt.

[The difficulty lies in the concept of *sequence*. If today's analysis even has an underlying point of view through which its propositions and proofs are understandable, then it is the following: the sequence comes about, in that the individual numbers are arbitrarily chosen one after the other. The result of these infinitely many choice acts lies completed before us, and in view of the completed infinite sequence, I can for example ask whether the number 1 occurs among its numbers. But this standpoint is nonsensical and untenable since inexhaustibility is part of the essence of infinity. A single *determined* (determined to infinity) sequence can only be defined by a *law*. If in contrast a sequence results step by step through free-choice acts, then it must be viewed as a *becoming* sequence and only such properties can be meaningfully asserted of a choice sequence, for which the decision "yes or no" (does the sequence have the property or not) can already be made when one has reached a certain point of the sequence such that any further development of the sequence beyond this point, whatever this development may be, cannot invalidate that decision. We thus may ask of a choice sequence whether or not the number 1 appears at its fourth place, but we may not ask whether the number 1 is completely absent from it.]

Weyl attributes to Brouwer the insight that *free-choice, becoming* sequences are possible objects for mathematical constructions. A simple example is given to illustrate the claim; however, many more detailed examples are presented later in connection with the concept of function.

Remark 59 Weyl does allow for real numbers that are determined by law-like sequences. Given the above discussion, such a real number could be determined by an initial interval i and by a recursively definable sequence $\phi: \mathbf{N} \to \mathbf{N}$. Such real numbers are referred to as single real numbers. These numbers are regarded as being in the continuum but as constituting only a small part of it. Paraphrasing Brouwer, Weyl (1921d, GA II, p. 153) says:

> Die Brouwersche Bemerkung ist einfach, aber tief: hier ersteht uns ein "Kontinuum", in welches wohl die einzelnen reellen Zahlen hineinfallen, das sich aber selbst keineswegs in eine Menge fertig seiender reeller Zahlen auflöst; vielmehr ein *Medium freien Werdens*.

> [Brouwer's remark is simple but deep; here arises a "continuum" in which no doubt the individual real numbers fall but which does not in any way dissolve itself into a set of pre-existing numbers; rather a *medium of free becoming*.]

Next, Weyl addresses the issue concerning the logical principle of the excluded middle. First, however, he notes that the domain of the real numbers, regarded as free-choice, becoming sequences of completely nested intervals, is uncountable. He points out that any map that generates an integer from a free-choice, becoming sequence of natural numbers, which he calls a *functio mixta*, must be such that the integer value is determined by some finite portion of the becoming sequence and remains unaffected by the subsequent development of the sequence; consequently, the map cannot possibly be injective. It follows that the domain of real numbers is extensionally indefinite; therefore, the question "Does there exist a real number that has a given property $\mathcal{E}$?" does not have a clear meaning. He notes that the question can only have a meaning if the domain of quantification is restricted to an extensionally definite subdomain of law-like reals such as the domain he used in *Das Kontinuum*. He concludes that one has the right to assert an existential statement $\exists \alpha \mathcal{E}(\alpha)$ only if one can construct a law that determines a real number that has the property $\mathcal{E}$. In addition, he insists that it is not enough to speak of the *possibility* of constructing such a real. Rather, one must *actually achieve* such a construction. Next, Weyl reformulates the denial of the existence claim, namely $\neg \exists \alpha \mathcal{E}(\alpha)$, as the positive assertion $\forall \alpha \neg \mathcal{E}(\alpha)$ or $\forall \alpha \bar{\mathcal{E}}(\alpha)$. He asserts that this statement has a meaning only insofar as the domain of quantification is the entire continuum, the domain of all free-choice, becoming sequences and that such a universal claim can only be justified on the basis of an *insight* regarding the essence of free-choice, becoming sequences. Weyl concludes that since there is not a sharp, well defined boundary that separates the cases for which either the one or the other assertion holds, it would be absurd to think of a complete or exhaustive disjunction. Thus, for an existential claim about real numbers, the law of the excluded middle does not hold.

Weyl subsequently points out that Brouwer goes even further in denying the principle of the excluded middle. Brouwer denies this principle not only for existential claims about sequences[117] of natural numbers but also for existential assertions about the natural numbers themselves despite the fact that the domain of natural numbers, in contrast with the domain of sequences of natural numbers, is extensionally definite. Noting that he formerly held that the principle of the excluded middle was valid for existential statements about natural numbers, Weyl critically examines the arguments that support this important principle. One is tempted, he says, to imagine that the natural numbers are spread out in a series $1, 2, 3, \ldots$, and to think that one can run through this series and test each number in turn as to whether or not it has a given property $\mathcal{E}$. If one encounters a natural number n such that $\mathcal{E}(n)$ holds, then $\exists n \mathcal{E}(n)$ holds. If one does not encounter such a natural number, then $\neg \exists n \mathcal{E}(n)$ holds. He says that the fallacy in this line of reasoning is the idea that one can actually *run through* the *infinite* series of natural numbers. Reporting on his own struggle with Brouwer's views on the law of the excluded middle, Weyl (1921d, p. 54, GA II, p. 156) says

> ... so wurde ich doch immer wieder auf meinen alten Standpunkt zurückgeworfen durch den Gedanken: Durchlaufe ich die Reihe der Zahlen und breche ab, falls ich eine Zahl von der Eigenschaft $\mathcal{E}$ finde, so tritt dieser Abbruch entweder einmal ein oder nicht; *es ist so, oder es ist nicht so,* ohne Wandel und Wank und ohne eine dritte Möglichkeit. Man muß solche Dinge nicht von außen erwägen, sondern sich innerlich ganz zusammenraffen und ringen um das "Gesicht", die Evidenz.

> [... I would always be thrown back to my old standpoint through the thought: If I run through the row of numbers and stop in case I find a number which has the property $\mathcal{E}$, then this breaking-off either occurs once or it does not: *it is so or it is not so,* purely and simply and without a third possibility. One must not consider such things from the outside, but one must internally pull oneself together and struggle for the "vision", the evidence.]

Weyl then says that he overcame his old standpoint through the realization that an existential assertion such as "there exists an even number" is not a judgement in the true sense which asserts a state of affairs, but is rather an "empty invention of the logicians". Weyl says

> Endlich fand ich für mich das erlösende Wort. *Ein Existentialsatz —* etwa "es gibt eine gerade Zahl" — *ist überhaupt kein Urteil im eigentlichen Sinne, das einen Sachverhalt behauptet;* Existential-Sachverhalte sind eine leere Erfindung der Logiker. "2 ist eine gerade Zahl": das ist ein wirkliches, einem Sachverhalt Ausdruck gebendes Urteil: "es gibt eine gerade Zahl" ist nur ein aus diesem Urteil gewonnenes *Urteilsabstrakt.* Bezeichne ich Erkenntnis als einen wertvollen Schatz, so ist das

---

[117]Recall that a real number is determined by an initial interval i together with a sequence of natural numbers.

Urteilsabstrakt ein Papier, welches das Vorhandensein eines Schatzes anzeigt, ohne jedoch zu verraten, an welchem Ort. Sein einziger Wert kann darin liegen, dass es mich antreibt, nach dem Schatze zu suchen. Das Papier ist wertlos, solange es nicht durch ein solches dahinter stehendes wirkliches Urteil wie "2 ist eine gerade Zahl" realisiert wird.

[Finally I found for me the magic word. *An existential sentence* — such as "there exists an even number" — *is not at all a judgement in the true sense which asserts a fact.* Existential facts are an empty discovery of the logicians. "2 is an even number": that is a genuine judgement which expresses a fact: "there exists an even number" is a *judgement-abstract* obtained from this judgement. If I characterize knowledge as a valuable treasure, then the judgement-abstract is a piece of paper which announces the existence of a treasure without however giving away its location. Its only value can lie in this, that it motivates me to look for the treasure. The piece of paper has no value as long as it is not backed up by a genuine judgement such as "2 is an even number".]

Having firmly rejected the validity of the law of the excluded middle for existential statements about the natural numbers, Weyl passes judgement on his previous work on the foundations of analysis presented in *Das Kontinuum*. He says that as long as one accepts the principle of the excluded middle for existential statements about natural numbers, Brouwer's analysis of the continuum and the analysis of the continuum presented in *Das Kontinuum*, must be regarded as equally possible even though Brouwer's analysis of the continuum might be preferred because it is more in accord with our intuitions about the essence of the continuum. However, as soon as the principle of the excluded middle is rejected for existential statements about natural numbers — and Weyl now insists on this rejection — the analysis presented by him in *Das Kontinuum* is radically impossible. Weyl (1921d, pp. 55–56, GA II, p. 157) says:

Sobald man aber den neuen Schritt tut — durch welchen, wie ich glaube, der Sinn des "es gibt" und "jeder" erst völlig klar wird — ist die erste Grundlegung [the analysis of *Das Kontinuum* (1918)] radikal unmöglich; denn die Verengerung des Gesetzesbegriffs auf den des $\mathcal{K}$-Gesetzes hilft uns dann nichts; die Frage nach der "Möglichkeit" stellt uns ebensowenig einem mit Ja oder Nein antwortenden Sachverhalt gegenüber, wenn sie gestellt wird mit Bezug auf die beliebig oft zu wiederholende Anwendung der Konstruktionsprinzipien wie mit Bezug auf die unendliche Zahlenreihe, d. i. den beliebig oft zu wiederholenden Prozess des Übergangs von einer Zahl zur nächstfolgenden. So gebe ich also jetzt meinen eigenen Versuch preis und schließe mich Brouwer an. In der drohenden Auflösung des Staatswesens der Analysis, die sich vorbereitet, wenn sie auch erst von wenigen erkannt wird, suchte ich festen Boden zu gewinnen, ohne die Ordnung, auf welcher es beruht, zu verlassen, indem ich ihr Grundprinzip rein und ehrlich durchführte; und ich glaube, das gelang — soweit es gelingen konnte. Denn *diese Ordunung ist nicht haltbar in sich*, wie ich mich jetzt überzeugt habe, und Brouwer — das ist die Revolution!

[The first foundation [the analysis of *Das Kontinuum* (1918)] is impossible, however, as soon as one takes the new step — through which, as I believe, the sense of "there exists" and "every" becomes completely clear; for the restriction of the concept of law to that of the $\mathcal{K}$-law will then not help us; the question about the "possibility" [the possibility of constructing a natural number that has a given property] does not present us with a state of affairs that provides a yes-or-no answer any more if it is posited with respect to the arbitrarily often repeated application of the principles of construction than if it is posited with respect to the infinite number sequence, that is, the arbitrarily repeated process of the transition of a number to its successor. I therefore give up my own attempt and join Brouwer. Faced with the advancing threat of the dissolution of the polity of analysis, a prospect which only a few recognized at first, I sought to gain solid ground, without abandoning the order on which the polity of analysis rests, by carrying out the fundamental principle of analysis purely and honestly; and I believe it was successful to the extent that it could have been successful. *For this order is not tenable in itself*, as I have now convinced myself, and Brouwer — that is the revolution!]

Weyl next considers the concept of function of which there are three types: *functio discreta, functio mixta* and *functio continua*. If the set of free-choice, becoming sequences of natural numbers is denoted by $\mathcal{B}(\mathrm{N},\mathrm{N})$, then in modern notation, the functions considered are of the following types: $f\colon \mathrm{N} \to \mathrm{N}$ (discreta); $f\colon \mathrm{N} \to \mathcal{B}(\mathrm{N},\mathrm{N})$ and $f\colon \mathcal{B}(\mathrm{N},\mathrm{N}) \to \mathrm{N}$ (mixta); and $f\colon \mathcal{B}(\mathrm{N},\mathrm{N}) \to \mathcal{B}(\mathrm{N},\mathrm{N})$ (continua).

Remark 60 It is not clear to what extent Weyl wants to identify a real number with an element of $\mathcal{B}(\mathrm{N},\mathrm{N})$. At the beginning of section 4, he suggests some possibilities. One might consider describing a sequence of completely nested, open intervals by means of a sequence of natural numbers n_i given by

$$m_1, k_1, m_2, k_2, m_3, k_3, \ldots, \tag{6.5.62}$$

where

$$n_{2j-1} = m_j \quad \text{and} \quad n_{2j} = k_j, \tag{6.5.63}$$

and the choice of the values of successive pairs is restricted only by the constraint

$$2^{k_{j+1}-k_j}(m_j - 1) + 1 < m_{j+1} < 2^{k_{j+1}-k_j}(m_j + 1) - 1, \tag{6.5.64}$$

in which $k_{j+1} > k_j > 0$. Alternatively, one can describe a real number by means of an initial interval i specified by (m_0, k_0), where $k_0 > 0$, and a free-choice, becoming sequence, that is, by a free-choice, becoming sequence of the form

$$m_0, k_0, n_1, n_2, n_3, \ldots \tag{6.5.65}$$

See remark 58 following (6.5.58) and the preceding discussion. On the other hand, when Weyl discusses real valued functions on the reals, that is, maps $f: \mathbf{R} \to \mathbf{R}$, he defines them in such a way that the focus is on the step-by-step mapping of each of the open intervals in the nested sequence of open intervals which constitutes the real number that is the argument of the function. For an arbitrary element of $\mathcal{B}(\mathbf{N}, \mathbf{N})$, there are no constraints imposed on the choice of the elements of the sequence; however, both (6.5.62) and (6.5.65) involve some constraints. It is true that in the case of the representation (6.5.65), the only constraint imposed is that $k_0 > 0$; however, Weyl's treatment of the maps $f: \mathbf{R} \to \mathbf{R}$ is more in accord with the representation (6.5.62) for which the choices must be made in pairs and for which the constraints are rather more substantial.

A *functio discreta* determines a *law-like* sequence. For any natural number n, a *functio discreta* generates a unique natural number according to a definite law, where the law is strictly inductive in character; for example, Weyl defines $O(n)$ which means 'n is odd' by

$$
\begin{aligned}
&O(1) = 1 \\
&\forall n(O(n) = 1 \implies O(n+1) = 2) \\
&\forall n(O(n) = 2 \implies O(n+1) = 1),
\end{aligned}
\tag{6.5.66}
$$

where for the output, '1' denotes 'true' and '2' denotes 'false'. The alternative definition

$$O(n) = \begin{cases} 1 & \neg\exists m(n = 2m) \quad \text{or} \quad \forall m(n \neq 2m) \\ 2 & \exists m(n = 2m) \end{cases} \tag{6.5.67}$$

is ruled out because the quantifier '$\exists m$' appears in the *definiens*. Weyl explicitly states that it is the form of (6.5.67) (ihrem Wortlaut nach) that makes it unacceptable as a law. Weyl also considers double sequences which are maps of the form $f: \mathbf{N} \times \mathbf{N} \to \mathbf{N}$. In particular, he shows how to define addition and multiplication on $\mathbf{N}$ inductively.

A *functio mixta* of the form $f: \mathbf{N} \to \mathcal{B}(\mathbf{N}, \mathbf{N})$ generates for each $m \in \mathbf{N}$ a *law-like* real, that is, a law-like *functio discreta* because the output of a function must be well defined. Thus, the function f reduces to a *functio discreta* of the form $g: \mathbf{N} \times \mathbf{N} \to \mathbf{N}$ as follows. Noting that $\forall m(f(m): \mathbf{N} \to \mathbf{N})$, define

$$\forall m \forall n \, (f(m)(n) = g(m, n)). \tag{6.5.68}$$

The definition of a *functio mixta* of the form $f: \mathcal{B}(\mathbf{N}, \mathbf{N}) \to \mathbf{N}$ is a more subtle matter because the domain of definition consists of all free-choice, becoming sequences. The essential characteristic of a function $f: \mathcal{B}(\mathbf{N}, \mathbf{N}) \to \mathbf{N}$, Weyl says, is that the natural number that corresponds to a free-choice, becoming sequence must depend only on a finite number of terms of that sequence in such a manner that the output cannot be affected by the subsequent development of the sequence whatever it may be. As noted above, Weyl used this characteristic to establish that the reals are not countable. Weyl gives a few examples of maps $f: \mathcal{B}(\mathbf{N}, \mathbf{N}) \to \mathbf{N}$ including examples in which the number of terms required to determine the output can itself be iteratively determined by some finite portion of the input sequence. Weyl does not, however, attempt to give a general characterization of all the possible principles of construction for such functions.

Weyl begins his discussion of *functiones continuae* by referring to the following example. For each element of $\mathcal{B}(\mathbf{N}, \mathbf{N})$, that is, for each free-choice, becoming sequence

$$m_1, m_2, m_3, \ldots, m_k, \ldots, \tag{6.5.69}$$

the output of $f: \mathcal{B}(\mathbf{N}, \mathbf{N}) \to \mathcal{B}(\mathbf{N}, \mathbf{N})$ is the sequence

$$n_1, n_2, n_3, \ldots, n_k, \ldots \tag{6.5.70}$$

defined by

$$n_1 = m_1 \quad \text{and} \quad n_k = n_{k-1} + m_k; \tag{6.5.71}$$

that is, n_k is the sum of the first k terms of the sequence (6.5.69). As each term in the sequence (6.5.69) is (freely) chosen, the rule (6.5.71) generates the corresponding term in the sequence (6.5.70).

Setting aside the peculiarities of this particular example, Weyl characterizes a *functio continua*, that is, a function $f: \mathcal{B}(\mathbf{N}, \mathbf{N}) \to \mathcal{B}(\mathbf{N}, \mathbf{N})$ as follows:

1. As each term m_k of the free-choice, becoming sequence $m_1, m_2, \ldots$ that constitutes the argument of f arises, either nothing or a definite natural number n_k is generated by f.

2. If at a given step a natural number n_k is generated at the k-th step, then n_k depends *in general* not only on m_k but also on all m_j for $j \leq k$.

3. The nature of f is such that for any input sequence, there is a point beyond which either one knows that for *all* subsequent m_j nothing is generated in which case f is undefined for the given argument sequence or one knows that an *infinite* number of elements of the output sequence that constitutes the value of f will be generated even though 'nothing' may be generated an arbitrary large but finite of number of times between each natural number that is generated.

In section 4 of the paper, Weyl points out that real numbers can be characterized in terms of free-choice, becoming sequences of natural numbers in ways that we have discussed above, in particular following (6.5.52) and in remark 60. He then discusses what it means for two numbers to 'fall together' or to 'be separate'. Weyl notes that these two possibilities do not constitute a complete dichotomy because of the becoming nature of the sequences that determine the real numbers. Essentially, one can never *know* that two real numbers fall together because it is always possible that with the next free choice they will fall apart. Weyl asserts on the basis of his concept of a real number that the continuum cannot be decomposed into three disjoint sets consisting of one of the points of the continuum, the set of elements to its left and the set of elements to its right. Weyl (1921d, p. 73, GA II, p. 173) says:

> Greifen wir auf der Zahlgeraden C, dem Variabilitätsgebiet einer reellen Variablen x, einen bestimmten Punkt heraus, z. B. $x = 0$, so kann man, wie wir sahen, keinesfalls behaupten, dass jeder Punkt entweder mit ihm zusammenfällt oder von ihm getrenntliegt. Der Punkt $x = 0$ zerlegt also das Kontinuum C durchaus nicht in zwei Teile $C^- : x < 0$ und $C^+ : x > 0$, in dem Sinne, dass C aus der Vereinigung von C^-, C^+ und dem einen Punkte 0 bestünde (jeder Punkt entweder mit 0 zusammenfiele oder zu C^- oder zu C^+ gehörte). Erscheint dies dem heutigen Mathematiker mit seiner atomistischen Denkgewöhnung anstössig, so war es in früheren Zeiten eine allen selbstverständliche Ansicht: innerhalb eines Kontinuums lassen sich wohl durch Grenzsetzung Teilkontinuen erzeugen; es ist aber unvernünftig, zu behaupten, dass das totale Kontinuum aus der Grenze und jenen Teilkontinuen zusammengesetzt sei. *Ein wahrhaftes Kontinuum* ist eben ein in sich Zusammenhängendes und *kann nicht in getrennte Bruchstücke aufgeteilt werden*; das widerstreitet seinem Wesen.

> [If we pick a definite point, e.g., $x = 0$, on the real line C, the domain of variability of the variable x, then, as we saw, one cannot assert that every point either coincides with the point or lies separate from it. The point $x = 0$ definitely does not decompose C into two parts $C^- : x < 0$ and $C^+ : x > 0$ in the sense, that C would consist of the union of C^-, C^+ and the one point 0 (every point either coincides with 0 or belongs to C^- or C^+). If this appears repugnant to the atomistic ways of thinking of the modern mathematician, it was in earlier times regarded by everyone as obvious: one may generate partial continua by imposing demarcations within the continuum; but it is foolish to assert that the total continuum consists of the boundary and the partial continua. *A true continuum* is one that hangs together and *cannot be partitioned into parts*; that goes against its essence.]

Next, Weyl discusses the usual operations for real numbers. Throughout the analysis, a real number is regarded as a free-choice, becoming sequence of completely nested binary intervals each of the form (6.5.49). It is required that a function $f: \mathbf{R} \to \mathbf{R}$ map each binary interval in the sequence that describes an argument α into a binary interval of the sequence that describes the value $\beta = f(\alpha)$ of the function.

Weyl uses the following generalization of the usual notation for denoting open intervals. Let $a_1, a_2, \ldots$ be a finite list of binary fractions. Then, $(a_1, a_2, \ldots)$ denotes the binary interval (b_1, b_2) where b_1 is the least of the a_i and b_2 is the greatest of the a_i.

As a simple example, he first considers the function x^2. Let

$$i = (a_1, a_2) \tag{6.5.72}$$

be one of the binary intervals of the real argument. Then, the corresponding output interval, denoted by i_{x^2}, is given by

$$i_{x^2} = (a_1^2, a_1 a_2, a_2^2). \tag{6.5.73}$$

There are several cases to consider. If $a_1 < a_2 < 0$, then $i_{x^2} = (a_2^2, a_1^2)$. If $0 < a_1 < a_2$, then $i_{x^2} = (a_1^2, a_2^2)$. Finally, if $a_1 < 0 < a_2$, then $i_{x^2} = (a_1 a_2, a_1^2)$ if $|a_1| \geq |a_2|$ and $i_{x^2} = (a_1 a_2, a_2^2)$ if $|a_2| > |a_1|$. The output sequence of completely nested, open intervals is generated from the input sequence of completely nested, open intervals on an *interval-by-interval* basis.

The binary operation xy of multiplication is considered next. In this case, two input binary intervals are considered, one for x and one for y.

$$i_x = (a_1, a_2) = \left(\frac{m-1}{2^k}, \frac{m+1}{2^k}\right), \quad i_y = (b_1, b_2) = \left(\frac{n-1}{2^k}, \frac{n+1}{2^k}\right). \tag{6.5.74}$$

The output interval

$$i_{xy} = (a_1 b_1, a_1 b_2, a_2 b_1, a_2 b_2). \tag{6.5.75}$$

Which of the numbers listed is the least and which is the greatest depends on the signs and relative magnitudes of the binary fractions a_1, a_2, b_1, b_2. There are many cases; for example, if $0 < a_1 < a_2$ and $0 < b_1 < b_2$ then

$$i_{xy} = (a_1 b_1, a_2 b_2). \tag{6.5.76}$$

Of course, for $x = y$, i_{xy} is just i_{x^2}. Although Weyl does not give the corresponding formulas for division, addition and subtraction, they are given here for completeness. For division x/y the output interval is

$$i_{x/y} = (a_1/b_1, a_1/b_2, a_2/b_1, a_2/b_2). \tag{6.5.77}$$

Even though it is assumed that $y \neq 0$, some of the intervals (b_1, b_2) at the beginning of the sequence for y may contain 0. For these intervals, the output interval (6.5.77) does not in general make sense; however, one of the two conditions

$$b_1 < b_2 < 0 \quad \text{or} \quad 0 < b_1 < b_2 \tag{6.5.78}$$

will eventually be satisfied by an interval and therefore for every subsequent interval, and 6.5.77 then yields appropriate output intervals. The start of the output sequence is delayed. For addition and subtraction the output intervals are given by

$$i_{x+y} = (a_1 + b_1, a_2 + b_2) \tag{6.5.79}$$

and

$$i_{x-y} = (a_1 - b_2, a_2 - b_1) \tag{6.5.80}$$

respectively. By combining these rules, one can develop a rule that will generate an output sequence for any rational function of any finite number of real variables.

As a final example, Weyl discusses in detail what it means for the identity

$$(x + y)(x - y) = x^2 - y^2 \tag{6.5.81}$$

to be valid. The output sequence generated for the left side will in general differ from the output sequence generated for the right side, but at each level, the corresponding intervals will overlap.

Following the discussion of particular examples, Weyl gives a general characterization of the concept of continuous function (stetige Funktion). Fundamentally, it is a law ϕ which assigns to each binary interval i a binary interval $\phi(i)$ in such a way that if a sequence of intervals $i_1, i_2, i_3, \ldots$ is completely nested then so is the sequence $\phi(i_1), \phi(i_2), \phi(i_3), \ldots$. The definition is somewhat complicated for several reasons. For some intervals i, there may be no output interval as in the case of division discussed above. In addition, Weyl allows for the possibility that in the sequence of output intervals, some of them may not lie *completely* inside the preceding one. A rather complicated condition ensures that it is always possible to select an *infinite* subsequence of completely nested, open intervals as the value of the function.

Remark 61 Note that for a continuous function, the output sequence is generated on an interval-by-interval basis. The k-th output interval depends only on the k-th input interval. For a *functio continua*, on the other hand, the k-th output integer n_k may depend on all of the members m_j of the input sequence up to and including the k-th element m_k. Clearly, there are *functio continua* that are not continuous functions (stetige Funktionen). Given the relationship discussed in remark 60 between real numbers and free-choice, becoming sequences of natural numbers, however, it would appear that every continuous function is a *functio continua*.

Weyl then goes on to assert that on a continuum, only *continuous* functions exist. He adds that the fact that the old analysis allowed for the possibility of discontinuous functions showed in the clearest way possible just how far it deviated from the essence of the continuum. Weyl considers the problem of pasting together several continuous functions on several continua. He denotes the real continuum by C and the subcontinua of positive and negative reals by C^+ and C^- respectively. First, he considers the continuous function $f_+(x) = x$ defined on C^+ by assigning to each binary interval (a_1, a_2) with $0 < a_1 < a_2$ the interval (a_1, a_2) and the continuous function $f_-(x) = -x$ defined on C^- by assigning to each binary interval (a_1, a_2) with $a_1 < a_2 < 0$ the interval $(-a_2, -a_1)$. To these two functions there corresponds a single function $f(x) = |x|$ defined on all of C such that f coincides with f_- on C^- and with f_+ on C^+. For the binary intervals that lie totally within C^- or C^+, one makes the same assignments as before. To an interval (a_1, a_2) such that $a_1 < 0 < a_2$, one assigns the interval

$$(a_1, -a_2, -a_1, a_2), \tag{6.5.82}$$

that is $(a_1, -a_1)$ or $(-a_2, a_2)$ whichever is the larger of the two intervals. On the other hand, consider the function $f_{+1}(x) = 1$ defined on C^+ by assigning to each binary interval (a_1, a_2) with $0 < a_1 < a_2$ the interval

$$\left(1 - \frac{(a_1 + a_2)}{2}, 1 + \frac{(a_1 + a_2)}{2}\right) \tag{6.5.83}$$

and the function $f_{-1}(x) = -1$ defined on C^- by assigning to each binary interval (a_1, a_2) with $a_1 < a_2 < 0$ the interval

$$\left(-1 + \frac{(a_1 + a_2)}{2}, -1 - \frac{(a_1 + a_2)}{2}\right). \tag{6.5.84}$$

In this case, however, there does not exist a function f defined on all of C which coincides with f_{-1} on C^- and with f_{+1} on C^+ because any binary interval (a_1, a_2) with $a_1 < 0 < a_2$ must be assigned to a binary interval that contains both -1 and $+1$. A function value for any free-choice, becoming sequence such that every element of the sequence contains zero cannot have a sequence of completely nested open intervals with widths that converge to zero as a value.

Bibliography

Anderson, J. L.: 1967, *Principles of Relativity Physics*, Academic Press, New York.

Bäuerle, G. G. A. and de Kerf, E. A.: 1990, *Lie Algebras*, Vol. 1 of *Studies in Mathematical Physics*, North Holland, Amsterdam.

Benacerraf, P. and Putnam, H. (eds): 1983, *Philosophy of Mathematics*, 2 edn, Cambridge University Press, Cambridge.

Bohr, N. and Rosenfeld, L.: 1933, Zur Frage der Messbarkeit der elektromagnetischen Feldgrössen, *Kgl. Danske Videnshab. Selskb, Mat.-Phys. Medd* **12**(8).

Brouwer, L. E. J.: 1911a, Beweis der Invarianz der Dimensionenzahl, *Mathematische Annalen* **70**, 161–165. Reprinted in (Brouwer 1976b, pp. 430–434).

Brouwer, L. E. J.: 1911b, Beweis der Invarianz des n-dimensionalen Gebiets, *Mathematische Annalen* **71**, 305–313. Reprinted in (Brouwer 1976b, pp. 477–485).

Brouwer, L. E. J.: 1911c, Über Abbildung von Mannigfaltigkeiten, *Mathematische Annalen* **71**, 97–115. Reprinted in (Brouwer 1976b, pp. 454–472).

Brouwer, L. E. J.: 1912, Zur Invarianz des n-dimensionalen Gebiets, *Mathematische Annalen* **72**, 55–56. Reprinted in (Brouwer 1976b, pp. 509–510).

Brouwer, L. E. J.: 1921, Besitzt jede reelle Zahl eine Dezimalbruch-Entwickelung?, *KNAW Verslagen* **29**, 803–812. Also: Math. Annalen **83**, pp. 201–210. Reprinted in (Brouwer 1976a, pp. 236–245).

Brouwer, L. E. J.: 1924, Über die Bedeutung des Satzes vom ausgeschlossenen Dritten in der Mathematik, insbesondere in der Funktionentheorie, *J. reine angew. Math.* **154**, 1–7. For an English translation, see (Brouwer 1976a, pp. 268–274).

Brouwer, L. E. J.: 1976a, *Collected Works*, Vol. 1, North Holland, New York. A. Heyting (ed.).

Brouwer, L. E. J.: 1976b, *Collected Works*, Vol. 2, North Holland, New York. Hans Freudenthal (ed.).

Cartan, E.: 1894, *Sur la structure des groupes de transformations finis et continus*, PhD thesis, Paris. Reprinted in (Cartan 1952–1955, 1.1, 137–287).

Cartan, E.: 1913, Les groupes projectifs qui ne laissent invariante aucune multiplicité plane, *Bull. Soc. Math. France* **41**, 53–96. Reprinted in (Cartan 1952–1955, 1.1, 355–398).

Cartan, E.: 1914, Les groupes projectifs continus réels qui ne laissent invariante aucune multiplicité plane, *J. Math. Pures Appl.* **10**(6), 149–186. Reprinted in (Cartan 1952–1955, 1.1, 493–530).

Cartan, E.: 1922, Sur un théorème fondamental de M. H. Weyl dan la théorie de l'espace métrique, *Comptes Rendus de l'Academie des Sciences* **175**, 82–85. Reprinted in (Cartan 1952–1955, 3.1, 629–632).

Cartan, E.: 1923a, Sur les variétés à connexion affine et la théorie de la relativité générallisée, *Annales de l' École Normale Supérieure* **40**, 325–412. Reprinted in (Cartan 1952–1955, 3.1, 659–746).

Cartan, E.: 1923b, Sur un théorème fondamental de M. H. Weyl, *Journal de Mathématique* II(2), 167–192. Reprinted in (Cartan 1952–1955, 3.1, 633–658).

Cartan, E.: 1952–1955, *Oeuvres Complètes*, Vol. 1–3, Gauthiers-Villars, Paris.

Cartan, E.: 1986, *On Manifolds with an Affine Connection and the Theory of General Relativity*, Monographs and Textbooks in Physical Science, Bibliopolis, Napoli.

Castagnino, M.: 1968, Some remarks on the Marzke-Wheeler method of measurement, *Nuovo Cimento* **B 54**, 149–150.

Castagnino, M.: 1971, The Riemannian structure of space-time as a consequence of a measurement method, *Journal of Mathematical Physics* **12**, 2203–2211.

Chern, S.-S.: 1996, Finsler geometry is just Riemannian geometry without the quadratic restriction, *Notices of the American Mathematical Society* **43**(9), 959–963.

Chevalley, C. and Weil, A.: 1968, Hermann Weyl (1885–1955), *in* K. Chandrasekharan (ed.), *Hermann Weyl Gesammelte Abhandlungen*, Vol. IV, Springer Verlag, Berlin, pp. 655–685. Extrait de L'Enseignement Mathématique, tome III, fasc. 3 (1957).

Church, A.: 1976, Comparison of Russell's resolution of the semantical antinomies with that of Tarski, *Journal of Symbolic Logic* **41**, 747–760.

Coleman, R. A. and Korté, H.: 1980, Jet bundles and path structures, *The Journal of Mathematical Physics* **21**(6), 1340–1351.

Coleman, R. A. and Korté, H.: 1981, Spacetime G-structures and their prolongations, *The Journal of Mathematical Physics* **22**(11), 2598–2611.

Coleman, R. A. and Korté, H.: 1982, The status and meaning of the laws of inertia, *The Proceedings of the Biennial Meeting of the Philosophy of Science Association*, Philadelphia, pp. 257–274.

Coleman, R. A. and Korté, H.: 1984a, Constraints on the nature of inertial motion arising from the universality of free fall and the conformal causal structure of spacetime, *The Journal of Mathematical Physics* **25**(12), 3513–3526.

Coleman, R. A. and Korté, H.: 1984b, A realist field ontology of the causal-inertial structure (the refutation of geometric conventionalism), University of Regina Preprint, 192 pages.

Coleman, R. A. and Korté, H.: 1987, Any physical, monopole, equation-of-motion structure uniquely determines a projective inertial structure and an $(n-1)$-force, *The Journal of Mathematical Physics* **28**(7), 1492–1498.

Coleman, R. A. and Korté, H.: 1989, All directing fields that are polynomial in the $(n-1)$-velocity are geodesic, *The Journal of Mathematical Physics* **30**(5), 1030–1033.

Coleman, R. A. and Korté, H.: 1990, Harmonic analysis of directing fields, *The Journal of Mathematical Physics* **31**(1), 127–130.

Coleman, R. A. and Korté, H.: 1992a, On attempts to rescue the conventionality thesis of distant simultaneity in STR, *Foundations of Physics Letters* **5**(6), 535–571.

Coleman, R. A. and Korté, H.: 1992b, The relation between the measurement and Cauchy problems of GTR, *in* H. Sato and T. Nakamura (eds), *The Sixth Marcel Grossmann Meeting on General Relativity*, World Scientific, pp. 97–119. Printed version of an invited talk presented at the meeting held in Kyoto, Japan, 23–29 June 1991.

Coleman, R. A. and Korté, H.: 1994, A semantic analysis of model and symmetry diffeomorphisms in modern spacetime theories, *in* U. Majer and H.-J. Schmidt (eds), *Semantical Aspects of Spacetime Theories*, Wissenschaftsverlag, pp. 83–94.

Coleman, R. A. and Korté, H.: 1995a, A new semantics for the epistemology of geometry I, Modeling spacetime structure, *Erkenntnis* **42**, 141–160.

Coleman, R. A. and Korté, H.: 1995b, A new semantics for the epistemology of geometry II, Epistemological completeness of Newton-Galilei and Einstein-Maxwell theory, *Erkenntnis* **42**, 161–189.

Coleman, R. A. and Korté, H.: 1999, Geometry and forces in relativistic and pre-relativistic theories, *Foundations of Physics Letters* **12**(2), 147–163.

Coleman, R. A. and Mann, R. B.: 1982, Spacetime metric and Yang Mills fields unified in a Galilean subspace structure, *Journal of Mathematical Physics* **23**, 2475–2487.

Deppert, W. and Hübner, K. (eds): 1988, *Exact Sciences and Their Philosophical Foundations: Exakte Wissenschaften und ihre philosophische Grundlegung. Vorträge des internationalen Hermann-Weyl-Kongresses*, Peter Lang Verlag, Frankfurt am Main. Vorträge des Internationalen Hermann-Weyl-Kongresses, Kiel 1985.

Detlefsen, M.: 1986, *Hilbert's Program; An Essay on Mathematical Instrumentalism*, Vol. 182 of *Synthese Library*, Reidel, Dordrecht.

Dieudonné, J.: 1981, *History of Functional Analysis*, North-Holland, New York.

Dirac, P.: 1973, Long range forces and broken symmetries, *Proceedings of the Royal Society* **333A**, 403–418.

Dirac, P. A. M.: 1928a, The quantum theory of the electron I, *Proceedings of the Royal Society (London) A* **117**, 610–624.

Dirac, P. A. M.: 1928b, The quantum theory of the electron II, *Proceedings of the Royal Society (London) A* **118**, 351–361.

Dirac, P. A. M.: 1930, A theory of electrons and protons, *Proceedings of the Royal Society (London) A* **126**, 360–365.

Dirac, P. A. M.: 1931, Quantised singularities in the electromagnetic field, *Proceedings of the Royal Society (London) A* **133**, 60–72.

Dirac, P. A. M.: 1971, *The Development of Quantum Theory*, Gordon and Breach Science Publishers, New York. J. Robert Oppenheimer Memorial Prize Acceptance Speech.

Dirac, P. A. M.: 1977, Recollections of an exciting era, *in* C. Weiner (ed.), *History of Twentieth Century Physics*, Vol. LVII of *Proceedings of the International School of Physics "Enrico Fermi"*, Italian Physical Society, Academic Press, pp. 109–146. The summer school on the history of twentieth century of physics took place from July 31 to August 12, 1972.

Dunford, N. and Schwartz, J. T.: 1963, *Linear Operators Part* II, Interscience.

Dyson, J. D.: 1983, Unfashionable pursuits, *Alexander von Humboldt Stiftung Mitteilung* **41**, 12–18.

Ehlers, J.: 1973a, The nature and structure of spacetime, *in* J. Mehra (ed.), *The Physicist's Conception of Nature*, Reidel, pp. 71–91.

Ehlers, J.: 1973b, Survey of general relativity theory, *in* W. Israel (ed.), *Relativity, Astrophysik and Cosmology*, Reidel, chapter 1, pp. 1–125.

Ehlers, J.: 1988, Hermann Weyl's contributions to the General Theory of Relativity, *in* W. Deppert and K. Hübner (eds), *Exact Sciences and Their Philosophical Foundations: Exakte Wissenschaften und ihre philosophische Grundlegung. Vorträge des internationalen Hermann-Weyl-Kongresses*, Peter Lang Verlag, Frankfurt/M - Bern etc., pp. 83–105.

Ehlers, J. and Köhler, E.: 1977, Path structures on manifolds, *The Journal of Mathematical Physics* **18**(10), 2014–2018.

Ehlers, J., Pirani, R. A. E. and Schild, A.: 1972, The geometry of free fall and light propagation, *in* L. O' Raifeartaigh (ed.), *General Relativity, Papers in Honour of J. L. Synge*, Clarendon Press, Oxford, pp. 63–84.

Ehresmann, C.: 1951a, Les prolongements d' une variété différentiable I, *Comptes rendus des séances de l' Académie des Sciences* **233**, 598–600. Reprinted in (Ehresmann 1983, pp. 343–345).

Ehresmann, C.: 1951b, Les prolongements d' une variété différentiable II, *Comptes rendus des séances de l' Académie des Sciences* **233**, 777–779. Reprinted in (Ehresmann 1983, pp. 346–348).

Ehresmann, C.: 1951c, Les prolongements d' une variété différentiable III, *Comptes rendus des séances de l' Académie des Sciences* **233**, 1081–1083. Reprinted in (Ehresmann 1983, pp. 349–351).

Ehresmann, C.: 1952a, Les prolongements d' une variété différentiable IV, *Comptes rendus des séances de l' Académie des Sciences* **234**, 1028–1030. Reprinted in (Ehresmann 1983, pp. 355–357).

Ehresmann, C.: 1952b, Les prolongements d' une variété différentiable V, *Comptes rendus des séances de l' Académie des Sciences* **234**, 1424–1425. Reprinted in (Ehresmann 1983, pp. 358–360).

Ehresmann, C.: 1983, Charles Ehresmann œuvres complètes et commentées, *in* A. C. Ehresmann (ed.), *Topologie Algébrique et Géométrie Différentielle*, number Suppléments #1 et #2 of Vol. 24 in *Cahiers de Topologie et Géométrie Différentielle*, Evrard, Amiens.

Einstein, A.: 1914, Prinzipielles zur verallgemeinerten Relativitätstheorie und Gravitationstheorie, *Physikalische Zeitschrift* **15**, 176–180.

Einstein, A.: 1916, Die Grundlage der allgemeinen Relativitätstheorie, *Annalen der Physik* **49**(7), 769–822. English translation "The Foundation of the General Theory of Relativity" in (Lorentz, Einstein, Minkowski and Weyl 1952).

Einstein, A.: 1918, Prinzipielles zur allgemeinen Relativitätstheorie, *Annalen der Physik* **55**, 241–244.

Einstein, A.: 1928, Riemann-Geometrie mit Aufrechterhaltung des Begriffes des Fernparallelismus, *Sitzungsberichte der Preussischen Akademie der Wissenschaften, Physikalisch-Mathematische Klasse* **7**, 217–221.

Einstein, A. and Grossmann, M.: 1914, Kovarianzeigenschaften der Feldgleichungen der auf die verallgemeinerte Relativitätstheorie gegründeten Gravitationstheorie, *Zeitschrift für Mathematik und Physik* **63**, 215–225.

Eisenhart, L. P.: 1927, *Non-Riemannian Geometry*, Vol. VIII of *American Mathematical Society Colloquium Publications*, American Mathematical Society, New York.

Elliott, J. P. and Dawber, P. G.: 1987, *Symmetry in Physics*, Vol. 2, The MacMillan Press, New York.

Feferman, S.: 1964, Systems of predicative analysis, *Journal of Symbolic Logic* **29**, 1–30. Reprinted in (Hintikka 1969).

Feferman, S.: 1985, A theory of variable types, *Rev. Colombiana de Matemáticas* **XIX**, 95–106.

Feferman, S.: 1988, Weyl vindicated: "Das Kontinuum" 70 years later, *in* C. Cellucci and G. Sambin (eds), Temi e prospettive della logica e della filosofia della scienza contemporanee, Vol. 1, CLUEB, Bologna Italy, pp. 59–93. The Congress was held in Cesena 7–10 January 1987. Reprinted in Feferman, S. (1998), *In the Light of Logic*, Oxford University Press.

Fock, V.: 1933, Zur Theorie des Positrons, *Dokl. Akad. Nauk. USSR* **1**, 267–278.

Frei, G. and Stammbach, U.: 1992, *Hermann Weyl und die Mathematik an der ETH Zürich 1913–1930*, Birkhäuser Verlag, Basel.

Freudenthal, H.: 1960, Zu den Weyl-Cartanschen Raumproblemen, *Archiv der Mathematik* XI, 107–115.

Friedman, M.: 1983, *Foundations of Space-Time Theories*, Princeton University Press, Princeton.

Fulton, W. and Harris, J.: 1991, *Representation Theory*, Vol. 129 of *Graduate Texts in Mathematics*, Springer-Verlag, Berlin.

Furry, W. H. and Oppenheimer, J. R.: 1934, On the theory of the electron and positron, *Physical Review* 45, 245–262.

Gilmore, R.: 1974, *Lie Groups, Lie Algebras, and Some of Their Applications*, John Wiley and Sons, New York.

Glymour, C.: 1980, *Theory and Evidence*, Princeton University Press, Princeton.

Gödel, K.: 1944, Russell's mathematical logic, *in* P. A. Schilpp (ed.), *The Philosophy of Bertrand Russell*, 3 edn, Northwestern University Press, Chicago. Tudor, New York, 1951. Reprinted in (Benacerraf and Putnam 1983).

Grünbaum, A.: 1973, *Philosophical Problems of Space and Time*, Vol. XII of *Boston Studies in the Philosophy of Science*, 2 edn, Reidel, Dordrecht.

Grzegorczyck, A.: 1954, Elementary definable analysis, *Fundamenta Mathematica* 41, 311–338.

Guillemin, V. and Sternberg, S.: 1964, An algebraic model of transitive differential geometry, *Bull. Amer. Math. Soc.* 70, 16–47.

Guillemin, V. and Sternberg, S.: 1966, Deformation theory of pseudogroup structures, *Mem. Amer. Math. Soc.*.

Hausdorff, F.: 1914, *Grundzüge der Mengenlehre*, Veit, Leipzig. Reprint: Chelsea, New York, 1949, 1965.

Hehl, F., von der Heyde, P., Kerlick, G. and Nester, J.: 1976, General relativity with spin and torsion: Foundations and prospects, *Reviews of Modern Physics* 48, 393–416.

Hehl, F. W.: 1985, On the kinematics of the torsion tensor of space-time, *Foundations of Physics* 15, 451–471.

Heisenberg, W.: 1934, Bemerkung zur Diracschen Theorie des Positrons, *Zeitschrift für Physik* 90, 209–231.

Hellinger, E.: 1965, Hilberts Arbeiten über Integralgleichungen und unendliche Gleichungssysteme, *David Hilbert Gesammelte Abhandlungen III*, Chelsea Publishing Company, New York, pp. 94–145. The collected papers of David Hilbert were originally published in three volumes (1932, 1933 and 1935) in Berlin.

Hilbert, D.: 1899, *Grundlagen der Geometrie*, Teubner, Stuttgard. Second English edition (1971) translated by Leo Unger from the 10th German edition. Second impression (1980). Open Court.

Hilbert, D.: 1902, Über die Grundlagen der Geometrie, *Math. Ann.* 56, 381–422.

Hilbert, D.: 1922, Neubegründung der Mathematik. Erste Mitteilung, *Abhandlungen aus dem Mathematischen Seminar der Hamburgischen Universität* 1, 157–177. Reprinted in (Hilbert 1932–1935), vol. III, pp. 157–177.

Hilbert, D.: 1926, Die Grundlagen der Mathematik, *Abhandlungen aus dem Mathematischen Seminar der Hamburgischen Universität* 6, 65–85. Reprinted with abridgements in *Grundlagen der Geometrie*, 7 edn, pp. 289–312 (1930). See also translation in (van Heijenoort 1967).

Hilbert, D.: 1932–1935, *Gesammelte Abhandlungen*, Vol. I–III, Springer, Berlin. Reprint Chelsea 1965.

Hilbert, D. and Ackermann, W.: 1950, *Principles of Mathematical Logic*, Chelsea Publishing Company, New York. This work is a translation into English of the 2 edn published in 1938 by Julius Springer in Berlin (1 edn, 1928) of the *Grundzüge der Theoretischen Logik* by Hilbert and Ackermann with revisions, corrections and added notes by Robert E. Luce.

Hilbert, D., von Neumann, J. and Nordheim, L.: 1928, Über die Grundlagen der Quantenmechanik, *Mathematische Annalen* 98, 1–30. Received 6 April 1927.

Hintikka, J. (ed.): 1969, *The Philosophy of Mathematics*, Oxford Readings in Philosophy, Oxford University Press, London.

Hurwitz, A.: 1897, Über die Erzeugung der Invarianten durch Integration, *Göttinger Nachrichten* p. 71.

Kibble, T. W. B.: 1961, Lorentz invariance and the gravitational field, *Journal of Mathematical Physics* 2, 212–221.

Kleene, S. C.: 1967, *Introduction to Metamathematics*, North-Holland, Amsterdam.

Klein, F.: 1893, Vergleichende Betrachtungen über neuere geometrische Forschungen, *Math. Ann.* 43, 63–100.

Klingenberg, W.: 1959, Eine Kennzeichnung der Riemannschen sowie der Hermiteschen Mannigfaltigkeiten, *Mathematische Zeitschrift* 70, 300–309.

Kodaira, K.: 1950, On ordinary differential equations of any even order and the corresponding eigenfunction expansions, *American Journal of Mathematics* 72, 502–544.

Korté, H.: 1981, *A Realist Interpretation of the Causal-Inertial Structure of Spacetime*, PhD thesis, University of Western Ontario, London, Ontario.

Kragh, H.: 1990, *Dirac A Scientific Biography*, Cambridge University Press.

Kramers, H. A.: 1930, Théorie générale de la rotation paramagnétique dans les cristaux, *Proceedings of the Koninklijke Nederlandse Akademie van Weten-schappen* **33**, 959–972. Communicated at the meeting of November 29, 1930.

Kramers, H. A.: 1937, The use of charge-conjugated wave-functions in the hole-theory of the electron, *Proceedings of the Koninklijke Nederlandse Akademie van Wetenschappen* **40**, 814–823.

Kretschmann, E.: 1917, Über den physikalischen Sinn der Relativitätstheorie, *Annalen der Physik* **53**(16), 576–614.

Kronig, R. and Weisskopf, V. F. (eds): 1964, *Collected Scientific Papers by Wolfgang Pauli*, Vol. II, John Wiley & Sons, New York.

Kundt, W. and Hoffmann, B.: 1962, Determination of gravitational standard time, *in* Editorial Committe (ed.), *Recent Developments in General Relativity*, Pergamon, New York, pp. 303–336. This collection of articles is dedicated to Leopold Infeld in connection with his 60th birthday.

Laugwitz, D.: 1958, Über eine Vermutung von Hermann Weyl zum Raumproblem, *Archiv der Mathematik* IX, 128–133.

Lie, S.: 1886/1935, Bemerkungen zu v. Helmholtzs Arbeit: Ueber die Tatsachen, die der Geometrie zu Grunde liegen, *Lie, Gesammelte Abhandlungen*, Vol. II, Teubner, Leipzig, pp. 374–379. Originally published in *Berichte über die Abhandlungen der Kgl. Sächsischen Gesellschaft der Wissenschaften in Leipzig, Math.-Phys. Klasse*, Supplement, abgeliefert am 21.2.1887, pp. 337–342.

Lie, S.: 1893, *Theorie der Transformationsgruppen*, Vol. 3, Teubner, Leipzig. Unter Mitwirkung von Prof. Dr. Friedrich Engel. Unveränderter Neudruck, 1930.

Lorentz, H. A.: 1923, The determination of the potentials in the general theory of relativity, with some remarks about the measurement of length and intervals of time and about the theories of Weyl and Eddington, *Proc. Acad. Amsterdam* **29**, 363–382.

Lorentz, H. A., Einstein, A., Minkowski, H. and Weyl, H.: 1952, *The Principle of Relativity: A Collection of Original Memoirs on the Special and General Theory of Relativity*, Dover Publications, Inc., New York. Translated from the third and enlarged German edition of 1923 "Das Relativitätsprinzip, eine Sammlung von Abhandlungen" (Leibzig: Teubner) by W. Perrett and G. B. Jeffrey.

Lorenzen, P.: 1969, *Einführung in die operative Logik und Mathematik*, 2 edn, Springer Verlag, Heidelberg.

Lorenzen, P.: 1971, *Differential and Integral; a Constructive Introduction to Classical Analysis*, University of Texas Press, Austin.

Majer, U.: 1988, Zu einer bemerkenswerten Differenz zwischen Brouwer und Weyl, *in* W. Deppert and K. Hübner (eds), *Exact Sciences and Their Philosophical Foundations: Exakte Wissenschaften und ihre philosophische Grundlegung. Vorträge des internationalen Hermann-Weyl-Kongresses*, Peter Lang Verlag, Frankfurt/M - Bern etc., pp. 543–551.

Mancosu, P.: 1998, *From Brouwer to Hilbert*, Oxford University Press, Oxford.

Marzke, R. F. and Wheeler, J. A.: 1964, Gravitation as geometry, I: The geometry of space-time and the geometrical standard meter, *in* H.-Y. Chiu and W. F. Hoffmann (eds), *Gravitation and Relativity*, Benjamin, Amsterdam, pp. 40–64.

Mehrtens, H.: 1990, *Moderne Sprache Mathematik*, Suhrkamp Verlag, Frankfurt am Main.

Nester, J. M.: 1977, Effective equivalence of the Einstein-Cartan and Einstein theories of gravity, *The Physical Review* **D16**, 2395–2401.

Newman, M. H. A.: 1957, Hermann Weyl, *Bibliographical Memoirs of Fellows of The Royal Society* **3**. A new series in continuation of *Obituary Notices of Fellows of the Royal Society*, volumes 1–9 (1932–1934).

Oppenheimer, J. R.: 1930a, On the theory of electrons and protons, *Phys. Rev.* **35** Second Series, 562–563. Letter to the editor.

Oppenheimer, J. R.: 1930b, Two notes on the probability of radiative transition, *Phys. Rev.* **35**, 939–947.

Pais, A.: 1986, *Inward Bound; Of Matter and Forces in the Physical World*, Clarendon Press, Oxford, New York.

Pauli, W.: 1941, Relativistic field theories of elementary particles, *Rev. Mod. Phys.* **13**, 203–232. Reprinted in (Kronig and Weisskopf 1964).

Peckhaus, V.: 1990, *Hilbertprogramm und Kritische Philosophie*, Vol. 7 of *Studien zur Wissenschafts-, Sozial- und Bildungsgeschichte*, Vandenhoeck & Ruprecht, Göttingen. Eds Michael Otte and Ivo Schneider and Hans-Georg Steiner.

Pirani, F. A. E.: 1973, Building space-time from light rays and free particles, *Symposia Mathematica* **XII**, 67–83.

Poincaré, H.: 1906, Les Mathématiques et Logique, *Revue de Métaphysique et de Morale* pp. 13–14.

Reichenbach, H.: 1924, *Axiomatik der relativistischen Raum-Zeit-Lehre*, Vieweg, Braunschweig. Reprinted in *Hans Reichenbach Gesammelte Werke*, volume 3, edited by Andreas Kamlah and Maria Reichenbach.

Reichenbach, H.: 1925, Über die physikalischen Konsequenzen der relativistischen Axiomatik, *Zeitschrift für Physik* pp. 32–48.

Riemann, B.: 1854, Ueber die Hypothesen, welche der Geometrie zu Grunde liegen, *Abhandlungen der Königlichen Gesellschaft der Wissenschaften zu Göttingen* **13**. Reproduced in (Riemann 1953).

Riemann, B.: 1953, *Gesammelte Mathematische Werke*, 2 edn, Dover, New York. Edited by Heinrich Weber with the assistance of Richard Dedekind; with a supplement edited by M. Noether and W. Wirtinger and with a new introduction by Professor Hans Lewy.

Rosenfeld, B. A.: 1988, *A History of Non-Euclidean Geometry*, Springer-Verlag. Translated by Abe Shenitzer with the editorial assistance of Hardy Grant. The Russion edition of this book appeared in 1976.

Russell, B.: 1903, *The Principles of Mathematics*, Cambridge University Press, Cambridge.

Russell, B.: 1908, Mathematical logic as based on the theory of types, *American Journal of Mathematics* **30**, 222–262. Reprinted in (Russell 1956) and (van Heijenoort 1967) pp. 150–182.

Russell, B.: 1956, *Logic and Knowledge*, Allen and Unwin, London. Edited by R. C. Marsh.

Salmon, W. C.: 1969, The conventionality of simultaneity, *Philosophy of Science* **36**, 44–63.

Salmon, W. C.: 1977, The philosophical significance of the one-way velocity of light, *Noûs* **11**, 253–292.

Scheibe, E.: 1957, Über das Weylsche Raumproblem, *Journal für Mathematik* **197**(3/4), 162–207. Dissertation Göttingen 1955.

Scheibe, E.: 1988, Hermann Weyl and the nature of spacetime, *in* W. Deppert and K. Hübner (eds), *Exact Sciences and Their Philosophical Foundations: Exakte Wissenschaften und ihre philosophische Grundlegung. Vorträge des internationalen Hermann-Weyl-Kongresses*, Peter Lang Verlag, Frankfurt/M - Bern etc., pp. 61–82.

Scholz, E.: 1996, Logische Ordnung im Chaos: Hausdorffs frühe Beiträge zur Mengenlehre, *in* E. Brieskorn (ed.), *Felix Hausdorff zum Gedächtnis*, Vol. 1 Aspekte seines Werkes, Vieweg, pp. 107–134.

Scholz, E.: 1999a, The concept of manifold, 1850–1950, *in* I. M. James (ed.), *History of Topology*, Elsevier Science B. V., chapter 2, pp. 25–64.

Scholz, E.: 1999b, Weyl and the theory of connections, *in* J. Gray (ed.), *The Symbolical Universe*, Oxford University Press, pp. 260–284.

Schur, I.: 1924, Neue Anwendungen der Integralrechnung auf Probleme der Invariantentheorie, *Sitzungsber. d. Berl. Akad. d. Wiss.* pp. 189–208, 297–321, 346–355.

Schweber, S. S.: 1994, *QED and the Men who made it: Dyson, Feynman, Schwinger and Tomonaga*, Princeton University Press, Princeton, New Jersey.

Schwinger, J.: 1988, Hermann Weyl and quantum kinematics, *in* W. Deppert and K. Hübner (eds), *Exact Sciences and Their Philosophical Foundations: Exakte Wissenschaften und ihre philosophische Grundlegung. Vorträge des internationalen Hermann-Weyl-Kongresses*, Peter Lang Verlag, Frankfurt/M - Bern etc., pp. 107–129.

Sciama, D. W.: 1961, On the interpretation of the Einstein-Schrödinger unified field theory, *Journal of Mathematical Physics* **2**, 472–477.

Sciama, D. W.: 1962, On the analogy between charge and spin in general relativity, *Recent Developments in General Relativity*, Pergamon Press, pp. 415–439.

Sigurdsson, S.: 1991, *Hermann Weyl, Mathematics and Physics, 1900–1927*, Ph.D., Harvard University, Cambridge, Massachusetts. Department of the History of Science.

Sklar, L.: 1977, Facts, conventions and assumptions, *in* J. Earman, C. N. Glymour and J. J. Stachel (eds), *Foundations of Space-Time Theories*, Vol. VIII of *Minnesota Studies in the Philosophy of Science*, University of Minnesota Press, Minneapolis, pp. 206–274.

Stone, M. H.: 1932, Linear transformations in Hilbert Space, *Amer. Math. Soc. Colloquium Publications* **15**.

Synge, J. L.: 1960, *The General Theory*, North-Holland, Amsterdam.

Tamm, I.: 1930, Über die Wechselwirkung der freien Elektronen mit der Strahlung nach der Diracschen Theorie des Elektrons und nach der Quantenelektrodynamik, *Zeitschrift für Physik* **62**, 545–568.

Thomas, T. Y.: 1925, On the projective and equi-projective geometries of paths, *Proceedings of the National Academy of Sciences* **2**(4), 199–209.

Thomas, T. Y.: 1926, A projective theory of affinely connected manifolds, *Mathematische Zeitschrift* **25**, 723.

Titchmarsh, E. C.: 1958, *Eigenfunction expansions associated with second order differential equations*, Clarendon Press. Revised edition 1962.

Trautman, A.: 1965, Foundations and current problems of general relativity, *in* S. Deser and K. W. Ford (eds), *Lectures on General Relativity*, Prentice-Hall, Englewood Cliffs.

Trautman, A.: 1966, Comparison of Newtonian and Relativistic Theories of Space-Time, *in* B. Hoffmann (ed.), *Perspectives in Geometry and Relativity*, Indiana University Press, Bloomington.

Utiyama, R.: 1956, Invariant theoretical interpretation of interaction, *The Physical Review* **101**, 1597–1607.

van Dalen, D.: 1995, Hermann Weyl's intuitionistic mathematics, *The Bulletin of Symbolic Logic* **1**(2).

van Heijenoort, J. (ed.): 1967, *From Frege to Gödel, A Source Book in Mathematical Logic, 1879–1931*, Harvard University Press, Cambridge Massachusetts.

Varadarajan, V. S.: 1974, *Lie Groups, Lie Algebras, and their Representations*, Prentice-Hall, Englewood Cliffs.

Veblen, O.: 1928a, Conformal tensors and connections, *Proceedings of the National Academy* **14**, 735.

Veblen, O.: 1928b, Projective tensors and connections, *Proceedings of the National Academy* **14**, 154.

Veblen, O. and Thomas, J. M.: 1926, Projective invariants of the affine geometry of paths, *Annals of Mathematics* **27**, 279–296.

Vinogradov, I. M. (ed.): 1993, *Encyclopaedia of Mathematics*, Vol. 1–10, Kluwer Academic Publishers, Dordrecht. An updated and annotated translation of the Soviet 'Mathematical Encyclopaedia'.

von Helmholtz, H.: 1866, Über die thatsächlichen Grundlagen der Geometrie, *Verhandlungen des naturhistorisch-medizinischen Vereins zu Heidelberg* **4**, 197–202. Reprinted in Wissenschaftliche Abhandlungen (1883) vol. II, pp. 610–617.

von Helmholtz, H.: 1868, Über die Thatsachen, die der Geometrie zum Grunde liegen, *Nachrichten von der Königlichen Gesellschaft der Wissenschaften zu Göttingen* pp. 192–222. Reprinted in Wissenschaftliche Abhandlungen (1883) vol. II, pp. 618–639.

Weinberg, S.: 1995, *The Quantum Theory of Fields*, Vol. I Foundations, Cambridge University Press, Cambridge.

Wigner, E. P.: 1931, *Gruppentheorie und ihre Anwendung auf die Quantenmechanik der Atomspektren*, Vieweg, Braunschweig.

Wigner, E. P.: 1932, Über die Operation der Zeitumkehr in der Quantenmechanik, *Nachrichten Akademie der Wissenschaften in Göttingen Mathematisch-Physikalische Klasse* pp. 546–559.

Wigner, E. P.: 1939, On the unitary representations of the inhomogeneous Lorentz group, *Ann. Math.* **40**, 149–204.

Wigner, E. P.: 1987, Remembering Paul Dirac, *in* B. N. Kursunoglu and E. P. Wigner (eds), *Reminiscences about a great Physicist: Paul Adrien Maurice Dirac*, Cambridge University Press, Cambridge.

Winnie, J. A.: 1970, Special relativity without one-way velocity assumptions, *Philosophy of Science* **37**, 81–99, 223–38.

Woodhouse, N. M. J.: 1972, The differentiable and causal structures of space-time, *Journal of Mathematical Physics* **14**(4), 495–501.

Yang, C. N.: 1982, The discrete symmetries *P*, *T* and *C*, *Journal de Physique* **43**(12), C8–439–C8–451.

Yang, C. N.: 1986, Hermann Weyl's Contribution to Physics, *in* K. Chandrasekharan (ed.), *Hermann Weyl*, Springer-Verlag, Berlin, pp. 7–21. Centenary Lectures delivered by C. N. Yang, R. Penrose, and A. Borel at the ETH Zürich.

Yang, C. N. and Mills, R. L.: 1954, Conservation of isotopic spin and isotopic gauge invariance, *The Physical Review* **96**, 191–195.

Young, A.: 1901, On quantitative substitutional analysis, *Proc. London Math. Soc.* **33**, 97–146.

Young, A.: 1902, On quantitative substitutional analysis, *Proc. London Math. Soc.* **34**, 361–397.

Part III

Appendices

Works by Hermann Weyl
Cited in this Volume

Weyl, H.: 1908a, Singuläre Integralgleichungen, *Mathematische Annalen* **66**, 273–324. GA I, 102–153, [3].

Weyl, H.: 1908b, *Singuläre Integralgleichungen mit besonderer Berücksichtigung des Fourierschen Integraltheorems*, Dissertation, Göttingen. GA I, 1–87, [1].

Weyl, H.: 1909a, Über beschränkte quadratische Formen, deren Differenz vollstetig ist, *Rendiconti del Circolo Matematico di Palermo* **27**, 373–392. GA I, 175–194, [5].

Weyl, H.: 1909b, Über die Konvergenz von Reihen, die nach Orthogonalfunktionen fortschreiten, *Mathematische Annalen* **67**, 225–245. GA I, 154–174, [4].

Weyl, H.: 1909c, Über gewöhnliche lineare Differentialgleichungen mit singulären Stellen und ihre Eigenfunktionen, *Nachrichten der Königlichen Gesellschaft der Wissenschaften zu Göttingen; Mathematisch-physikalische Klasse* pp. 37–63. GA I, 195–221, [6].

Weyl, H.: 1910a, Die Gibbsche Erscheinung in der Theorie der Kugelfunktionen, *Rendiconti del Circolo Matematico di Palermo* **29**, 308–323. GA I, 305–320, [10].

Weyl, H.: 1910b, Über die Definitionen der mathematischen Grundbegriffe, *Mathematisch-naturwissenschaftliche Blätter* **7**, 93–95 and 109–113. GA I, 298–304, [9].

Weyl, H.: 1910c, Über die Gibbsche Erscheinung und verwandte Konvergenzphänomene, *Rendiconti del Circolo Matematico di Palermo* **30**, 377–407. GA I, 321–353, [11].

Weyl, H.: 1910d, Über gewöhnliche Differentialgleichungen mit Singularitäten und die zugehörigen Entwicklungen willkürlicher Funktionen, *Mathematische Annalen* **68**, 220–269. GA I, 248–297, [8].

Weyl, H.: 1910e, Über gewöhnliche lineare Differentialgleichungen mit singulären Stellen und ihre Eigenfunktionen (2. Note), *Nachrichten der Königlichen Gesellschaft der Wissenschaften zu Göttingen; Mathematisch-physikalische Klasse* pp. 442–467. GA I, 222–247, [7].

Weyl, H.: 1911a, Konvergenzcharakter der Laplaceschen Reihe in der Umgebung eines Windungspunktes, *Rendiconti del Circolo Matematico di Palermo* **32**, 118–131. GA I, 376–389, [14].

Weyl, H.: 1911b, Über die asymptotische Verteilung der Eigenwerte, *Nachrichten der Königlichen Gesellschaft der Wissenschaften zu Göttingen; Mathematisch-physikalische Klasse* pp. 110–117. GA I, 368–375, [13].

Weyl, H.: 1911c, Zwei Bemerkungen über das Fouriersche Integraltheorem, *Jahresbericht der Deutschen Mathematikervereinigung* **20**, 129–141. GA I, 354–366, [12]. Correction p. 367.

Weyl, H.: 1912a, Das asymptotische Verteilungsgesetz der Eigenwerte linearer partieller Differentialgleichungen (mit einer Anwendung auf die Theorie der Hohlraumstrahlung), *Mathematische Annalen* **71**, 441–479. GA I, 393–430, [16].

Weyl, H.: 1912b, Über das Spektrum der Hohlraumstrahlung, *Journal für die reine und angewandte Mathematik* **141**, 163–181. GA I, 442–460, [18].

Weyl, H.: 1912c, Über die Abhängigkeit der Eigenschwingungen einer Membran von deren Begrenzung, *Journal für die reine und angewandte Mathematik* **141**, 1–11. GA I, 431–441, [17].

Weyl, H.: 1913a, *Die Idee der Riemannschen Fläche*, 1 edn, B. G. Teubner, Leipzig. 2 edn, B. G. Teubner, Leipzig, 1923; Reprint of 2 edn, Chelsea Co., New York, 1951; 3 edn, revised, B. G. Teubner, Leipzig, 1955. English translation of 3 edn, *The Concept of a Riemann Surface*, Addison-Wesley, 1964.

Weyl, H.: 1913b, Über die Randwertaufgabe der Strahlungstheorie und asymptotische Spektralgesetze, *Journal für die reine und angewandte Mathematik* **143**, 177–202. GA I, 461–486, [19].

Weyl, H.: 1914a, Sur une application de la théorie des nombres à la mécanique statistique et la théorie des perturbations, *L'Enseignement mathématique* **16**, 455–467. GA I, 498–510, [21].

Weyl, H.: 1914b, Über ein Problem aus dem Gebiete der diophantischen Approximationen, *Nachrichten der Königlichen Gesellschaft der Wissenschaften zu Göttingen; Mathematisch-physikalische Klasse* pp. 234–244. GA I, 487–497, [20].

Weyl, H.: 1915, Das asymptotische Verteilungsgesetz der Eigenschwingungen eines beliebig gestalteten elastischen Körpers, *Rendiconti del Circolo Matematico di Palermo* **39**, 1–50. GA I, 511–562, [22].

Weyl, H.: 1916a, Über die Bestimmung einer geschlossenen konvexen Fläche durch ihr Linienelement, *Vierteljahrsschrift der naturforschenden Gesellschaft in Zürich* **61**, 40–72. Russische Übersetzung: Uspehi Matem. Nauk (N. S) 3, no. 2 (24), 159–190 (1948). GA I, 614–644, [25].

Weyl, H.: 1916b, Über die Gleichverteilung von Zahlen mod. Eins, *Mathematische Annalen* **77**, 313–352. GA I, 563–599, [23].

Weyl, H.: 1917, Über die Starrheit der Eiflächen und konvexer Polyeder, *Sitzungsberichte der Königlich Preußischen Akademie der Wissenschaften zu Berlin* pp. 250–266. GA I, 646–662, [27].

Weyl, H.: 1918/1994, *The Continuum*, Dover, Mineola N.Y. An unabridged and corrected republication of the English translation, edited by Stephan Pollard and Thomas Bole, of *Das Kontinuum*, first published by The Thomas Jefferson University Press, Kirksville, Missouri, in 1987.

Weyl, H.: 1918a, *Das Kontinuum*, Veit & Co., Leipzig. 2 edn, de Gryter & Co., Berlin, 1932.

Weyl, H.: 1918b, Gravitation und Elektrizität, *Sitzungsberichte der Königlich Preußischen Akademie der Wissenschaften zu Berlin* pp. 465–480. GA II, 29–42, [31].

Weyl, H.: 1918c, *Raum, Zeit, Materie*, J. Springer, Berlin. 3 edn, essentially revised, J. Springer, Berlin 1919; 4 edn, essentially revised, J. Springer, Berlin 1921; 5 edn, revised, J. Springer, Berlin, 1923; 7 edn, edited (with notes) by J. Ehlers, Springer, Berlin 1988; *Temps, espace, matière* (from the 4th German edn), A. Blanchard, Paris, 1922; *Space, Time, Matter*, (from the 4th German edn), Methuen, London, 1922.

Weyl, H.: 1918d, Reine Infinitesimalgeometrie, *Mathematische Zeitschrift* **2**, 384–411. GA II, 1–28, [30].

Weyl, H.: 1919a, Der circulus vitiosus in der heutigen Begründung der Analysis, *Jahresbericht der Deutschen Mathematikervereinigung* **28**, 85–92. GA II, 43–50, [32].

Weyl, H.: 1919b, Eine neue Erweiterung der Relativitätstheorie, *Annalen der Physik* **59**, 101–133. GA II, 55–87, [34].

Weyl, H.: 1919c, *Kommentar zu Riemanns "Über die Hypothesen, welche der Geometrie zu Grunde liegen"*, 2 edn, J. Springer, Berlin. 3 edn, J. Springer, Berlin, 1923.

Weyl, H.: 1919d, Über die statischen kugelsymmetrischen Lösungen von Einsteins "kosmologischen" Gravitationsgleichungen, *Physikalische Zeitschrift* **20**, 31–34. GA II, 51–54, [33].

Weyl, H.: 1920a, Das Verhältnis der kausalen zur statistischen Betrachtungsweise in der Physik, *Schweizerische Medizinische Wochenschrift* **50**, 737–741. GA II, 113–122, [38].

Weyl, H.: 1920b, Die Einsteinsche Relativitätstheorie, *Schweizerland (1920)*. Schweizerische Bauzeitung (1921). GA II, 123–140, [39].

Weyl, H.: 1920c, Elektrizität und Gravitation, *Physikalische Zeitschrift* **21**, 649–650. GA II, 141–142, [40].

Weyl, H.: 1921a, Das Raumproblem, *Jahresbericht DMV* **30**, 92ff. GA II, 212–228, [45].

Weyl, H.: 1921b, Electricity and gravitation, *Nature* **106**, 800–802. GA II, 260–262, [48].

Weyl, H.: 1921c, Feld und Materie, *Annalen der Physik* **65**, 541–563. GA II, 237–259, [47].

Weyl, H.: 1921d, Über die neue Grundlagenkrise der Mathematik, *Mathematische Zeitschrift* **10**, 39–79. GA II, 143–180, [41]. Reprinted by Wissenschaftliche Buchgesellschaft, Darmstadt, 1965.

Weyl, H.: 1921e, Über die physikalischen Grundlagen der erweiterten Relativitätstheorie, *Physikalische Zeitschrift* **22**, 473–480. GA II, 229–236, [46].

Weyl, H.: 1921f, Zur Infinitesimalgeometrie: Einordnung der projektiven und konformen Auffassung, *Nachrichten der Königlichen Gesellschaft der Wissenschaften zu Göttingen; Mathematisch-physikalische Klasse* pp. 99–112. GA II, 195–207, [43].

Weyl, H.: 1922a, Das Raumproblem, *Jahresbericht der Deutschen Mathematikervereinigung* **31**, 205–221. GA II, 328–344, [53].

Weyl, H.: 1922b, Die Einzigartigkeit der Pythagoreischen Maßbestimmung, *Mathematische Zeitschrift* **12**, 114–146. GA II, 263–295, [49].

Weyl, H.: 1922c, Die Relativitätstheorie auf der Naturforscherversammlung, *Jahresbericht der Deutschen Mathematikervereinigung* **31**, 51–63. GA II, 315–327, [52].

Weyl, H.: 1922d, Letter to E. Bovet, 27. 7. 1922. In: E. Bovet (ed.), Die Physiker Einstein und Weyl antworten auf eine metaphysische Frage, *Wissen und Leben* **15**, 901–906 (not included in GA).

Weyl, H.: 1923a, Análisis situs combinatorio, *Revista Matematica Hispano-Americana* **5**, 43. GA II, 390–415, [58].

Weyl, H.: 1923b, Entgegnung auf die Bemerkungen von Herrn Lanczos über die de Sittersche Welt, *Physikalische Zeitschrift* **24**, 130–131. GA II, 373–374, [55].

Weyl, H.: 1923c, *Mathematische Analyse des Raumproblems*, J. Springer, Berlin.

Weyl, H.: 1923d, Repartición de corriente en una red conductora. (Introducción al análisis combinatorio), *Revista Matematica Hispano-Americana* **5**, 153–164. English translation: George Washington University Logistics Research Project (1951). GA II, 378–389, [57].

Weyl, H.: 1923e, Zur allgemeinen Relativitätstheorie, *Physikalische Zeitschrift* **24**, 230–232. GA II, 375–377, [56].

Weyl, H.: 1923f, Zur Charakterisierung der Drehungsgruppe, *Mathematische Zeitschrift* **17**, 293–320. GA II, 345–372, [54].

Weyl, H.: 1924a, Análisis situs combinatorio (continuación), *Revista Matematica Hispano-Americana* **6**, 1–9 and 33–41. GA II, 416–432, [59].

Weyl, H.: 1924b, Das gruppentheoretische Fundament der Tensorrechnung, *Nachrichten der Gesellschaft der Wissenschaften zu Göttingen. Mathematisch-physikalische Klasse* pp. 218–224. GA II, 461–467, [62].

Weyl, H.: 1924c, Massenträgheit und Kosmos. Ein Dialog, *Die Naturwissenschaften* **12**, 197–204. GA II, 478–485, [65].

Weyl, H.: 1924d, Observations on the note of Dr. L. Silberstein: determination of the curvature invariant of space-time, *The London, Edinburgh and Dublin philosophical Magazine and Journal of Science* **48**, 348–349. GA II, 476–477, [64].

Weyl, H.: 1924e, Randbemerkungen zu Hauptproblemen der Mathematik, *Mathematische Zeitschrift* **20**, 131–150. GA II, 433–452, [60].

Weyl, H.: 1924f, Was ist Materie?, *Die Naturwissenschaften* **12**, 561–568, 585–593, and 604–611. GA II, 486–510, [66]. Reprinted by J. Springer, Berlin, 1924 and Wissenschaftliche Buchgesellschaft, Darmstadt, 1963.

Weyl, H.: 1924g, Zur Theorie der Darstellung der einfachen kontinuierlichen Gruppen. (Aus einem Schreiben an Herrn I. Schur), *Sitzungsberichte der Preußischen Akademie der Wissenschaften zu Berlin* pp. 338–345. GA II, 453–460, [61].

Weyl, H.: 1925/1988, *Riemanns geometrische Ideen, ihre Auswirkungen und ihre Verknüpfung mit der Gruppentheorie*, Springer, Berlin. K. Chandrasekharan, ed.

Weyl, H.: 1925a, Die heutige Erkenntnislage in der Mathematik, *Symposion* **1**, 1–32. GA II, 511–542, [67].

Weyl, H.: 1925b, Theorie der Darstellung kontinuierlicher halbeinfacher Gruppen durch lineare Transformationen I, *Mathematische Zeitschrift* **23**, 271–309. GA II, 543–579, [68].

Weyl, H.: 1926a, Elementare Sätze über die Komplex- und die Drehungs-gruppe, *Nachrichten der Gesellschaft der Wissenschaften zu Göttingen. Mathematisch-physikalische Klasse* pp. 235–243. GA III, 25–33, [70].

Weyl, H.: 1926b, Theorie der Darstellung kontinuierlicher halbeinfacher Gruppen durch lineare Transformationen II, *Mathematische Zeitschrift* 24, 328–376. GA II, 580–605, [68].

Weyl, H.: 1926c, Theorie der Darstellung kontinuierlicher halbeinfacher Gruppen durch lineare Transformationen III, *Mathematische Zeitschrift* 24, 377–395. GA II, 606–645, [68].

Weyl, H.: 1926d, Theorie der Darstellung kontinuierlicher halbeinfacher Gruppen durch lineare Transformationen (Nachtrag), *Mathematische Zeitschrift* 24, 789–791. GA II, 645–647, [68].

Weyl, H.: 1926e, Universe, modern conceptions of, *The Encyclopedia Britannica*, 13 edn, pp. 908–911.

Weyl, H.: 1926f, Zur Darstellungstheorie und Invariantenabzählung der projektiven, der Komplex- und der Drehungsgruppe, *Acta Mathematica* 48, 255–278. GA III, 1–24, [69].

Weyl, H.: 1927a, Quantenmechanik und Gruppentheorie, *Zeitschrift für Physik* 46, 1–46. GA III, 90–135, [75].

Weyl, H.: 1927b, Zeitverhältnisse im Kosmos, Eigenzeit, gelebte Zeit und metaphysische Zeit, *in* E. S. Brightman (ed.), *Proceedings of the Sixth International Congress of Philosophy*, Harvard University Cambridge/Massachusetss, pp. 54–58. September 13th to 17th, 1926.

Weyl, H.: 1927[sic]/1966, *Philosophie der Mathematik und Naturwissenschaft*, 1 edn, Handbuch der Philosophie, Abt. 2A, Verlag R. Oldenbourg, München. Second unchanged edition, 1948. Third edition re-translated from the extended English edition (?), Verlag R. Oldenbourg, München, 1966.

Weyl, H.: 1928a, Diskussionsbemerkungen zu dem zweiten Hilbertschen Vortrag über die Grundlagen der Mathematik, *Abhandlungen aus dem mathematischen Seminar der Hamburgischen Universität* 6, 86–88. GA III, 147–149, [77].

Weyl, H.: 1928b, *Gruppentheorie und Quantenmechanik*, S. Hirzel, Leipzig. (a) 2nd reworked edition, S. Hirzel, Leipzig 1931. (b) English translation: The Theory of groups and quantum mechanics, Dutten, New York, 1932. (c) Reprinting of (b): Dover Publications, New York, 1949.

Weyl, H.: 1929a, Consistency in mathematics, *The Rice Institute Pamphlet* 16, 245–265. GA III, 150–170, [78].

Weyl, H.: 1929b, Elektron und Gravitation, *Zeitschrift für Physik* **56**, 330–352. GA III, 245–267, [85].

Weyl, H.: 1929c, Gravitation and the electron, *The Rice Institute Pamphlet* **16**, 280–295. GA III, 229–244, [84].

Weyl, H.: 1929d, Gravitation and the electron, *Proceedings of the National Academy of Sciences of the United States of America* **15**, 323–334. GA III, 217–228, [83].

Weyl, H.: 1929e, On the foundations of infinitesimal geometry, *Bulletin of the American Mathematical Society* **35**, 716–725. GA III, 207–216, [82].

Weyl, H.: 1930, Redshift and relativistic cosmology, *The London, Edinburgh and Dublin philosophical Magazine and Journal of Science* **9**, 936–943. GA III, 300–307, [89].

Weyl, H.: 1931a, *Die Stufen des Unendlichen*, G. Fischer, Jena.

Weyl, H.: 1931b, Geometrie und Physik, *Die Naturwissenschaften* **19**, 49–58. GA III, 336–345, [93].

Weyl, H.: 1932, Zu David Hilberts siebzigstem Geburtstag, *Die Naturwissenschaften* **20**, 57–58. GA III, 346–347, [94].

Weyl, H.: 1934a, Harmonics on homogeneous manifolds, *Annals of Mathematics* **35**, 486–499. GA III, 386–399, [98].

Weyl, H.: 1934b, On generalized Riemann matrices, *Annals of Mathematics* **35**, 714–729. GA III, 400–415, [99].

Weyl, H.: 1934c, Universum und Atom, *Die Naturwissenschaften* **22**, 145–149. GA III, 420–424, [101].

Weyl, H.: 1935, Elementare Theorie der konvexen Polyeder, *Commentarii mathematici Helvetici* **7**, 290–306. English translation in: Contributions to the theory of games I, Annals of Mathematics Studies, Princeton University Press **24**, 3–18 (1950). GA III, 517–533, [106].

Weyl, H.: 1936, Generalized Riemann Matrices and factor sets, *Annals of Mathematics* **37**, 709–745. GA III, 534–570, [107].

Weyl, H.: 1937a, Note on matric algebras, *Annals of Mathematics* **38**, 477–483. GA III, 572–578, [109].

Weyl, H.: 1937b, Riemannsche Matrizen und Faktorensysteme, *Comptes Rendus du Congrès International des Mathématiciens Oslo* **2**, 3. GA III, 571, [108].

Weyl, H.: 1938a, Mean motion, *American Journal of Mathematics* **60**, 889–896. GA III, 634–641 , [113].

Weyl, H.: 1938b, Symmetry, *Journal of the Washington Academy of Sciences* **28**, 253–271. GA III, 592–610, [111].

Weyl, H.: 1939a, *The classical groups, their invariants and representations*, Princeton University Press; Oxford University Press; H. Milford, London. 2 edn, Princeton University Press; Oxford University Press; H. Milford, London, 1946.

Weyl, H.: 1939b, Invariants, *Duke Mathematical Journal* **5**, 489–502. GA III, 670–683, [117].

Weyl, H.: 1939c, Mean motion II, *American Journal of Mathematics* **61**, 143–148. GA III, 642–647, [114].

Weyl, H.: 1940, Theory of reduction for arithmetical equivalence, *Transactions of the American Mathematical Society* **48**, 126–164. GA III, 719–757, [120].

Weyl, H.: 1942a, On geometry of numbers, *Proceedings of the London Mathematical Society* **47**, 268–289. GA IV, 75–96, [127].

Weyl, H.: 1942b, Theory of reduction for arithmetical equivalence II, *Transactions of the American Mathematical Society* **51**, 203–231. GA IV, 46–74, [126].

Weyl, H.: 1943, *Meromorphic Functions and Analytic Curves*, Vol. 12 of *Annals of Mathematics Studies*, Princeton University Press and Oxford University Press, Princeton and London. In collaboration with F. Joachim Weyl; Reprinted by Kraus Reprint Corporation, New York, 1965.

Weyl, H.: 1944a, David Hilbert and his mathematical work, *Bulletin of the American Mathematical Society* **50**, 612–654. Portugiesische Übersetzung: Boletin da Sociedade de Matemática de São Paulo **1**, 76–104 (1946), and **2**, 37–60 (1947). GA IV, 130–172, [132].

Weyl, H.: 1944b, Obituary: David Hilbert 1862–1943, *Obituary Notices of Fellows of the Royal Society* **4**, 547–553. American Philosophical Society Year Book 387–395 (1944). GA IV, 121–129, [131].

Weyl, H.: 1945, Fundamental domains for lattice groups in division algebras I, II, *I:Festschrift zum 60. Geburtstag von Prof. Dr. A. Speiser*, Orell Füssli, Zürich, pp. 218–232. II: Commentarii mathematici Helvetici **17**, 283–306 (1944/45). GA IV, 232–264, [136].

Weyl, H.: 1946, Mathematics and logic. A brief survey serving as a preface to a review of "The Philosophy of Bertrand Russell", *The American Mathematical Monthly* **53**, 2–13. GA IV, 268–279, [138].

Weyl, H.: 1948, Wissenschaft als symbolische Konstruktion des Menschen, *Eranos-Jahrbuch* pp. 375–431. GA IV, 289–345, [142].

Weyl, H.: 1949a, *Philosophy of Mathematics and Natural Science*, 1 edn, Princeton University Press. 2 edn, Princeton University Press, 1950.

Weyl, H.: 1949b, Relativity theory as a stimulus in mathematical research, *Proceedings of the American Philosophical Society* **93**, 535–541. GA IV, 394–400, [147].

Weyl, H.: 1950a, 50 Jahre Relativitätstheorie, *Die Naturwissenschaften* **38**, 73–83. GA IV, 421–431, [149].

Weyl, H.: 1950b, Ramifications, old and new, of the eigenvalue problem, *Bulletin of the American Mathematical Society* **56**, 115–139. GA IV, 432–456, [150].

Weyl, H.: 1950c, A remark on the coupling of gravitation and electron, *The Physical Review* **77**, 699–701. GA IV, 286–288, [141].

Weyl, H.: 1954, Erkenntnis und Besinnung (Ein Lebensrückblick), *Studia Philosophica*. GA IV, 631–649, [166]. A talk given at the University of Lausanne, May 1954.

Weyl, H.: 1955, *Selecta Hermann Weyl*, Birkhäuser, Basel.

Weyl, H.: 1968, *Gesammelte Abhandlungen*, Vol. I–IV, Springer Verlag, Berlin. Edited by K. Chandrasekharan.

Weyl, H.: 1985, Axiomatic versus constructive procedures in mathematics, *The Mathematical Intelligencer* **7**, 12–17. A posthumous publication, edited by Tito Tonietti.

Weyl, H. and Weyl, J.: 1938, Meromorphic curves, *Annals of Mathematics* **39**, 516–538. GA III, 611–633, [112].

Weyl, H. and Weyl, J.: 1942, On the theory of analytic curves, *Proceedings of the National Academy of Sciences of the United States of America* **28**, 417–421. GA IV, 111–114, [129].

Brauer, R. and Weyl, H.: 1935, Spinors in n dimensions, *American Journal of Mathematics* **57**, 425–449.

Jerosch, F. and Weyl, H.: 1908, Über die Konvergenz von Reihen, die nach periodischen Funktionen fortschreiten, *Mathematische Annalen* **66**, 67–80. GA I, 88–101, [2].

Peter, F. and Weyl, H.: 1927, Die Vollständigkeit der primitiven Darstellungen einer geschlossenen kontinuierlichen Gruppe, *Mathematische Annalen* **97**, 737–755. GA III, 58–75, [73].

Robertson, H. P. and Weyl, H.: 1929, On a problem in the theory of groups arising in the foundations of infinitesimal geometry, *Bulletin of the American
Mathematical Society* **35**, 686–690. GA III, 203–206, [81].

Authors

Robert Coleman
Department of Physics
University of Regina
Regina, Sasketchewan
Canada S4S 0A2
coleman@trillium.phys.uregina.ca

Hubert Goenner
Institut für Theoretische Physik
Universität Göttingen
Bunsenstr. 9
D–37073 Göttingen
Deutschland — Germany
goenner@theorie.physik.uni-goettingen.de

Herbert Korté
Department of Philosophy
University of Regina
Regina, Sasketchewan
Canada S4S 0A2
korte@trillium.phys.uregina.ca

Erhard Scholz
Fachbereich Mathematik
Universität–Gesamthochschule Wuppertal
Gaußstr. 20
D–42097 Wuppertal
Deutschland — Germany
scholz@math.uni-wuppertal.de

Skúli Sigurdsson
Science Institute
University of Iceland
Dunhaga 3
IS-107 Reykjavik
Iceland/Island
sksi@raunvis.hi.is

Norbert Straumann
Institut für Theoretische Physik,
Universität Zürich
Winterthurer Str. 190
CH–8057 Zürich
Schweiz — Switzerland
norbert@physik.unizh.ch

Robert Coleman is a physicist with active interests in the philosophy of science, Hubert Goenner physicst and historian of physics, Herbert Korté is a professional philosopher of science with research interests in foundations of mathematics and physics including their history, Erhard Scholz is a historian of mathematics, Skúli Sigurdsson historian of science with broad interests in cultural and intellectual history, Norbert Straumann physicist with active interests in the history of physics.

Index